THEORETICAL AND APPLIED MECHANICS

THEORETICAL AND APPLIED MECHANICS

Proceedings of the XVIth International Congress of Theoretical and Applied Mechanics held in Lyngby, Denmark 19-25 August, 1984

Edited by

Frithiof I. NIORDSON

and

Niels OLHOFF

The Technical University of Denmark
Lyngby
Denmark

1985

NORTH-HOLLAND
AMSTERDAM • NEW YORK • OXFORD

ISBN: 0 444 87707 X

Published by:
Elsevier Science Publishers B.V.
P.O. Box 1991
1000 BZ Amsterdam
The Netherlands

Sole distributors for the U.S.A. and Canada:
Elsevier Science Publishing Company, Inc.
52 Vanderbilt Avenue
New York, N.Y. 10017
U.S.A.

Library of Congress Cataloging in Publication Data

International Congress of Theoretical and Applied
Mechanics (16th : 1984 : Lyngby, Denmark)
Theoretical and applied mechanics.

Bibliography: p.
1. Mechanics--Congresses. 2. Mechanics, Applied--
Congresses. I. Niordson, Frithiof I. II. Olhoff, Niels.
III. Title.
QA801.I39 1984 531 85-1457
ISBN 0-444-87707-X (U.S.)

PRINTED IN THE NETHERLANDS

PREFACE

This book contains the Proceedings of the XVIth International Congress of Applied Mechanics, held at the Technical University of Denmark, Lyngby, Denmark, August 19-25. The Congress was organized by the International Union of Theoretical and Applied Mechanics (IUTAM) and held in Lyngby by invitation from the Royal Danish Academy of Sciences and Letters, the Technical University of Denmark, and the Danish Center for Applied Mathematics and Mechanics (DCAMM).

The full texts of the two General Lectures and the sectional lectures according to the list on page xv are included in this volume and will not appear elsewhere. The contributed papers are listed by author and title; most of them will be published in appropriate technical journals.

The publication of these Proceedings has been handled very promptly and capably by North-Holland and their editors, to whom we are very grateful.

Lyngby
October 1984 — Frithiof Niordson — Niels Olhoff

Registration during the Early Bird Reception held Sunday evening.

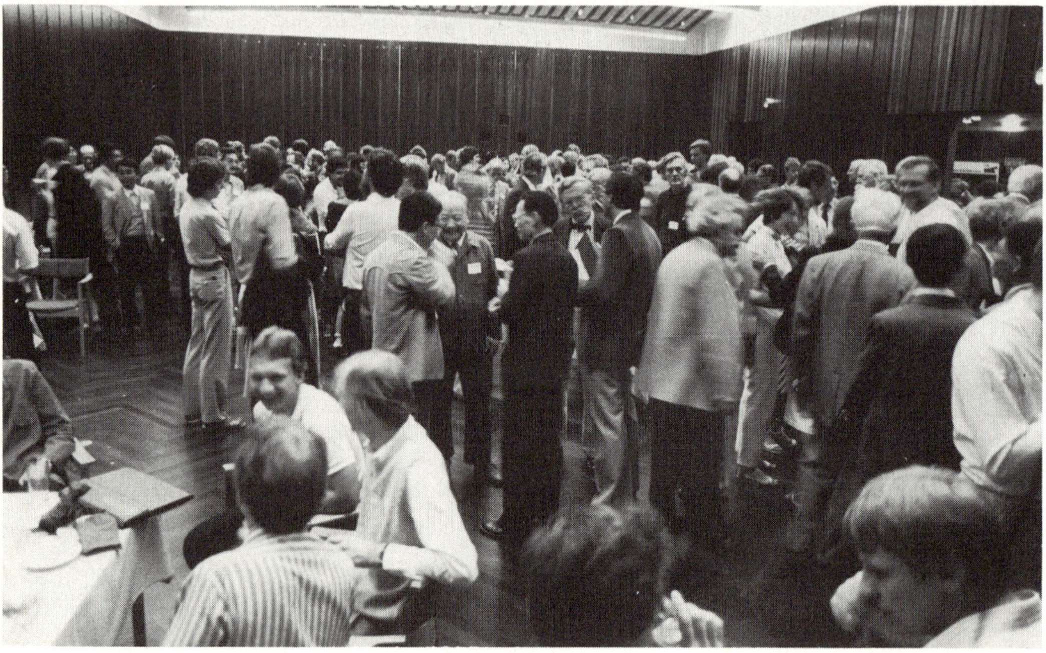

Meeting friends and colleagues at the Early Bird Reception in the Society of Engineers of Denmark.

The entrance to the Assembly Hall and Congress Area in Building 101 of the Technical University.

Arrival of the Danish Minister of Education prior to the Opening of the Congress on Monday morning.

"Lur" blowers opening the Congress.

From the Opening Ceremony. From the left: Dr. Niels Olhoff (Executive Secretary of ICTAM Lyngby), Mrs. Ann Drucker, Professor Daniel Drucker (President of IUTAM), Mr. Bertel Haarder (Minister of Education), Dr. Peter Lawætz (Rector of the Technical University), and Professor Frithiof Niordson (President of ICTAM Lyngby).

Discussions during one of the Poster Sessions.

From the Information Desk.

During the Tuesday Night Reception in the City Hall of Copenhagen.

Mood from the courtyard of building 101 during one of the lunch-breaks.

CONTENTS

Preface v

Congress Committees xiii

List of Authors xv

List of Participants xvii

Report on the Congress xxvii

OPENING AND CLOSING LECTURES

H. Alfvén and F. Čech: Space Research and the New Approach to the Mechanics of Fluid Media in Cosmos 1

J.B. Keller: Computers and Chaos in Mechanics 31

SECTIONAL LECTURES

H. Aref: Chaos in the Dynamics of a Few Vortices - Fundamentals and Applications 43

V.V. Beletsky: Resonant Phenomena in Rotational Motions of Artificial and Natural Celestial Bodies 69

M.V. Berry: Quantum, Classical and Semiclassical Adiabaticity 83

L. Bjørnø: Aspects of Nonlinear Acoustics 97

O.M. Faltinsen: Hydrodynamic Loads on Marine Structures 117

E.J. Hinch: The Recovery of Oil from Underground Reservoirs 135

K. Hutter and T. Alts: Ice and Snow Mechanics, A Challenge to Theoretical and Applied Mechanics 163

K. Kirchgässner: Nonlinear Wave Motion and Homoclinic Bifurcation 219

A. Libchaber: From Periodicity to Chaos in Hydrodynamic Systems 233

J.J. McCoy: Continuum Formulations for Stochastic Multi-component Systems 237

H. Mitsuyasu: Recent Studies on Ocean Wave Spectra 249

D.W. Moore: Numerical and Analytical Aspects of Helmholtz Instability 263

J.R. Paulling: Hydrodynamic Synthesis of Marine Structures 275

P. Terndrup Pedersen: Structural Design of Marine Structures 293

W.G. Price and Y. Wu: Hydroelasticity of Marine Structures 311

D. Rogula: Non-Classical Material Continua 339

P.G. Saffman: Three Dimensional Stability and Bifurcation of Steady Water Waves 355

J. Salençon: Yield-Strength of Anisotropic Soils 369

W.O. Schiehlen: Vehicle System Dynamics 387

S. Taneda: Flow Field Visualization 399

List of Contributed Papers 411

CONGRESS COMMITTEE OF IUTAM

A. Acrivos (USA)
E.R. Arantes e Oliveira (Portugal)
G.K. Batchelor (UK)
E. Becker (BRD)
B.A. Boley*, Secretary (USA)
J. Brilla (CSSR)
S. Dhawan (India)
D.C. Drucker*, Chairman (USA)
W. Fiszdon* (Poland)
N.J. Hoff (USA)
S. Kaliszky (Hungary)
W.T. Koiter (Netherlands)
Y.H. Ku (USA)
T.C. Lin (People's Republic of China)
G. Maier (Italy)
H.K. Moffatt* (UK)
R. Moreau (France)
F.I. Niordson (Denmark)
F.P.J. Rimrott* (Canada)
M. Roy (France)
W. Schiehlen (BRD)
W. Schneider (Austria)
L.I. Sedov (USSR)
J. Singer (Israel)
I. Tani (Japan)
L. van Wijngaarden* (Netherlands)
J.R. Willis (UK)
N.N. Yanenko (USSR)
P.-Y. Zhou (People's Republic of China)

*) Members of the Executive Committee

LOCAL ORGANIZING COMMITTEE

F.I. Niordson*, Chairman	- Dept. Solid Mechanics
N. Olhoff*, Secretary	- Dept. Solid Mechanics
E. Byskov	- Dept. Structural Engrg.
J. Christoffersen	- Dept. Solid Mechanics
E. Hansen	- Lab. Applied Math. Physics
J. Jensen	- Dept. Solid Mechanics
J. Juncher Jensen	- Dept. Ocean Engrg.
P. Scheel Larsen*	- Dept. Fluid Mechanics
A. Gudmann Nielsen	- Dept. Solid Mechanics
P. Pedersen*	- Dept. Solid Mechanics
O. Skovgaard	- Lab. Applied Math. Physics
H. True	- Lab. Applied Math. Physics
P. Tryde	- Inst. Hydrodynamics and Hydraulic Engrg.
R. Zetterlund*	- Dept. Solid Mechanics

*) Members of the Executive Committee

LIST OF AUTHORS OF PAPERS IN THIS VOLUME

Prof. ALFVÉN, H., The Royal Institute of Technology, Department of Plasma Physics, S-100 44 Stockholm, Sweden.

Prof. AREF, H., Division of Engineering, Box D, Brown University, Providence, Rhode Island 02912, USA.

Prof. BELETSKY, V.V., Keldysh Institute of Applied Mathematics, USSR Academy of Sciences, Moscow, USSR.

Prof. BERRY, M.V., H.H. Wills Physics Laboratory, Tyndall Avenue, Bristol BS8 1TL, UK.

Prof. BJØRNØ, L., Technical University of Denmark, Building 352, DK-2800 Lyngby, Denmark.

Prof. FALTINSEN, O.M., Division of Marine Hydrodynamics, Norwegian Institute of Technology, Trondheim, Norway.

Prof. HINCH, E.J., Department of Applied Mathematics and Theoretical Physics, University of Cambridge, Silver Street, Cambridge CB3 9EW, England.

Prof. HUTTER, K., Laboratory of Hydraulics, Hydrology and Glaciology, ETH, Gloriastrasse 37-39, CH-8092 Zürich, Switzerland.

Prof. KELLER, J.B., Departments of Mathematics and Mechanical Engineering, Stanford University, Stanford, California, USA.

Prof. KIRCHGÄSSNER, K., Mathematische Institut A, Universität Stuttgart, D-7000 Stuttgart, BRD.

Prof. LIBCHABER, A., James Franck and Enrico Fermi Institutes and Department of Physics, The University of Chicago, Chicago, IL 60637, USA.

Prof. McCOY, J.J., The Catholic University of America, Washington, D.C. 20064, USA.

Prof. MITSUYASU, H., Research Institute for Applied Mechanics, Kyushu University, Kasuga, 816, Japan.

Prof. MOORE, D.W., Department of Mathematics, Imperial College, London, S.W.7, England.

Prof. PAULLING, J.R., Department of Naval Architecture and Offshore Engineering, University of California, Berkeley, California 94720, USA.

Prof. PEDERSEN, P. TERNDRUP, Institute of Ocean Engineering, The Technical University of Denmark, DK-2800 Lyngby, Denmark.

Prof. PRICE, W.G., Department of Mechanical Engineering, Brunel University, Uxbridge, Middlesex, UK.

Prof. ROGULA, D., Polish Academy of Sciences, Institute of Fundamental Technological Research, ul. Swietokrzyska 21, 00-049 Warsaw, Poland.

Prof. SAFFMAN, P.G., Applied Mathematics 217-50, California Institute of Technology, Pasadena, California 91125, USA.

Prof. SALENCON, J., Ecole Polytechnique, Ecole Nationale des Ponts et Chaussées, Laboratoire de Mécanique des Solides, 91128 Palaiseau Cedex, France.

Prof. SCHIEHLEN, W.O., Institute B of Mechanics, University of Stuttgart, Stuttgart, FRG.

Prof. TANEDA, S., Research Institute for Applied Mechanics, Kyushu University, Kasuga, Fukuoka 816, Japan.

LIST OF PARTICIPANTS

(Authors of papers presented at the Congress are indicated by an asterisk)

*** ALGERIA
(1 participant)

* Boulala, M.

*** AUSTRALIA
(6 participants)

* Grimshaw, Roger H.J.
* Karihaloo, Bhushan L.
McElwain, Sean
* Philip, John R.
* Rozvany, George I.N.
Tuck, Ernest O.

*** AUSTRIA
(4 participants)

* Kluwick, Alfred
* Schneider, Wilhelm
* Zauner, Erwin
* Ziegler, Franz

*** BELGIUM
(11 participants)

Boucher, Serge
Janssens, Paul A.
Kestens, Jean
* Ladriere, Patrice
* Lebon, Georgy
Lefeber, Dirk
* Mertens, Robert A.
* Ottoy, Jean-Pierre
Phmelevskaja-Plotnikova, G.
* Platten, Jean
Save, Marcel

*** BRAZIL
(3 participants)

* Bevilacqua, Luiz
* Koiller, Jair
* Weber, Hans I.

*** BRD
(43 participants)

Besdo, Dieter
* Boer, Reint de
Bohme, Gert
Das, Arabindo
* Detemple, Elisabeth
* Eckelmann, Helmut
* Ehlers, Wolfgang
* Ellermeier, Wolfgang
Emmerling, Franz A.
* Esslinger, Maria
Fazio, Claudio
* Gaul, Lothar
* Geier, Bodo
Grabe, Gunter
* Grabitz, Georg
Gravert, Peter
Hornung, Hans
Kienzler, Reinhold
* Kirchgassner, K.
Kosinski, Witold
* Krawietz, Arnold
* Kreuzer, Edwin J.
* Labisch, Franz K.
* Lehmann, Theodor
* Lippmann, Horst
* Mahrenholtz, Oskar H.G.
Matthies, Hermann
* Meier, Gerd E.A.
Muller, Ernst-August
* Muller, Ulrich
* Obermeier, Frank
* Pecherski, Ryszard B.
Popp, Karl
* Rohwer, Klaus
* Schiehlen, W.
Schlegel, Volker
* Schmidt, Herbert
* Schwieger, Horst
* Stumpf, Helmut
* Vasanta-Ram, V.I.
* Wauer, Jorg
* Weber, Herbert
* Weichert, Dieter

*** BULGARIA
(2 participants)

* Ivanov, Tsolo P.
* Kozarov, Marin M.

*** CANADA
(32 participants)

Blackwell, John H.
Bojadziev, George N.
* Camotim, Dinar
Chater, Elie
Curran, John
Davies, Hun G.
* Ellyin, Fernand
* Epstein, Marcelo
* Gladwell, Graham
* Glass, Irvine I.
* Haddad, Yehia M.
* Hinchey, Michael J.
* Huseyin, Koncay
* Krasinski, Joseph de
* Kujawski, Daniel
* Leutheusser, Hans J.
Mandl, Paul
* Neale, Ken
* Newman, Barry G.
* Novak, Milos
* Parkinson, Geoffrey V.
* Pindera, Jerzy T.
Rasmussen, Henning
* Rimrott, Fred
* Savage, Stuart B.
* Selvadurai, Patrick
* Tabarrok, Bez
* Tennyson, Rod
Tory, Elmer M.
Vandamme, Luc
* Vaughan, Henry
* Vinogradov, A.M.

*** CHINA
(37 participants)

* Bai, Yilong
* Chao, Ching Cheng
Chen, Zhi-da
* Chen, Sixiong
* Cheng, Gengdong
Cheng, Pei-Ji
Cheng, Chi-Ta
* Chien, Wei Zang
* Fan, Jinghong
Gai, Bing Zheng
* Gao, Yu-chen
* He, Fubao
* He, You-Sheng
* Huang, Yanzhang
* Jin, Wenlu
* Li, Hao
Lin, T.C.
* Ling, Fuhua
* Liu, Chengqun
* Liu, Feng
Qiu, Ji bao
* Shen, Hui-li
* Shen, Kuo-Shan
Shi, Guang-yi
Sun, Fang-Toh
Sun, Mingru
* Wang, Ren
Xu, Bo Hou
Yang, Gui-tong
* Ye, Kaiyuan
Ye, Shanghui
* Yu, Mao-Hong
* Zhang, Ruqing
* Zhang, Jin
* Zhang, Wen
* Zhao, Ling-cheng
Zhou, Pei-Yuan

*** CSSR
(8 participants)

Curev, A.G.
Danek, Milan
* Hyca, Milan
* Kafka, Vratislav
Kolar, Vaclav
Krupka, Vlastimil
Nemec, Jaroslaw A.C.
* Pichal, M.

*** DDR
(3 participants)

* Altenbach, Johannes
Hennig, Klaus
* Hoffmeister, Manfred

*** DENMARK
(75 participants)

Aage, Christian
Andersen, Poul
Andersen, Sven Ib
* Behrendt, Lars
* Bendsoe, Martin P.
Birker, Bertel
* Bjorno, Leif
* Braestrup, Mikael W.
* Brons, Morten
Byskov, Esben

Cederkvist, Jan
Christiansen, Edmund
Christoffersen, Jes
Damkilde, Lars
Ditlevsen, Ove
Dyrbye, Claes
Fabian, Ole
Fenger, Niels Peter
* Fredsoe, Jorgen
Goltermann, Per
* Gunneskov, Ole
Haahr, Henrik
Hansen, Erik B.
Hansen, John M.
Hartig, Arne
Jakobsen, Tom B.
Jakobsen, Stig Baungaard
Jensen, Niels O.
Jensen, Jorgen J.
Jensen, Jarl
Jensen, Henrik M.
Jensen, John Korsgaard
Jensen, Henrik
* Jonsson, Ivar G.
Jorgensen, Henrik K.
* Justesen, Peter
* Kaas-Petersen, Christian
Kildegaard, Arne
* Knudsen, Thomas S.
* Krenk, Steen
* Kristensen, Hans Saustrup
Ladefoged, Troels
* Larsen, Jesper
* Larsen, Poul S.
Larsen, Gunner
Laursen, Mogens
Madsen, Per J.
Maegaard, Karl Aa.
Matulewicz, Zofia Maria
Mollmann, Hugo
* Morch, Knud Aa.
Nielsen, Arne Gudmann
Niordson, Johan
Niordson, Frithiof I.
* Olhoff, Niels
* Pedersen, Preben T.
Pedersen, Pauli
Pedersen, B. Maribo
Pommer, Christian
* Pyrz, Ryszard
* Ravn-Jensen, Kim
Rickelt, Jens
Schonfeldt, Henrik
Simonsen, Jens
* Skovgaard, Ove
Sorensen, Jens Norkaer
Svendsen, Ib A.
* Talreja, Ramesh
Thomsen, Soren Gyde
True, Hans
Tryde, Per
Tuttle, Mark E.
* Tvergaard, Viggo
West, Michael P.
Zhang, Hong-Zhao

*** EGYPT
(1 participant)

* El Chazly, Nihad M.

*** FINLAND
(5 participants)

* Boehm, Juhani von
* Parland, Herman
* Pylkkanen, Jaakko V.
Ranta, Matti A.
Raty, Raimo

*** FRANCE
(63 participants)

* Adler, Pierre M.
Ando, Kohei
* Atten, Pierre
* Azouni, Maherzia A.
* Barthes-Biesel, Dominique
Brousse, Pierre
Brun, Louis C.J.
Cabannes, Henri
* Calamote, Jacqueline
* Caputo, Jean G.
* Chomaz, Jean-Marc
* Chossat, P.
Coirier, Jean
* Collet, Bernard
* Couder, Yves
* Dahan, Marc
* Debieve, J.F.
* Devienne, Rene
Dodu, Jacques
Duvaut, Georges
* El Fatmi, Rached
* Fauve, Stephan
Favre, Alexandre J.A.
Finifter, Daniel
* Frisch, Uriel
Germain, Paul
Guiraud, Jean-Pierre
* Iooss, G.
* Kida, Shigeo
* Klepaczko, Janusz

Lagache, Jean-Marie
Lagarde, Alexis
Lanchon, Helene
* Lebouche, Michel
* Lemaitre, Jean
* Lene, Francoise
* Lenoir, Marc
Levyn, Cecile
* Libchaber, A.
* Malraison, Bernard
* Maugin, Gerard A.
Moreau, Rene J.
Muller, Patrick
* Nguyen, Quoc Son
* Ohayon, Roger
Orsero, Pierre
Piau, Monique
Plotard, M.
* Pouget, Joel
* Rabaud, Marc
Roseau, Maurice
Roussel, Michel
Roy, Maurice
* Salencon, Jean
* Sommeria, Joel
Souchet, Rene
* Suguet, Pierre
Thiry, Yves R.
* Touratier, Maurice
* Tuckerman, Laurette
* Valid, Roger
* Verron, Jacques
* Vullierme-Ledard, Martine

*** GREECE
(4 participants)

* Baniotopoulos, C.C.
Georgantopoulos, George A.
Kikeras, Ph. J.
Stogianidis, P.

*** HOLLAND
(39 participants)

* Arbocz, Johann
Bakker, P.J.M. de
* Bakker, Peter G.
* Beek, Peter van
Besseling, J.F.
* Biesheuvel, Arie
Blaauwendraad, J.
* Bos, Hans
Bruin, Gerrit J. de
Campen, Dick H. van
* Coene, Rene
* Dieterman, Harm A.
Geerlings, Joseph J.P.
Giessen, Erik van der
* Gorissen, Wim C.M.
Grootenboer, Henk J.
Heijst, Gert-Jan van
Heugten, Petrus C.M. van
Heyden, Arnold van der
* Jansen, A.J.M.
Janssen, J.D.
* Jonker, J. Ben
Koiter, Warner T.
Koster, Willem G.
Kuypers, Wim
Meijers, Pieter
Ommen, Klaas van
Omta, Roel
* Radder, A.C.
* Schouten, Gerrit
* Schwab, Arend L.
Sevenster, Arjen
Soerjadi, R.
* Steketee, Jakob A.
Velthuizen, Hendricus G.M.
Ven, Alphons A.F. van de
Vossers, Gerrit
Weisenborn, Toon
Wijngaarden, Leen van

*** HUNGARY
(3 participants)

* Bosznay, Ádam
* Pomazi, Lajos
* Tarnai, Tibor

*** INDIA
(3 participants)

* Arya, Vinod K.
* Narasimha, Roddam
* Sharma, R.N.

*** IRAN
(3 participants)

Eslami, M. Reza
Shadman, Dariush
Vafai, Abolhassen

*** IRELAND
(2 participants)

Hayes, Michael
Kelly, George V.

*** ISRAEL
(28 participants)

* Aboudi, Jacob
* Banks-Sills, Leslie
Bar-Yoseph, Pinchas
Baruch, Menachem
* Beltzer, Abraham
* Benveniste, Yakov
* Bodner, Sol R.
* Durban, David
* Elishakoff, Isaac
Gan-Mor, Samuel
* Hashin, Zvi
Kochavi, Eytan
Leibowitz, Moti
Libai, Avinoam
* Librescu, Liviu
Lifshitz, J.
Lottati, I.
Nissim, E.
* Parnes, Raymond
Partom, Yehuda
Pnueli, David
Rautu, S.
Rotem, A.
Rubin, Miles
* Schmidt, Rudiger
Shilkrut, Dov
Singer, Josef
Siton, Givon

*** ITALY
(26 participants)

Amabile, Tatone
* Bergamaschi, Silvio
Bianchi, Giovanni
Boffi, Vinicio
Cattaneo, Gianpiero
* Cercignani, Carlo
* Cinquini, Carlo
* Como, Mario
* Corradi, Leone
Germano, Massimo
* Grimaldi, Antonio
Grioli, Giuseppe
* Lucchetti, Fabrizio
Maffioli, Giancarlo
Maier, Giulio
* Martelli, Francesco
* Matteis, Guido de
* Orlandi, Paolo
Pantano, Pietro
* Pignataro, Marcello
Prosperetti, Andrea
* Schmidt, Henrik
Sinopoli, Anna
Spinelli, Giancarlo
* Stagni, Luigi
* Vestroni, Fabrizio

*** JAPAN
(43 participants)

Abe, Takeji
* Arai, Makoto
* Fukasawa, Toichi
* Hasimoto, Hidenori
Ikeda, Tamon
Imai, Isao
* Ishii, Katsuya
* Ishikawa, Hiromasa
* Kambe, Tsutomu
Kawai, Tadakiko
* Kawamura, Tetuya
* Kawashima, Koichiro
* Kitagawa, Hiroshi
* Kiya, Masaru
* Kobayashi, Yasunori
* Koga, Tatsuzo
Kozo, Kawata
* Kuwahara, Kunio
* Maruo, Hajime
* Mitsuyasu, Hisashi
Miya, Kenzo
* Miyazaki, Takeshi
* Mizushima, Jiro
* Nishimura, Tetsu
Okamura, Hiroyuki
* Onishi, Yoshimoto
* Ooka, Masahiro
* Oshima, Nobunori
* Oshima, Koichi
* Saito, Yoshio
* Sakai, Yasuhiko
* Sano, Masaki
* Seguchi, Yasuyuki
* Sone, Yoshio
* Sumi, Yoichi
* Tada, Yukio
* Takahashi, Kazuo
* Tanaka, Eiichi
* Tanaka, Masao
* Taneda, Sadatoshi
Tani, Itiro

* Toda, Susumu
* Yamamoto, Yoshiyuki

*** MEXICO
(2 participants)

* Peralta-Fabi, Ricardo
* Sabina, Federico J.

*** NEW ZEALAND
(1 participant)

* Hosking, Roger J.

*** NIGERIA
(1 participant)

* Mahanti, P.K.

*** NORWAY
(4 participants)

* Bergan, Pal G.
* Faltinsen, Odd M.
* Papatzacos, Paul
* Tveitereid, Morten

*** POLAND
(17 participants)

* Bauer, Jacek M.
* Bogacz, Roman
  Fiszdon, W.
* Grysa, Krzysztof
* Grzedzinski, Janusz
* Kowalewski, Tomasz A.
* Lekszycki, Tomasz
* Litewka, Andrzej
* Mroz, Zenon
* Perzyna, Piotr
  Pietraszkiewicz, Wojciech
* Rakowski, Jerzy
* Rogula, Dominik
* Sokolowski, Jan
* Sygulski, Ryszard
* Walenta, Zbigniew A.
* Zyczkowski, Michal

*** PORTUGAL
(1 participant)

* Campos, L.M.B.C.

*** ROUMANIA
(4 participants)

* Atanasiu, Nicolae
* Chiriacescu, Sergiu
* Isarie, I.
* Sarbu, Nicolae

*** SAUDI ARABIA
(2 participants)

  Badr, Hassan
* Doyle, John F.

*** SINGAPORE
(1 participant)

* Wang, Chien Ming

*** SOUTH AFRICA
(1 participant)

  Moes, Johannes

*** SPAIN
(3 participants)

* Casas-Vazquez, Jose
* Diaz, Francesc
* Jimenez, Javier

*** SWEDEN
(38 participants)

* Alfven, Hannes
* Amberg, Gustav
* Andersson, Lars-Erik
Aronsson, Gunnar
Bark, Goran
* Bark, Fritz
Bodelind, Bertil
Broberg, Leif
Broberg, K. Bertram
Carlsson, Claes-Goran
* Carlsson, Janne
* Dahlkild, Anders
Digby, Peter J.
Dillstrom, Peter
* Enflo, Bengt
* Hoegfors, Christian
Holmberg, Bengt R.
* Hsieh, R.K.T.
Huang, Yuanzhong
Hult, Jan
Jansson, Per-Ake
Jansson, Stefan
* Jonsson, Mikael
* Karlsson, Lennart
Karlsson, Lars
* Klarbring, Anders
Klasen, Bjorn
* Lundberg, Bengt
Magi, Mart
Nilsson, J.E.V.
Nordlund, Erling
* Soderholm, Lars H.
Stahle, Per
Stigh, Ulf
Storakers, Bertil
Sundin, K-G
Wihlborg, Goran
* Zhou, Shu-Ang

*** SWITZERLAND
(6 participants)

* Deger, Yasar
Hauviller, Claude
* Hutter, Kolumban
* Moser, Alfred
Partl, Manfred
Sayir, Mahir

*** TUNISIA
(1 participant)

Abdelhadi, A.

*** TURKEY
(3 participants)

* Boduroglu, Hasan
* Demiray, Hilmi
Dikmen, Murat

*** UK
(54 participants)

Ansell, Jim
Austin, David M.
Batchelor, George K.
Benjamin, T. Brooke
* Berry, Micheal
* Brindley, John
Britter, Rex E.
* Brown, Eric H.
* Burdess, James
* Burgoyne, Chris
Byatt-Smith, John G.
* Calladine, Christopher R.
* Cowell, Robert G.
* Daniels, Peter G.
* Davies, John T.
* Downie, Martin J.
* Glendinning, Paul
* Greated, Clive Alan
* Gregory, R. Douglas
Grundy, Robert E.
* Hall, Philip
* Harris, David
Helliwell, John B.
* Hinch, Edward J.
Holland, Charles
Hunt, Julian C.R.
* Hunt, Giles W.
Johannesen, Niels H.
Johnson, Kenneth L.
* Jones, Norman
Lighthill, James
Michael, David H.
* Mobbs, Stephen D.
Moffatt, H. Keith
* Moore, Derek W.
* Pellegrino, Sergio
* Peregrine, D.H.
* Price, W.G.
* Roberts, John W.
Robinson, Peter D.
Shail, Ronald
* Shanmugasundaram, V.
Smith, Michael R.
Smith, Frank I.P.
* Spence, David
Stephen, Neil
* Talbot, David

* Thornton, Colin
* Ursell, Fritz
  Walpole, Leslie J.
  Wang, Peiji
* Willis, John
  Wills, John
  Wittrick, William H.

*** USA
(183 participants)

  Achenbach, Jan D.
  Acrivos, Andreas
  Adams, George G.
* Advani, Sunder H.
* Akylas, Triantaphyllos
* Antar, Basil N.
* Aref, Hassan
* Babcock, Charles
* Bajaj, Anil K.
* Bammann, Douglas J.
* Barber, James R.
* Bassani, John
* Bendiksen, Oddvar O.
* Berger, Stanley A.
* Bogy, David B.
  Boley, Bruno A.
* Bolton, Edward W.
* Brady, John F.
* Brenner, Howard
  Broeck, Jean-Marc van den
* Buckmaster, John
* Budiansky, Bernard
* Burton, Thomas D.
  Buskirk, William van
* Caflisch, Russel
* Carroll, Michael M.
* Casey, James
  Chen, Hwei-Ju
  Chen, Yu
* Chen, C.F.
* Cheng, Hsien K.
  Chou, Shun-Chin
* Christensen, Richard M.
  Chulay, Steven
* Clifton, Rodney J.
* Comninou, Maria
* Cowin, Stephen C.
* Cramer, M.S.
  Cranch, Edmund
* Crandall, Stephen H.
  Dally, James W.
* Datta, Subhendu K.
* Dolgopolsky, Alexander
* Dowell, Earl
  Drucker, Daniel C.
* Dunayevsky, Victor A.
* Dvorak, George J.
* Dybbs, Alexander
* Feng, William W.
* Fisher, Martin J.
  Flugge, Wilhelm
* Freund, L. B.
* Gautesen, Arthur K.
* Gorlov, Alexander
  Gupta, K.K.
* Hall, Mary S.
* Hanagud, S.V.
* Harris, John G.
* Havner, Kerry S.
* Hendriks, Ferdinand
* Henriksen, Mogens
  Herakovich, Carl T.
* Herrmann, George
  Hetnarski, Richard B.
  Hodge, Philip
  Hoff, Nicholas J.
* Holmes, Philip J.
* Hsu, C.S.
* Hsu, Chen-Chi
* Huang, T.C.
* Hutchinson, John W.
* Hutchinson, James R.
* Ibrahim, Raouf A.
* Inman, Daniel
* Iversen, James D.
* Jenkins, James T.
* Johnson, Robert E.
* Johnson, Millard W.
* Johnson, George C.
* Jones, Frederick L.
  Juhasz, Stephen
* Kachanov, Mark
* Kant, Rishi
* Karim-Panahi, Khosrow
* Katsube, Noriko
* Keller, Joseph B.
* Kerr, Arnold D.
  Kirmser, Philip G.
* Kohn, Robert V.
* Koschmieder, Lothar
* Krajcinovic, Dusan
  Ku, Y.H.
* Kwatny, Harry
* Kyriakides, Stelios
* Lakin, William D.
* Lee, Erastus H.
* Lee, Jon
* Leissa, Arthur W.
* Levinson, Mark
* Libove, Charles
* Lin, S.R.
* Lin, T.H.
* Mac Sithigh, Gearoid P.
* Magnuson, Allen H.
* Majerus, John
* Margolis, Stephen B.
* Markenscoff, Xanthippi

* McCoy, J.J.
* Meecham, William C.
* Melville, W. Kendall
* Miksis, Michael J.
* Miller, Gregory R.
* Mitchell, Thomas P.
* Mook, Dean
* Moon, Francis C.
* Mukherjee, Subrata
* Muller, Michael R.
* Naghdi, Paul M.
Nash, William A.
Ndefo, Ejike D.
* Needleman, Alan
* Newman, J.N.
Nied, Herman F.
* Nunziato, Jace W.
* Ochoa, Ozden
* Pariseanu, George
* Paulling, J. Randolph
* Pearlstein, Arne J.
Pence, Thomas J.
* Pian, Theodore H.H.
* Plaut, Raymond H.
* Ramaprian, Belakavadi
* Rashed, Ahmed
* Reifsnider, Kenneth L.
* Reinhall, Per
* Renouard, Dominique
* Rice, Jim
* Richman, Mark W.
Richmond, Owen
Rogers, Arthur W.
* Sachse, Wolfgang
* Saffman, Philip
* Sarpkaya, Turgut
Schaffers, W.J.
* Schreyer, Howard L.
* Sclavounos, Paul
Senechal, Lester J.
Sethna, P.R.
* Shaw, Steven W.
* Siginer, Aydeniz
* Simmonds, James G.
Skalak, Richard
* Smith, C.W.
* Stadler, Wolfram
Steele, Charles R.
* Storti, Duane
* Strang, Gilbert
* Sture, Stein
* Su, T.C. Joe
* Subrahmanyam, K. Bala
* Sumner Jr., Eric
Talke, Frank E.
* Taylor, John E.
Thompson, Philip D.
* Ting, T.C.T.
* Traugott, S.C.
* Trevino, George
* Triantafyllidis, Nicolas
Tsai, Stephen W.
Ungarish, Marius
Vogelius, Michael
* Voyiadjis, George Z.
* Walton, Otis R.
* Wan, Frederic Y.M.
* Wang, Chang-Yi
Wang, John D.
* Wang, S.S.
* Weaver, Richard
Weitsman, Yechiel
* Weng, George J.
Williams, Harry E.
* Wilson, James F.
* Wu, Han-Chin

*** USSR
(11 participants)

* Ambartsumian, Sergey A.
* Beletsky, V.V.
Ishlinsky, Alexandr
Khristianovich, A.
* Lyubimov, Grigory
Mikhailov, Gleb K.
* Mossakovsky, Vladimir I.
Portnykh, S.
Sedov, Leonid
Shilov, Alexandr
* Yavorskaya, Inna M.

*** YUGOSLAVIA
(2 participants)

* Djukic, Djordje S.
* Picuga, Alija

REPORT ON THE CONGRESS

Frithiof I. Niordson

The decision to accept the invitation from Denmark to hold the XVI'th International Congress at the Technical University in Lyngby was taken by the Congress Committee of IUTAM during its meeting in August 1980 in Toronto, Canada. As before, it was decided that the Congress would cover the entire field of mechanics: analytical, solid, and fluid mechanics, including applications. However, the urge to try new ways and improve the scientific value of such a meeting led the Congress Committee to decide on several changes of earlier policies and the introduction of some new activities. Thus, a few topics were to be selected for special emphasis and furthermore, for the first time, poster-sessions would appear.

The decisions of the Congress Committee were implemented by its Executive Committee, which worked in close cooperation with the local organizers. Among a number of proposals the Executive Committee finally selected three topics for special emphasis, namely

* Marine-structure wave interaction
* Micromechanics of multicomponent media
* Development of chaotic behaviour in dynamical systems

Convenors were appointed to select lecturers for an introductory session with an instructional element intended for non-specialists and to coordinate the sessions with invited sectional lectures and contributed papers.

The decision to hold poster-sessions was preceeded by thorough discussions in the Congress Committee, *pro et contra*. When finally the decision was taken it was with the firm intention to give contributed papers accepted as posters and as lectures the same status. The unquestionable success of the poster-sessions as witnessed by an overwhelming majority of the participants, was certainly partly due to the principle that the posters should not be rated as papers of secondary quality.

The Congress Committee had decided that there would be only two general lectures, and invited the Nobel laureate professor Hannes Alfvén (Sweden) to give the opening lecture and professor Joseph Keller (USA) to give the closing lecture. In addition, 16 sectional lecturers were invited.

The number of contributed papers submitted to the congress was 787. Of these, 418 were presented, 235 as lectures and 183 as posters. The distribution according to countries was as follows:

Country	Submitted	Presented	Lectures	Posters
Algeria	2	1	1	0
Australia	6	4	4	0
Austria	3	2	2	0
Belgium	9	5	2	3
Brazil	8	3	1	2
BRD	34	21	10	11
Bulgaria	6	2	1	1
Canada	39	23	17	6
Chile	1	1	0	1
China	99	22	9	13
CSSR	11	3	1	2
DDR	2	2	1	1
Denmark	20	17	8	9
Egypt	8	1	0	1
Finland	6	3	1	2
France	52	29	17	12
Greece	4	1	0	1
Holland	17	12	7	5
Hungary	5	3	2	1
India	14	3	3	0
Israel	15	9	6	3
Italy	18	11	6	5
Japan	35	28	14	14
Mexico	2	2	1	1
New Zealand	1	1	1	0
Nigeria	4	1	0	1
Norway	3	3	2	1
Poland	27	13	7	6
Portugal	4	1	0	1
Roumania	14	3	1	2
Saudi Arabia	2	1	0	1
Singapore	1	1	1	0
Spain	3	2	2	0
Sweden	12	10	4	6
Switzerland	2	2	1	1
Turkey	11	2	1	1
UK	36	23	13	10
USA	231	139	82	57
USSR	17	6	5	1
Yugoslavia	3	2	1	1
All	787	418	235	183

In BRD, Canada, France, Japan, Poland, UK, USA, and USSR National Committees reviewed papers from within their own country and made recommendations to the International Papers Committee, which was authorized by the Congress Committee to select papers for presentation at the meeting. The selection was based on summaries and abstracts as announced in the call for papers. The contributed papers presented at the congress are listed below on pp. 411-435.

The Congress was based on pre-registration and 820 scientists registered before and during the meeting. 39 late cancellations reduced the number of active participants to 781, not including 218 accompanying persons. In all, 43 countries were represented and the distributions by country is given below on pages xvii-xxvi.

*

During the Congress the weather in Denmark was sunny and the air comfortably warm. On the Sunday evening preceeding the Congress a reception was arranged in down town Copenhagen at "Den Gyldne Fortun" where about 600 early comers met and received their briefcases.

The official opening of the Congress took place in the Assembly Hall of the Technical University on Monday at 9.15 hours. Two musicians from the Royal Orchestra blew a fanfare on ancient Danish horns. The session was opened by Professor FRITHIOF NIORDSON, President of the Congress with the following words:

"Mr. Minister,
Mr. Rector,
Highly Honoured Guests,
Ladies and Gentlemen.

With the sound of these horns I call this meeting to order, and on behalf of the organizing committee I welcome you to the XVIth International Congress on Theoretical and Applied Mechanics.

To those of you, who come from abroad, we danes extend a hearty welcome to Denmark.

The logo of the Congress has been interpreted by some of you as a ball-bearing, as boiling water, as solid particles suspended in fluid, and what not... Indeed, it may be all that, but for us it is a special trade mark, the front end of those bronze horns, we just heard, and which were used in this country some three thousand years ago, when a meeting was summoned. More than thirty of these musical instruments have been found in Denmark, always in symmetrical pairs.

We thought it would be appropriate to use them today for the first International Congress in Mechanics ever held in Denmark. I should add that - of course - we do not know what kind of meetings were inagurated with the sound of these horns in the Bronze Age, but they were probably not congresses in Mechanics.

In 1930, more than 50 years ago, the Third International Congress was held in Stockholm - the first and until now - the only one ever held in Scandinavia. I have compared the list of participants from that meeting with ours.

The 1930 list is printed on glossy paper, and you need no magnifying glass to read the names. Indeed, it contains an oval photografic portrait of every participant; many names of which are now famous: von Karman, G.I. Taylor, Prandtl, Oseen, Timoshenko and many, many more. Not many of them are here today, but I do recognize No. 189, Maurice Roy, Ingénieur au Corps de Mines. Professor Roy, who was former President of this Union, is now one of the honoured guests of the Congress."

The Minister of education, Mr. BERTEL HAARDER then welcomed the congress participants and said:

"Mr. Chairman,
Ladies and Gentlemen.

I am very pleased with this opportunity to welcome a congress of this dimension. The topics of this IUTAM Congress are of crucial importance to the development of science and industry. The professional level of the IUTAM Congresses is considered very high. We therefore take the Union's decision of placing the Congress in Denmark as an acknowledgement of Danish research in this area.

I am sure that the Danish organizing committee will respond to this challenge. You are, Ladies and Gentlemen, attending this Congress in an institution founded by the discoverer of electromagnetism and aluminum, the Danish scientist H.C. Ørsted, and you have just heard the sound of the oldest Danish information technology, although the horns were most likely designed for warfare or religious rituals.

In this Congress there are many interesting themes, which will be discussed. They are outstanding examples of cooperation between theory and application, between science and industry. In a recent address to the Danish Parliament I have stressed the need for basic research as a support for technological development. The Danish Government has launched 2 new programs totalling 180 million Dollars to cover development and research, and the Government has also urged both the public and the private sector to build up an open-minded cooperation. I should add, that the additional funds I mentioned are on top of the regular funds, so it is an increase.

On this background it is encouraging to host a Congress persuing the goals of cooperation between the public and the private sector. Furthermore, international cooperation in the research and development area has a high priority in Danish research policy. Therefore, we are very pleased to have this opportunity to enlarge and further our international scientific relations especially because of the high quality of work performed in your area. I feel convinced that the scientific research presented here will have a long term effect on the international cooperation, and in particular, Danish scientists will take the opportunity to further their international contacts. As a small country we have long time ago decided to participate as much as possible in international cooperation within research and science. We simply have no alternative, if we want to keep up with the international level in the future as we have done successfully in several fields in the past.

This Congress must be some sort of an Anniversary Congress, since I know that the first Congress was held in 1924. I would like to congratulate you on this Anniversary, and I would like to thank you once again for placing this very important Congress in Denmark in this place, and finally, Ladies and Gentlemen, on behalf of the Danish Government I once again welcome you to this country, and I hope that you will have some fruitful and good days in Copenhagen. Thank you."

The Rector of the Technical University, Dr. PETER LAWÆTZ spoke on behalf of the host institution. He said:

"Mr. President,
Mr. Chairman,
Ladies and Gentlemen.

As Rector of the Technical University of Denmark it is a great pleasure for me to welcome you on our campus as participants of this XVIth International Congress of Theoretical and Applied Mechanics.

In the early 19th Century, when this university was created as an independent institute of applied sciences, Mechanics was one of the predominant subjects. The word applied had two senses at that time. One was application to practical life and thus engineering, the other was connected with the idea of performing experiments. Something not really esteemed at the pure university at the time. It is obvious that Mechanics has come a long way since that initial period. The present stand of the subject is well illustrated in the programme of your Conference. However, the two old features of experimentation and practical connection are well preserved as facets of applied science. Even in the theoretical area a new development of experimentation has more recently started up by the use of large computers. I hope that this feature will continue, and will prosper. The Technical University is proud to host such a distinguished scientific conference; it gives a place and atmosphere of internationalism, which in our daily life could need some improvement. It also stresses the point that scientific research is an integral part of an institution of higher education. I am happy to observe that many members of our staff are participants in the conference.

Every time I visit another university, I see very few students. Of course, there are many good explanations for that. If you lack them here, it is because the fall term has not yet begun. On the other hand, if you had seen the number of students that we are expecting in the near future, due to a political expansion of the intake at this university, I think that you would rather not have come. The idea is, of course, that you use the facilities normally available to them, and since they are expected to be quite crowded, there would be no space for you. I hope that you will have good use for the facilities now available for you, the auditoria, the canteen, and even this hall, which, as you can see, is programmed for sports, if not used for written examinations. I also hope that the good weather along with other things will support the scientific and social programme of your conference, and I thus wish you a profitable stay in Copenhagen and in Lyngby. Thank you."

Finally Professor DANIEL C. DRUCKER, President of IUTAM officially opened the Congress with the following words:

"Thank you Professor Niordson,

It is indeed my priviledge and pleasure to officially open this Congress as President of your organization, IUTAM. We are deeply grateful to the organizing committee, Professor Niordson and Professor Olhoff in particular, and their many colleagues, who spent so many days, so many weeks, so many months in long hours of preparation for us today. Our choice of this location was, indeed, a recognition of the very high level of research conducted here in Denmark, and at the Technical University in particular. We are honoured that we have had both the Minister of Education, and the Rector of the Technical University of Denmark here to greet us, and in turn express our appreciation to them and to all of Denmark for acting as hosts. I think, it is particularly appropriate to have had the musical introduction that we did have, although Professor Niordson suggested that perhaps the first use of these instruments was not to call a congress of applied mechanics to order. I would point out that Bronze Age instruments and the roots of our subject Mechanics do, indeed, have much in common.

The instruments are certainly symbolic of roots extending far back into antiquity, and at the same time, as you have heard today, they are able to project modern music, and I presume, the music of the future. And so to our subject, which goes far back into History. It is an active subject at present, and will be an active subject, and in fact the foreteller of much of the technology of the future. As the Minister said: Basic research provides tools, on which our civilization does depend. So it is again my pleasure to officially open the Congress and return the meeting back again to Professor Niordson."

*

A summary survey of the scientific programme is given below.

MONDAY, 20 AUGUST

09.15 - 09.45	Opening (Assembly Hall)
09.45 - 10.45	Opening lecture (Assembly Hall) H. ALFVÉN: Space Research and the New Approach to the Mechanics of Fluid Media in Cosmos
11.15 - 12.15	Sectional lectures
13.45 - 14.45	Parallel sessions - Initial lecture of the special session "Marine-structure Wave Interaction" - Contributed papers, lectures - Contributed papers, posters General display
15.00 - 17.00	Parallel sessions - Initial lectures of the special session "Marine-structure Wave Interaction" - Contributed papers, lectures - Contributed papers, posters Discussions with authors 15.00 - 16.00 General open discussions 16.00 - 17.00.

TUESDAY, 21 AUGUST

09.00 - 10.00	Parallel sessions - Initial lecture of the special session "Micromechanics of Multi-component Media" - Contributed papers, lectures
10.15 - 12.15	Parallel sessions - Initial lectures of the special session "Micromechanics of Multi-component Media" - Contributed papers, lectures
13.45 - 14.45	Parallel sessions - Sectional lectures - Contributed papers, posters General display
15.15 - 16.45	Parallel sessions - Contributed papers, lectures - Contributed papers, posters Discussions with authors 14.45 - 15.45 General open discussions 15.45 - 16.45.

WEDNESDAY, 22 AUGUST

09.00 - 10.00	Parallel sessions - Initial lecture of the special session "Development of Chaotic Behaviour in Dynamical Systems" - Contributed papers, lectures
10.15 - 12.15	Parallel sessions - Initial lectures of the special session "Development of Chaotic Behaviour in Dynamical Systems" - Contributed papers, lectures

THURSDAY, 23 AUGUST

09.00 - 10.00	Parallel sessions - Sectional lectures - Contributed papers, posters General display
10.15 - 12.15	Parallel sessions - Contributed papers, lectures

- Contributed papers, posters
  Discussions with authors 10.15 - 11.15
  General open discussions 11.15 - 12.15

13.45 - 14.45 Parallel sessions
- Contributed papers, lectures
- Contributed papers, posters
  General display

15.00 - 17.00 Parallel sessions
- Contributed papers, lectures
- Contributed papers, posters
  Discussions with authors 15.00 - 16.00
  General open discussions 16.00 - 17.00.

FRIDAY, 24 AUGUST

09.00 - 10.00 Parallel sessions
- Sectional lectures
- Contributed papers, posters
  General display

10.15 - 12.15 Parallel sessions
- Contributed papers, lectures
- Contributed papers, posters
  Discussions with authors 10.15 - 11.15
  General open discussions 11.15 - 12.15

13.45 - 14.45 Parallel sessions
- Contributed papers, lectures
- Contributed papers, posters
  General display

15.15 - 16.45 Parallel sessions
- Contributed papers, lectures
- Contributed papers, posters
  Discussions with authors 14.45 - 15.45
  General open discussions 15.45 - 16.45.

SATURDAY, 25 AUGUST

09.00 - 10.30 Contributed papers, lectures
11.00 - 12.00 Closing lecture (Assembly Hall)
J. KELLER:
Computers and Chaos in Mechanics
12.00 - 12.15 Closing (Assembly Hall).

*

The Lord Mayor of Copenhagen had cordially invited the participants and accompanying persons of the Congress to a reception and a buffet style dinner at the City Hall of Copenhagen on Tuesday, 21 August at 7 p.m.

An excursion to Frederiksborg Castle in Hillerød and Kronborg Castle in Elsinore was arranged for participants and accompanying persons on Wednesday afternoon.

The Congress Banquet was held at Hotel Marienlyst in Elsinore on Friday, 24 August. About 650 participated in the dinner. After dinner dancing to a jazz orchestra continued until well after midnight.

Theoretical and Applied Mechanics
F.I. Niordson and N. Olhoff
Elsevier Science Publishers B.V. (North-Holland)

SPACE RESEARCH AND THE NEW APPROACH TO THE MECHANICS OF FLUID MEDIA IN COSMOS

Hannes Alfvén[1,2] and Franz Čech[1]

1) The Royal Institute of Technology
Department of Plasma Physics
S-100 44 Stockholm
Sweden

2) University of California
San Diego, Mail Code C-014
La Jolla, California 92037
USA

Space research has increased our knowledge of fluid media in space in two ways.
1. It has made it possible to SEE our cosmic environment not only in visual light and radio waves but also in infrared, ultraviolett, X-rays and gamma rays. The picture we get in these now accessible wavelengths is drastically different from the pre-space-age view. One of the results is that (at least by volume) our cosmic environment consists to more than 99% of magnetized, often dusty, plasma.
2. In situ measurements in the magnetospheres - including heliosphere - combined with laboratory experiments have given us new information about the properties of cosmic plasmas. This makes it possible to UNDERSTAND what we see. It turns out that a drastic revision of the classical theory of plasmas is necessary.
3. The new results are APPLIED to cosmogony (origin and evolutionary history of the solar system). Magnetospheric physics has matured to such an extent that it is possible to treat certain aspects of cosmogony as an extrapolation of magnetospheric physics. It is shown that certain events 4-5 billion years ago can be reconstructed with an accuracy of a few percent.

1.0 NEW OBSERVATIONS

1.1 Is Hardware or Theory Deciding the Progress of Science?

The progress of science down through the centuries is usually illustrated by describing the important new ideas which have changed whole fields of science. The milestones in our way to increased understanding of the macroscopic world are the replacement of the Ptolemeian geocentric cosmology with the Copernican heliocentric one, and the work by Kepler and Newton, who built the foundation for celestial mechanics. More recently we have witnessed how Einstein's theory of general relativity has been the fundament of an at present rather generally accepted cosmology (big bang).

However, there is a complementary way of describing progress, viz., by focussing our attention on the development of the "hardware", the instruments which produced the data on which the new ideas were based. According to this "materialistic" view the scientific breakthrough during the 17th century was due to Galileo's introduction of the telescope. This was a more decisive factor in the "Copernican revolution" than the heliocentric theory of Copernicus. In fact, his system had been proposed by Aristarchos 2000 years earlier, but in the absence of telescopes it could not be proved. In other words, it was primarily a new instrument, not a new idea, which smashed the crystal spheres.

Similarly, it was the building of giant telescopes which made the heliocentric cosmology obsolete. The use of them made it possible to map our galaxy. It became inevitable to discover an increasing number of foreign galaxies, thus degrading the sun to an unimportant celestial body in a "peripheral" region of the universe. The introduction of radio telescopes has been another important technical achievement.

The leading role which the development of instruments plays is of course not confined to the field of cosmology. It was the construction of the spectroscope, the Geiger-Müller counter and the Wilson chamber which made it possible to understand the shell structure of the atom; it was the accelerators which gave us nuclear physics. Similarly, it was the microscope which introduced the modern era in biology and medicine. To claim that invisibly small bacteria could kill large animals, including men, was ridiculous before we had instruments to observe them.

We are now witnessing a new breakthrough of a similar nature. We are entering a new epoque in our understanding of our cosmic environment. This new era was introduced by Korolev, who launched the first spacecraft, and by van Allen, who introduced the technique of high quality in situ measurements in space - a technique which has been further developed by a large number of his brilliant students, including Krimigis and McIlwain. These two new types of "hardware" are changing our views of the macroscopic world as drastically as Galileo's telescope 350 years ago. Indeed spacecraft are now making it possible to observe the universe not only in visual light and at radio frequencies, but also in infrared, ultraviolet, X-rays, and γ-rays, thus about doubling the number of observable octaves (Fig. 1). For every new octave which is added to the observable electromagnetic spectrum we see the universe in a new light.

Moreover, spacecraft make in situ measurements possible in the magnetospheres, including the solar magnetosphere (solar wind region). Such measurements have shown that the properties of cosmic plasma are drastically different from what we

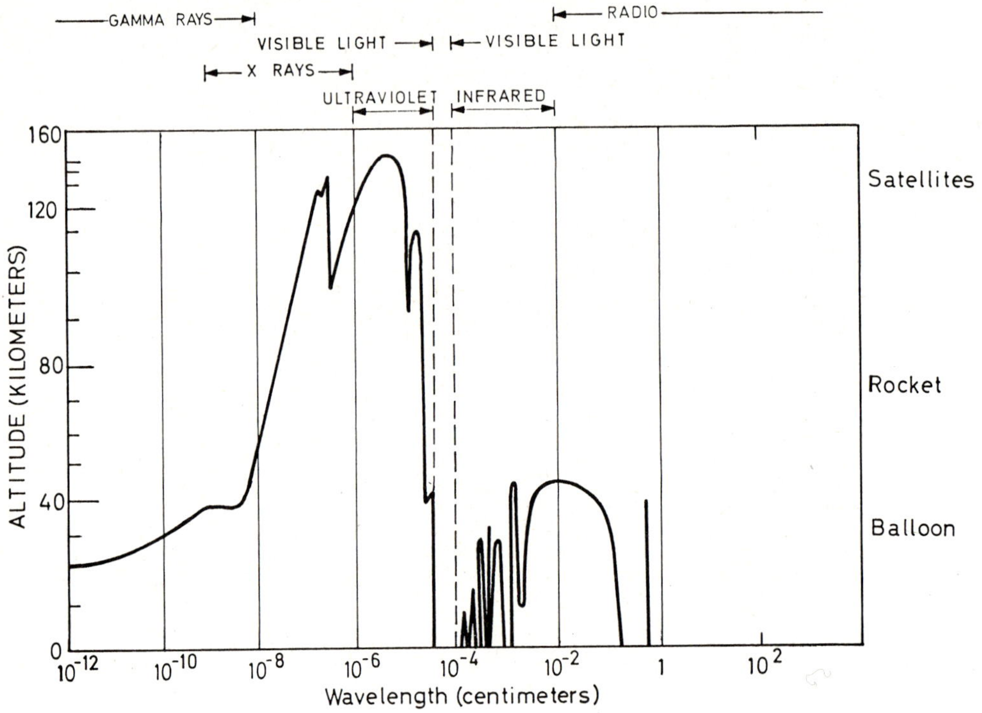

Fig. 1. Opacity of Earth's atmosphere to incoming radiation. The curve shows the altitude below which the radiation is invisible.

thought earlier. This is even more important than the increase in spectral range, because it is basic to the understanding of many of the astrophysical phenomena we observe.

The result is a "paradigm transition" in astrophysics. It is not a question of modifying one or two or three old theories, but of rewriting major chapters of cosmic plasma physics. Certainly the in situ measurements do not extend further than to the outer reach of spacecraft. However, as it is probable that the basic properties of plasmas are the same in more distant regions, large parts of astrophysics, including the origin and evolution of the solar system, and cosmology, must be revised.

The instrument designers have good reasons to be even more proud of the fact that they have ushered in a new era in the history of astronomy and astrophysics, including the history of the evolution of an interstellar cloud into a solar system of the present structure, and also of a new approach to cosmology.

1.2 The Cataract Operation

Fig. 1 shows that before the space age only the visual and the radio octaves were available for observations. High orbit rockets and spacecraft have added the infrared, ultraviolet, X-ray and gamma ray regions. Our X-ray eye which earlier was

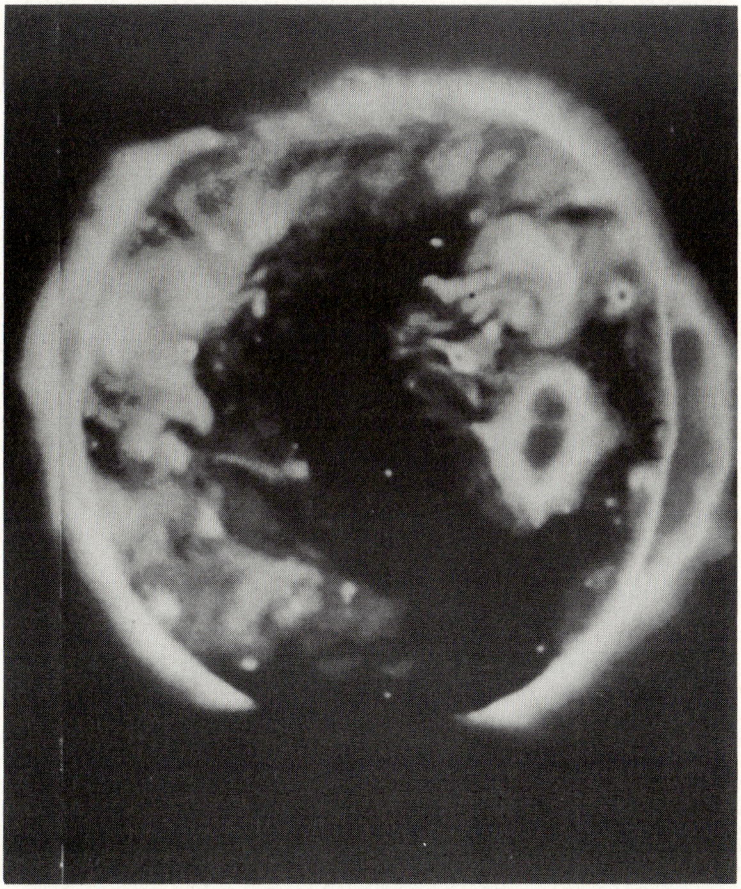

Fig. 2. X-ray picture of the Sun. Dark features called "coronal holes" are one of the largest and one of the most important features of the Sun.

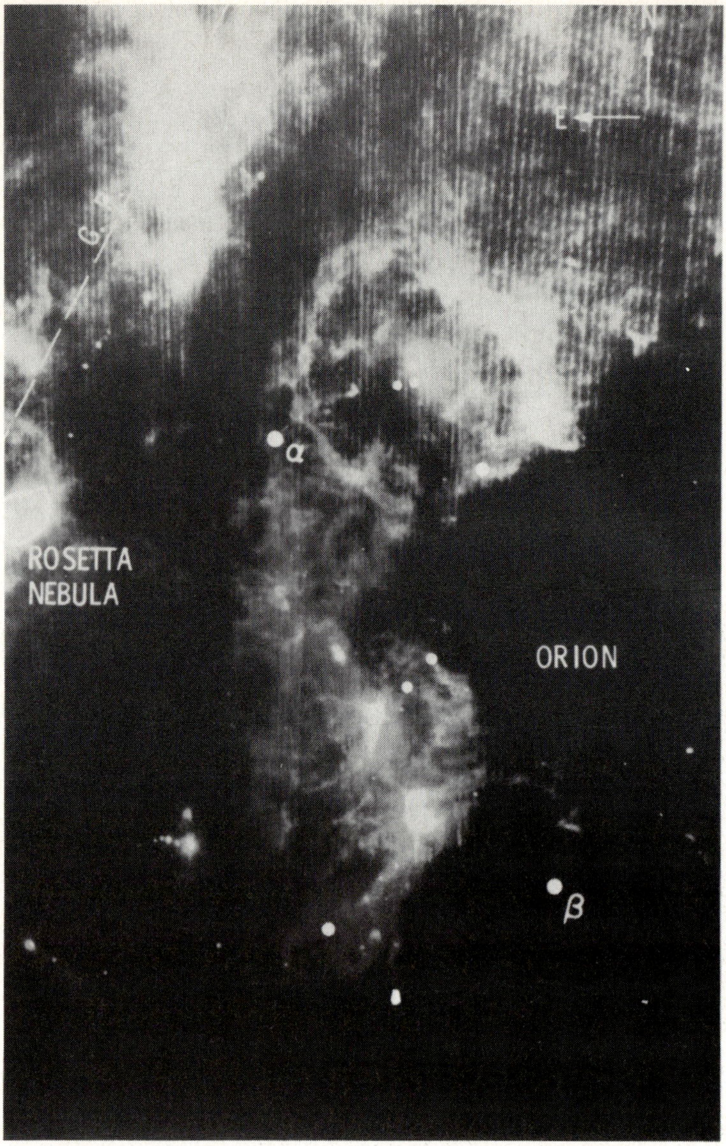

Fig. 3. IR - picture of the Orion region. What is seen is cosmic plasma. Its emission dominates in this wavelength region.

blind, has undergone a cataract operation, so that we can see our cosmic environment also in this new light. This has changed - or is beginning to change - our views of the universe.

Almost wherever we look we see drastically new unexpected phenomena. Compare our ordinary picture of the sun, an almost uniform disc (with the exception of a few tiny sunspots) with the X-ray picture which shows that often half of the sun is black (coronal holes) (Fig. 2). Or look at the interstellar plasma which seen in IR - ray sometimes is brighter than the stars (Fig. 3). A still more striking contrast is obtained if we consider the majectic calmness of the star-spangled night sky which changes with time constants of months, years or centuries or millienia (except for the Moon) and compare this with a gamma ray burst, in which an enourmous amount of energy is released in less than a minute with order of magnitude variation in milliseconds (Fig. 4).

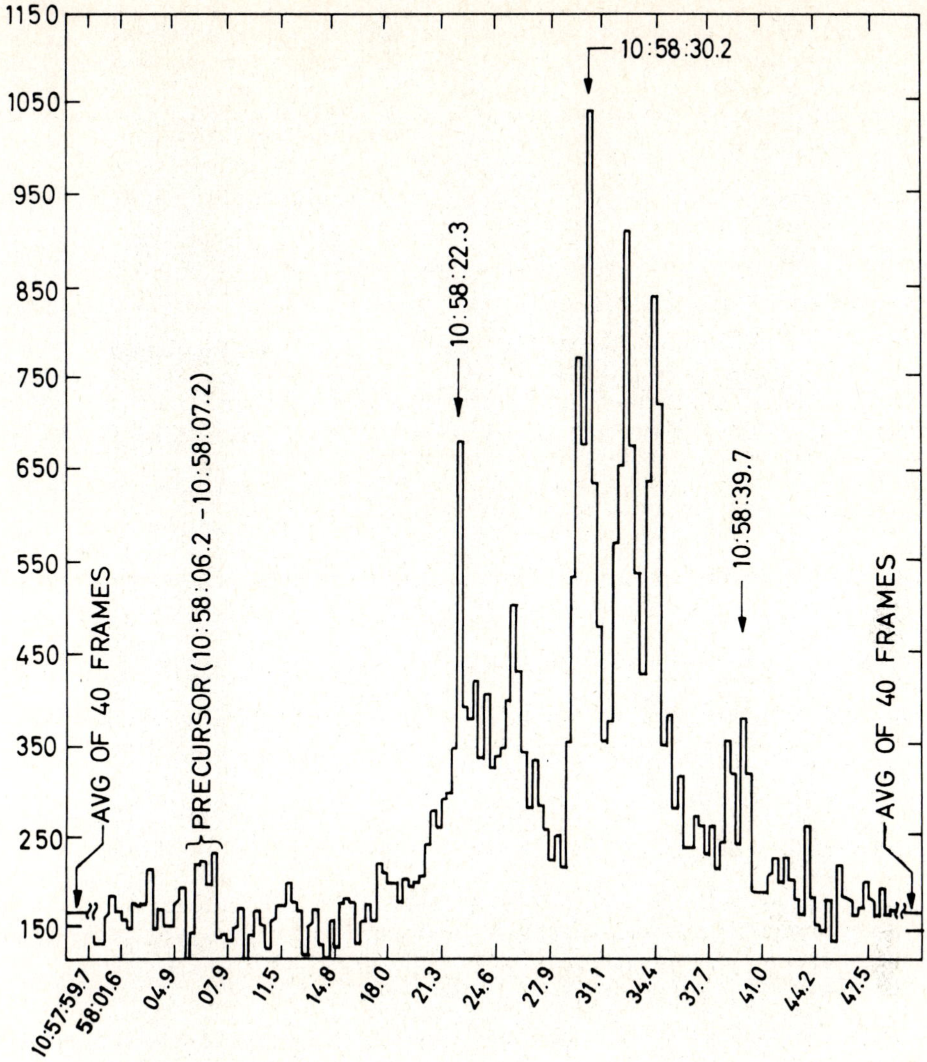

Fig. 4. Typical emission in a gamma ray burst. Total duration 50 s .

We can expect that this new information will introduce a new era in our views of the cosmic environment. Indeed it does.

Moreover, the increased sophistication of our measurements in the "old" octaves adds much new information.

2.0 NEW UNDERSTANDING

2.1 Limitations of Classical Mechanics

As basically cosmic physics is an empirical science, new and unexpected observations will have a tendency to shake its foundations. Obviously the observations of e.g. gamma ray burst demonstrate that the venerable Newtonian (or Einsteinian) universe could not represent the final truth. There must be other phenomena which appear unimportant in the visual universe but definitely are dominating the X-ray - gamma ray universe. It turns out that most if not all of the X-ray - gamma ray phenomena must be due to plasma processes. Indeed, many if not all should be described by relativistic plasma physics.

Already visual and radio observations demonstrated that, at least by volume, our cosmic environment consisted to more than 99% of plasma. The new observations indicate that more than 99.99...% is plasma. This means that it is no longer Newton alone who governs the mechanics in space. He has to share the power with Maxwell and Langmuir.

Hence classical (Newtonian) mechanics which was sacresanct up to the beginning of this century is now subject to three limitations:

1. Atomic phenomena required the introduction of quantum mechanics.

2. At velocities comparable to the velocity of light special relativity effects were decisive. (We are not discussing the general theory of relativity).

3. In magnetized low density plasmas, hydromagnetics and plasma phenomena are often orders of magnitude larger than the mechanical effects. This is usually the case in most of the interplanetary, interstellar, and intergalactic space. Also in stars these phenomena are often very important.

2.2 Theoretical or Experimental-observational Approach?

When it became obvious that (what we now call) plasma phenomena were decisive in many geophysical and astrophysical problems (first perhaps for magnetic storms and aurora) two different approaches were made:

Birkeland wanted to base the study on observations and on laboratory experiments.

Chapman and Cowling claimed that the properties of plasmas ("ionised gases" in their terminology) could be derived from basic physical principles by a sophisticated mathematical analysis.

Hence the approach to an understanding of low density cosmic plasmas has followed two different main lines.

(1) In the Chapman-Cowling approach, the plasma is treated essentially as a fluid. The general properties of non-ionized media (liquids or gases) are generalized by introducing electrically charged particles instead of (or together with) molecules. This lead to the classical Chapman-Cowling approach.

(2) The other alternative is to treat the motion of each individual electrically charged particle under the action of electric and magnetic fields which often are produced mainly by the particle population. This is basically the Birkeland-Störmer approach. To everyone who has studied the closest representative of a cosmic plasma, viz. the aurora with its extreme degree of variability both in space and time, it is not very natural to consider it as a fluid (Fig. 5). The plasma in Birkeland's terrella experiments did not either give the impression of being a fluid. The relative merits of the two approaches are discussed in Cosmic Plasma (Alfvén, 1981a; in the following referred to as CP), Chapter I:1, p. 1.

The classical approach is useful for treating waves in plasmas etc. However, many of the conclusions drawn from it have turned out to be in conflict with observations. Science is basically empirical. That this hold also for plasma physics is demonstrated by the list of new effects which we now know must be considered (see Table I). They are mainly due to experimental-observational discoveries. Very few of them have been predicted by the classical theory, and even post factum it is difficult to account for them using the classical formalism (see CP, Chapter I:4, p. 8). As a summary, the classical theory has been rather sterile and quite often greatly misleading. This is due to the fact that a plasma is so complicated that in order to make a mathematically elegant treatment it is necessary to make a number of simplifying assumptions of which not all have turned out to be realistic.

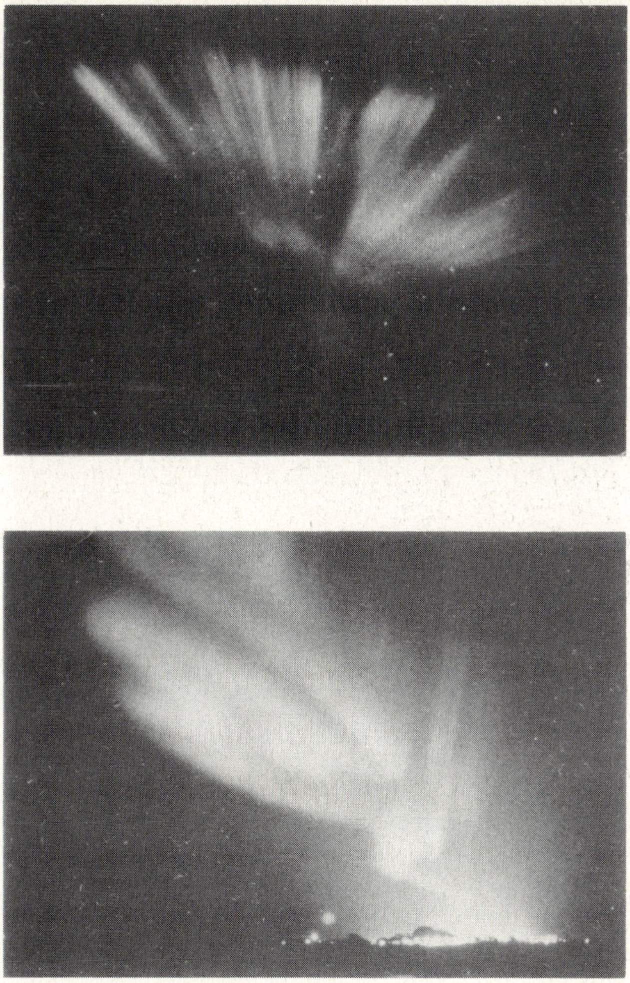

Fig. 5. When a cosmic plasma invades the atmosphere aurorae are observed. They show the true nature of a cosmic plasma. It is very inhomogeneous and the aurora changes its shape within a timescale of down to minutes.

The change in cosmic plasma physics is essentially the result of three factors:

(a) In situ measurements in the magnetospheres (including the heliosphere)

(b) Laboratory studies of phenomena of interest in cosmic plasma physics

(c) Increased understanding of how to extrapolate results obtained in one field to other fields of plasma physics.

A survey of some of the "paradigm transitions" which this has caused or is c sing has been published in a monograph (Alfvén, 1981a). Summaries of this have be presented in Alfvén, 1982, 1983a, 1983b. Table I gives a very brief summary.

TABLE I: MAGNETOSPHERIC RESEARCH

is causing a paradigm transition in geophysics and astrophysics for the following reasons:

§1. Electric double layers are realized to be very important.

§2. The often misleading "magnetic merging" theories of energy transfer should be replaced by an electric current description, including the circuits in which the currents flow.

§3. Homogeneous models often are found to be misleading and should be extensively replaced by inhomogeneous models.

§4. It is realized that inhomogeneities are produced by filamentary currents,

§5. and by surface currents, dividing space into cells.

§6. It is concluded that space in general has a cellular structure.

§7. The introduction of the current-circuit description makes it impossible to neglect the pinch effect term in the pressure equation

$$\nabla(p + B^2/2\mu_o) - (\underline{B}\cdot\nabla)\underline{B}/\mu_o = 0 \ .$$

§8. It is doubtful whether large-scale turbulence is of importance in diffuse media.

§9. In a space plasma, electric currents may produce chemical separation.

§10. In a dusty plasma, gravito-electromagnetic effects are often important.

§11. The "critical velocity" is often decisive to the interaction of neutral gas and magnetized plasma.

Cosmological consequences will not be discussed here.

A few of these items will be discussed in 2.2.1.

2.2.1 Double Layers. The Present State Seems To Be -

(a) Electric currents produce double layers if the drift velocity of the electrons exceeds the thermal velocity (Iizuka et al., 1979; Iizuka et al., 1982).

(b) The properties of a double layer does not only depend on the plasma conditions but also on the outer circuit. If a double layer configuration has a negative volt-current characteristic ("negative resistance" R_i) it can still be stationary if the outer circuit contains a positive resistance (R_e) which is sufficiently large, e.g. numerically larger than $|R_i|$. Also a stationary double layer is usually "noisy", i.e. it produces erratic high frequency (sometimes also low frequency) oscillations (Torvén and Lindberg, 1980; Carpenter et al., 1984).

(c) The voltage drop ΔV over a double layer is normally 3-5 kT/e (volt equivalent of thermal energy). This holds in such laboratory experiments where the ionisation is mainly produced by one plasma source (Torvén and Lindberg, 1980).

(d) If there is one plasma source at each side of the double layer and there is little ionisation produced near the double layer ΔV may be several orders of magnitude larger (Torvén, 1982; Lindberg and Torvén, 1983; Sato et al., 1980).

(e) A double layer in an inductive circuit has a tendency to explode. This occurs for example if $|R_i| > R_e$. At such an explosion practically all the inductive energy $1/2\,L\,I^2$ (L = inductance, mainly in the outer circuit, I = total current) of the circuit is released in the double layer (Lindberg and Torvén, 1983).

(f) In the magnetosphere normally stationary double layers accelerate the auroral electrons to energies of 3-5 keV . It is a matter of controversy whether the acceleration occurs in one single double layer or in a series of double layers. Magnetic mirror produced electric fields may also be important. Exploding double layers in the magnetotail may be the cause of magnetic substorms.

(g) Exploding double layers produced by the prominence currents in sun are likely to be the basic phenomenon in solar flares. Also in small scale "flashes" double layers may be important (Carlqvist, 1982).

(h) Double layers in galactic circuits may generate the intense energy release observed as double radio sources. It is possible that they accelerate particles up to at least 10^{14} eV and may be responsible for generation of high energy cosmic radiation (CP, Chapter III:4.4, p. 56; Carlqvist, 1982).

Reviews of double layers have been given at the Symposium at Risø 1982 (Michelsen et al.) and at Innsbruck 1984 (Carpenter et al.).

§2. The switch from the magnetic field to the electric current-particle description was favoured by the Armstrong and Zmuda mapping of the magnetospheric current system followed up by Potemra, Bostrom and several others. This settled the half a century old dispute between the Birkeland school and the Chapman school to Birkeland's favour. (Dessler has appropriately called the field aligned currents "Birkeland currents"). Further, at Potemra's Conference on Magnetospheric currents 1983 the advantages of the latter description were demonstrated. It is a most important treatise of how the theory of the magnetosphere should be based on the particle description. The relation between the electric current (particle) formalism of energy transfer and the "magnetic merging" formalism seems now to be:

(a) When currents in a plasma change and hence the magnetic fields associated with them vary magnetic field lines merge (reconnect). There is no objection to using "merging" and "reconnection" as a purely geometrical description.

(b) The merging theories cannot be used generally for energy transfer calculations in cosmical physics. One of the reasons is that they do not take adequate account of boundary conditions (see CP Chapter II:5, p. 26).

Of the other items §4 and 5 are rather well-known. §6 is of course very important for astrophysics in general and perhaps especially for cosmology. §7, 8, 10 and 11 will be discussed in the next section.

3.0 NEW APPROACH TO COSMOGONY

3.1 Cosmogony as an Extrapolation of Magnetospheric Research

From what has been said in Sections 1 and 2 it is obvious that introduction of the new paradigm means a revision of large parts of astrophysics. We shall here confine ourselves to one of the fields viz. the origin and evolution of the solar system (cosmogony). Besides cosmology which we shall not discuss here (see CP Chapter VI and Alfvén, 1983d) this is probably the field of most far-reaching interest. Indeed it has a central position for geology, oceanography, paleobiology, evolution of interstellar clouds, the formation of stars and "solar nebulae", and for astrophysics in general.

We shall begin by a brief review of an approach to cosmogony which started in 1942. (See Table II). The new idea was that electromagnetic (or hydromagnetic) effects were of decisive importance for understanding how the solar system got to its present state. Because previous cosmogonies since Laplace considered mechanical forces alone, this was not reconcilable with the generally accepted types of cosmogonies. Certainly, these have changed drastically during the ages, but almost all of them neglected hydromagnetic and plasma effects. Few cosmogonists had more than a superficial knowledge of hydromagnetics and plasma

TABLE II: HISTORY OF THE SOLAR SYSTEM WITH ELECTROMAGNETIC EFFECTS INTRODUCED

1942	It was shown that a cosmogonic model of this type required that three mechanisms be postulated: a. Electromagnetic transfer of angular momentum. b. The existence of a phenomenon which later was called "critical velocity". c. A plasma-planetesimal transition associated with a 2:3 contraction. This produced "cosmogonic shadows".
1954	A survey of the theory was published as a monograph On the Origin of the Solar System (Alfvén, 1954). It included a development of the theory of Saturnian rings leading to a correction in the contraction factor by a few percent.
1960	Laboratory confirmation of critical velocity (Fahleson, 1961), essentially based on a technique developed by Bratenahl (Anderson et al., 1959).
1974	Zmuda and Armstrong (1974) (and Iijima and Potemra, 1978; for a survey, see Potemra, 1979) map the magnetospheric current system which gives the needed transfer of angular velocity.
1975, 1976	Cosmogonic theory systematically developed in two monographs by Alfvén and Arrhenius (1975, 1976).
1980	Space research calls for a "paradigm transition". A brief review is given in Alfvén, 1981, and further developed in 1982, 1983 (summarized in Table I).
1982	Critical velocity effect in space demonstrated by space experiment by Haerendel (1982).
1983	Holberg's treatments of Voyager results (Holberg et al., 1982) make possible a further confirmation of 2:3 contraction and cosmogonic shadow effect. This is also supported by earlier investigations of the asteroidal belt. All together the 2:3 fall-down ratio is found in seven cases, four in the Saturnian ring and three in the asteroidal belt. This is an encouragement to the further pursuit of this cosmogonic approach.

physics, with the result that the decisive importance of the 2:3 contraction and the band structure have not been appreciated.

Space research has now changed the situation by giving us new information about electromagnetic and plasma effects in space as has been discussed in Sections 1 and 2. From in situ measurements in the magnetospheres we know the properties of plasmas over five or ten orders of magnitude in density, in magnetization, in temperature, etc. and we also begin to understand what processes are possible and which are not. This has introduced or is introducing a new climate in cosmical physics which may be more favorable for a serious discussion about the evolutionary history of the solar system. In fact, we do not need to base cosmogonic theory on more or less reasonable assumptions about conditions at the time when the solar system was formed (probably 4-5 Gyears ago), or on uncertain interpretation of distant, marginally observable phenomena. We can instead treat cosmogony as an extrapolation of reasonably well-established processes from space research, often derived from in situ measurements (see Fig. 6). The result is an approach in which the evolutionary history is decided by a combination of mechanical and electromagnetic (plasma) effects.

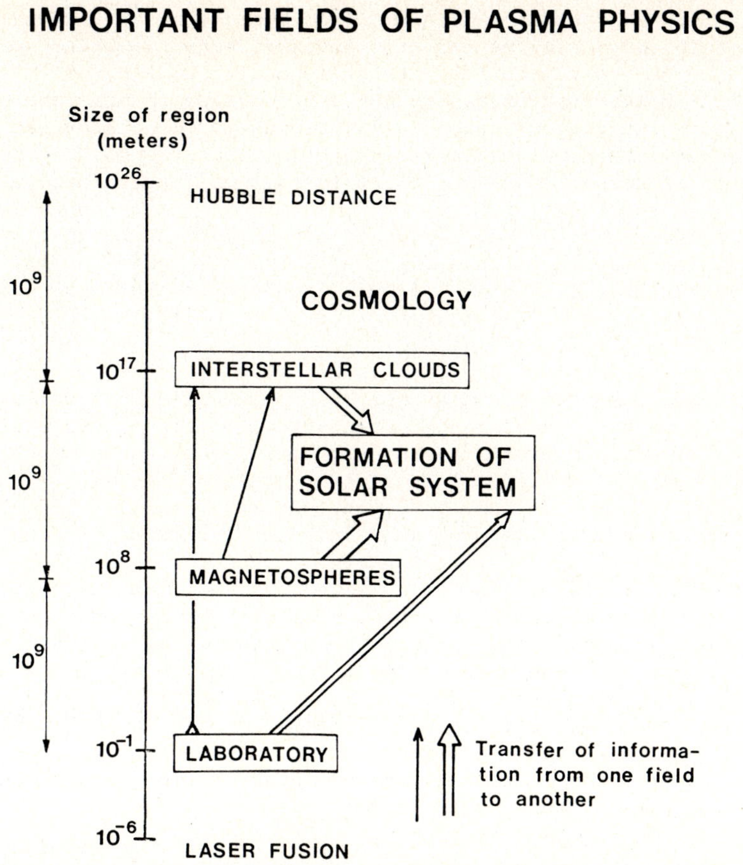

Fig. 6. Magnetospheric research has matured to such an extent that it is possible to treat essential parts of the evolutionary history of the solar system as an extrapolation of magnetospheric research. Laboratory experiments also form an important basis for this. Further, extrapolation from both magnetospheric and laboratory results contribute to a revision of our view of interstellar clouds, and hence influence also the way in which we approach cosmogony. The transfer of information from one field to another is shown in the figure. The thickness of the arrows is proportional to the flow of information.

What has been summarized above leads to an evolutionary history of the solar system, which in some respects is similar to what is described in the Alfvén-Arrhenius monographs. This is not unexpected, because, as seen in Table II, the evolution of these theories has from the beginning been coordinated with the development in cosmic plasma physics (Alfvén, 1983b). A brief survey will be given here (compare Table II).

3.2 Electric Currents in Interstellar Clouds

There are good reasons for the general view that stars and solar systems are born out of an interstellar cloud of dusty plasma. However, the theory of the origin and evolution of such clouds and the formation of stars and solar nebulae is a field which must now be revised for the following reasons:

As we have seen in Section 2 in situ measurements in magnetospheric plasmas (including the solar wind) have caused drastic changes in our views of the properties of cosmic plasmas. What was considered sacrosanct ten or even five years ago is now hopelessly obsolete. This theoretical paradigm transition, which is summarized in Section 2.2 and Table I has penetrated as far our as in situ measurements are made; i.e. as far as spacecraft have travelled. Outside this limit the paradigm transition has not yet taken place. Plasmas in interstellar space are still being treated according to the old paradigm. This means in reality that the present theories of interstellar clouds and of the formation of stars and solar nebulae are based on the tacit assumption that the basic properties of cosmic plasmas change at the outer reach of spacecraft.

It is obvious that astrophysics cannot remain in this unstable state (indeed, a "universal instability" in plasma physics!). The new paradigm will sooner or later be extended to interstellar space. It will cause a revolutionary change in our view of the evolution of interstellar clouds, in the following respects.

(a) According to CP Chapter II:3 etc. cosmic plasmas cannot be described by the magnetic field picture alone. This must be supplemented by an electric current description. Astrophysicists are often reluctant to accept the existence and importance of electric currents in interstellar space, but none of them claims that the magnetic fields are curl-free. As a non-curlfree magnetic field means electric currents, they implicitly accept that interstellar space is penetrated by electric currents. However, there is an immense difference between an implicit acceptance and an explicit description of the phenomena in terms of electric currents. The latter description calls immediately for models of the circuits in which the currents flow, and models of the dynamo which produces the currents. Such currents may transfer energy from one region to another, sometimes over distances comparable to the size of the whole galaxy. With regard to the circuit description, it has been objected that "there are no wires in space". But "circuits" do not necessarily mean an aggregate of simple linear elements. Especially in the "computer age", circuits often contain non-linear distributed elements. An example of a circuit with non-linear elements is given by Boström (1974) in his circuit of a magnetic substorm.

(b) As soon as electric currents are introduced explicitly, attention is focussed on the pinch effect, which is discussed in CP Chapter IV:8, p. 93.

In a typical Bennett pinch both the pressure and the magnetic field are large inside the pinch but zero outside. A typical Bennett pinch is produced when

$$\frac{\mu_o}{4\pi} I_z^2 > 2Nk(T_e + T_i)$$

(I_z = current, T_e, T_i are electron and ion temperatures, N = number of particles per unit length). Fig. 7 illustrates three typical cases of stationary and cylindrically symmetric currents (i) and magnetic field line (B) configurations. In most treatments of the evolution of an interstellar gas cloud, it is assumed that electromagnetic forces oppose the contraction, as in (a), whereas they just as well may assist or cause the contraction, as in (c). The intermediate case (b) may be a first approximation of a model of filamentary currents (e.g. see CP Chapter IV:8.1, p. 95).

(c) According to §3, homogeneous models of plasmas are now increasingly replaced by inhomogeneous models. When a new field is opened, it is natural to approach it by making homogeneous models, in the belief that these will in any case be a reasonable first-order approximation to a final theory. In plasma physics we have the sad experience that this is very often not true. When a field has matured to such an extent that it is obvious that homogeneous models are no longer sufficient, it is often evident that inhomogeneous models give a drastically different description of the phenomena. The homogeneous model was of no use. Instead, it led the modeling into a dead-end from which it often is very difficult to turn back because a powerful establishment committed to the homogeneous model has already been formed.

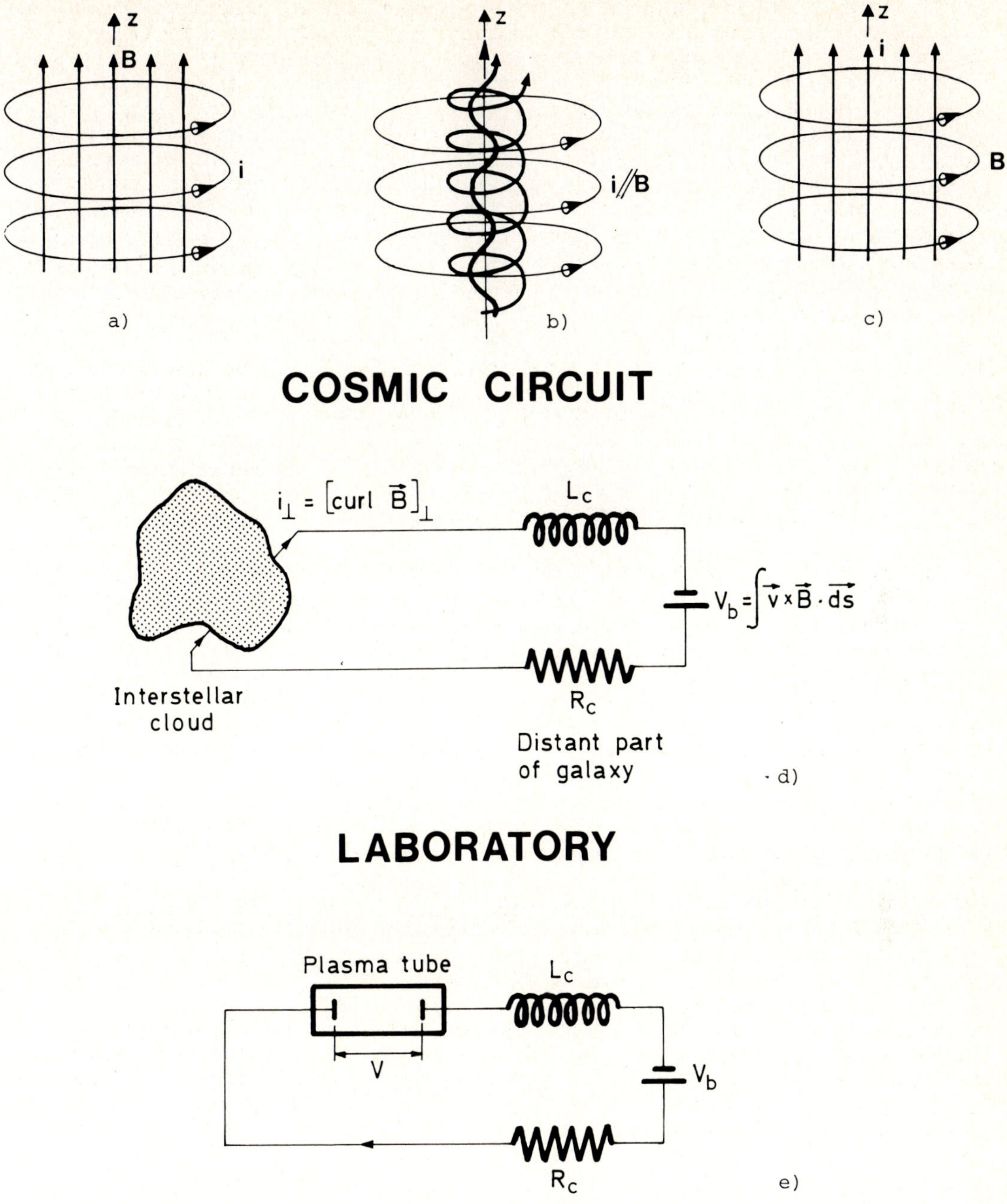

Fig. 7. Three special cases of stationary and cylindrically symmetric current (i) and magnetic field (B) configurations. (a) A toroidal current and an axial magnetic field leading to a force opposing contraction. (b) A force-free configuration with i and B parallel. (c) The Bennett pinch with an axial current and a toroidal magnetic field. Electromagnetic effects aid and even start contraction. (d) If curl $\underline{B}_\perp \neq 0$ at any point on surface of the cloud its evolution depends on the total circuit. (Magnetic merging theories not applicable). (e) The behaviour of plasma in a tube depends not only on local plasma parameters but also on the total circuit, including R and L . Energy transfer cannot be calculated by magnetic merging.

(d) According to §4, there is often an association between electric currents and observed filaments. Examples of this in our close vicinity are auroral rays (probably) associated with filametary currents, the filamentary structure of the solar corona, and the filamentary currents in the ionosphere of Venus (CP Chapter II:4.7, p. 26). In interstellar clouds, there are often observed filamenatary structures (especially in contrast-enhanced photographs). Such observations support our conclusion that interstellar space, and not the least interstellar clouds, are penetrated by a network of electric currents. Concerning clouds in which no filamenatary structure is observed, it is an open question whether this depends on an absence of them or the inadequacy of observational methods to detect them. From the general picture of the new paradigm the latter interpretation seems to be preferable.

(e) §5 and §6: it is not obvious that these are decisive for the present development of cosmogony.

(f) §7: this has already been discussed in the beginning of this section.

(g) §8: turbulence is generally believed to be decisive for the evolution of interstellar clouds and the formation of the solar system. There seems to be no convincing observational evidence for this (see CP Chapter IV:4, p. 84).

3.3 New Approach to the Evolution of Interstellar Clouds

As stated above, sooner or later the new paradigm will penetrate also the field of the evolution of interstellar clouds. The theory of interstellar clouds should be treated as an extrapolation of magnetospheric research (CP Chapter IV:8, p. 93; see Fig. 6).

Very much work will be required for this transition, and it is difficult to predict in detail what the result will be. As a reasonable guess as to what a future model of the formation and evolution of interstellar clouds should be, we may suggest the following.

a. Electric currents in "void" interstellar space assist gravitation in collecting matter by the pinch effect, so that interstellar clouds are formed.

b. These develop under the combined action of mechanical and electromagnetic forces. The volume occupied by currents may constitute a very small fraction of the total volume, so that the plasma regions are not evident in the averages of measurements with insufficient resolution. Still, a network of filamentary currents may be decisive to the evolution of the clouds. It is correct to treat the evolution of an interstellar cloud independent of its surroundings only if there is no current connecting it with the surrounding (cf. Fig. 7e).

c. As stated above (cf. Fig. 7), the general belief that electromagnetic forces oppose the contraction of a cloud is not necessarily correct. Pinch effects may contribute to the contraction and, indeed, cause a collapse of clouds with a mass that is orders of magnitude smaller than the Jeans mass.

d. A "stellesimal" star formation out of a dusty cloud seems possible (cf. CP Chapter V, p. 110).

3.4 Properties of the Solar Nebula

When the sun is formed it will be surrounded by a dusty plasma penetrated by a network of currents which partially support it (Fig. 8). This "solar nebula" is drastically different from the Laplacean nebula. It is possible that Oort's cometary cloud is a relic of this. The cloud is strongly inhomogeneous and contains regions of different chemical composition (compare §9). From this primeval cloud there rain cloudlets of different composition down towards the sun. Moreover, there is a rain of cosmic dust, perhaps similar to the present rain of

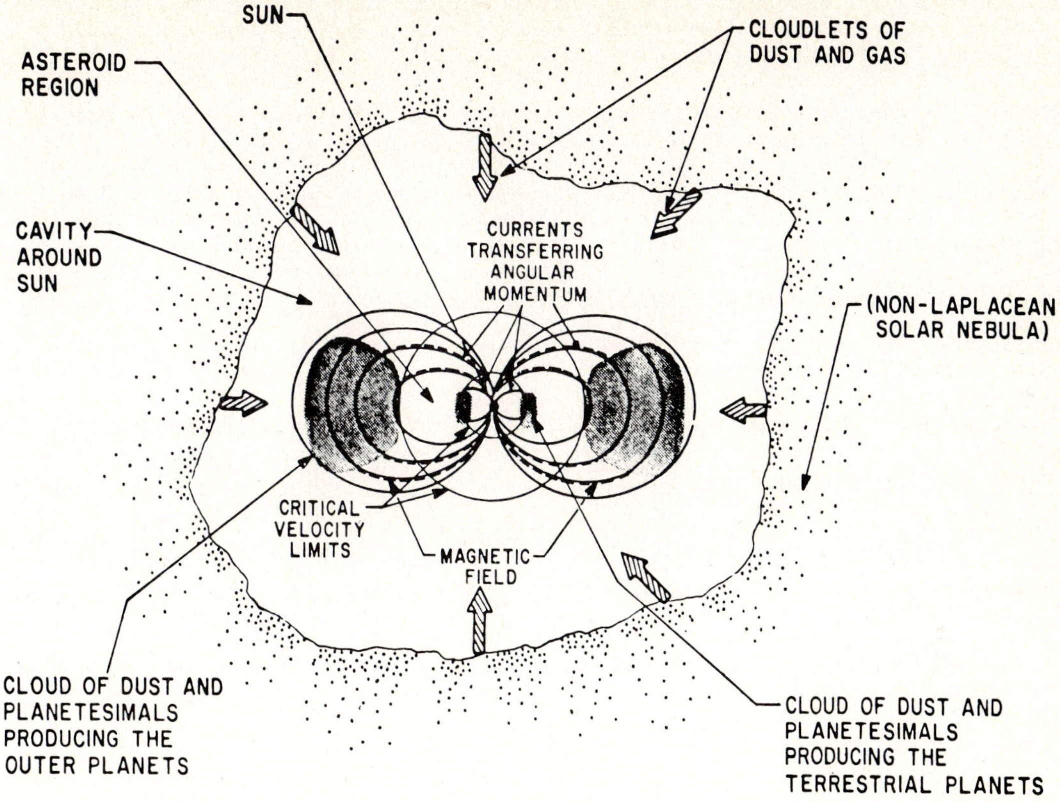

Fig. 8. After the formation of the sun and its magnetic field, remnants of the (non-Laplacean) solar nebula (formed according to 3.3 and 3.4) fall towards the sun in cloudlets of dust and gas. When a falling cloudlet reaches its critical velocity, it is stopped. Currents from the spinning sun support the cloudlet for a short period of time and transfer angular momentum to it. At the plasma-planetesimal transition, grains in Kepler orbit are produced and plasma processes are then no longer important.

meteoroids from the cometary cloud. For possible models, see Alfvén and Arrhenius, 1975, 1976.

From now on our attempt to reconstruct the evolutionary history of the solar system enters a new phase in two respects:

a. Up to this point we have discussed the evolution of an interstellar cloud as a unity, even if the cloud is very inhomogeneous. From now on we have to discuss the evolution of individual cloudlets as distinct from the evolution of the whole solar nebula. The latter process is an integration of the processing of cloudlets, and consists essentially of a slow transformation of the solar nebula into planetesimals (and later planets) during a long period of time.

b. The second reason why we are entering another phase is analytical. The process of planet formation around the sun is similar to the formation of satellites around some planets, especially Jupiter, Saturn and Uranus, which have well-developed satellite systems. Hence we should aim at a general theory of formation of secondary bodies around a central gravitating, rotating and magnetized body. This requirement has been referred to as the "hetegonic principle" (see Alfvén and Arrhenius, 1975, 1976. The arguments for this view are discussed in some detail there). The principle is related to what now is usually called "comparative planetology" but should include also the formation of bodies around the sun. Galileo, when discovering the Jovian satellites, already called this system a "solar system in miniature".

As we have four well-developed systems to base our conclusions upon, we can speak with more confidence about the processes than for the earlier states of development. Hence, our model for all four systems should start from the assumption that a magnetized central body was formed already and surrounded by a dusty plasma from which cloudlets of different chemical composition, together with dust grains, fall in towards it.

3.5 Basic Processes in Evolution of Solar Nebula

In this state there are three processes which were decisive for the present structure of the solar system (cf. Table II).

a. The transfer of angular momentum from the central body to the surrounding plasma. The transferred angular momentum is now found in the orbital moment of the secondary bodies.

There is a rather obvious candidate for this process, viz. the auroral current system, which is known to transfer angular momentum between a rotating central body and a surrounding plasma (Fig. 9). Essentially the same process is well known from the Jupiter-Io system (Hill et al., 1983). One of the processes which is claimed to account for the loss of momentum from the sun is turbulence, but with reference to §8, and especially what will be demonstrated later, this is not an acceptable cosmogonic process.

Another process to account for the loss of solar momentum is the solar wind. This is an interesting and perhaps partially correct suggestion. However, it is not clear how it can incorporate the band structure and the cosmogonic shadow effects.

With our present knowledge there seems to be no serious objection to accepting the electromagnetic transfer qualitatively as the basic process. When we come to a quantitative evaluation of the process there seems to be no serious objection to the view that the satellite systems were formed by such a transfer from the mother planet. However, for the planetary system the situation looks more difficult.

If we assume that the formation of the solar system was a very rapid chaotic process with a time constant of less than a million years, we would run into

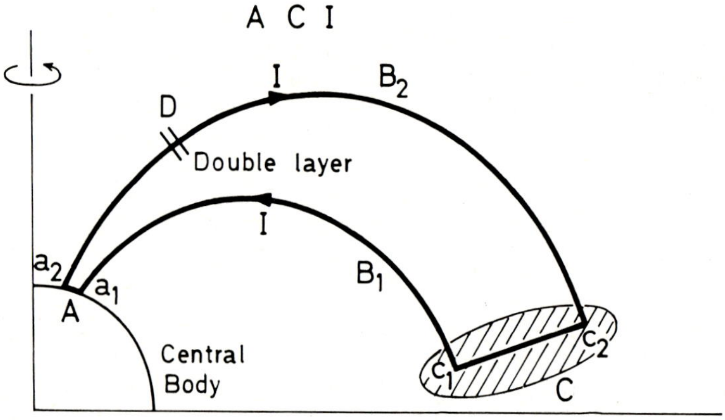

Fig. 9. Simplified picture of the auroral circuit transferring angular momentum between a magnetized rotating body A and a surrounding plasma cloud C . Current I may produce double layers D .

difficulties. The magnetic field must first support all the matter and then transfer angular momentum to it. This would require an enormously strong solar magnetic field. We can avoid such a difficulty by assuming that the planetary system was formed by cloudlets going through the momentum transfer process with a short time constant T_c while this processing led to a slow buildup of planetesimals and planets with a time constant T_i which may be many orders of magnitude larger than T_c. The result is that the density ρ of the plasma supported at a certain moment by the magnetic field need only be a small fraction of the total smeared-out density ρ_s of the produced planetesimals $\rho = \rho_s T_c/T_i$.

The distinction between the rapid processing of cloudlets and the slow integrated buildup means that ρ was so small that the momentum transfer processes could take place in a low-density plasma (collisionless plasma). This is fortunate because much of the study of magnetospheric plasmas has been and is concentrated on such plasmas. Hence we should be able to treat the transition from plasma to planetesimals as an extrapolation of present-day magnetospheric results (for details, see Alfvén and Arrhenius, 1976).

b. Band structure of the solar system. The second basic plasma process is the critical velocity.

When developing a tentative early theory of the possible importance of electromagnetic processes in solar system evolution, it was necessary to postulate the existence of "the critical velocity" in order to explain the band structure of the solar system (see Table II). Such a process was unknown at that time, but the cosmogonic evidence for its existence was considered so compelling that laboratory experiments to demonstrate it were started as soon as possible. These were successful, and there exists now a literature of some hundred papers regarding this phenomenon (Axnäs et al., 1982).

The existence of such a process in space is now confirmed by space experiments (Haerendel, 1982). As there are very few phenomena which have been discovered from a cosmogonic theory, this gives some confidence that this is the process responsible for the band structure.

However, in spite of all this, the problem of planetary formation remains difficult. The critical velocity is a phenomenon in pure gases, but how a dusty plasma behaves is not clear. The most serious problem is to understand how the planets and satellites have acquired their present chemical composition. Possible chains of processes have been discussed but a convincing solution has not yet been found.

3.6 Scenario of the Cosmogonic Process

The general scenario of the cosmogonic processes is shown in Table III. Plasma effects were of considerable importance for the evolutionary history of the solar system from the formation and evolution of cosmic coulds to the formation of the sun and a surrounding solar nebula. In the solar nebula the plasma effects were of decisive importance in two respects.

a) They transferred angular momentum from the sun to the plasma out of which later the planets (and asteroids) were formed.

b) The critical velocity produced the band structure of the solar system. The basic plasma processes which cause the critical velocity are still not very well clarified theoretically, but the phenomenon is extensively studied in the laboratory, and space experiments have demonstrated its importance for cosmic plasma physics.

c) After the plasma phase of the solar nebula came the plasma-planetesimal transition (PPT). This was not a sudden and violent turbulent transition, but a slow,

TABLE III: FORMATION OF PLANETS/SATELLITES FROM INTERSTELLAR CLOUDS

State of matter which is located at present in planets/satellites	Evolutionary Process	Main Evolutionary Mechanism	References
Dusty Plasma	Evolution of Interstellar Cloud	Gravitation	3.2
	Formation of Sun and Solar Nebula	Pinch Effect	
	Evolution of Solar Nebula	Electro-Magnetic Transfer of Angular Momentum	3.5(a),3.6
		Critical Velocity	3.5(b),3.6
	Plasma-Planetesimal transition	2/3 Contraction Cosmogonic Shadow Effect Rosseland Field	4.0 Fig.10.4.3
Planetesimals	Accretion of Planetesimals to Planets	Mechanical Effects Plasma Processes not Important	
Planets			
Satellites	Formation of Satellites around Planets occurs by a Repetition in Miniature of these Processes (starting with formation of nebula around planet).		

continuous process working for a very long time, perhaps 10 - 100 million years. It discontinuously but persistently transformed in-falling matter of a dusty plasma into a planetesimal state. (However, the processing of individual cloudlets was a rapid process, see 3:5a). In the pre-planetary period this process worked in the planetary system. At a later period (when the planets were formed) it worked around the planets, producing "satellitesimals".

d) The mass of matter in the planetesimal state increased slowly, until the planetesimals began to aggregate to planets. Later, similar processes led the "satellitesimals" to aggregate to satellites. Plasma processes are of negligible importance for these processes.

In one region in the planetary system, viz. the asteroidal region, and in one region in the satellite systems, viz. the Saturnian ring, the aggregation to planets or satellites has not taken place. The reasons for this are the low density in the asteroid region and the location of the Saturnian ring inside the Roche limit (Alfvén and Arrhenius, 1975,1976). Hence, at least in certain respects, the state in these two regions represents the planetesimal state. This makes them of decisive importance for our attempts to reconstruct the plasma-planetesimal transition. Because much - or rather most - of the cosmogonic information initially stored in the planetesimal state is obliterated in the planetesimal-planet transition, they are of unique value for clarifying the early evolutionary history of the solar system.

In the following, our approach to cosmogony is to a large extent based on a study of the plasma-planetesimal transition (PPT). It turns out that the Saturnian ring and the asteroidal belt contain information which is decisive for our reconstruction of the history of the solar system. In this paper we concentrate our attention on the Saturnian ring.

4.0 SCENARIO OF THE SATURNIAN RING FORMATION

The Pioneer and Voyager exploration of the Saturnian rings has given us most valuable material about its structure. For reasons given in Alfvén 1983b we concentrate our attention here on the bulk structure, which seems to give information of decisive value for clarifying the origin of the rings. That paper relied on Holberg's curves (Holberg et al., 1982; Holberg, 1983). Now similar curves (Esposito et al., 1983) have also been published. In the points that are of interest to us the newly published curves agree very well with those we have used.

What has been said above leads to the following scenario for the formation of the ring system (see Alfvén and Arrhenius, 1975, 1976; Alfvén, 1983b,c).

1. Saturn was already formed, with approximately its present mass and present spin. Its magnetic field may also have had the present shape (close to a dipole field) but we do not know its strength.

2. Cloudlets of gas and dust from interplanetary space fell in towards Saturn. They became ionized, which led to current of the same type as mapped in the planetary magnetosphere by Zmuda and Armstrong (1974) and later by Iijima and Potemra (1978). The Jupiter-Io circuit is similar (see Hill et al., 1983). This current system transferred angular momentum from the planet to the plasma (Alfvén and Arrhenius, 1976, Chapter 16; Alfvén, 1981a, p. 52,120).

3. This brought the plasma into a state of partial corotation so that Saturn's gravitation was compensated to two-thirds by the centrifugal force and to one-third by electromagnetic forces (from the Saturnian magnetic field; see Fig. 10).

4. At the transition from the plasma to the planetesimal phase the electromagnetic forces vanished, which caused a contraction by a factor $\Gamma = 2/3$. (This factor is given by the geometry of Saturn's magnetic dipole field, see Fig. 10). Early Voyager results have already demonstrated that there is strong evidence for this process in the present structure of the Saturnian ring (Alfvén, 1983b).

Further development of this approach has shown that it is possible to account for more than 30 signatures in the Voyager diagrams.

4.1 Negative Diffusion and Stability

The first question we have to answer must be: is it reasonable that essential parts of the present ring structure are "fossils" from cosmogonic times (4-5 billion years ago)?

Baxter and Thompson (1971, 1973) have demonstrated that under certain conditions the diffusion in a population of grains in Kepler orbits is "negative" (see Fig. 11). This result is confirmed by Lin and Bodenheimer (1981). The present ring system consists of 1000, if not 10,000, ringlets (cf. Fig. 12). This indicates that the negative diffusion mechanism is active today. There seems to be no obvious objection to the assumption that the same mechanism was active in the past. This means that the present structure may derive from cosmogonic times (Alfvén, 1983b). Hence, it is meaningful to try to reconstruct essential events in the evolutionary history from "fossils" stored in the Saturnian ring. There are reasons to believe that this holds also for the asteroidal belt.

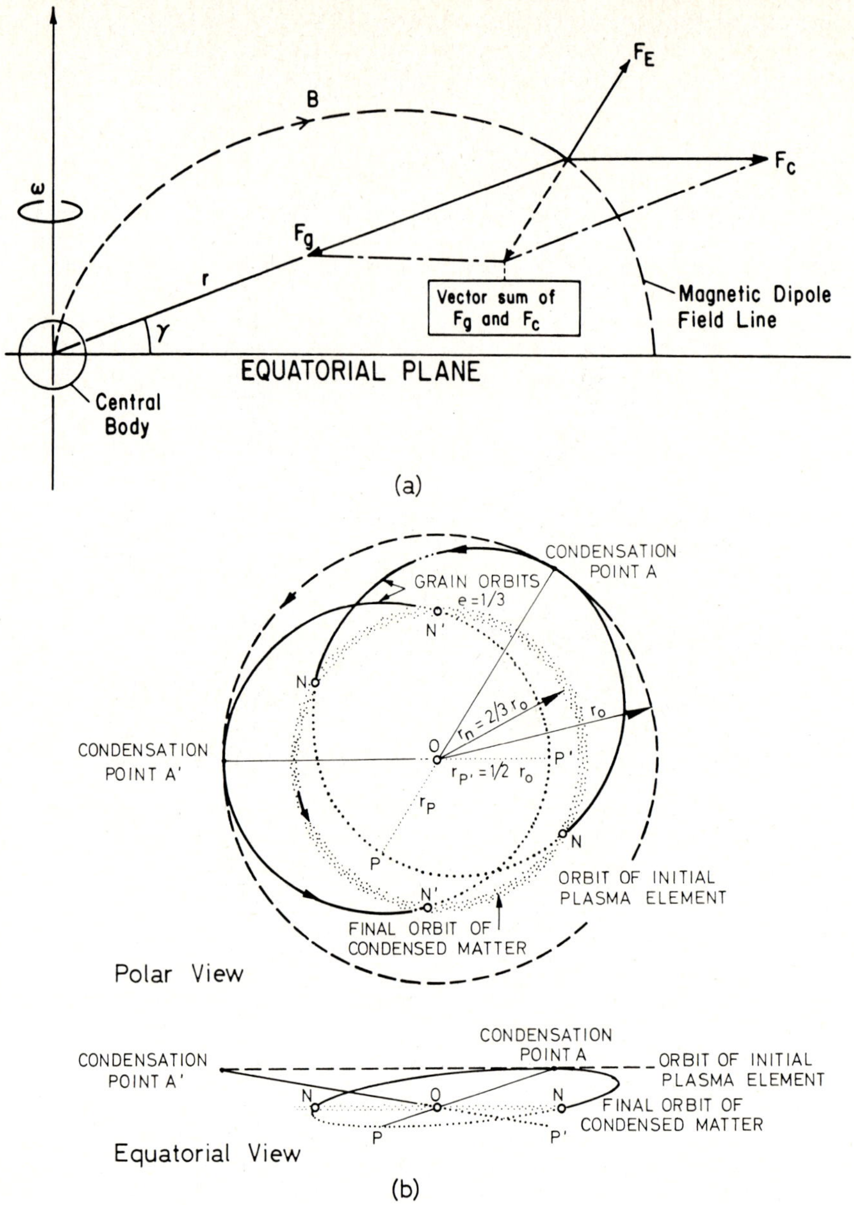

Fig. 10. (a) Charged particles (plasma, charged dust, or cloudlets) in an axisymmetric magnetic dipole field around a gravitating rotating body. If their motion is magnetic-field dominated, a quasistationary motion requires that the vector sum of gravitation and centrifugal force be perpendicular to the magnetic field line. As shown by Alfvén and Arrhenius (1975, 1976), this means $v = (2/3)^{1/2} v_k$, where v is the rotational velocity and v_k is the Kepler velocity. (b) Vanishing magnetic forces give a transfer into elliptic orbits. If the magnetic field or the particle charge suddenly disappears, the particles at the central distance a_0 will orbit in ellipses with semimajor axis $a = (3/4)\, a_0$, and eccentricity $e = 1/3$. They will collide mutually when they reach the nodes in the equatorial plane with $a = (2/3)\, a_0$, and after collisions move in a circle at $a = (2/3)\, a_0$ with $\Gamma = 2{:}3$.

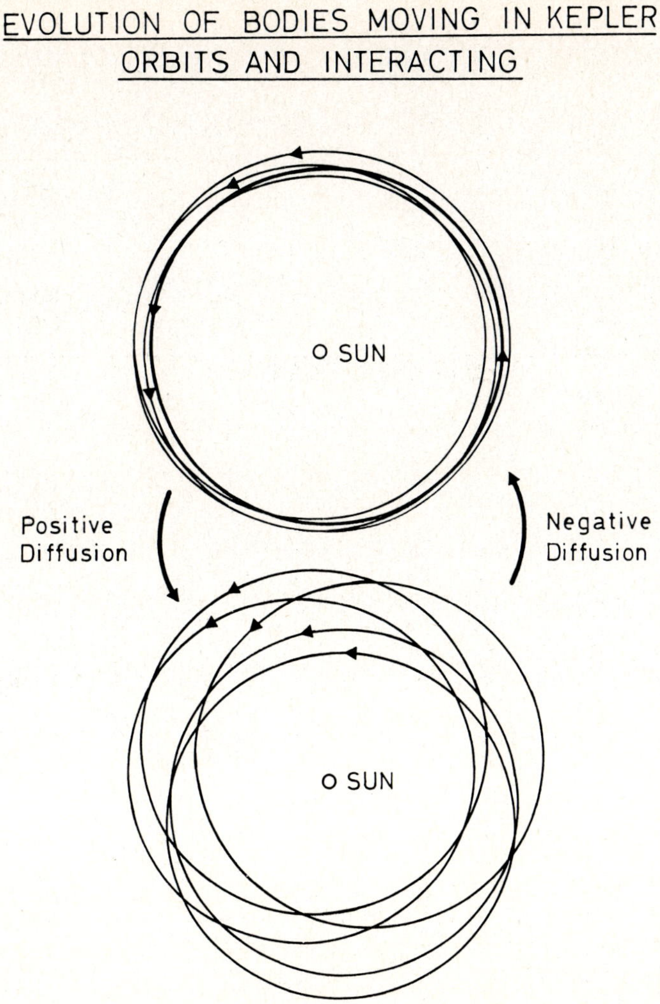

Fig. 11. Interaction of a large number of particles in Kepler orbits. In the discussion of collisions between particles in interplanetary space (e.g. evolution of the asteroidal belt or meteor streams) it is usually taken for granted that state a will evolve into state b (positive diffusion). This is usually not correct. Collisions between the particles will not spread the orbits since the diffusion coefficient is negative (Baxter and Thompson, 1971, 1973). Instead, collisions will lead to equalization of the orbital elements, leading from state b to state c so that a jet stream is formed.

4.2 Non-Catastrophic Formation

The general scenario must be, consequently, that the Saturnian satellites and ring system was formed not as a result of a sudden event, but by a slow injection of diffuse matter during a period of many millions of years. Very early during this period a ring-satellite system was already formed which was qualitatively similar to the present one, but with only a small fraction of its present mass. When more mass accumulated the bodies became more massive, but the same structure was retained. Hence, during most of the time of accretion shadow-producing bodies were located at the same places as today.

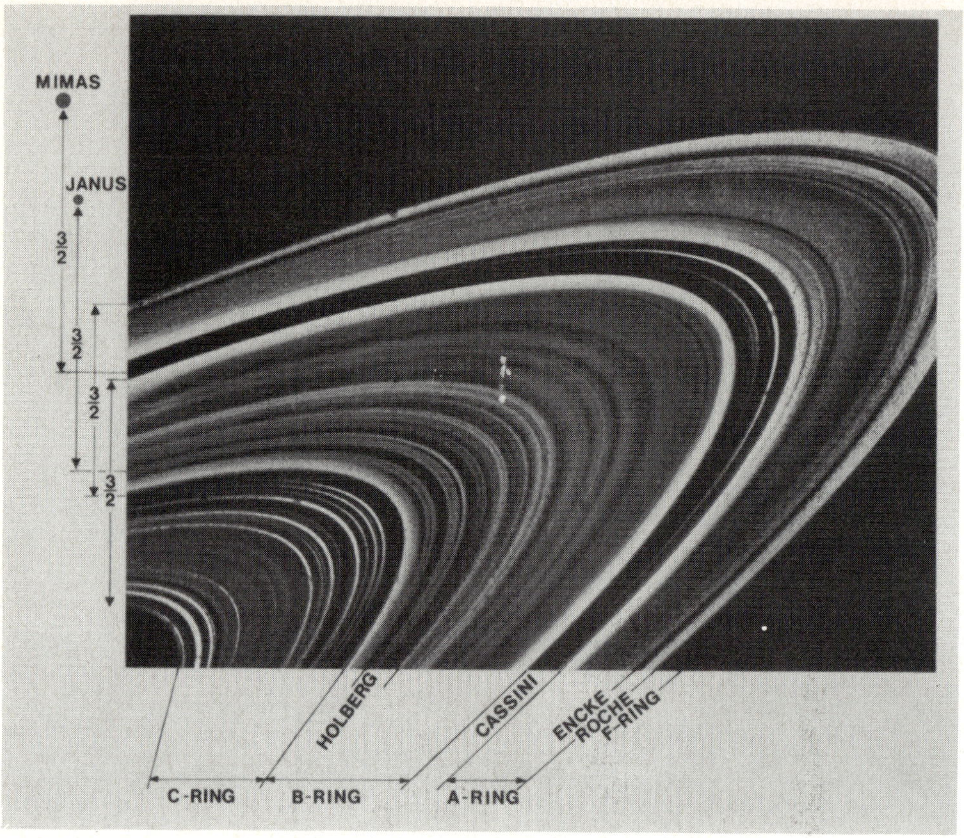

Fig. 12. This photo of the Saturnian Ring shows some of its fine structure. In order to maintain a large number of ringlets, a negative diffusion process is indicated.

4.3 Cosmogonic Shadow

In the cosmogonic model we discuss it was assumed that in the Saturnian magnetosphere there was a dusty plasma which, to some extent, was concentrated at the euqatorial plane. At the Saturnian distance of Mimas, this satellite (or the jet stream out of which it was formed) swept the plasma so that a "hole" - in reality an empty ring - was produced.

During the Pioneer mission, Fillius and McIlwain (1980) actually observed this kind of phenomenon. In fact, at the distance of Mimas, the counting rate went down by orders of magnitude (see Fig. 13). Janus produced a similar although more narrow and shallower "hole". The shepherd satellites and the A ring gave an almost complete cut-off in the plasma density. Hence, actual measurements have shown that the "hole" formation is not an ad hoc assumption of a new process, but was actually observed in the magnetospheres. (Other observations in different energy regions show similar phenomena, although sometimes blurred, presumably by radial electric field drifts).

When, at the plasma-planetesimal transition the 2:3 contraction takes place, the hole should be transferred to 2/3 of the distance of Mimas. This is indeed the location of Cassini's division. Similarly, Janus should produce a marked minimum which now should be found at 2/3 of their distance. This can be identified with the Holberg minimum* in the B ring (Holberg et al., 1982). The Shepherd

* A proposed name for the minimum at 1.58 Saturnian radii, motivated by the fact that J. Holberg has drawn attention to its importance in this connection.

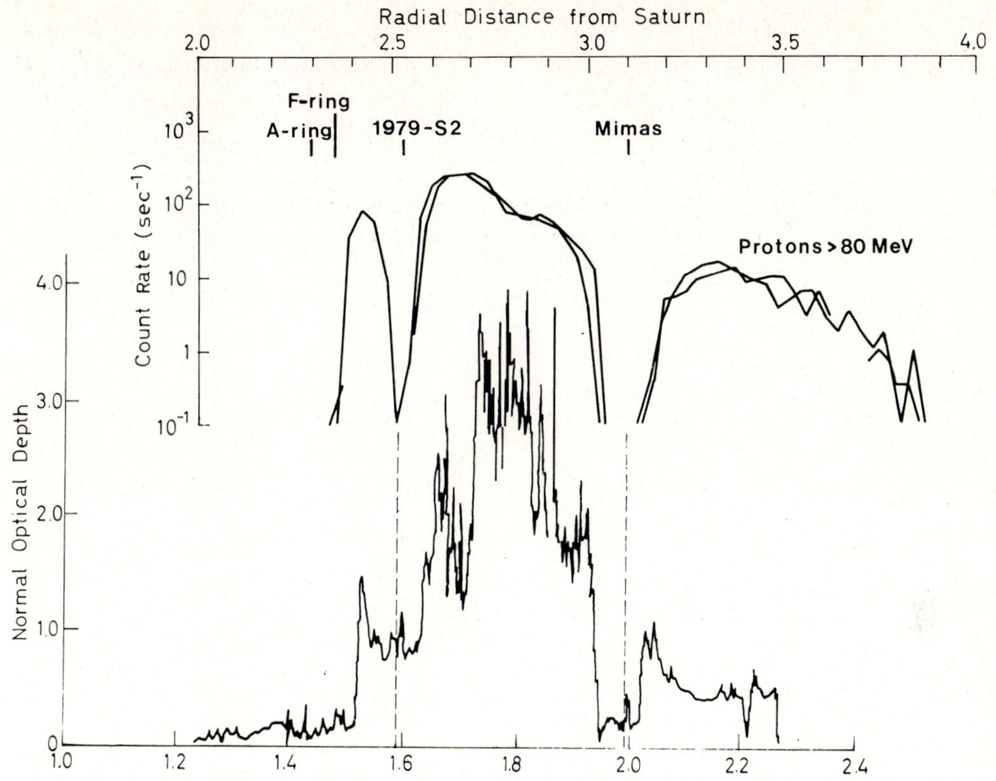

Fig. 13. Comparison between present magnetospheric plasma distribution and mass distribution in the rings. Present-day charged particle distribution in the Saturnian magnetosphere often shows void regions produced by absorption by the satellites. The upper curves are obtained by Fillius and McIlwain (1980). It is argued that the plasma distribution was qualitatively the same in cosmogonic times. The contraction by a factor $\Gamma = 2:3$ at the transition from plasma to planetesimal should result in a somewhat similar mass distribution at a Saturnian distance of 2:3 of the plasma distribution. The lower curve shows the present mass distribution in the Saturnian rings (Holberg, 1983). It is compared with the Fillius-McIlwain curve reduced by a factor $\Gamma = 0.64$. The "cosmogonic shadows" of Mimas, the co-orbital satellites and the Shepherds are identified with Cassini's division, the deepest minimum in the B ring, and the inner limit of the B ring.

satellites should produce a hole in a similar way. The A ring also produces a shadow which extends so far out that it joins the shadow produced by the shepherd satellites.

The result of the combined action of the shepherd satellites and the A ring (and perhaps also of the F ring) is an extended region of low intensity which accounts for the C ring. Hence the rapid decrease in intensity at the border between the B and the C ring should be 2/3 of the position of the shepherds. Finally, the outer edge of the very massive B ring should give the strong decrease in intensity which marks the inner edge of the C ring, which is located at 2/3 of the outer limit of the B ring.

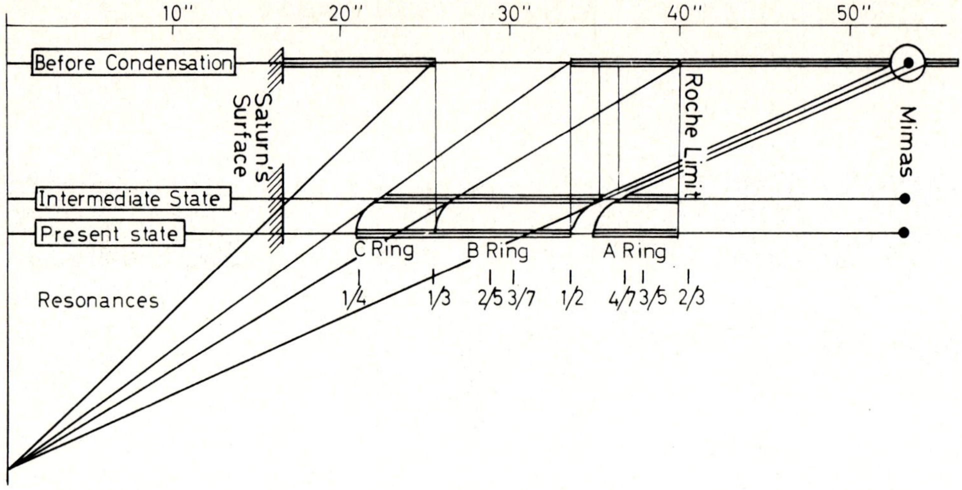

Fig. 14. Correction of the fall-down ratio. The upper line, marked "Before Condensation", shows the plasma density in the equatorial plane of Saturn, the center of which is at the left end. Next line, marked "Intermediate State" represents the density of the grains produced by condensation of the gas. As these move at 2:3 of the distance of the gas out of which they are produced, the density distribution is obtained by a geometrical construction reducing the central distances to 2:3. The lower line, marked "Present State", represents the density distribution into which the "Intermediate State" is transformed by the "shadow reaction", or "shadow load" (see Alfvén, 1954).

The combined actions of these four cosmogonic shadows give the bulk structure of the Saturnian ring. (See Fig. 14; for details see Alfvén, 1983b).

When the ring particles produce the shadows, it implies that they absorb plasma with a smaller angular momentum. This leads to a decrease in their distance. Hence, the contraction factor $\Gamma = 2/3 = 0.667$ should decrease by a few per cent. Theoretically we should expect the value to go down to $0.63 - 0.65$ (Alfvén, 1954; 1981b). In the following we shall use the value 0.64 instead of 0.67. This 4% difference is theoretically motivated, but since we never claim a higher accuracy than a few percent, it is not very important for the main discussion.

4.4 Comparison with the Asteroidal Belt

The hetegonic principle states that all systems of secondary bodies should follow a general theory. This means that our result for the Saturnian rings should also be applicable to the asteroid belt around the sun, which is the other important case where the initial planetesimals have not accreted to larger bodies (see Fig. 15; Alfvén, 1984).

A similar analysis of the asteroidal belt has given three identifications of cosmogonic shadows (see Table IV). The small correction should not be applied to the asteroidal belt (see Alfvén and Arrhenius, 1976, Chapter 11.8, 18.5 and 18.8; Alfvén, 1983b).

Also in this case the mutual agreement and the agreement with theory is surprisingly good.

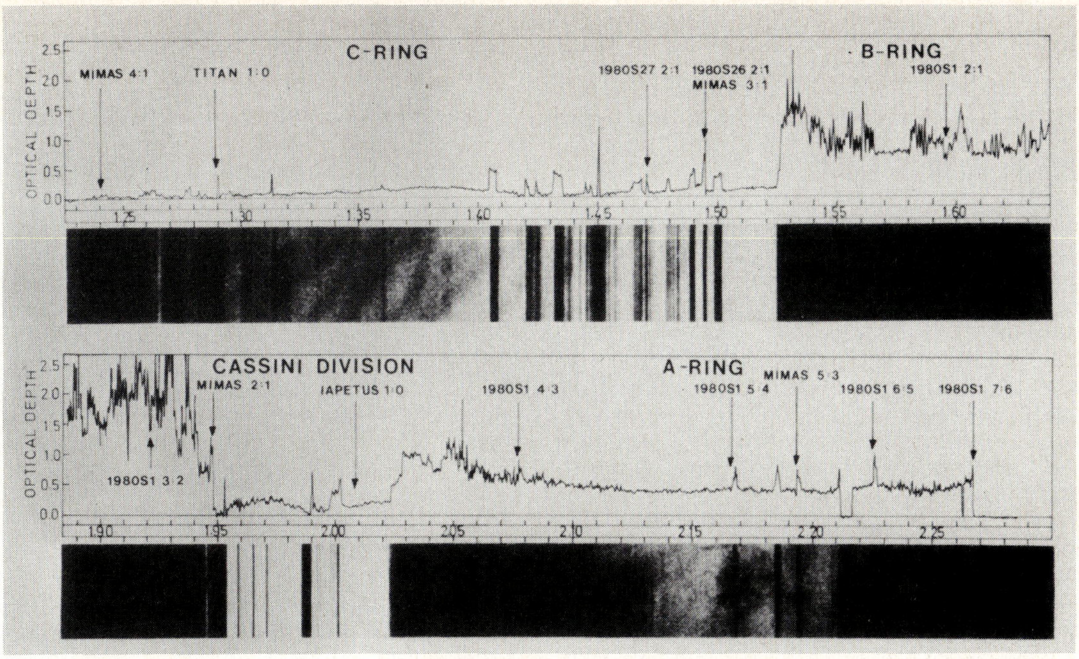

Fig. 15. The normal optical depth of the rings from Voyager 2 UVS ring occultation data. Below is the brightness of the rings in transparent light (Holberg *et al.*, 1982).

TABLE IV

Cosmogonic shadows

Saturnian ring from Holberg's data	a	Γ
Mimas	3.075	
Co-orbitals	2.510	0.646
Shepherds	2.349	0.63
	2.310	
Cassini Center	1.984	0.655 (0.650-0.660)
Outer B	1.945	
Holberg min	1.58	0.635
Inner B	1.525	
Inner C	1.235	
Average 0.642 ± 2%		

Asteroidal region	a	Γ
Jupiter	5.18	
Main belt		0.676
outer limit	3.50	
High density		0.674
outer limit	3.22	
High density		0.683
inner limit	2.36	
Main belt		
inner limit	2.20	
Theoretical value		0.667

(Alfvén, 1981b)

4.5 Remarks on the Larmor Radius of the Accreting Grains

The basis for the derivation of the 2:3 contraction has been that electrically charged particles to a first approximation are bound to move along a magnetic field line (seen from a coordinate system in which the electric field perpendicular to B is zero). This is correct as long as their Larmor radius is small compared to the relevant size parameter. However, we have changed the model from the early simple picture of an electron-ion plasma to a treatment of the behaviour of a charged dust in a dusty plasma. The dust grains have q/m (charge to mass) ratios, which are orders of magnitude smaller than for electrons or protons. In fact, their Larmor periods may exceed the Kepler period and their Larmor radii may be large compared to the characteristic parameters of the dipole field. This seems to make it difficult to treat their motions as we have done. However, this difficulty is often fictitious, as is shown in some detail in Alfvén (1984). A further analysis of this important problem is in progress (Carlqvist, personal communication).

5.0 CONCLUSIONS ABOUT THE PLASMA-PLANETESIMAL TRANSITION

We have shown that essential features of the bulk structure of the Saturnian rings can be understood as a result of the cosmogonic shadow effect produced by a 2/3 contraction which probably took place at the transition from the plasma to the planetesimal phase (PPT), presumably 4-5 billion years ago. The 2:3 ratio appears in four cases. Adding to this the three cases from the asteroidal belt, we have no less than seven identification. This means that we can state with considerable confidence that the cosmogonic shadow effect must have been essential at the formation of the solar system.

Besides the cosmogonic shadow effect, gravitational resonances were important, especially in the asteroid belt, where they produce the Kirkwood gaps. Due to the very small ratio of Mimas/Saturn in comparison to Jupiter/Sun, corresponding resonance effects in the Saturnian rings are small but clearly identifiable (Holberg et al., 1982). (Also, some other effects are important (see Alfvén, 1983b)). Hence, a 2:3 contraction should be characteristic for the plasma-planetesimal transition (PPT). It seems difficult to interpret the above results as being due to any other effect. The surprisingly high degree of agreement between the observational and theoretic values means that we have a possibility of reconstructing certain features of the PPT with an accuracy within a few percent.

The work on doing this is in progress. A straightforward development of the theoretical model shows that the "holes" produced by plasma absorption by the satellites must be associated with radial electric fields both at the inside and outside of the depleted region. These electric fields will change the Γ-values so that both the Cassini division and the Holberg minimum would exhibit "banks" (i.e. maxima) both inside and outside of the minima. Such structures are actually visible in the Holberg curves (Figs. 13 and 15).

Further, important features in the bulk structure of the rings (discussed in Alfvén, 1983b) are analyzed, including why the center of the B ring is so massive, why there is a structure inside the Cassini division, and how the several maxima in the C ring are produced. Also, the importance of the Northrop and Hill (1982) instability and the possibility of explaining the Encke division as produced by an adiabatic circularization of the initial orbits of the grains are being studied. In this way it may be possible to reconstruct the state of the Saturnian environment at cosmogonic times.

With a similar analysis of the asteroidal region, we will have the possibility of reconstructing how the planetesimal state developed into planets/satellites.

Further, we can conclude:

Since the PPT, the structure of the Saturnian rings and the asteroid belt cannot have undergone any violent large-scale disruptions. A slow evolution, resulting in concentration of mass in a few bodies, is indicated. In the equatorial region there could have been no "solar gale" strong enough to disrupt the basic pattern. However, there could very well be a solar wind (or solar gale) in the polar regions. Indeed the transfer of angular momentum from the sun to the solar nebula must be accompanied by a large release of energy (rotational energy of the sun) which must produce an intense heating partically in the close neighbourhood of the sun, resulting in a solar wind emission (see Alfvén and Arrhenius 1976 Chapter 5.5 p. 78). There will be little to prevent this in the axial directions, but the infall of matter accreting in the equatorial plane will make it difficult to proceed very far. This is in agreement with the results of Axon and Taylor (1984), who give a model of a biconical protostellar wind for a star in the Orion nebula.

Further during the PPT and later there could not have been any strong, large-scale turbulence.

Violent events before the PPT cannot be excluded, but so far there seems to be no decisive arguments in favor of such phenomena. On the contrary, what has been said in Section 3 speaks against it. The large-scale evolution of the solar system, from an interplanetary cloud to the present structure, could very well have been a slow, quiet process (but consisting of rapid, consecutive processing of cloudlets) which, during many millions of years, built up the present structure.

6.0 SUMMARY OF THE PROCESSES NECESSARY TO UNDERSTAND THE EVOLUTION OF THE SOLAR SYSTEM

As we stated in Table II, a cosmogony taking account of electromagnetic effects requires these processes:

(a) Electromagnetic transfer of angular momentum

(b) The existence of a phenomenon later called "critical velocity"

(c) A plasma-planetesimal transition associated with a 2:3 contraction which produced "cosmogonic shadows".

The existence of all three predicted processes has been confirmed. When elaborating the theory in different respects it was found necessary to introduce one more process:

(d) Local pinch effects in the interstellar source cloud are required to form a non-Laplacean solar nebula of a kind which could give the required initial condition for the cosmogonic processes.

Of these, (b) and (c) have been discovered after being predicted from the cosmogonic theory, whereas (a) has been discovered independently.

Our claim that our result could not be uncertain by more than a few percent seems to be legitimate.

The usefulness of a theory is often judged by the number of earlier unknown phenomena it predicts. If we apply this criterion to the introduction of electromagnetic effects in the cosmogonic processes, the electromagnetic approach seems to be not too bad.

Acknowledgements - We wish to thank Professor C.-G. Fälthammar for placing the resources of the Department of Plasma Physics to our disposal. The editing has been done by Mrs. E. Florman and Mrs. K. Vikbladh, the figures have been drawn by Mrs. K. Forsberg. Support has been obtained from the Swedish Natural Science Research Council.

7.0 REFERENCES

Alfvén, H., On the Cosmogony of the Solar System I, Stockholms Observatoriums Annaler I, 14, No. 2, 1942.

Alfvén, H., On the Origin of the Solar System, Oxford University Press, 1954.

Alfvén, H., Cosmic Plasma, D. Reidel Publ. Co., Dordrecht, Holland, 1981a.

Alfvén, H., The Voyager 1/Saturn Encounter and the Cosmogonic Shadow Effect, Astrophys. Space Sci. 79, 491, 1981b.

Alfvén, H., Paradigm Transition in Cosmic Plasma Physics, Physica Scripta T2/1, 10, 1982.

Alfvén, H., Paradigm Transition in Cosmic Plasma Physics, Geophys. Res. Letters, 10, 487, 1983a.

Alfvén, H., Solar System History as Recorded in the Saturnian Ring Structure, Astrophys. Space Sci. 97, 79, 1983b.

Alfvén, H., Space Research and Cosmic Plasma Physics, TRITA-EPP-83-06, The Royal Institute of Technology, Stockholm. Lecture at the IUGG Meeting, Hamburg, Aug. 22, 1983, 1983c.

Alfvén, H., On Hierarchical Cosmology, Astrophys. Space Sci. 89, 313, 1983d.

Alfvén, H., Cosmogony as an Extrapolation of Magnetospheric Research, TRITA-EPP-84-02, The Royal Institute of Technology, Stockholm (in print: Space Sci. Rev. Oct. 1984), 1984.

Alfvén, H. and Arrhenius, G., Structure and Evolutionary History of the Solar System, D. Reidel Pub. Co., Dordrecht, Holland, 1975.

Alfvén, H. and Arrhenius, G., Evolution of the Solar System, NASA SP-345, US Government Printing Office, Washington, D.C., 1976.

Anderson, O.A., Baker, W.R., Bratenahl, A., Further, H.P., Kunkel, W.B. and Stone, J.A., Hydromagnetic Capacitor, J. Appl. Phys. 30, 188, 1959.

Axnäs, I., Brenning, N. and Raadu, M.A., The Critical Ionization Velocity: A Bibliography, in Haerendel, G. and Möbius, E., (Eds.) Proceedings of the Workshop on Alfvén's Critical Velocity (Oct. 11-13, 1982), Max-Planck-Institut für Extraterrestrische Physik, Garching, FRD, 1982.

Axon, D.J. and Taylor, K., Discovery of a Family of Herbig-Haro Objects in M42: Implications for the Geometry of the High Velocity Molecular Flow?, Mon. Not. R. Astr. Soc. 207, 241, 1984.

Baxter, D. and Thompson, W.B., Jetstream Formation Through Inelastic Collisions, in Physical Studies of Minor Planets, NASA SP-267, Ed. T. Gehrels, US Government Printing Office, Washington, D.C., 319, 1971.

Baxter, D. and Thompson, W.B., Elastic and Inelastic Scattering in Orbital Clustering, Astrophys. J., 183, 323, 1973.

Boström, R., Ionosphere-Magnetosphere Coupling, in Magnetospheric Physics, Ed. McCormac, B.M., D. Reidel Publ. Co., Dordrecht, Holland, p. 45, 1974.

Carlqvist, P., Double Layers in Space, Report Risø-R-472, Risø National Laboratory, DK-4000 Roskilde, Denmark, Eds. P. Michelsen and J. Juul Rasmussen, 255-273, 1982.

Carpenter, R.T., Torvén, S. and Lindberg, L., Potential Oscillations of Double Layers in Inductive Current Circuits, Proceedings of Second Symposium on Plasma Double Layers and Related Topics, 5-6 July 1984, Innsbruck, Austria, 1984.

Esposito, L.W., O'Callaghan, M., Simmons, K.E., Hord, C.W., West, R.A., Lane, A.L., Pomphrey, R.B., Coffeen, D.L. and Sato, M., Voyager Photopolarimeter Stellar Occultation of Saturn's Rings, J. Geophys. Res. 88, 8643, 1983.

Fahleson, U.V., Experiments with Plasma Moving Through Neutral Gas, Phys. Fluids 123, 1961.

Fillius, W. and McIlwain, C.E., Very Energetic Protons in Saturn's Radiation Belt, J. Geophys. Res. 85, 5803, 1980.

Haerendel, G., Alfvén's Critical Velocity Effect Tested in Space, Z. f. Naturforschung, 37a, 728, 1982.

Hill, T.W., Dessler, A.J. and Goertz, C.K., Magnetospheric Models, in Physics of the Jovian Magnetosphere, ed. A.J. Dessler, Cambridge University Press, Cambridge, p. 353, 1983.

Holberg, J., Forrester, W.T. and Lissauer, J.J., Identification of Resonance Features within the Rings of Saturn, Nature 297, 115, 1982.

Holberg, J., (Private communication), 1983.

Iijima, T. and Potemra, T.A., Large-Scale Characteristics of Field-Aligned Currents Associated with Substorms, J. Geophys. Res. 83, 599, 1978.

Iizuka, S., Saeki, K., Sato, N. and Hatta, Y., Buneman Instability and Double-Layer Formation in a Collisionless Plasma, Phys. Rev. Lett. 43, 1404, 1979.

Iizuka, S., Michelsen, P., Rasmussen, J.J. and Schrittwieser, R., Dynamics of a Potential Barrier Formed on the Tail of a Moving Double Layer in a Collisionless Plasma, Phys. Rev. Lett. 48, 145, 1982.

Lin, R. and Bodenheimer, P., On the Stability of Saturn's Rings, Ap. J. 248, L83, 1981.

Lindberg, L. and Torvén, S., Spontaneous Transfer of Magnetically Stored Energy to Kinetic Energy by Electric Double Layers, TRITA-EPP-83-05, The Royal Institute of Technology, Stockholm, 1983.

Michelsen, P. and Rasmussen, J.J. (Ed.), Symposium on Plasma Double Layers, Risø-R-472, Risø National Laboratory, DK-4000 Roskilde, Denmark, June 1982, 1982.

Northrop, T.G. and Hill, J.R., Stability of Negatively Charged Dust Grains in Saturn's Ring Plane, J. Geophys. Res. 87, 6045, 1982.

Potemra, T., Current Systems in the Earth's Magnetosphere, Rev. Geophys. Space Phys., 1979.

Potemra, T., Magnetospheric Currents, Chapman Conference on Magnetospheric Currents, Irvington, VA, April 5-8, 1983, Geophysical Monograph 28, American Geophysical Union, Washington D.C., 1984.

Torvén, S., High-Voltage Double Layers in a Magnetized Plasma Column, J. Phys. D: Appl. Phys. 15, 1943, 1982.

Torvén, S. and Lindberg, L., Properties of a Fluctuating Double Layer in a Magnetized Plasma Column, J. Phys. D: Appl. Phys, 13, 2285, 1980.

Zmuda, A.J. and Armstrong, J.C., The Diurnal Flow Pattern of Field-Aligned Currents, J. Geophys. Res. 79, 4611, 1974.

Theoretical and Applied Mechanics
F.I. Niordson and N. Olhoff
Elsevier Science Publishers B.V. (North-Holland)

COMPUTERS AND CHAOS IN MECHANICS

Joseph B. Keller*

Departments of Mathematics and Mechanical Engineering
Stanford University
Stanford, California
USA

Much recent progress in mechanics is related to the use of computers in the solution of problems, in the control of experiments, and in the analysis of data. These tools have made certain kinds of calculations and measurements easier. But they have also revealed the widespread occurrence of chaotic and stochastic behavior of mechanical systems, and have shown that there are certain regularities in this behavior. The understanding of the chaotic behavior and its regularities, and of how to analyze and control it, are unsolved problems. Many theoretical attacks have been made on these problems, but they have only touched the surface.

1. INTRODUCTION

It is a pleasure and an honor for me to present the closing lecture of the Sixteenth International Congress of Theoretical and Applied Mechanics. I want to thank Professor Frithiof Niordson, the Chairman of the Organizing Committee, Dr. Niels Olhoff, the Executive Secretary of the Committee, and all the other members of this Committee and of the Congress Committee for having invited me. In addition I want to thank them for having organized the Congress so efficiently. They have enabled all of us to attend a meeting on the highest technical level and at the same time to enjoy the hospitality for which Denmark is so well known.

From the papers presented at this Congress, and from the lively discussions in the lecture rooms and in the halls, it is clear that mechanics is thriving. A quantitative indication of this is shown in Figure 1, which shows the number of mechanical engineers employed in the United States as a function of time from 1900 to 1970. The number seems to have grown exponentially, doubling every ten years until 1950. Then the growth rate slowed down considerably, as is to be expected when the number approaches the saturation level. This behavior is in accordance with the logistic equation governing a population $N(t)$ in a limited environment, such as flies in a bottle:

$$\frac{dN(t)}{dt} = \alpha N(t)\left[1 - \frac{N(t)}{K}\right] .$$

---

* This work was supported in part by the National Science Foundation, the Office of Naval Research and the Air Force Office of Scientific Research.

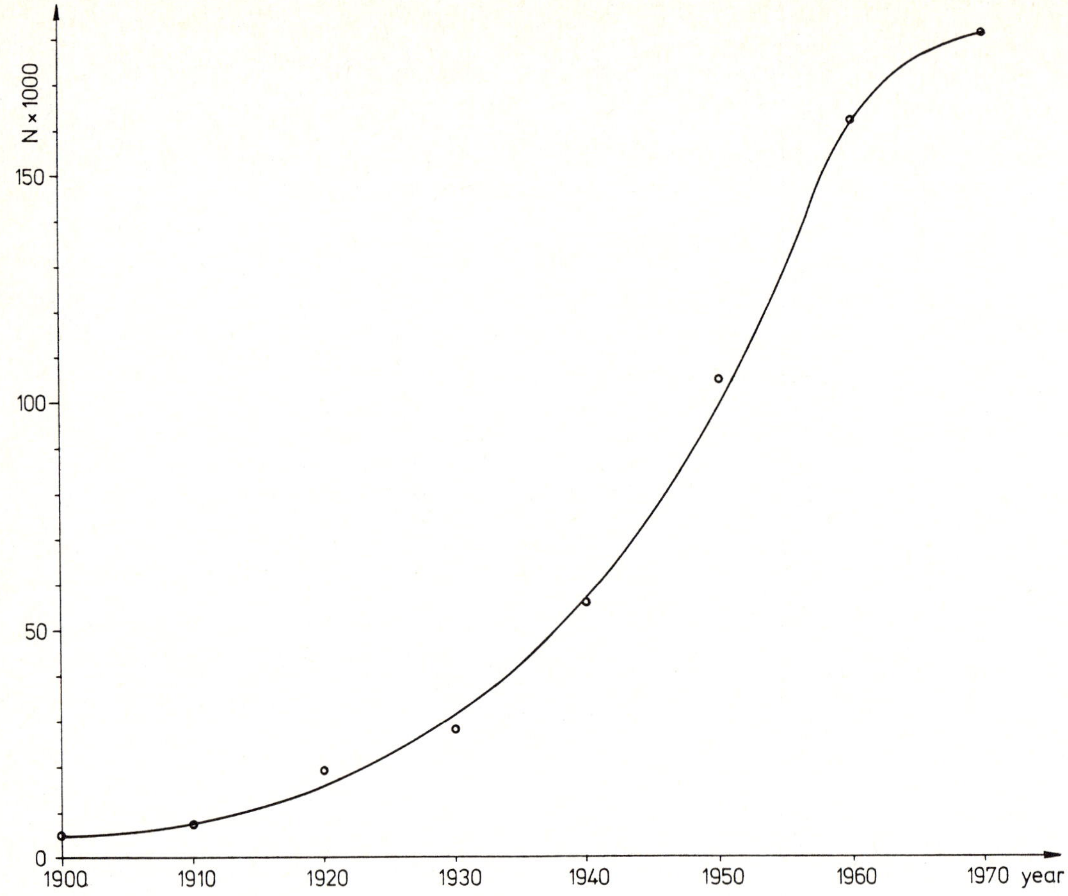

Figure 1. The number of mechanical engineers employed in the United States as a function of time. Based upon data on page 140 of "Historical Statistics of the United States, Colonial Times to 1970. Part 1", U.S. Department of Commerce, Bureau of Census, 1975. For the earlier years, the combined figures for various categories of engineers were pro rated in accordance with the ratios for later years. The solid curve was drawn freehand to indicate the trend.

Here α is the growth rate and K is the maximum possible population.

Figure 2 shows the number of journals of interest to workers in mechanics. This number appears to be doubling every 20 years and does not seem to have begun to level off yet.

To find out what all these mechanical engineers are doing, we can use the results of a questionnaire prepared by Professor Charles Steele of Stanford University. It was sent to members of the Applied Mechanics Division of the American Society of Mechanical Engineers. The first question concerned the primary responsibilities of the members, and the results are shown in Table 1. Since each member could give more than one responsibility, the percentages add up to nearly 200%. The table shows that research and development are the major activities, followed by education and analysis.

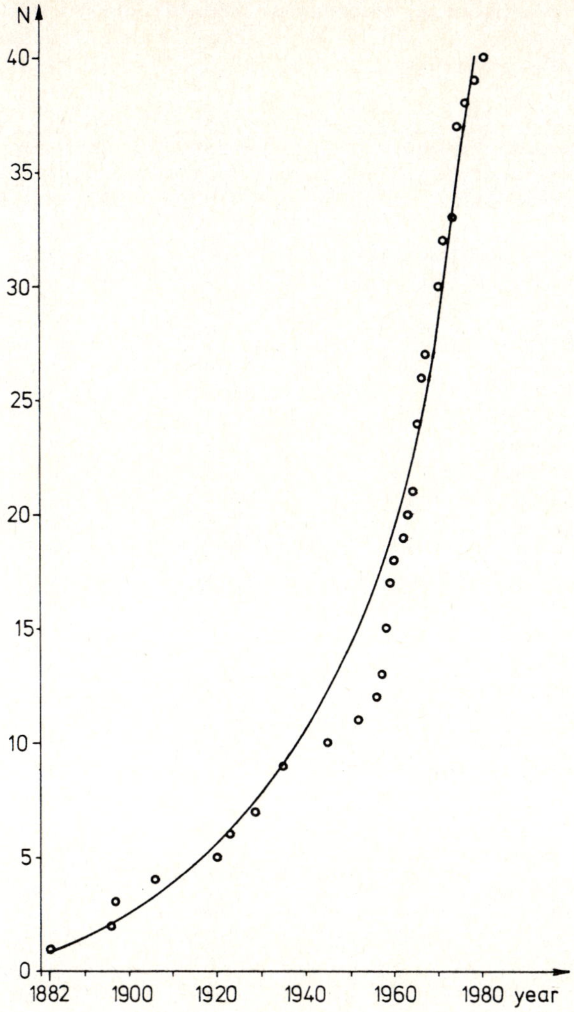

Figure 2. The number of journals pertaining to mechanical engineering as a function of time. Based upon various sources. The solid curve was drawn freehand to indicate the trend.

Table 1. Primary responsibilities, members of ASME's Applied Mechanics Division

1. Research & Development	=	54.7%
2. Education	=	42.9%
3. Analysis	=	32.4%
4. Consulting	=	15.5%
5. Design	=	11.1%
6. Product Development	=	10.8%
7. Testing	=	10.4%
8. Management	=	7.0%
9. Maintenence	=	1.3%
10. Sales	=	.6%

The second question, inquiring about the areas of interest of the members, yielded the results shown in Table 2. Here also each member could list more than one area. The three most popular areas are

Table 2. Areas of Interest, Members of ASME's Applied Mechanics Division

			total	in industry	in education
1.	Finite Element Methods	=	35.8%	42.6%	26.8%
2.	Computing	=	29.1%	31.4%	26.0%
3.	Analytical Methods	=	28.4%	29.0%	27.6%
4.	Fracture (crack) Mechanics	=	24.3%	26.6%	21.3%
5.	Vibrations (structures)	=	23.6%	24.3%	22.8%
6.	Composite Materials	=	19.9%	20.1%	19.7%
7.	Plasticity	=	18.9%	18.9%	18.9%
8.	Plates, Shells and Membranes	=	18.6%	16.6%	21.3%
9.	Vibrations (struct. elements)	=	18.2%	17.8%	18.9%
10.	Elasticity	=	17.2%	17.2%	17.3%
11.	Machine Design	=	15.9%	17.8%	13.4%
12.	Heat Transfer	=	15.9%	17.8%	13.4%
13.	Continuum Mechanics	=	15.5%	13.0%	18.9%
14.	Fracture Processes	=	15.5%	20.7%	8.7%
15.	Structures (applied)	=	15.2%	18.9%	10.2%
16.	Kinematics & Dynamics	=	14.9%	12.4%	18.1%
17.	Experimental Stress Analysis	=	14.9%	16.0%	13.4%
18.	Impact on Solids	=	13.5%	18.3%	7.1%
19.	Incompressible Flow	=	13.5%	11.8%	15.7%
20.	Vibrations of Solids (basic)	=	12.8%	10.1%	16.5%
21.	Structural Stability	=	12.8%	10.7%	15.7%
22.	Materials Test Techniques	=	12.5%	14.2%	10.2%
23.	Viscoelasticity	=	11.1%	9.5%	13.4%
24.	Machine Elements	=	11.1%	11.2%	11.0%
25.	Wave Motion in Solids	=	10.8%	8.9%	13.4%
26.	Structures (basic)	=	10.8%	8.9%	13.4%
27.	Strings, Ropes, Beams	=	9.5%	7.1%	12.6%
28.	Thermal Effects in Solids	=	9.5%	9.5%	9.4%
29.	Fastening & Joining	=	8.1%	13.0%	1.6%
30.	Compressible Flow	=	8.1%	8.9%	7.1%
31.	Aerodynamics	=	8.1%	8.9%	7.1%
32.	Thermodynamics	=	7.4%	7.7%	7.1%
33.	Acoustics	=	7.4%	5.9%	9.4%
34.	Systems & Controls	=	7.1%	7.7%	6.3%
35.	Materials Processing	=	7.1%	8.9%	4.7%
36.	Hydraulics	=	7.1%	7.1%	7.1%
37.	Internal Flow	=	7.1%	5.3%	9.4%
38.	Structures (containment)	=	6.8%	8.3%	4.7%
39.	Flow Measurements	=	6.8%	6.5%	7.1%
40.	Explosions & Ballistics	=	6.8%	10.1%	2.4%
41.	Multiphase Flows	=	6.4%	5.3%	7.9%
42.	Turbulence	=	6.1%	5.9%	6.3%
43.	Wave Motion	=	6.1%	4.1%	8.7%
44.	Machine Fluid Dynamics	=	6.1%	8.9%	2.4%
45.	Flow Stability	=	5.7%	3.6%	8.7%
46.	Soil Mechanics	=	5.1%	5.9%	3.9%
47.	Rheology	=	5.1%	3.0%	7.9%
48.	Aeroelasticity	=	4.4%	4.1%	4.7%
49.	Combustion	=	4.1%	4.1%	3.9%
50.	Wall Layers	=	4.1%	3.0%	5.5%
51.	Mass Transfer	=	3.7%	3.0%	4.7%
52.	Prime Movers & Propusion	=	3.0%	4.1%	1.6%
53.	Free Shear Layers	=	2.7%	1.2%	4.7%
54.	Biocontrol	=	2.0%	1.2%	3.1%
55.	Astronautics	=	1.7%	1.8%	1.6%
56.	Electro-Magneto-Fluids	=	1.4%	1.2%	1.6%
57.	Mechanics of Plasma	=	.7%	1.2%	.0%
58.	Rarefied Flow	=	.0%	.0%	.0%
59.	Flight Test Techniques	=	.0%	.0%	.0%

finite element methods, computing, and analytical methods. They are followed by fracture, vibrations, composites, and plasticity, which are all aspects of solid mechanics. The first topic in fluid dynamics is incompressible flow, which is nineteenth in the list. The next fluid dynamics topics are compressible flow, aerodynamics, hydraulics and internal flow, which are numbers 30, 31, 36 and 37 respectively. Turbulence, the outstanding unsolved problem of classical physics, is number 42 on the list, and is of interest to six percent of the members. On the other hand, fracture, the major open problem of solid mechanics, is fourth and is of interest to one quarter of the members.

The two most popular areas of interest shown in Table 2 both involve computers. Thus computers seem to have become the most commonly used tool of workers in mechanics. The table also shows that engineers in industry have more interest in both finite element methods and in computing than do those in education.

Another indication of the activities of research workers in mechanics is provided by an unpublished study I made in 1972 with the help of Andrew Callegari and Maurice Machover. We studied the type of mathematics used in various research journals by recording the highest level of mathematics used on each page of a given journal. This showed which kinds of mathematics were used and how frequently each was used. For all the engineering and physics journals which we examined, including the journal of Applied Mechanics, the most commonly occurring branches of mathematics were parts of mathematical analysis. This included calculus, elementary ordinary differential equations, complex variables, special functions, elementary partial differential equations, perturbation methods and numerical calculation. From 1950 to 1970 there was a general increase in the level of mathematics and its frequency of occurrence. There was also much greater use of partial differential equations.

However the most noticeable change was the increase in numerical results. In 1970 almost every article contained graphs and tables obtained with the help of a computer, whereas in 1950 there were many fewer such results. At the same time there was a decrease in the discussion of numerical analysis because of the availability of packaged programs to solve ordinary differential equations, to do matrix algebra, etc.

The questionnaire results and the journal study both confirm the obvious fact that computers have become a most important instrument for research in mechanics. We shall now consider some consequences of their use.

2. SOME PROGRESS DUE TO COMPUTERS

Computers have enabled us to make progress in many different ways in mechanics. For example, they have been used to collect experimental data, to analyse data and to control experiments. They have made possible computer aided design (CAD) and computer aided manufacturing (CAM). They have also made possible the construction of the science citation index, which is a valuable research tool.

In addition to all these roles, computers have been used to compute, that is to calculate numerically the solutions of mathematical problems. In fact that is the use to which they have been put by all those engineers whose interest is in the finite element method and in computing. It is also the way in which they were used to construct the tables and graphs which appear in the research journals.

There now are numerous computer programs available to solve various classes of problems. There are ordinary differential equation solvers, random number generators, elliptic equation solvers, minimization programs, etc. There are special programs for the analysis of structures, for stress calculations, etc.

In view of all these developments in problem solving by computer, what role is there for human problem solvers? It seems to me that there are essentially two kinds of roles. One is to "join them", that is to use computers, to devise methods whereby problems can be solved with computers, to formulate new problems to be solved by them, etc. The other role is to "fight them", that is to do what computers cannot do. This includes studying singularities of solutions, determining asymptotic behavior, deducing general results, etc.

These two roles are not antagonistic but are actually complementary to one another. This is shown very clearly in the study of non-integrable Hamiltonian mechanical systems, which has been advanced considerably in recent years by a combination of analytical and computational methods. Therefore we shall describe some of the results of this study, beginning with the analytical ones.

3. ANALYTICAL RESULTS ON NON-INTEGRABLE SYSTEMS

A mechanical system is called a Hamiltonian system of n degrees of freedom if it has n coordinates $q(t) = [q_1(t), \dots, q_n(t)]$, n conjugate momenta $p(t) = [p_1(t), \dots, p_n(t)]$ and a Hamiltonian function $H[p, q]$. The equations of motion are

$$\frac{dp_i}{dt} = -\frac{\partial H}{\partial q_i}, \quad \frac{dq_i}{dt} = \frac{\partial H}{\partial p_i}, \quad i = 1, \dots, n .$$

Such a system is said to be integrable if it has n new angle coordinates $\theta(t) = [\theta_1(t), \dots, \theta_n(t)]$ and n new action variables $J(t) = [J_1(t), \dots, J_n(t)]$ with the following three properties:

1. $p = p(\theta, J)$ and $q = q(\theta, J)$ are periodic in each θ_i with period 2π,

2. $H[p(\theta, J), q(\theta, J)] = \mathcal{H}(J)$ is independent of θ,

3. $$\frac{dJ_i}{dt} = -\frac{\partial \mathcal{H}(J)}{\partial \theta_i} = 0, \quad \frac{d\theta_i}{dt} = \frac{\partial \mathcal{H}(J)}{\partial J_i} \equiv \omega_i(J), \quad i = 1, \dots, n .$$

From the equations of motion (3) for J_i, it follows that each J_i is a constant, so the vector J is constant in time. Then from the equations of motion (3) for θ_i, it follows that each θ_i increases linearly in t:

4. $$\theta_i(t) = \omega_i(J)t + \theta_{i0}, \quad i = 1, \dots, n .$$

Now from (1) and (4) we see that p and q are multiply-periodic in t with the periods $2\pi/\omega_i(J)$, $i=1,\dots,n$.

We have now seen that an integrable Hamiltonian system can be integrated, i.e. that the equations of motion for $\theta(t)$ and $J(t)$ can be solved explicitly. Then from (1) the original variables $p(t)$ and $q(t)$ can be obtained, and they are multiply-periodic, or quasi-periodic, functions of time.

The motion can be described simply in the $2n$ dimensional phase space of θ and J , in which each θ_i ranges from 0 to 2π . Then the surface J = constant is a torus, and the path $\theta(t) = \omega(J)t + \theta_0$ winds around it. This torus is said to be invariant because every orbit which starts at any point on it stays on it forever. Thus the whole phase space consists of invariant tori.

Now we can pose the following basic problem. Suppose that $H_0(J)$ is the Hamiltonian of an integrable system and that $H_1(J,\theta)$ is a smooth function which is periodic in each θ_i with period 2π . Let ε be a small parameter, and consider the perturbed Hamiltonian $H(J,\theta)$ defined by

$$H(J,\theta) = H_0(J) + \varepsilon H_1(J,\theta) \ .$$

Is the system with this new Hamiltonian $H(J,\theta)$ integrable for ε sufficiently small?

In order to answer this question Poincaré in 1892 and Von Zeipel in 1916 developed the classical perturbation theory of Hamiltonian mechanics. In that theory one seeks new action and angle variables $\bar{J}$ and $\bar{\theta}$, and a transformed Hamiltonian $\bar{H}(\bar{J},\varepsilon)$ which is independent of $\bar{\theta}$. The transformation from the old variables to the new ones depends upon the small parameter ε . To find this transformation, one represents $\bar{H}$ and the transformation as power series in ε . Then by formal calculations, one determines the coefficients in those power series.

It turns out that the denominators of these coefficients involve products and powers of the quantities $n\cdot\omega$, defined by

$$n\cdot\omega \equiv n_1\omega_1 + n_2\omega_2 + \dots + n_n\omega_n \ .$$

Here n is the vector with integer components $n_1, n_2, \dots, n_n$, which may be positive, negative or zero. When $n\cdot\omega = 0$, the frequencies $\omega(J)$ of the unperturbed system are said to be in resonance. Then some of the coefficients in the power series blow up, i.e. they are not defined, and the perturbation expansion fails to yield the desired transformation. Even when the $\omega(J)$ are incommensurate, so that there is no exact resonance, some of the denominators still become very small. The occurrence of these small denominators has prevented anyone from proving that the perturbation series converge.

In 1954 Kolmogoroff proposed that the transformation should be determined by a rapidly converging iteration method, rather than by the traditional power series method. His proposal was carried out by Arnold (1961) and by Moser (1962). They proved that when the $\omega(J)$ are sufficiently incommensurable, the torus J = constant is transformed into another invariant torus for the perturbed system, and in fact a finite fraction of all the tori are transformed in this way. This result is called the KAM theorem after Kolmogoroff, Arnold and Moser. "Sufficiently incommensurable" means that for all n,

$$|n \cdot \omega| \geq \gamma |n|^{-\tau} .$$

Here γ and τ are some constants independent of n. The theorem also requires various other conditions, and especially that ε be sufficiently small.

This theorem shows that a system which is close enough to an integrable system has a large set of invariant tori. It does not describe what happens to the other tori which are not preserved by the perturbation. Nor does it tell how the fraction of tori which remain invariant varies with ε as ε increases.

4. COMPUTERS AND CHAOS

Computed solutions of the equations of motion for particular systems have provided answers to these questions for those systems. The results which stimulated many of these calculations were those of the meteorologist Lorenz in 1963. He studied Benard convection, the motion which occurs in a horizontal layer of liquid which is heated from below. Due to thermal expansion, the hot fluid rises from the bottom and thereby generates convective motion.

To analyze this motion, Lorenz described it by a few modes. Then he derived equations of motion for the time-dependent amplitudes $x(t)$, $y(t)$ and $z(t)$ of those modes. In dimensionless variables these equations are

$$\dot{x} = -\sigma x + \sigma y ,$$

$$\dot{y} = -xz + rx - y ,$$

$$\dot{z} = xy - bz .$$

Numerical solution of these equations showed that for certain values of the parameters σ, r and b, the solutions are chaotic. For other parameter values the solutions are quite regular.

There is no precise definition of chaotic, but it certainly means that the solution is not constant, not periodic, not multiply periodic, nor does it approach a motion of any of these regular kinds. Instead the solution continually oscillates about in an irregular fashion. As a consequence it has a continuous frequency spectrum, rather than a spectrum consisting of discrete frequencies, as is the case for regular motions.

The Lorenz system is not a Hamiltonian system but rather a dissipative one. However the same behavior - the occurrence of chaotic solutions for certain parameter values and regular solutions for other values - was soon found by Henon and Heiles (1964) for a Hamiltonian system. They considered a system with two degrees of freedom, with coordinates $x(t)$ and $y(t)$ and conjugate momenta $p_x(t)$ and $p_y(t)$. Their Hamiltonian is

$$H = \frac{1}{2}(p_x^2 + p_y^2 + x^2 + y^2) + x^2 y - \frac{1}{3}y^3 .$$

They found that for total energy E less than $1/12$, the solutions were regular for most initial conditions. As E increased the fraction of initial conditions leading to chaotic solutions increased.

For $E \geq 1/6$ most of the solutions were chaotic.

Since then chaotic solutions have been looked for and found in a great variety of systems - mechanical, electrical, electro-mechanical, thermo-mechanical, chemical, etc. They have been found by computer solution of the governing equations, and by experimental measurements on actual systems.

These results suggest the following conclusion: Most systems with two or more degrees of freedom have some chaotic motions and some regular motions. The regular motions occur when the system is far away from resonances of low order, i.e. resonances for which $|n|$ is small. This is in accord with the conclusion of the KAM theorem. On the other hand, when the system is near such resonances, chaotic motions occur.

The size of ε determines how near a system must be to resonance in order that chaotic motions occur. This resonance "width" increases as ε increases. Therefore when ε is so large that neighboring resonances overlap, i.e. when their separation is less than the sum of their widths, we may expect "all" motions to be chaotic. This is the criterion of Chirikov (1960), which is useful for some systems.

Accounts of the numerous investigations of chaotic motions are presented in the three recent books by Lichtenberg and Lieberman [1], Sparrow [2], and Guckenheimer and Holmes [3]. The first presents a broad survey of various results, the second gives a detailed account of the solutions of the Lorenz equations, and the third presents a very clear account of the mathematical theory. A short survey by Holmes and Moon [4] is also very informative.

5. CONCLUSION

A number of general questions are raised by the discovery of the widespread occurrence of chaotic motions, such as the following:

a) How does the occurrence of chaotic motions fit in with our previous experience, in which such motions were either absent or unimportant?
b) Why is there so much interest in chaos now?
c) What significance does chaos have in mechanics?

Some more technical questions are:

d) When does chaotic motion occur?
e) What mathematical properties do chaotic solutions have?
f) How can we tell if a system is integrable or not?
g) Can the extensive verbal and graphical descriptions of chaotic solutions be condensed into a few general principles?

These technical questions will be answered by future research. Let me therefore consider the more general questions.

It is helpful to recall that our understanding of, and intuition about, any branch of science are based upon the study of simple systems. Since these systems are special, there is the possibility that they may not be typical of all systems. This has turned out to be the case in mechanics, where our intuition is based upon linear and nonlinear systems with one degree of freedom, linear systems with many degrees of freedom, and general integrable systems.

All of these systems have only regular motions - i.e. periodic or multiply periodic oscillations, etc. Therefore we have come to expect this behavior. Thus it has been a shock to learn from computed solutions, and from some theoretical results, that very simple non-integrable systems have chaotic motions.

These results require that we revise our view of the behavior of mechanical systems, and recognize that generally such systems will have both regular and chaotic motions. Previously experimental apparatus was adjusted to yield regular motions. Noisy data were unwanted and were often disregarded. Now it is realized that noise is an essential feature of many motions, so the noise is being studied. In fact we might say that the noise is the signal!

Chaotic motions were known to occur in the past, but they were believed to occur only in systems with very many degrees of freedom. For example, the motion of all the molecules in a gas or liquid was believed to be chaotic. In fact, Boltzmann's ergodic hypothesis was a conjecture about the nature of that motion. Similarly the turbulent motion of fluids has long been recognized as chaotic, and it provides the most common instance of such motion.

One suggestion about the way in which turbulence occurs in fluids was proposed by Landau (1941) and developed by Hopf (1950). It is that the regular or laminar motion undergoes a bifurcation as some parameter, such as Reynolds number or Rayleigh number, increases through a critical value. As the parameter increases further the new solution undergoes bifurcation, and this process continues repeatedly. After a large number of bifurcations the motion is turbulent. This scheme requires a large number of degrees of freedom.

An alternative proposal was made by Ruelle and Takens in 1971. They suggested that after a small number of bifurcations, say three or four, the motion might already be turbulent. This suggestion was based on the mathematical property that in a phase space of three or four dimensions, there can be an attracting set with very peculiar properties toward which the motion can tend. Such "strange attractors", as they are called, have been observed since then in many systems in chaotic motion, including fluid systems. The Landau-Hopf behavior of very many bifurcations has been observed also.

The possibility that chaotic motion can occur in systems with just a few degrees of freedom had in fact been proved before. Cartwright and Littlewood in the 1940's, and Levinson in about 1950, showed that the forced Van der Pol equation has chaotic solutions. Smale (1965) generalized this result to a class of transformations called "horseshoe maps". But it had not been realized before Ruelle and Takens that turbulent motion might occur in fluids when only a few modes have been excited. Many investigators hope that by understanding chaotic motion in mechanical systems with a few degrees of freedom, they may gain understanding of the turbulent motion of fluids.

The preceding comments provide partial answers to questions a and b. Concerning question c, about the significance of chaotic motions in mechanics, we can expect them to play a role as important as that of regular motions. How soon they assume this role remains to be seen.

REFERENCES

1. A.J. Lichtenberg and M.A. Lieberman, Regular and Stochastic Motion, Springer-Verlag, New York, 1983.

2. C. Sparrow, The Lorenz Equation, Springer-Verlag, New York, 1982.

3. J. Guckenheimer and P.J. Holmes, Nonlinear Oscillations, Dynamical Systems and Bifurcations of Vector Fields, Springer-Verlag, New York, 1983.

4. P.J. Holmes and F.C. Moon, Strange Attractors and Chaos in Nonlinear Mechanics, Journal of Applied Mechanics, 1984.

Theoretical and Applied Mechanics
F.I. Niordson and N. Olhoff
Elsevier Science Publishers B.V. (North-Holland)

CHAOS IN THE DYNAMICS OF A FEW VORTICES - FUNDAMENTALS AND APPLICATIONS

Hassan Aref

Division of Engineering, Box D
Brown University
Providence, Rhode Island 02912
U.S.A.

The emergence for small N of chaotic dynamics in the N-vortex problem is reviewed. Vortex systems on the unbounded plane, vortex systems confined by boundaries and various "restricted" problems are discussed. Results of relevance to applications are highlighted.

1. ON VORTICES AND VORTICISTS

In analytical theories of the mechanics of fluids [1,2] the fluid is usually viewed as a continuum and the mechanics written as a field theory, wherein the basic entity, a velocity field dependent on space and time, is most often thought of in the so-called Eulerian representation, viz. the velocity $\underline{u}$ at space-time point $(\underline{x},t)$ equals the velocity of whatever fluid particle happens to occupy that space-time point [3]. In the conventional view used for most of classical mechanics, on the other hand, the evolution in time of identifiable, distinguishable particles of matter is the object of study [4]. A prime example is the system of gravitationally interacting point particles used to model stellar dynamics in celestial mechanics. Whenever this latter, often-times useful point of view is adopted in fluid mechanics, we speak of a Lagrangian representation of the theory [3] (although, as the textbooks are usually quick to point out, both representations were known to and extensively used by Euler.)

Vortex dynamics turns out to be an area of fluid mechanics in which the Lagrangian viewpoint is particularly useful. This stems ultimately from the result of Helmholtz [5], that, in the idealized case of dissipationless flow of a constant density fluid, vortex lines are material lines. Hence, in keeping track of the evolution of vorticity one is led immediately to a Lagrangian description. And keeping track of the stretching, bending and flexing of vortex lines in a flow is typically tantamount to a mechanistic description [6]. Küchemann in a frequently cited passage refers to vortices as "the sinews and muscles of fluid motions" [7].

If we restrict ourselves to two-dimensional flows and consider a model in which only a finite number of point particles carry vorticity, a state of affairs that by Helmholtz' theorems [5] is preserved in time, we arrive at a system which is to fluid mechanics what the system of point masses is to celestial dynamics [8]. True, many important effects are being glossed over in this simplified model, just as many important effects are completely ignored when modelling stars and planets as point masses. Let me, nevertheless, first describe the model and some of its properties and then return, briefly, to comment on apparent limitations when comparing to actual flow situations.

I should mention that the point vortex model, which I am about to describe, is of such great formal beauty that Nature could not resist providing at least one close realization of it in the vortex dynamics of superfluid Helium [9]. Furthermore, let me remind you that the great French natural philosopher René Descartes sought an explanation for the mechanics of the Heavens in a model where stars, planets and comets were centers of vorticity [10]. This theory was thoroughly refuted by Newtonian mechanics. And another great mind, William Thomson, the later Lord Kelvin, sought to model atomic structure in terms of steady configurations of concentrated vortices [11]. This theory was refuted by quantum mechanics. But I digress.

2. THE DYNAMICS OF POINT VORTICES

Thus, consider two-dimensional or plane flow. For notational convenience think of the flow plane as the complex z-plane. A point vortex is a singularity of the two-component velocity field (u,v) about which the flow is circumferential with the flow speed decreasing inversely with distance from the vortex. In symbols

$$u + iv = i \frac{\Gamma}{2\pi} \frac{z}{|z|^2} \tag{1}$$

for a point vortex of strength Γ (also called circulation) located at the origin z=0. The velocity field from an assembly of vortices is given by superposition:

$$u + iv = \frac{i}{2\pi} \sum_{\alpha=1}^{N} \Gamma_\alpha \frac{z-z_\alpha}{|z-z_\alpha|^2}, \tag{2}$$

where $z_\alpha = x_\alpha + iy_\alpha$ is the position in the complex plane of vortex $\alpha=1,\ldots,N$.

Now by Helmholtz' theorem [5] any one of the vortices will move with the fluid velocity at its position (excluding, of course, the infinite contribution to the velocity from the vortex itself.) Thus, in the absence of other background potential flows and in the absence of any boundaries that would limit the validity of the velocity law (1), we have

$$\dot{z}_\alpha = \frac{i}{2\pi} \sum_{\substack{\beta=1 \\ \beta\neq\alpha}}^{N} \Gamma_\beta \frac{z_\alpha-z_\beta}{|z_\alpha-z_\beta|^2}, \tag{3}$$

which is also conveniently written

$$\dot{z}^*_\alpha = \frac{1}{2\pi i} \sum_{\substack{\beta=1 \\ \beta\neq\alpha}}^{N} \frac{\Gamma_\beta}{z_\alpha-z_\beta} \tag{3'}$$

Apart from the complex-variable notation these are the equations of motion for point vortices as they have been known since Helmholtz' seminal 1858 paper [5].

Let me pause briefly to bring out similarities and differences between the point vortex problem and the point mass dynamics of celestial mechanics. Equation (1) gives the velocity field about a point vortex and corresponds in the analogy to the formula of Newton giving the gravitational force field about a point mass [12]. A parameter, the strength of the vortex, enters this formula just as the mass of the point particle enters the expression for the gravitational force. Any vortex in the system responds to the velocity field produced by all the others by moving in it according to Helmholtz' theorem [5]. Analogously, the point masses in a gravitational system respond to the combined gravitational attraction of all the others by accelerating according to Newton's second law [12]. So far the analogy seems rather close, although the vortex circulations can be of either sign whereas particle masses are intrinsically positive. Also, the point vortex problem is restricted to two dimensions.

Let me next turn to some important differences. These are brought out most clearly when we consider the Hamiltonian formulation [4,13] of our two models. For point masses the state of each particle is specified by giving both the position coordinates and the components of momentum. And each coordinate, momentum-component pair defines a set of canonically conjugate variables and thus a dynamical degree of freedom. Hence, for N point particles in d-dimensional space there are Nd degrees of freedom. For point vortices, on the other hand, the Hamiltonian formulation, which in this case is due to Kirchhoff [14], takes the following curious form: Eqs. (3) may be written as

$$\Gamma_\alpha \dot{x}_\alpha = \frac{\partial H}{\partial y_\alpha} \quad , \quad \Gamma_\alpha \dot{y}_\alpha = - \frac{\partial H}{\partial x_\alpha} \quad , \tag{4}$$

where

$$H = - \frac{1}{2\pi} \sum_{1 \leq \alpha < \beta \leq N} \Gamma_\alpha \Gamma_\beta \log|z_\alpha - z_\beta| . \tag{5}$$

Thus for vortices $(x_\alpha, \Gamma_\alpha y_\alpha)$ constitutes a pair of canonically conjugate variables and there are exactly as many degrees of freedom as there are point vortices in the system. Also note, and this will be important later, that the "phase plane" $(x_\alpha, \Gamma_\alpha y_\alpha)$ of any vortex coincides (apart from scale and, possibly, reversal of orientation) with the actual plane of flow (x,y) [15]. Thus, in this problem all the intricate phase space structure that we have come to associate with the transition to chaos (and that will be illustrated for point vortex dynamics as we proceed) is accessible to immediate visual observation. For vortices orbits in phase space are trajectories in real space.

3. INTEGRABILITY OR CHAOS

As is well known the problem of two gravitationally interacting point masses yields to solution giving the familiar elliptic orbits of Kepler [4]. Also, those who have been exposed to the ideas of statistical mechanics appreciate that very complicated behaviour, usually labelled "ergodic", presumably emerges in a many-body problem [16]. What is generally less well known, and this is in essence the topic of chaos, is that vestiges of this complex behaviour (that one has been taught to expect only in a many-body system) are discernible as soon as problems, which in their formulation are just a shade more complicated than the Kepler problem are investigated [17-19]. This fact was already apparent to workers in celestial mechanics at the turn of the century, most notably among them H. Poincaré, and many of the ideas that are in use today in the description of chaotic behaviour for Hamiltonian systems stem directly from Poincaré's work [20]. For point masses the three-body problem cannot be solved by quadratures [21], not because we currently lack the skill, but because there do not exist solutions expressing the coordinates and momenta in terms of the time, with global integrals such as energy, total momentum and total angular momentum entering as the only parameters. Indeed, what apparently happens as the number of particles is increased, is that the known integrals are insufficient in number to determine the solutions and no other general integrals emerge to aid them.

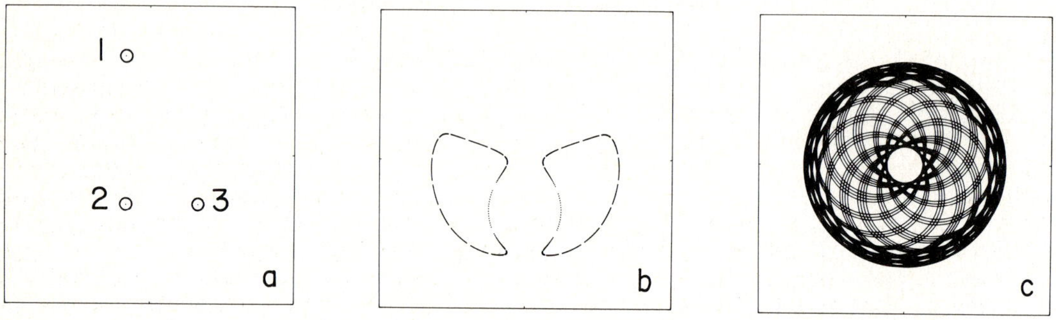

Figure 1: Integrable motion of three identical point vortices on the unbounded plane. (a) Initial configuration; (b) Poincaré section; (c) Trajectory of vortex #2. Units of length are chosen such that the distance between vortices 2 and 3 is one length unit.

Precisely this same issue of integrability arises in point vortex dynamics. The N-vortex problems with N=1, 2 and 3 are all integrable (in an unbounded domain). The four-vortex problem is not [8]. A proof of integrability for N=1,2 is trivially given by constructing a solution: A pair of opposite vortices translates uniformly (providing a two-dimensional analog of a vortex ring). Two vortices with circulations that do not sum to zero revolve around the centroid of vorticity [2]. For N=3 a formal proof of integrability goes as follows [8,22,23]: We start from the well-known integrals [2]

$$\sum_{\alpha=1}^{N} \Gamma_\alpha z_\alpha \equiv Q + iP \quad , \tag{6}$$

and

$$\sum_{\alpha=1}^{N} \Gamma_\alpha |z_\alpha|^2 \equiv I \quad , \tag{7}$$

and, of course, H itself, Eq. (5), and then investigate the Poisson bracket algebra of these entities. The Poisson bracket is defined as usual [4] by

$$[f,g] \equiv \sum_{\alpha=1}^{N} \frac{1}{\Gamma_\alpha} \left(\frac{\partial f}{\partial x_\alpha} \frac{\partial g}{\partial y_\alpha} - \frac{\partial f}{\partial y_\alpha} \frac{\partial g}{\partial x_\alpha} \right) , \tag{8}$$

since in this problem x_α and $\Gamma_\alpha y_\alpha$ are canonically conjugate (for each $\alpha=1,\ldots,N$). We find that

$$[Q, P] = \sum_{\alpha=1}^{N} \Gamma_\alpha , \tag{9a}$$

$$[Q, I] = 2P , \tag{9b}$$

and

$$[P, I] = -2Q , \tag{9c}$$

so the list of integrals cannot be lengthened by the operation of taking Poisson brackets [4]. However, we also see that

$$[P^2 + Q^2, I] = 0 \tag{10}$$

and so the three quantities H, I and $P^2 + Q^2$ are integrals in involution (regardless of the values of the circulations). It then follows from Hamiltonian mechanics that the system with three degrees of freedom, i.e. the three-vortex problem, is integrable [13]. This much essentially appears already in a short treatise by Poincaré on vortex motion published in 1893 [24] and in fact even earlier in an 1877 thesis by W. Gröbli [25] on which I shall have more to say in a moment.

Although integrable the three-vortex problem still provides the not uninteresting challenge to discover the nature of the motion as a function of the vortex strengths. This problem of elucidation was the main one addressed by Gröbli [25] in his thesis, and he succeeded in reducing the solution to expressions involving integrals of elliptic and hyperelliptic type for a few special (but as it turns out quite representative) choices of the vortex strengths. In particular, Gröbli obtained a solution for three identical vortices. This solution was rediscovered independently by E. A. Novikov almost exactly a century later [26]. While attempting to generalize Novikov's results I found a geometrical method that elucidates the complete solution for three vortices of arbitrary strengths [27]. That work in turn had been pre-empted by another forgotten paper, this one by J. L. Synge written in 1949, in which equivalent results were obtained using a very similar method [28].

I do not here want to discuss details of these analyses. Suffice it to say that the geometrical constructions accomplish essentially the same tasks for the three-vortex problem as does the well-known Poinsot construction [4] for the problem of force-free motion of a rigid body. In the rigid body case complicated solution formulae involving elliptic functions are being interpreted geometrically. In the three-vortex problem the solutions are not in general given in terms of known functions but for the "worst" cases contain integrals with implicitly given integrands [26]. Hence, in the three-vortex problem the geometrical representation of solutions is genuinely helpful. Recently, while considering the four-vortex problem (to which I want to turn momentarily) N. Pomphrey and I found a reduction scheme that uses a sequence of canonical transformations to reduce the problem of N identical point vortices to N-2 degrees of freedom [22,29]. A similar reduction procedure was found by Khanin [30] at about the same time. Applied to three identical vortices we obtain as the main problem to be solved a simple, nonlinear, ordinary differential equation for the area of the triangle spanned by the three vortices. The solution is given compactly in terms of Jacobi elliptic functions [22].

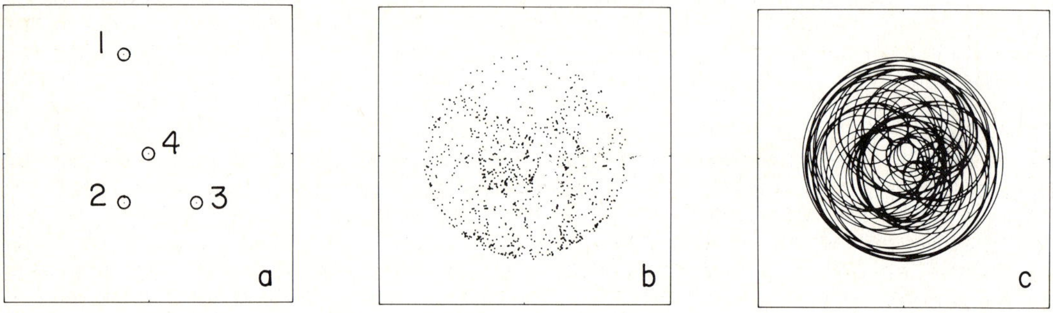

Figure 2: Chaotic motion of four identical point vortices on the unbounded plane. (a) Initial configuration; (b) Poincaré section; (c) Trajectory of vortex #2.

But we are really more interested in the cases where integrability fails and the solutions of our problem are not given entirely in terms of global integrals. We expect from the celestial mechanics precedent [21] that the resulting behaviour of the deterministic vortex system can be essentially stochastic and the main focus of the work that I shall now describe is to elucidate how, when and (to some extent) why chaotic behaviour enters point vortex dynamics. We want to know how to detect chaos in models that resemble flows of practical interest, sometimes because it is essential to avoid chaotic motion, sometimes because we wish to enhance the signal from chaos for study and sometimes because we can use this type of dynamics to practical advantage as I shall show later on.

4. SOME CONCESSIONS TO REALISM

Let me sound a few cautionary notes before embarking on a detailed description of point vortex results. First of all, as a model of real, extended vortices in ordinary fluids point vortices obviously leave something to be desired. However, let me remind you that the classical study by von Kármán [1] used point vortices to model the vortex street shed by a cylinder in a uniform flow. The obvious objection to such modelling is that all internal degrees of freedom usually associated with the "vortex core" have disappeared. A point vortex has no core. Even the next simplest model of two-dimensional vortex motion, in which finite regions of uniform vorticity are used, the so-called "vortex patches" [31,32], shows that interactions between neighbouring vortices excite dynamically significant modes in which the cores are deformed. These are the analogs for vortices of tidal forces for gravitationally interacting globes. Several illustrations of such effects may be found in the work of Zabusky and collaborators [33]. Although core deformation effects obviously modify the results for vortex interactions relative to those of a point vortex model, the notion that chaotic behaviour is entirely eliminated when using smooth distributions of vorticity in the two-dimensional Euler equation runs counter to currently accepted phenomenology.

A related question that frequently arises is whether the chaotic motion of a few vortices can be considered an example of turbulence. The answer is, I believe, a resounding "No". Turbulence in the usual meaning of the word in fluid mechanics is indeed a stochastic flow governed by deterministic equations (usually the Navier-Stokes equations) [34]. However, many aspects of turbulent flow, in particular such concepts as the inertial range or cascade [34,35], imply many excited degrees of freedom. And this is quite obviously not achieved in a four-vortex problem. We are, in my view, dealing here with a novel flow regime not properly envisioned in the traditional division into "laminar" and "turbulent". Currently this type of behaviour has not found its ultimate place in the general classification of flows. In this epistemological sense the status of chaotic vortex motion is oddly reminiscent of the status of organized vortex structures in turbulent shear flows [36]. Chaotic few-vortex systems are essentially laminar flows with stochastic properties. Organized vortex structures, on the other hand, are regular flow patterns in otherwise stochastic velocity fields. Neither fits comfortably into the inherited hierarchy of laminar versus turbulent.

5. CHAOTIC MOTION OF FOUR VORTICES

With these concessions made let me proceed to explain how one goes about detecting chaos in a few-vortex-system. The time-honored diagnostic (usually associated with the name of Poincaré) is the surface of section [17-19, 21]. In a Hamiltonian system of at least two degrees of freedom let (q_1, p_1) and (q_2, p_2) be two pairs of conjugate variables. The operational definition then goes as follows: Each time q_1 crosses some reference value (say $q_1 = 0$ for convenience) with $p_1 > 0$, record (q_2, p_2). The surface of section plot consists of the assembly of such points (q_2, p_2).

For identical point vortices this recipe is very simple to carry out on a digital computer. Assume units chosen so that the identical vortices all have strengths $\Gamma = 1$. Then follow the evolution in time from some initial configuration and every time x_1 crosses zero with $y_1 > 0$, i.e. vortex 1 crosses the positive y-axis, record the physical position of vortex 2 (or vortex 3, or vortex 4). The set of these recorded positions provides the surface of section. Usually section points from several initial conditions with the same values of known integrals such as Q, P, I and H are shown superimposed on the same plot in order to achieve some sampling of the system phase space. For N identical vortices this is very easily done by viewing any run as a simultaneous run of N! different possible labellings of the point vortices and recording all positive y-axis crossings. This enhancement has been performed in all the surface of section plots shown for identical vortices in this paper.

Figure 1(b) gives an example of such a section for the three-vortex system starting from the (rather arbitrarily chosen) initial configuration in Fig. 1(a). It is immediately apparent that the section points fall on a family of smooth curves. To appreciate this plot I show in Fig. 1(c) the complete trajectory of one of the vortices (all three are qualitatively similar). This trajectory consists of a precessing oval-shaped orbit and in general will eventually fill the annular region that bounds it. The points in Fig. 1(b) are selected from such a trajectory and so it is clear that if the selections were made at random, we would obtain a diffuse cloud of dots within the annular region. But the selection is not random. Indeed, if we think of Fig. 1(b) as an image of Fig. 1(c) taken with a particular method of stroboscopic illumination, the "stroboscope was finely tuned" to the revolution frequency of vortex 1 in its orbit. And what Fig. 1(b) then shows us is that corresponding to the y-axis crossings of vortex 1 a relationship exists between x_2 and y_2 (or x_3 and y_3). Now, clearly, a similar result would have emerged if instead of the positive y-axis we had used some other line x_1 = cst. Hence, what the result in Fig. 1(b) shows us graphically is that there exists a relation between x_2 and y_2 for each value of x_1. This is precisely the additional piece of information that we need to deduce that a solution exists in which x_1, y_1, x_2, y_2, x_3, y_3 can be expressed in terms of Q, P, I, H and the time t. From the general proof given earlier we know, of course, that such a solution exists. Fig. 1(b) provides a geometrical manifestation of this existence proof.

So let us complicate matters by adding yet another vortex, similar to the three that we already have, and let us again perform an initial value computation. The initial positions for vortices 1, 2 and 3 are the same as those considered above and vortex 4 was placed at the centroid of vorticity (Fig. 2(a)). The irregularity of the Poincaré section plot, Fig. 2(b), immediately reveals that this four-vortex problem is intrinsically different from the three-vortex problem. Apparently it is no longer true that the coordinates $x_1,\ldots,y_4$ of the vortices can be written as smooth functions of time and a set of global integrals. The integrals that we already know, i.e., Q, P, I and H, are insufficient in number to adequately constrain the motion, and what Fig. 2(b) suggests is that no additional integral has emerged to supplement them. The physical trajectories of the vortices are also much less regular than the trajectory of Fig. 1(c). It is almost obvious to the unaided eye that the trajectory in Fig. 1(c) corresponds to a motion with two (possibly) incommensurate frequencies, and thus is quasi-periodic, whereas the trajectory in Fig. 2(c) corresponds to a band of excited frequencies and thus is nonperiodic. Nonperiodicity is a characteristic "symptom" of chaotic behaviour [37].

I should comment here that although the conclusions stated are correct, one cannot in general deduce nonintegrability from a surface of section plot such as Fig. 2(b) when the number of degrees of freedom exceeds two. The reason is that even for an integrable system the region in a (q,p)-plane projection of the section, that is visited by the orbit, need not be simply a curve (as is the case for two degrees of freedom) but can be an annulus. Only if the system is separable, with (q,p) being one pair of separation variables, will the projection be a simple curve [19]. The reason that the conclusions drawn from Fig. 2(b) are, nevertheless, correct is that the four-vortex system is equivalent to a system with just two degrees of freedom when the integrals $Q^2 + P^2$ and I are taken into account. Formally this may be seen by the reduction schemes mentioned earlier [8,22,29,30], or one can appeal to the equations of motion for the variables $\ell_{\alpha\beta} = |z_\alpha - z_\beta|$, $1 \leq \alpha < \beta \leq N$, that give the separations of the vortices [8, 26, 29]. This latter technique was historically the first to be used in numerical experiments on integrability versus chaos [22,23,29,38]. Some very pretty sections obtained in this way appear in the recent thesis by Knauf [39]. The section plotted in Fig. 2(b) is just an analytic transformation of these other sections (in which integrability does correspond to smooth curves), and therefore can be used to "diagnose" chaotic behaviour.

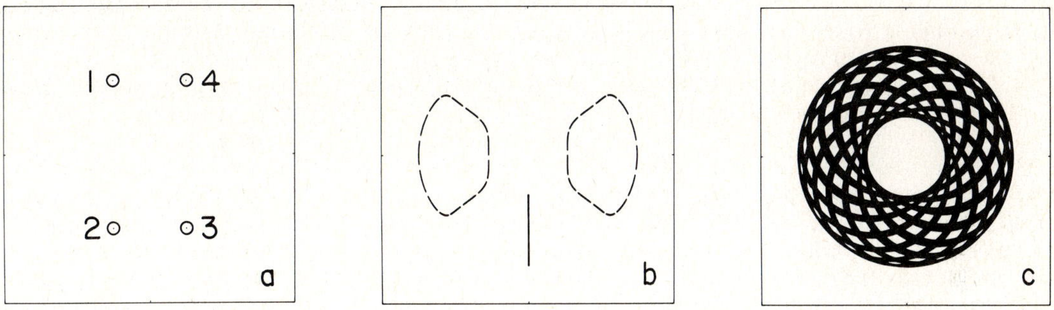

Figure 3: Integrable case of four-vortex motion due to the discrete symmetry (13). (a) Initial configuration; (b) Poincaré section; (c) Trajectory of vortex #2.

Let me digress once more to make a formal comment that shows how relatively subtle this property of chaos in the motion of four identical vortices is [8]. Reconsider Eqs. (3') with all $\Gamma_\beta = \Gamma$ (i.e. identical vortices), viz.

$$\dot{z}^*_\alpha = \frac{\Gamma}{2\pi i} \sum_{\substack{\beta=1 \\ \beta\neq\alpha}}^{N} (z_\alpha - z_\beta)^{-1} . \tag{11}$$

As we have just discussed this system of ordinary differential equations is solvable by quadratures for $N \leq 3$ but not for N=4 (and thus probably not for any $N > 4$). Now it may be shown that the system

$$\dot{z}_\alpha = \frac{\Gamma}{2\pi i} \sum_{\substack{\beta=1 \\ \beta\neq\alpha}}^{N} (z_\alpha - z_\beta)^{-1} , \tag{12}$$

which differs from (11) only by the absence of complex conjugation on the left hand side, can be solved (reduced to quadratures) for all values of N. In fact (12) are just the pole decomposition equations for Burgers' equation, intimately related to the well-known Calogero-Moser system [40].

It should be noted that imbedded in the generally chaotic regime of motion of four identical vortices are subsets of initial conditions yielding regular motion. One such subset arises by exploiting a discrete symmetry of the equations of motion, in this case a

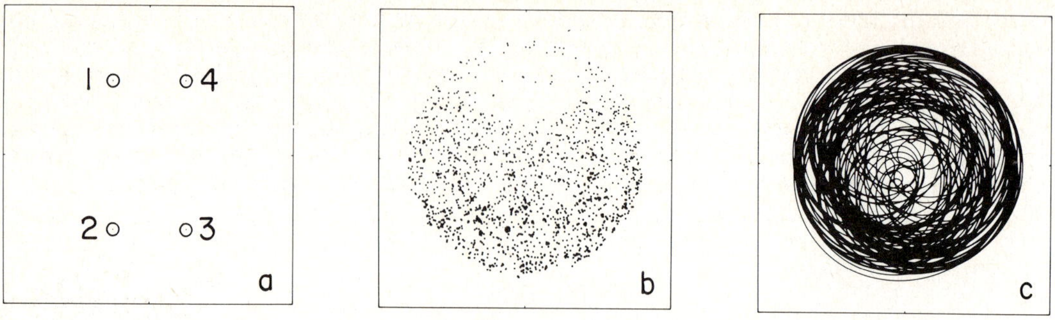

Figure 4: Chaotic four-vortex motion achieved by slightly disrupting the symmetry in Fig. 3. (a) Initial configuration; (b) Poincaré section; (c) Trajectory of vortex #2.

reflection symmetry about the centroid of vorticity (see [41] for a general discussion of point vortex motions with a center of symmetry). Thus, if we consider the initial state in Fig. 3(a), wherein the four identical vortices are situated at the vertices of a rectangle, it is not difficult to verify that the reflection symmetry

$$z_1 + z_3 = 0, \quad z_2 + z_4 = 0 \tag{13}$$

is preserved by Eqs. (3), effectively reducing them to a system of four coupled, ordinary differential equations. This system is again Hamiltonian and has the additional integral I (appropriately modified to take (13) into account). Thus it is integrable. Figure 3(b) illustrates the implications of the discrete symmetry (13) for the Poincaré section and Fig. 3(c) shows a corresponding vortex trajectory. Compare these results now to Fig. 4. In Fig. 4(a) we show a slightly different initial configuration, still essentially a rectangle but the discrete symmetry (13) has been destroyed. Indeed for Fig. 4(a) we have $z_1 + z_3 = .01$, $z_2 + z_4 = 0$. The full-fledged four-vortex problem is reintroduced, as manifested in the Poincaré section, Fig. 4(b), which again shows the characteristic random "splatter" of points, and in the nature of the vortex trajectory, Fig. 4(c). All the known analytical solutions with four or more vortices in unbounded fluid rely on some kind of discrete symmetry for their determination.

6. BOUNDARY-INDUCED CHAOS

So far we have seen an analogy between the dynamics of point masses in infinite space and the dynamics of point vortices in unbounded fluid with the three-vortex problem playing a role analogous to the two-body problem, roughly speaking. Nonintegrability sets in for four vortices and for three "bodies". I now wish to pursue the analogy somewhat further in order to arrive at certain extensions of the results obtained so far, extensions that probably are of greater importance to applications than the idealized case of infinite fluid which has occupied us till now.

The first extension concerns the possible effect of boundaries. How will the introduction of rigid bounding walls or other boundary conditions affect the transition to chaos in the few-vortex problem? The immediate suggestion, made already in 1980 by Novikov and Sedov [38], is that since the integrals that we know for the point vortex equations in infinite fluid hinge on the invariance of the Hamiltonian under spatial translations and rotations, and since such invariances will be at least partially destroyed by the imposition of other boundary conditions, chaotic behaviour should emerge for a smaller number of vortices in the case of bounded flow. But, of course, this heuristic argument must be checked.

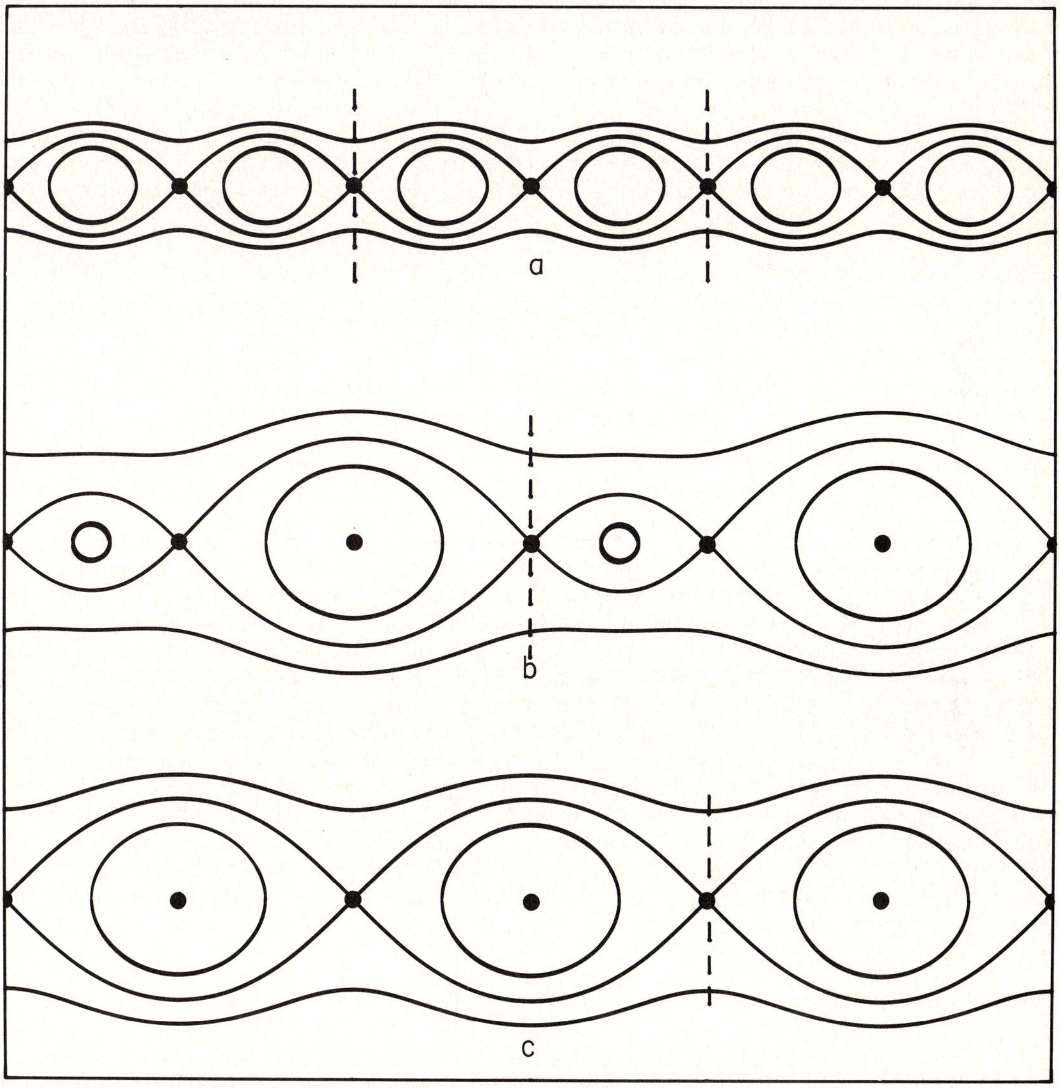

Figure 5: Exact solutions for a periodic system of identical vortices with (a) N=2, (b) N=3, (c) N=4 vortices in the "basic cell". The curves give possible trajectories for the vortices if perturbed from the equi-spaced steady state.

There are some very simple boundary conditions that may be expected to have the least impact on integrability. Consider, for example, the imposition of periodic boundary conditions in the x-direction with a certain spatial period L. This situation arises naturally in the discussion of infinite vortex rows. Shear layers after roll-up and vortex streets are frequently modelled by systems of this type [1,8]. One can show in a straight forward way that the Hamiltonian is now changed to

$$H = -\frac{1}{2\pi} \sum_{1\leq\alpha<\beta\leq N} \Gamma_\alpha \Gamma_\beta \log\left|\sin\left\{\frac{\pi}{L}(z_\alpha - z_\beta)\right\}\right| . \quad (14)$$

This Hamiltonian again depends only on relative positions of the vortices. It is thus invariant to all translations, even those with a component in the x-direction. Hence, both the integrals Q and P, Eq. (6), survive, an observation that was apparently first made by Birkhoff and Fisher in 1959 [42]. However, as we saw previously, Eq. (9a), Q and P in general do not commute (i.e. they are not integrals in involution), and so one only expects the three-vortex problem in a periodic strip to be integrable when the sum $\Gamma_1 + \Gamma_2 + \Gamma_3$ vanishes.

Vortex rows in which the periodically repeated cell consists of three vortices with net zero circulation is not a configuration readily met with in practice. Of greater relevance is a system with identical vortices, which has been used on several occasions to model the dynamics of a shear layer after roll-up [8,43-45]. Here the steady state of interest is a row of uniformly spaced vortices. It turns out very conveniently that the most unstable mode for this configuration, according to a linear stability analysis, is a mode with wavelength equal to twice the vortex spacing [1], which, thus, can be captured at arbitrary amplitudes by analyzing the integrable N=2 system. Other exact solutions that can be found by invoking discrete symmetries in systems with a basic cell of N=3 [46] or N=4 [47] vortices are illustrated in Fig. 5. In all cases the steady state is a saddle point in the system phase space, and the solutions in Fig. 5 show this saddle point to be connected to others, that differ from it only in the relabelling of the vortices along the row. For example, the N=2 mode clearly can be invoked to transpose the initial row ...abcdef... into ...badcfe... (albeit in an infinite time). This corresponds to the well known "vortex pairing" process for finite area vortices of Winant and Browand [43]. Similarly, the N=3 mode [46]can change ...abcdef... into ...cbafed...; the N=4 mode [47] can produce ...cbadgfeh... from ...abcdefgh... and so on. There are counterparts of these saddle connections for the three- and four-vortex problems on the unbounded plane, where in fact they play an important role in understanding the topology of the accessible phase space and the onset of chaos [22]. The significance of such "vortex permutations" was evident already to Synge [28] while analyzing the three-vortex problem: "These oscillations between configurations which differ only through interchange of vortices of equal strength appear rather interesting," he wrote ([28], p. 267).

At the present time a complete list of such perfectly periodic motions is not available but the motions that I have mentioned all have close counterparts in the recent experiments by Ho and Huang [48,49]on forced shear layers. One may ask why the pairing mode [43] is essentially the only one seen in freely developing (i.e. unforced) shear layers. Part of the answer is, of course, that this is the most unstable mode. But another part is, I believe, provided by the notion of chaotic vortex motion [8,49]: when the discrete symmetry of perfect periodicity is not enforced, the system becomes once again non-integrable and any phase incoherence along the row will amplify exponentially. Indeed, the region of phase space "close to" saddle connections is known to be particularly prone to chaos [17-19,21]. All modes with $N \geqslant 3$ are susceptible to this mechanism. Thus, such modes will enter with random phases and will not produce a coherent signal. They will also act to degrade phase coherence from one pair to the next along the shear layer. Spatially coherent forcing is a mechanism for suppressing this "phase-chaos", as shown in the experiments of Ho and Huang [48]. Although this application of the concept of chaotic vortex motion is still somewhat tenuous, it strikes me as a promising area for further investigations. It is thought-provoking that the largest number of vortices per cell for which vortex permutation modes have been found, viz. N=4 (Fig. 5), coincides with the number in the gravest mode of Ho & Huang's experiment [48]. This could be a temporary coincidence or it could point to a more basic aspect of vortex interactions.

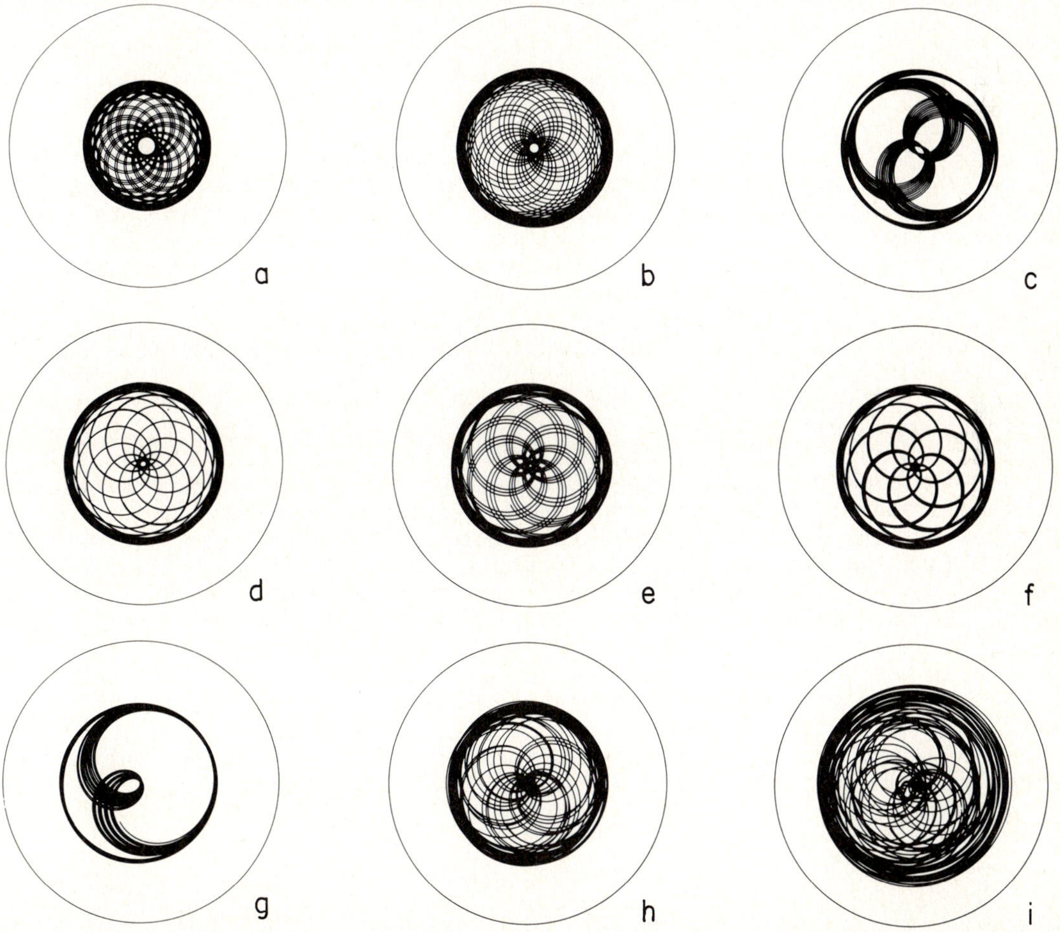

Figure 6: The three-vortex configuration of Fig. 1 was enclosed in a circle of radius R. The trajectory of vortex #2 is shown for (a) R = 3.0; (b) 2.45; (c) 2.4; (d) 2.39; (e) 2.37; (f) 2.36; (g) 2.35; (h) 2.3; (i) 2.2.

Other systems of interest to applications, which can be approached using vortices in a periodic strip, are the models of vortex streets initiated by von Kármán [1]. The N=2 system with $\Gamma_1 = -\Gamma_2$ essentially just succeeds in giving the velocity of propagation of the street as a function of configurational parameters. The essence of point vortex street stability is embodied in the N=4 system studied by Domm [50]. So far the notion of chaotic vortex motion has not been successfully applied to the problem of dissolution of a (point) vortex street [41] although there are suggestions that it is relevant [51].

In a doubly periodic domain the Hamiltonian for point vortices may be written

$$H = \frac{1}{8\pi^2} \sum_{m,n=-\infty}^{+\infty} \left(|\xi_{mn}|^2 - \sum_{\alpha=1}^{N} \Gamma_\alpha^2\right)/(m^2 + n^2) , \qquad (15)$$

an expression apparently first given by Kraichnan in 1975 [52]. Here the ξ_{mn} are Fourier components of the singular vorticity field

$$\xi_{mn} = \sum_{\alpha=1}^{N} \Gamma_\alpha \exp(i \frac{2\pi}{L}(mx_\alpha + ny_\alpha)) . \qquad (16)$$

Again H depends only on position differences and thus the integrals Q and P survive also the introduction of these doubly periodic boundaries. Since the doubly periodic boundary constrains the sum of the vortex circulations to vanish, the implication is that both the two- and three-vortex problems are integrable. The existence of chaotic motion for doubly periodic vortex arrays with sufficiently many vortices in a "basic cell" could have important consequences for the propagation of finite amplitude Tkachenko-waves [53].

Rigid boundaries also lead to a reduction in the number of point vortices sufficient for chaotic behaviour. A stationary rigid boundary requires that we associate with each point vortex a set of images in order to satisfy the condition that the normal velocity vanishes at the boundary. For the simplest boundary geometries the construction of images can be carried out by inspection. For example, if an infinite plane wall is introduced, dividing the flow plane in two, each vortex acquires a single mirror image of opposite circulation. In this case the N=2 system is still integrable but the three-vortex system is not. If the wall is along the x-axis, the only general integral (besides the Hamiltonian) is P, Eq. (6). Indeed, the solutions for N=2 in this case can also be thought of as solutions of the unbounded four-vortex problem with $\Gamma_1=\Gamma_2=-\Gamma_3=-\Gamma_4$ that satisfy a discrete symmetry, viz.

$$z_1 - z_4^* = 0, \; z_2 - z_3^* = 0. \qquad (17)$$

These solutions describe collisions of two vortex pairs with a common axis. They have been discovered and analyzed several times [25,44,54-56].

Another very simple image system arises if we enclose the vortices in a circle modelling the boundary of a cylindrical vessel. Again it is true that each vortex has just one image of opposite circulation. If the vortex is located at z, its image will be at R^2/z^*, where R is the radius of the circular boundary. Now, neither the integral Q nor P survives but I, Eq. (7), does and so the two-vortex problem inside a circle is integrable whereas the three-vortex problem is not. Somewhat inconclusive evidence of this transition can be found already in a 1970 paper by Murty and Rao [57] although the authors were concerned primarily with other issues.

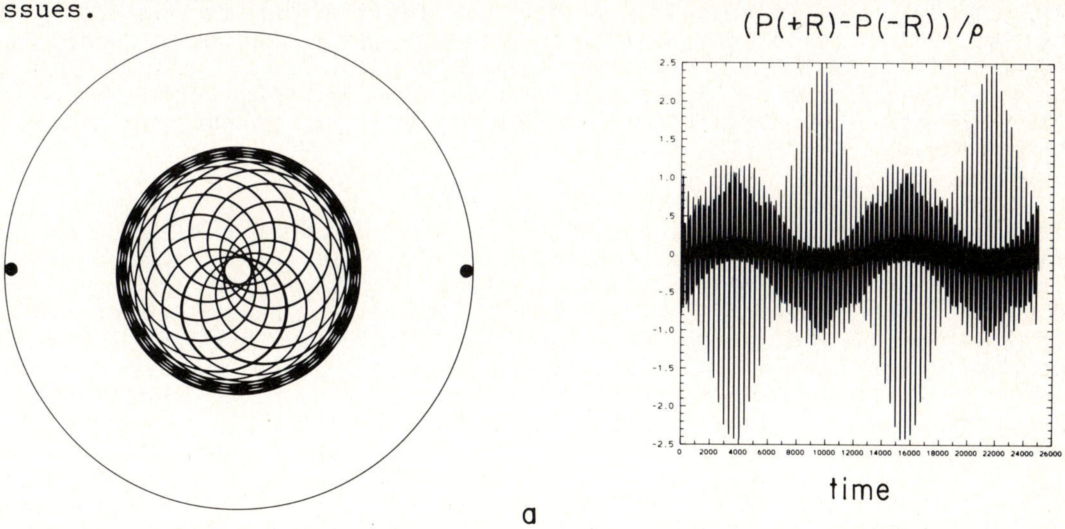

a

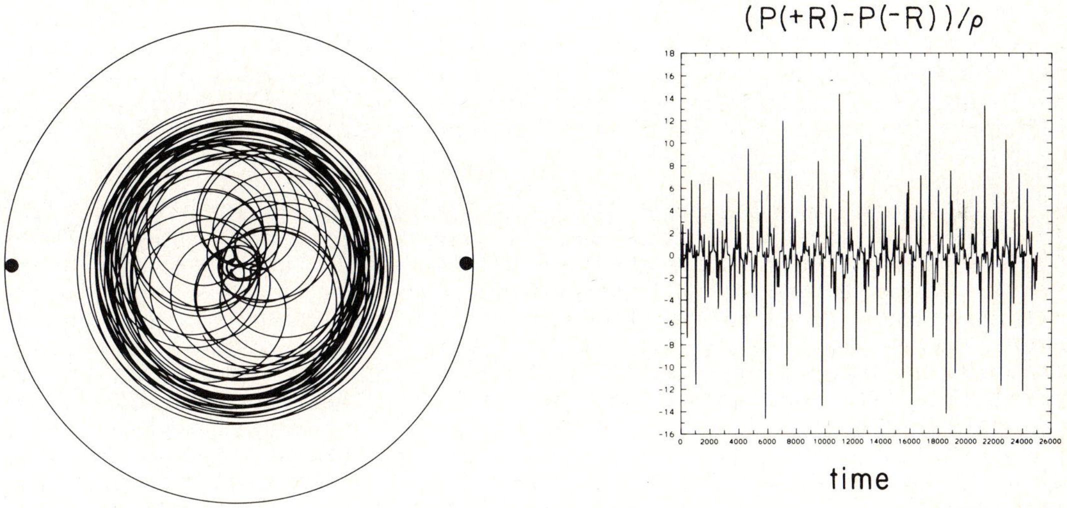

b

Figure 7: Two corresponding plots of a vortex trajectory and the boundary pressure signal for the system of Figure 6. In (a) R=2.65, in (b) R=2.25.

This particular case of boundary-induced chaos, as I suggest the phenomenon be called, can be easily illustrated. In Fig. 6 I have taken the same initial configuration of three identical vortices that was used for the unbounded fluid calculation in Fig. 1, but now I have enclosed the vortices in a circle (centered at the initial centroid of vorticity) by adding the image contributions. I am only displaying the trajectory of one of the vortices for different values of the circle radius, but Poincaré sections were also generated. We know from Fig. 1 what the trajectory will look like when the circle is of sufficiently large radius and, indeed, as Fig. 6(a) shows, when the radius R=3.0 (in the units of Fig. 1), the trajectory is similar to the one illustrated in Fig. 1(c). As we change the radius, i.e. "squeeze" the vortices, and repeat the calculation, interesting shifts occur in the ratio of the two controlling frequencies. These changes lead to the sequence of trajectory plots displayed in Fig. 6 terminating in the chaotic regime. The transition to "large scale" chaos is rather sharp in terms of R. In Fig. 6(h,i) the radius R=2.3 and 2.2, respectively, only slightly smaller than the values used for Fig. 6(b-g), where the Poincaré sections (not shown) reveal no trace of chaotic behaviour.

The qualitative changes apparent in the sequence of orbits shown in Fig. 6 have important implications for flow situations dominated by intense vortices. To emphasize this I show in Fig. 7 two corresponding instances of a vortex trajectory and the time series of the pressure difference between two points on the fixed, rigid boundary. The vortex calculations are of the type discussed already in Fig. 6. The pressure difference between the two diametrically opposite points indicated in Fig. 7(a) and 7(c) is given by the unsteady Bernoulli's equation [1-3]. (For this case of a flow induced by point vortices an expression in terms of instantaneous vortex positions results.) It is clearly seen that with regular vortex orbits a regular pressure signal emerges (Fig. 7(a,b)). However, as soon as we enter the chaotic regime in earnest, the pressure fluctuations on the wall become much more "noisy" (Fig. 7(b,c)). This simple illustration shows that the appearance of chaotic motion of confined vortices can have a profound influence on the nature of pressure forces on surrounding walls. At the very least one must pay attention to the number of vortex centers that are within the region of interest.

In the above developments I have implicitly assumed that the Hamiltonian formulation developed for unbounded point vortex systems carries over to the case of bounded flow. For the very simple examples mentioned so far this can be verified by direct calculation. For arbitrary boundaries there is a formal theory, initiated by Routh in 1881 [58] and completed by Lin in 1941 [59-61], which shows that the dynamics of point vortices is always Hamiltonian in the sense of Eqs. (4). The Hamiltonian is in general given in terms of Green's function for the Laplacian within the boundaries in question. For rigid boundaries the Hamiltonian invariably depends on more than the relative vortex positions. For vortices in a bounded domain without translational or rotational symmetry, e.g. vortices in a rectangle, none of the general integrals Q, P and I survive, and the one-vortex problem is expected to be the only integrable problem [38,62]. I shall return to further consequences of this plausible statement in the next subsection. There are no analytical proofs of these expectations at the present time.

Let me remark that this reduction in the number of point vortices necessary for chaos when rigid boundaries are introduced also has a counterpart in the more familiar dynamics of point particles with inertia. Indeed, if such a particle is confined inside a region with a closed boundary, from which it is reflected elastically upon collision, we arrive at the famous stadion or billiards problem. It is well known that one can obtain chaotic solutions for this type of problem, and that the issue of integrability or nonintegrability hinges on the actual geometry of the reflecting boundaries [63-66]. For billiards gravitational interaction with the boundaries is not required in order to produce chaotic motion. Point vortices, on the other hand, do not obey the "law of inertia" [67], and so for them the interaction with the boundary (through their induced flow field or, equivalently, a system of images) is necessary, and a single point vortex inside rigid boundaries always defines an integrable system.

7. THE RESTRICTED PROBLEM

So far I have dealt with systems of point vortices in which all constituents were on a similar if not equal footing. Frequently systems with identical vortices were chosen for discussion. Now I want to consider systems that arise as limiting cases of the previous analyses by letting the strength of one of the vortices tend to zero. Similar systems have been considered in point mass dynamics where the term "restricted problem" is commonly used to designate situations where one or more particles are passively swept along in the gravitational field created by the others, without affecting the motion of these "active" particles in any way.

In fluid mechanics this kind of restricted problem has a very natural interpretation as the problem of advection of a passive marker in a prescribed or evolving flow field [8,29]. Again the suggestion from the point mass analogy is that the equations of motion for the passively advected particle will be nonintegrable even though the vortex system responsible for the advecting flow is integrable. The investigations of Sitnikov [69], for example, showed that motion with an essentially arbitrary degree of randomness can be obtained by following a very light particle moving in the field of two massive "stars" orbiting one another in Kepler ellipses [21].

Figure 8 illustrates passive particles being advected by a system of three identical vortices. The initial configuration of the vortices and five passive marker particles is shown in Fig. 8(a). Figure 8(b) shows the trajectory of one of the vortices in order to emphasize that the vortex motion is just the integrable problem discussed earlier. Fig. 8(c) shows a stroboscopic Poincaré section for all the markers, corresponding (as before) to crossings by one of the vortices of the positive y-axis. Chaos is clearly in evidence for marker particles started within or close to the vortex triangle, but markers started at some distance orbit the entire vortex triple and have essentially regular motions. Similar illustrations appear already in a brief paper with N. Pomphrey published in 1980 [29]. In fact, of all the problems mentioned so far this "restricted four-vortex problem" has received most attention from the analytical side. Ziglin [70] published a proof of non-integrability, using methods usually associated with the name of Melnikov, which, however, has been questioned on technical grounds by Holmes and Marsden [71]. Recently Koiller and collaborators [72] have initiated a somewhat different analytical approach. As a curious historical aside I note that on p. 7 of his thesis Gröbli [25] explicitly states that he will not consider advection problems of this type ("Auf die Bestimmung der Bewegung von Flüssigkeitstheilchen welche sich in endlicher Entfernung von den Wirbelfäden befinden, werden wir nicht eingehen").

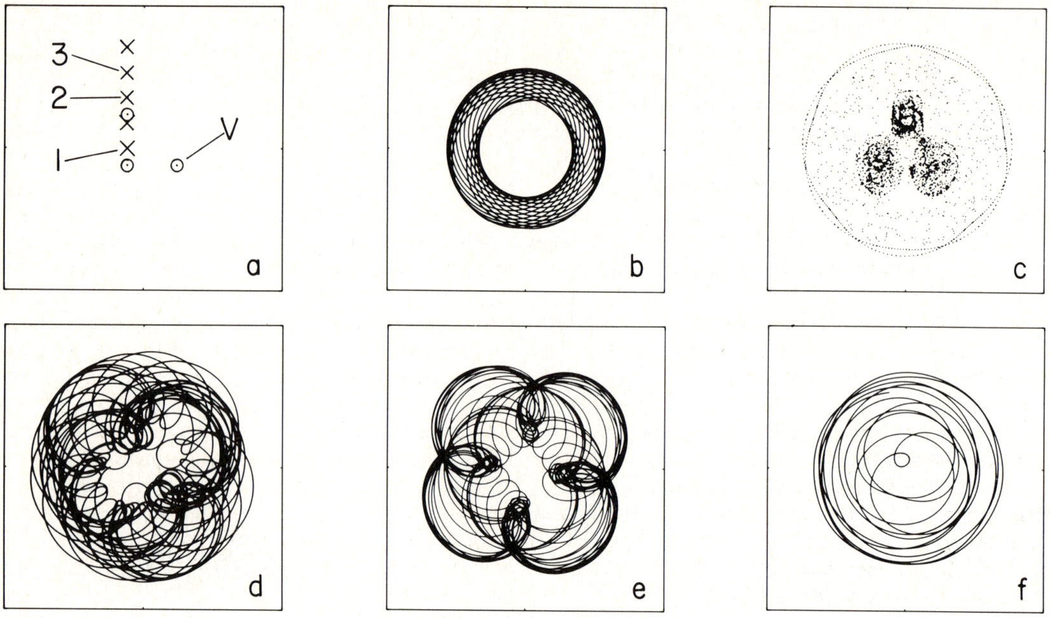

Figure 8: Passive advection by three vortices on the unbounded plane. (a) Initial configuration of vortices (o) and passive markers (x); (b) Trajectory of vortex V; (c) Poincaré section for marker particles. Marker trajectories (d) #1, (e) #2, (f) #3. The scale in (b), (d) and (e) is expanded by a factor 2 for added detail.

The results just mentioned are really only special cases of the more general proposition that flow fields, which are extremely simple in an Eulerian sense, can produce signals in a Lagrangian variable (such as a passively advected marker) of arbitrarily great complexity. The underlying reason is that "the advection equations"

$$\begin{aligned} \dot{x} &= u(x,y,z,t), \\ \dot{y} &= v(x,y,z,t), \\ \dot{z} &= w(x,y,z,t), \end{aligned} \tag{18}$$

where u, v, w are prescribed, encompass a broad range of dynamical systems, even when the field (u,v,w) is "constrained" to satisfy equations of hydrodynamics. And systems of the form (18) are adequately "complicated" that chaotic solutions can appear.

Chaotic advection, as I have proposed the phenomenon be called [68], was first shown numerically by Hénon in 1966 [73] for certain steady, three-dimensional flows (usually called Beltrami flows) which solve the Euler equation. In two dimensions it is necessary to use a time-dependent flow to produce chaotic particle motion.

In particular, if the fluid is incompressible, Eqs. (18) can be written in terms of a stream-function $\psi(x,y,t)$ thus:

$$\dot{x} = \frac{\partial\psi}{\partial y} \quad , \quad \dot{y} = - \frac{\partial\psi}{\partial x} \quad , \tag{19}$$

so the stream-function plays the role of Hamiltonian. For steady flow this one-degree-of-freedom system is autonomous and hence integrable [13]. This is the dynamical systems version of the familiar hydrodynamic argument that in steady flow path-lines coincide with stream-lines [1,2]. The stream-lines of the steady flow case would be called Kolmogorov-Arnold-Moser (KAM) tori [17-19,21] in the language of the theory of dynamical systems.

One can give elementary examples of chaotic advection (with some practical relevance) using unsteady potential flows. For example, a system of two bound vortices, that are switched on and off alternately, provides a model of the flow field in a bath of viscous fluid with two alternating, cylindrical rotors. For this example chaotic advection is readily produced, and one shows easily by numerical calculations that the efficiency of mixing is greatly enhanced by the chaotic particle motion [68]. The effects of chaotic advection on material interfaces are also easily illustrated. Particularly interesting is the signal from homoclinic oscillations, referred to as "tendrils" by Berry et al. [74], and the dramatic stretching of an interface to which this phenomenon may lead [68,75]. These ideas are currently gaining ground in chemical engineering applications [76]. In particular, the possibility of achieving efficient mixing at very low Reynolds numbers, where mixing mechanisms dependent on turbulent flow are not available, through chaotic advection is considered intriguing. We are currently pursuing the possibility of chaotic advection in a certain class of Stokes flows.

Advection by laminar flows is a very natural area for the application of these ideas from chaotic dynamics. The new lesson for fluid mechanics is that one can have fine-scale advection patterns in a flow field that instantaneously only has large-scale structure. The problem has great intuitive appeal because phase space coincides with configuration space, as I have already explained. Chaos unfolds "before our very eyes", and we now realize that the marvelous patterns with which we are familiar when cream spreads on coffee [74], or a drop of oil is drawn out in a puddle of water, are probably fractal sets [77] associated via chaos with the "dynamical system" called passive advection. Very similar ideas have been pursued for many years in plasma physics under the heading "magnetic field line flow" [19], although there the motivation is to avoid the chaos! The central role of Hamiltonian systems, which is essentially a manifestation of fluid incompressibility, is interesting. Recent work on transport in Hamiltonian systems (including the new concepts of "cantori" and "turnstiles", see [78]) seems destined to have practical impact.

It is important to note that in this application of chaotic dynamics we are dealing directly with the full fluid-mechanical advection problem as it stands. There are no modal expansions here followed by (sometimes ad hoc) truncation, which to some has been the "Achilles heel" of the work on transition to turbulence. It is interesting to speculate on extensions to compressible flows, where passive particles in principle should be able to advect onto attracting sets that are fractals (socalled "strange" attractors [79]). One can speculate on relations to cloud- and other pattern-formation processes. There are many exciting questions that derive from a dynamical systems interpretation of Eqs. (18) (see also [75]).

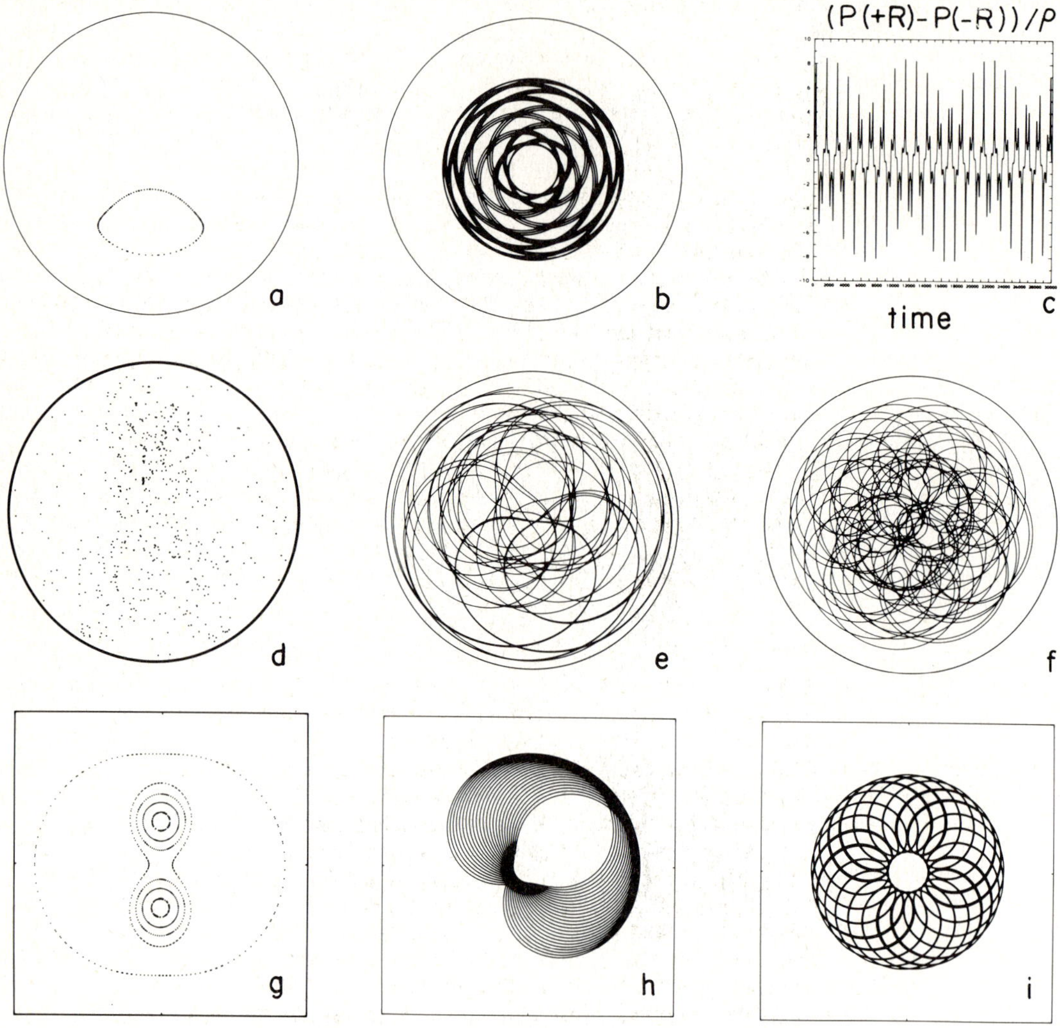

Figure 9: Two-vortex motion and passive advection with and without a circular boundary of radius R=2.5. (a) Poincaré section for bounded vortices; (b) Trajectory of vortex #2 (with boundary); (c) Boundary pressure signal; (d) Poincare section for passively advected markers (with boundary); (e),(f) Sample trajectories of markers (with boundary); (g) Poincaré section for markers (without boundary); (h),(i) Sample trajectories of markers (without boundary).

8. SYNOPSIS

In conclusion let us attempt to piece together a picture of the causes and effects of chaotic few-vortex motion (within the context of the point vortex model). Figure 9 was prepared to assist in this task. We have two identical point vortices in a cylindrical tank started at the positions labelled 1 and 2 in Fig. 1(a). In Fig. 9(b) I show the real space trajectory of vortex 2. It is a regular, quasi-periodic precession as we expect for this integrable problem. Integrability manifests itself also in the smooth, regular section, Fig. 9(a), constructed as described in Sec. 5. Finally, integrability manifests itself in the pressure signal shown in Fig. 9(c), constructed as outlined in Sec. 6. Note that the section and the orbit are pieces of Lagrangian information, the pressure time-series is an Eulerian signal. At this stage, as far as the vortex motion itself is concerned, the various diagnostics agree in predicting a regular Eulerian flow.

Now we seed the flow field with passive Lagrangian marker particles, all started on the y-axis. Because of the boundary this restricted problem is nonintegrable, a result that follows by combining the results of Secs. 6 and 7. We verify this from the section plot in Fig. 9(d) and (for the keen eye) directly from the orbits shown in Fig. 9(e,f). The observer doing flow visualization will disagree with the observer monitoring wall pressures on whether the flow should be called regular or irregular. In a very precise sense we see here "the relative orderliness of Eulerian representation over Lagrangian", to quote an apt phrase from Amsden & Harlow [84].

Finally, remove the boundary and we revert to an integrable problem. The section essentially produces a steady stream-line pattern in a rotating frame, Fig. 9(g), and the particle orbits are again quasi-periodic (Fig. 9(h,i)).

* * *

I have illustrated in this lecture how the notion of chaos can be applied to the Lagrangian dynamics of material points in fluid flow, in particular point vortices. I believe that it is important and frequently productive to think about problems in fluid mechanics in these terms. The concept of chaos is tailored to the description of deterministic processes that yield outputs with stochastic properties. Apart from the somewhat idealized examples discussed here there are myriads of such phenomena in fluid mechanics. I anticipate that we shall see tools and methods from the theory of dynamical systems emerge in many problems, and that for fluid mechanics these ideas will eventually gain the stature of more established methodologies such as stability analysis, the theory of random functions and perturbation methods.

The research reviewed here has been supported by the Fluid Mechanics Program of the National Science Foundation under grant MEA81-16910 to Brown University. Computer resources made available through the Scientific Computing Division of the National Center for Atmospheric Research have been invaluable. NCAR is sponsored by NSF.

REFERENCES

1. H. Lamb, Hydrodynamics, 6th ed. (Dover, New York, 1945).

2. G. K. Batchelor, An Introduction to Fluid Dynamics (Cambridge Univ. Press, 1967).

3. L. Prandtl and O. G. Tietjens, Fundamentals of Hydro- and Aeromechanics (Dover, New York, 1957).

4. L. D. Landau and E. M. Lifshitz, Mechanics, 2nd ed. (Pergamon Press, Oxford, 1969).

5. H. v. Helmholtz, Über Integrale der hydrodynamischen Gleichungen, welche den Wirbelbewegungen entsprechen, Crelles J. 55 (1858) 25. Transl., P. G. Tait, Phil. Mag. (4), 33 (1867) 485-512.

6. H. J. Lugt, Vortex Flow in Nature and Technology (Wiley, New York, 1983).

7. D. Küchemann, Report on the IUTAM Symposium on Concentrated Vortex Motions in Fluids, J. Fluid Mech. 21 (1965) 1-20.

8. H. Aref, Integrable, chaotic and turbulent vortex motion in two-dimensional flows, Ann. Rev. Fluid Mech. 15 (1983) 345-89.

9. R. J. Donnelly, Experimental Superfluidity (Univ. Chicago Press, 1967).

10. E. J. Aiton, The Vortex Theory of Planetary Motions (American Elsevier, New York, 1972).

11. W. Thomson (Lord Kelvin), Mathematical and Physical Papers, Vol. IV (Cambridge Univ. Press, 1910).

12. I. B. Cohen, Newton's discovery of gravity, Scient. Amer. 244 (1981) 166-79.

13. E. T. Whittaker, A Treatise on the Analytical Dynamics of Particles and Rigid Bodies, 4th ed. (Cambridge Univ. Press, 1937).

14. G. R. Kirchhoff, Vorlesungen über Mathematische Physik, Vol. I (Teubner, Leipzig, 1876).

15. L. Onsager, Statistical hydrodynamics, Nuovo Cimento (Suppl.) 6 (1949) 279-87.

16. N. Wiener, Cybernetics, 2nd ed. (MIT Press, Cambridge, 1961).

17. V. I. Arnold and A. Avez, Ergodic Problems of Classical Mechanics (Benjamin, New York, 1968).

18. M. V. Berry, Regular and irregular motion, AIP Conf. Proc. (S. Jorna, editor) No. 46 (1978) 16-120.

19. A. J. Lichtenberg and M. A. Lieberman, Regular and Stochastic Motion (Springer, Berlin, 1983).

20. H. Poincaré, Les méthodes nouvelles de la mécanique céleste (Gauthier-Villars, Paris, 1899).

21. J. Moser, Stable and Random Motions in Dynamical Systems (Princeton Univ. Press, 1973).

22. H. Aref and N. Pomphrey, Integrable and chaotic motions of four vortices. I. The case of identical vortices, Proc. Roy. Soc. (London) A380 (1982) 359-87.

23. E. A. Novikov and Yu. B. Sedov, Stochastic properties of a four-vortex system, Sov. Phys. JETP 48 (1978) 440-44.

24. H. Poincaré, Théorie des Tourbillons (Deslis Frères, Paris, 1983).

25. W. Gröbli, Specielle Probleme über die Bewegung geradliniger paralleler Wirbelfäden (Zürcher und Furrer, Zürich, 1877).

26. E. A. Novikov, Dynamics and statistics of a system of vortices, Sov. Phys. JETP 41 (1975) 937-43.

27. H. Aref, Motion of three vortices, Phys. Fluids 22 (1979) 393-400.

28. J. L. Synge, On the motion of three vortices, Can. J. Math. 1 (1949) 257-70.

29. H. Aref and N. Pomphrey, Integrable and chaotic motions of four vortices, Phys. Lett. A 78 (1980) 297-300.

30. K. M. Khanin, Quasi-periodic motions of vortex systems, Physica D 4 (1982) 261-9.

31. G. S. Deem and N. J. Zabusky, Vortex waves: Stationary 'V states', interactions, recurrence and breaking, Phys. Rev. Lett. 40 (1978) 859-62.

32. J. Burbea, On patches of uniform vorticity in a plane of irrotational flow, Arch. Ration. Mech. Anal. 77 (1982) 349-58.

33. N. J. Zabusky, Recent developments in contour dynamics for the Euler equations, Ann. N. Y. Acad. Sci. 373 (1981) 160-70.

34. G. K. Batchelor, The Theory of Homogeneous Turbulence (Cambridge Univ. Press, 1953).

35. A. N. Kolmogorov, The local structure of turbulence in incompressible viscous fluid for very large Reynolds numbers, C. R. Acad. Sci. URSS 30 (1941) 301.

36. A. K. M. F. Hussain, Coherent structures - reality and myth, Phys. Fluids 26 (1983) 2816-50.

37. E. N. Lorenz, Deterministic nonperiodic flow, J. Atmos. Sci. 20 (1963) 130-41.

38. E. A. Novikov and Yu. B. Sedov, Stochastization of vortices, JETP Lett. 29 (1979) 677-9.

39. A. Knauf, Das N-Vortex-System der Klassischen Mechanik, Diplomarbeit, Fachber. Physik, Freien Univ. Berlin (1983).

40. F. Calogero, Motion of poles and zeros of special solutions of nonlinear and linear partial differential equations and related "solvable" many-body problems, Nuovo Cimento B 43 (1978) 177-241.

41. H. Aref, Point vortex motions with a center of symmetry, Phys. Fluids 25 (1982) 2183-7.

42. G. Birkhoff and J. Fisher, Do vortex sheets roll up?, Rend. Circ. Mat. Palermo 8 (1959) 77-90.

43. C. D. Winant and F. K. Browand, Vortex pairing: A mechanism of turbulent mixing-layer growth at moderate Reynolds number, J. Fluid Mech. 63 (1974) 237-55.

44. E. Acton, The modelling of large eddies in a two-dimensional shear layer, J. Fluid Mech. 76 (1976) 561-92.

45. J. Jimenez, On the visual growth of a turbulent mixing layer, J. Fluid Mech. 96 (1980) 447-60.

46. I. Tengara, Some Mathematical Problems in Waves and Vortices, Thesis, Mathematics Department, Imperial College of Science and Technology, London (1981).

47. D. C. Blomberg, Point Vortex Models of a Forced Shear Layer, M.Sc. Thesis, Division of Engineering, Brown University (1984).

48. C-M. Ho and L-S. Huang, Subharmonics and vortex merging in mixing layers, J. Fluid Mech. 119 (1982) 443-73.

49. C-M. Ho and P. Huerre, Perturbed free shear layers, Ann. Rev. Fluid Mech. 16 (1984) 365-424.

50. U. Domm, Über Wirbelstrassen von geringster Instabilität, Z. Angew. Math. Mech. 36 (1956) 367-71.

51. P. G. Saffman and J. C. Schatzman, Stability of a vortex street of finite vortices, J. Fluid Mech. 117 (1982) 171-85.

52. R. H. Kraichnan, Statistical dynamics of two-dimensional flow, J. Fluid Mech. 67 (1975) 155-75.

53. V. K. Tkachenko, Stability of vortex lattices, Sov. Phys. JETP 23 (1966) 1049-56.

54. A. E. H. Love, On the motion of paired vortices with a common axis, Proc. London Math. Soc. 25 (1894) 185-94.

55. W. M. Hicks, On the mutual threading of vortex rings, Proc. Roy. Soc. (London) A 102 (1922) 111-31.

56. K. O. Friedrichs, Special Topics in Fluid Dynamics (Gordon & Breach, New York, 1966).

57. G. S. Murty and K. S. Rao, Numerical study of the behaviour of a system of parallel line vortices, J. Fluid Mech. 40 (1970) 595-602.

58. E. J. Routh, Some applications of conjugate functions, Proc. London Math. Soc. 12 (1881) 73-89.

59. C. C. Lin, On the motion of vortices in two dimensions - I. Existence of the Kirchhoff-Routh function, Proc. N.A.S. (USA) 27 (1941) 570-5.

60. C. C. Lin, On the motion of vortices in two dimensions - II. Some further investigations on the Kirchhoff-Routh function, Proc. N.A.S. (USA) 27 (1941) 575-7.

61. C. C. Lin, On the Motion of Vortices in Two Dimensions, (Toronto Univ. Press, 1943).

62. E. A. Novikov, Stochastization and collapse of vortex systems, Ann. N. Y. Acad. Sci. 357 (1980) 47-54.

63. L. A. Bunimovich, On the ergodic properties of nowhere dispersing billiards, Comm. Math. Phys. 65 (1979) 295-312.

64. L. A. Bunimovich, On ergodic properties of certain billiards, Funct. Anal. Appl. 8 (1974) 254-5.

65. M. Hénon, Le Billiard Ovale, C. R. Sem. Obs. Nice 2 (1982-3) XIII.1-9.

66. M. V. Berry, Regularity and chaos in classical mechanics, illustrated by three deformations of a circular 'billiard', Eur. J. Phys. 2 (1981) 91-102.

67. A. Sommerfeld, Mechanics of Deformable Bodies (Academic Press, New York, 1964).

68. H. Aref, Stirring by chaotic advection, J. Fluid Mech. 143 (1984) 1-21.

69. K. Sitnikov, The existence of oscillatory motions in the three-body problem, Sov. Phys. Dokl. 5 (1961) 647-50.

70. S. L. Ziglin, Nonintegrability of a problem on the motion of four point vortices, Sov. Math. Dokl. 21 (1980) 296-9.

71. P. J. Holmes and J. E. Marsden, Horseshoes in perturbations of Hamiltonian systems with two degrees of freedom, Comm. Math. Phys. 82 (1982) 523-44.

72. J. Koiller and S. P. Carvalho, Nonintegrability of a restricted problem of four vortices, J. Fluid Mech. (submitted).

73. M. Hénon, Sur la Topologie des Lignes de Courant dans un Cas Particulier, C. R. Acad. Sc. Paris. Ser. A 262 (1966) 312-4.

74. M. V. Berry, N. L. Balazs, M. Tabor and A. Voros, Quantum Maps, Ann. Phys. 122 (1979) 26-63.

75. H. Aref and G. Tryggvason, Vortex dynamics of passive and active interfaces, Physica D (in press).

76. D. V. Khakhar, R. Chella and J. M. Ottino, Stretching, chaotic motion and breakup of elongated droplets in time dependent flows, Dept. Chem. Engrng., Univ. Mass. Amherst (1984).

77. B. B. Mandelbrot, Fractals, Form, Chance and Dimension, (Freeman, San Francisco, 1977).

78. R. S. Mackay, J. D. Meiss and I. C. Percival, Stochasticity and transport in Hamiltonian systems, Phys. Rev. Lett. 52 (1984) 697-700.

79. O. E. Lanford III, The strange attractor theory of turbulence, Ann. Rev. Fluid Mech. 14 (1982) 347-64.

80. A. A. Amsden and F. H. Harlow, Slip instability, Phys. Fluids 7 (1964) 327-34.

Theoretical and Applied Mechanics
F.I. Niordson and N. Olhoff
Elsevier Science Publishers B.V. (North-Holland)

RESONANT PHENOMENA IN ROTATIONAL MOTIONS OF ARTIFICIAL AND NATURAL CELESTIAL BODIES

V.V.Beletsky

Keldysh Institute of Applied Mathematics
USSR Academy of Sciences
Moscow

The resonant rotations of the Moon, Mercury, Venus and other natural and artificial celestial bodies are considered. Gravitational, magnetic and tidal interactions are taken into account. The generalized laws of Cassini for the resonant rotations and extremal properties of the resonant motions are discussed.

1. INTRODUCTION

Let the motion under investigation have a set of frequencies ω_i. The motion is called resonant if

$$\sum n_i \omega_i \simeq 0 \qquad (1.1)$$

where n_i are integers, small by assumption. Let $\varkappa(t)$ be the phase shift of the motion from the resonance so that $\dot{\varkappa} = 0$ when the condition (1.1) is exactly satisfied. If the motion $\varkappa = \varkappa_0$, $\dot{\varkappa} = 0$ is stable then one may speak of the phase stability of the resonant motion.

It is now admitted that the resonant motions play a special role which is unlikely to be accidental. Such motions are frequent in nature and used in technology, including space technology. Examples of such use are:

(a) the synchronization phenomenon in rotors of different mechanisms and technical devices (Blekhman, 1981);
(b) passive stabilization systems in artificial celestial bodies (Grechko, Sarychev et al., 1983).

Some controlled motions may also be resonant. In particular, man's gait is a resonant process (Beletsky, 1984).

There is even a hypothesis published by A.M.Molchanov in 1968 that the complete resonance of the Solar System is a consequence of the evolution maturity of the system. Without going into detail we only note that the resonant phenomena are positively very common in the Solar System, including those with the phase stability. The latter evidences the fact that resonance is actually caused by physical reasons - "the resonant interaction". On the other hand, it does not necessarily mean that a given resonance is accidental if phase stability is absent (or unobserved).

Confident examples of orbital resonances with phase stability are given by motions of following celestial bodies (Alfven and Arrhenius, 1975):

1. The triple resonance $\omega_I - 3\omega_{EU} + 2\omega_G = 0$ between the orbital frequences of Jupiter's satellites Io, Europa, Ganymede.

The resonances in the Saturn satellite system:

2. $2\omega_D - \omega_E - \omega_{\pi E} = 0$ between the orbital frequences of Dione and Enceladus (the angular frequency of Enceladus' pericenter is involved in the resonance relationship).
3. $4\omega_T - 2\omega_M - \omega_{ST} - \omega_{SM} = 0$ between the orbital frequences of Tethys and Mimas (the angular frequencies ω_{ST} and ω_{SM} of the orbit nodes of these satellites are also involved).
4. $4\omega_H - 3\omega_{TI} - \omega_{\pi H} = 0$ between the orbital frequences of Titan and Hyperion (the angular frequency of Hyperion's orbit pericenter is also involved).
5. Trojans at the libration points of the Sun-Jupiter systen (resonance 1:1).
6. "Hilda group "asteroids with the mean period $T_M = \frac{2}{3}T_J$ (T_J is Jupiter's period).
7. The Neptune-Pluto system: the phase shift
$$\varkappa = 3\lambda_P - 2\lambda_N - \delta_{\pi P} - 180°$$
is limited, though not small ($|\Delta\varkappa| \leq 76°$). Here $\lambda_P, \lambda_N, \delta_{\pi P}$ are, respectively, longitudes of Pluto, Neptune and Pluto's orbit pericenter.

Other examples may also be given. Uranus' satellites Miranda-Ariel--Umbriel are in the Io-Europa-Ganymede type resonance but its phase shift slowly grows. Famous resonance 5:2 between Saturn's and Jupiter's periods is unlikely to be in the phase stability, but undoubtedly it affects the planet orbits.

Among various sets of celestial bodies one may find a few dozens of quasicommensurabilities in orbital motions and even show that the probability of their occurrence is very low in comparison with the observed cases (Goldreich, 1964: Molchanov, 1968). But an explicit physical interaction not always can be revealed in those cases.

A lot of stable resonances are encountered in rotational motions of celestial bodies. All of them are based on a clear physical picture of interactions.

This paper deals with the resonant rotational motion dynamics of natural and artificial celestial bodies. Investigations of this problem performed by the author and his students as well as other researchers are discussed.

2. RESONANCES IN ROTATIONAL MOTIONS OF CELESTIAL BODIES

The necessary conditions for the resonance are:

(a) the existence of a conservative factor in the system that generate the resonant "traps" - the stability regions in the vicinity of the resonance;

(b) the existence of a dissipative factor that generates the conditions for capturing the motion into the resonant "trap".

In the Solar System the gravitational interaction is the major conservative factor and the tidal deceleration is the major dissipative factor. A great variety of conservative and dissipative factors may be observed in the dynamics of artificial celectial bodies. Occurrence of the resonant motion considerably depends on the nature of conservative and dissipative forces. The modern theory of passive stabilization of artificial celestial bodies is mainly based on the study of resonant rotations (Okhotsimsky and Sarychev, 1963).

Let us indicate some types of resonance in the rotational motions (Ω is the angular velocity of the body's axial rotation, ω is the orbital angular velocity):

1. $\Omega - \omega \simeq 0$ the Moon type resonance (1:1). The angular velocities of the axial and rotational motions are coincide. Such a motion is also called "the relative equilibrium": a body rests in a rotating orbital system of coordinates and, hence, one of its sides always faces the center of attraction like the Moon and the Earth. The gravity gradient systems of passive stabilization are based on this type of resonance.

2. $2\Omega - 3\omega \simeq 0$ the Mercury type resonance (3:2). Mercury makes exactly three revolutions about its axis for exactly two orbital periods.

3. $\Omega - 2\omega \simeq 0$ the magnetic stabilization resonance (2:1). Following the magnetic field line of the Earth a magnetized satellite makes two revolutions in its rotational motion for one orbital period.

The extraordinary resonance in the rotational motion of Venus will be discussed below.

The unique feature of resonance 1:1 is that it lays on the surface of the theory of rotational motions. It means that there is an explicit finite solution of the sufficiently strictly formulated problem. The fact was known by the classics of celestial mechanics (Lagrange, Laplace) who studied the stability of such a motion in a linear statement. A strict, sufficiently general, nonlinear study of the relative equilibrium was carried out when time came for the Earth's artificial satellite launches. The main results of the study may be formulated as follows (Beletsky, 1959).

A free solid body in a central Newtonian force field performs a motion with a circular orbit of its center of mass and principal central axes of inertia directing along the orbital (rotating) axes. For the motion to be Lyapunov stable it is sufficient that the major axis of inertia ellipsoid is directed along the radius vector of the orbit, and the minor axis along the normal to the orbit plane.

Numerous artificial satellites with gravity gradient attitude control systems as well as many natural satellites satisfy the above criterion. Among 33 natural satellites of the Solar System ten satellites perform nonresonant rotational motion, another ten - unspecified motion, and the rest thirteen - the resonant one in resonance 1:1. The latter are the Moon, four Jupiter's satellites (Io, Europa, Ganymede, Callisto), five Saturn's satellites (Enceladus, Iapetus, Rhea, Tethys, Dione), Neptune's satellite (Triton), both Mars' satellites (Phobos and Deimos).

3. ON CASSINI'S LAWS OF THE MOON MOTION

The above discussion of the stable relative equilibrium does not fully describe resonance 1:1. Indeed, perturbations due to other celestial bodies result in more complex dynamical effects. For example, in the Moon case "the third body" - the Sun - introduces strong perturbations into the Moon's orbit. The Moon's orbital plane precesses with the period T_{Ω} = 18.6 years. In its turn this causes the perturbations in the rotational motion of the Moon.

In 1693 D.D.Cassini discovered empirically the laws of the rotational motion of the Moon. We do not quote them here since below "the generalized Cassini's laws" will be formulated.

Cassini's laws are empirical. They do not follow from exact motion equations in the form of an exact final solution. The classical theory of the Moon's libration and its further developments were based on the linearization of the motion equations in the vicinity of the empirical Cassini's laws. Therefore the problem of a more strict theoretical justification of Cassini's laws as the real laws of nature had been open for nearly 300 years. Only in the present time it has become possible to solve this problem due to theoretical advances and development of space research. Now we shall discuss it in detail.

4. ON MERCURY'S ROTATION

In 1889 Sciaparelli has interpreted a series of his observations of Mercury declaring that Mercury rotates about its axis with the period T = 88 days that is equal to the orbital period. However, using the radar techniques (1965) established that Mercury's rotation period is close to two thirds of the orbital period (58.65 days).Later it was found that the optical observations by Sciaparelli allow nonunique interpretation, and the period of about 59 days well agrees with these observations. Resonance 3:2 in Mercury's rotation in now widely admitted.

The plane rotation of a celestial body whose center of mass moves in an orbit with the eccentricity e is described by the equation (Beletsky, 1959):

$$(1+e)\cos\nu \, \frac{d^2\delta}{d\nu^2} - 2e\sin\nu \, \frac{d\delta}{d\nu} + 3\,\frac{A-C}{B}\sin\delta = 4e\sin\nu, \quad \delta = 2\theta \qquad (4.1)$$

Here ν is the true anomaly; θ is the angle between the body's axis of inertia and the radius vector of the orbit; A, B, C are the principal axes of inertia.

Even before the discovery of resonance 3:2 in Mercury's rotation Chernous'ko (1963) showed that equation (4.1) allows resonant solutions including those of the Mercury type (3:2). The phase stability zone of such a resonance has the size of about e. Since the eccentricity of Mercury's orbit is not small ($\sim$0.206) the existence of the (3:2) - type resonance becomes even more possible.

The investigation of the exact resonant (3:2) solutions of equation (4.1) was carried out by Beletsky and Lavrovsky (1975). The stability regions of these solutions are shown as the shaded zones in Fig. 1 in the plane of parameters e, $n^2 = 3(A-C)/B$.

In reality, however, Mercury moves in the perturbed non-Keplerian orbit; the plane model (4.1) is only a step in the study of the general case. Thus the necessity has become imminent for the development of a generalized theory of resonant rotations of celestial bodies with consideration of systematic perturbations of their orbits.

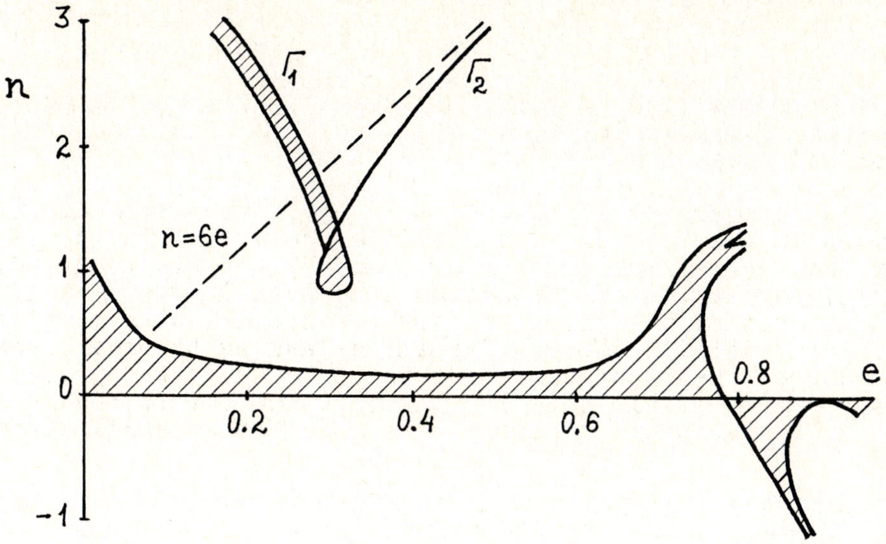

Figure 1
Stability Regions of the Resonant Solutions

Advances in mathematics of the XXth century - the theory of periodic solutions of Poincare, asymptotic methods of nonlinear oscillations - allowed to construct such a theory. Launches of space vehides and the necessity to meet the requirements of spaceflights, have promoted a new stage in the development of the theory of rotational motions of celestial bodies.

5. THE GENERALIZED LAWS OF CASSINI

As a result of those investigations the theory of the so called "generalized laws of Cassini" (Beletsky, 1972) has been constructed. Successive (sometimes, parallel) steps of the analysis have been published since 1958 by Beletsky, Chernous'ko and Torzhevsky (1963), Colombo (1966), Peale (1974) and others. The discussion of these publications may be found, for example, in the books by Beletsky (1965, 1975, 1977).

Cassini's laws are determined by the superposition of the stabilizing effect of the gravitational field for the resonance described in pp. 2, 4 and the evolution laws for the angular momentum of the body's rotational motion with respect to the precessing orbit (Beletsky, 1958).

The generalized laws of Cassini may be represented by the stationary points of nonlinear autonomous equations obtained from the general motion equations by the procedure of asymptotic technique in the resonance case (Beletsky, 1972). Extrema of the average Hamiltonian take place at these points.

Below these laws are given in a formal representation, using the following designations: K_ω, K_Ω are, respectively, the precession velocities of the body's orbit pericenter and node; i is the orbital inclination.

1. A body rotates regularly about its principal central axis of inertia with the angular velocity

$$\Omega_0 = \Omega + K_\omega + K_\Omega \operatorname{Cos}(i \pm \rho_0) \qquad (5.1)$$

that is close to one of the resonant values: $\Omega = \omega$ (the Moon), $\Omega = 3\omega/2$ (Mercury), and so on. Here ω is the orbital angular velocity.

2. The axis of the body's angular rotation, the normal to the orbit plane and the axis of the orbit precession lie in the same plane.

3. The axis of the body's angular rotation and the normal to the orbit plane form the constant angle ρ_0 given by the equation

$$\operatorname{Cos}\rho^* \mp \operatorname{Sin}\rho^* \operatorname{Ctg}\rho_0 + \chi \operatorname{Cos}\rho_0 = 0 \qquad (5.2)$$

Here the parameters ρ^*, χ are uniquely determined by the orbital parameters, the moments of inertia of the body, and Ω_0. The equation (5.2) has 2 or 4 solutions.

4. The phases of the rotational and orbital motions concide at the passage through pericenter: the angle between the radius vector and the node line coincide with the angle between the axis of inertia and the node line.

From the second law it follows that the axis of the body's angular rotation precesses in space with the same period as the orbital axis. Thus, the generalized laws of Cassini (the first and the second one) include two resonances.

The main stability condition for the generalized laws of Cassini is the position of inertia axes described in p. 2 (the Moon case). In the Mercury case such a position of axes must be at the passage through the orbital pericenter.

Lidov and Neishtadt (1975) extended the above theory by taking into account the tidal effects and the asymptotic stability of Cassini's laws. Finally, it was shown (Barkin, 1978) that the generalized laws of Cassini are Poincare's generating motions for the exact periodic solutions of the initial equations.

6. VENUS

In 1962 radar measurements showed that Venus has retrograde rotation. By modern data the period of this rotation is $T_V = 243.0 \pm 0.03$ days, which is close to the resonance value of 243.16 days. For the period of conjunction with the Earth $\tau = 583.92$ days Venus performs exactly five revolutions about its axis with respect to the Sun-Venus line and exactly four revolutions with respect to the Sun-Earth line. At each conjunction Venus face the Earth by the same side. The angular velocity of Venus' axial rotation relates the orbital angular velocities of Venus and the Earth, ω_V and ω_E, as

$$\Omega = 4\omega_V - 5\omega_E .$$

This extraordinary resonance was found in the plane model (Goldreich and Peale, 1968). However, spatial effects are of the most interest in this problem.

The theory of the spatial resonant rotation of Venus was developed in a series of publications (Beletsky, Levin and Pogorelov, 1979--1981). The theory makes allowance for the gravitational interaction between the Sun and Venus, and the Earth and Venus, the evolution of Venus' orbit, the tidal effects in the rotational motion of Venus. The resonance zone in the phase space is formed, on the average, by the gravitational field of the Earth. The gravitational torque produced by the Sun does not overagedly influence the formation of this zone.

It is appropriate first to discuss the evolution of the celestial body rotation under the torque of tidal forces (Beletsky, 1981). In this epoch the tidal effects are minor. The major tidal evolution had occurred in a cosmogonically short time of 10^4 to 10^6 years at the first stage of the planets formation. The possible result of the tidal evolution is a capture into the resonant rotation with phase stability.

The picture of the tidal evolution in the plane of parameters ρ, Ω is given in Fig. 2. Here ρ is the angle between the vector

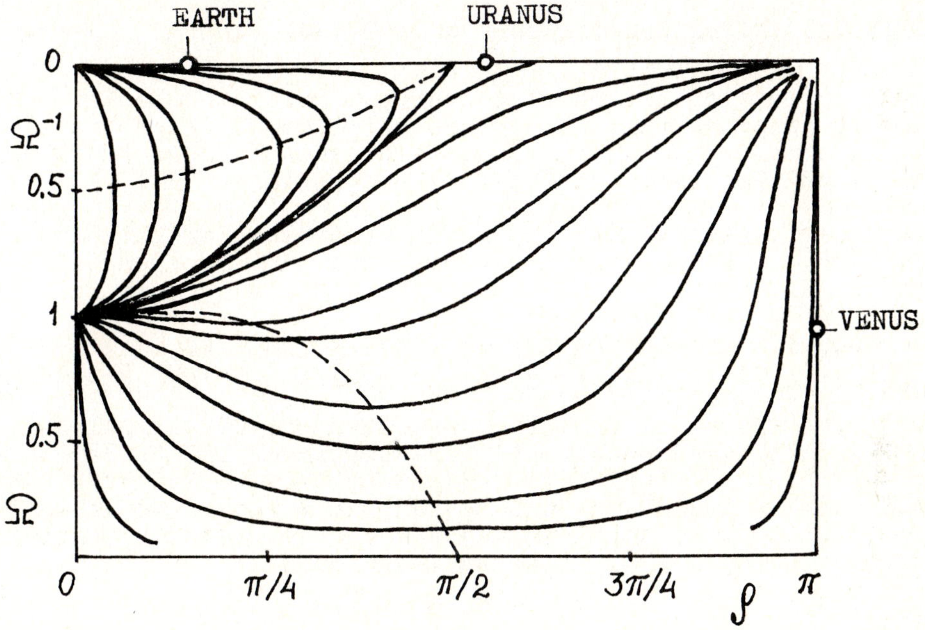

Figure 2
Tidal Evolution

of angular momentum and the normal to the orbit plane, Ω is the normalized value of this vector. The case $\rho = 0$ corresponds to the direct rotation and $\rho = \pi$ to the retrograde rotation. It can be seen that all motions tend to be direct rotation with zero inclination and a certain angular velocity. The process depends on the initial conditions. Extreme versions of the first stage of the evolution are: (a) strong inclination evolution at nearly constant rotation (Uranus); (b) rotation evolution at nearly constant inclination (Venus).

In orbits close to circular ones the limiting angular velocity is near the orbital value (the Moon, etc.). For the orbit eccentricity $e \sim 0.2$ the limiting angular velocity is about one and a half of the orbital one(Mercury).

The tidal instability of the retrograde rotation and the existence of the retrograde rotation of Venus are, on the face of it, in contradiction. It gave birth to a hypothetic theory on the existence in the past of a retrograde satellite of Venus that stabilized it in the retrograde rotation (Khentov, 1974).

The conclusion of the new cosmogonic theory that the retrograde rotation of Venus has a cosmogonical origin seems to be more natural (Eneev and Kozlov, 1981).

The theory of Venus rotation shows that in this case Venus could not manage to tilt over for the time of tidal evolution (less than 10^6 years). The tilting process runs much slower than the process of capture into the resonant rotation.

The steady-state resonant rotation of Venus is described by the generalized Cassini type laws. It should be noted, however, that the probability of Venus' capture into the resonance is quite small in comparison with capturing the Moon or Mercury into their resonances.

7. EXTREMAL PROPERTIES OF RESONANT MOTIONS

Among a variety of motions the resonant ones are special. Therefore it may be supposed that the characteristics of the force fields are extremal at the resonances. There are several theoretical and empirical principles that confirm this suggestion.

Developing ideas and methods of Poincaré Blekhman in 1960 proposed an extremal principle which was discussed and applied to specific problems (Blekhman, 1981). One of its formulations is following.

Let consider a generating motion with frequencies ω_S. Let introduce the value $\sigma_S = \mathrm{Sgn}\,(d\omega_S/dh_S)$. Here h_S is the energy corresponding to the frequency ω_S. The object is called strictly anisochronous if $\sigma_S = 1$ and nonstrictly anisochronous if $\sigma_S = -1$.

Consider the value Λ of the Lagrangian L, averaged in the generating motion. Assume that all objects of the system are similarly anisochronous ($\sigma_S = \sigma$). Then the functional

$$D = -\sigma\Lambda = \min \qquad (7.1)$$

i.e. it has a minimum on stable resonant modes of the motion.

For the orbital cases $\sigma < 0$. Under certain conditions one may ignore in Λ the terms owing to the kinetic energy, and then the condition (7.1) turns into $\langle U_0 \rangle = \min$. Here $\langle U_0 \rangle$ is the value of the gravitational force function, averaged on generating motions. This formulation is equivalent, in a sense, to the known heuristic principle of "the least interaction" (Ovenden et al., 1973).

In the case of rotational motions of celestial bodies $\sigma > 0$, and from (7.1) the opposite principle $\langle U_0 \rangle = \max$ may be obtained for stable resonant modes.

The principle (7.1) is formulated for generating motions, and allows to determine the motion phases on stable modes. Beletsky and Shlyakhtin (1976) proposed an extremal principle for stable resonant motions in the following form:

$$\bar{U}(x_0,\dot{x}_0) = \lim_{t\to\infty} \frac{1}{t}\int_0^t U\,(x\ (x_0,\dot{x}_0,t),t)dt = \max \qquad (7.2)$$

Here the averaged value of the force function U on real, not "generating", motions is involved. Equation (7.2) is based rather on the Lyapunov functions than the small parameter.

The principle (7.2) allows one to find the initial conditions x_0, $\dot{x}_0$ for the stable resonant modes.

The numerical experiment testing this principle was carried out mainly for the equation (4.1), though other equations were also considered. Some results are given in the figures below. Instead of the average force function $\bar{U}$ the average dimensionless potential energy $\bar{u} = -K\bar{U}$, $K > 0$ was computed. Minima of the function $\bar{u}$ correspond to maxima of the function $\bar{U}$.

The minima for resonances 1:1, 3:2, 2:1 are shown in Fig. 3. Fig. 4 shows the orbital period map in the phase plane (θ, θ'). It convin-

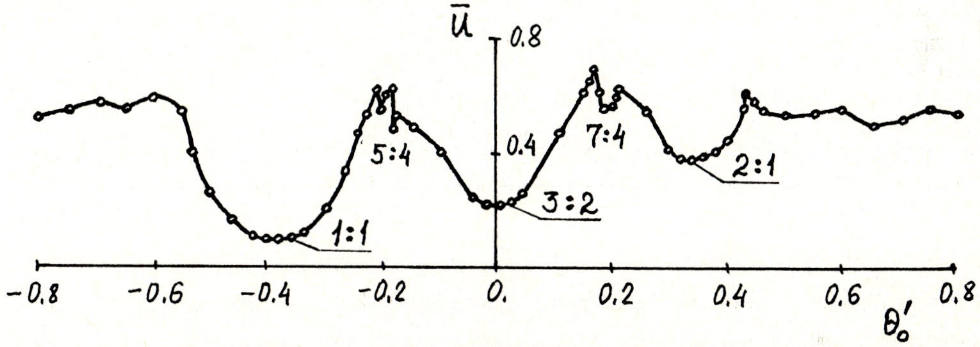

Figure 3
Average Potential Energy

ces us in the stability of resonances mentioned above.

An analogous situation (Fig. 5a) takes place in the problem of rotation of a magnetized satellite in a polar circular orbit. Two stable resonant rotations are found: 2:1 (stabilization with respect to a local magnetic field line): 0:1 (stabilization with respect to the fixed direction). If a gravitational torque is involved, another stable resonance 1:1 (Fig. 5b) appears.

Resonances of the type 0:1 may be used as a basis for an analytical theory of slow rotations of celestial bodies (Beletsky, 1981).

Extremal properties of resonant motions were investigated by Beletsky and Kasatkin (1980), Kozlov (1982) and others.

8. CONCLUDING REMARKS

The resonances in the rotational motions of celestial bodies observed in nature are of the low order (1:1, 3:2). Khentov (1982) showed that it is inevitable in the system of two gravitating bodies. Resonances of higher order may be due to the interaction of more than two bodies (Venus) or to the additional force fields (magnetic re-

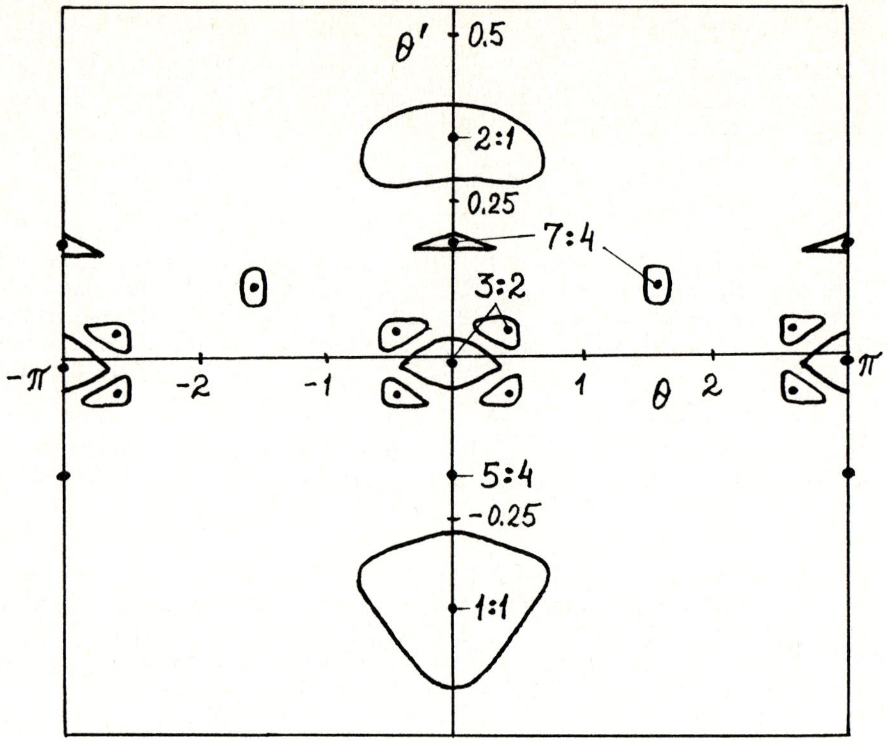

Figure 4
Orbital Period Map

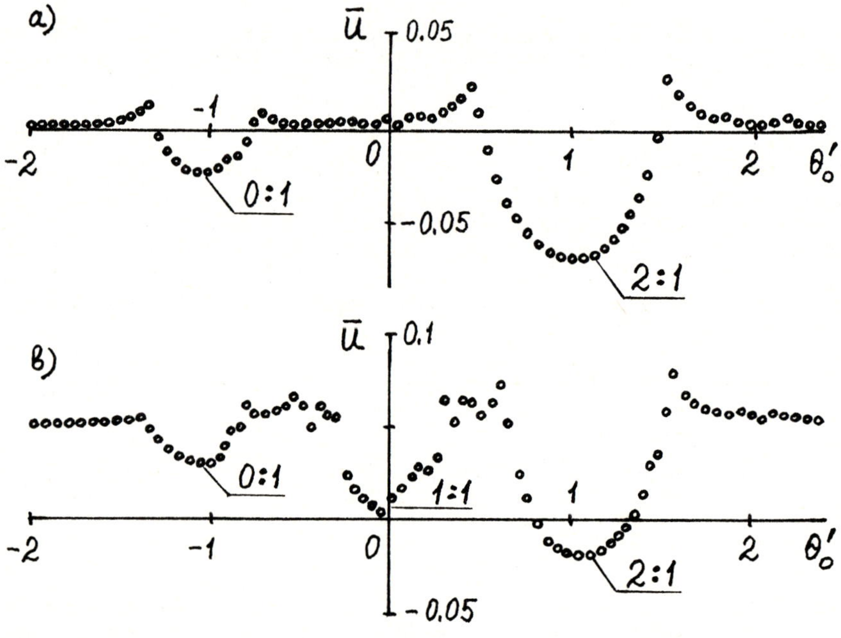

Figure 5
Average Potential Energy for a Magnetized Satellite

sonance).

Nature has set the resonance traps along the paths of motion of celestial bodies. But a trap is not yet a beast in the trap. To trap the beast one must have patience and good luck.

9. ACKNOWLEDGEMENT

The author is grateful to Yu.V.Bolotin and E.M.Levin for their assistance during the preparation of this report.

REFERENCES:

[1] Alfven H., Arrhenius G., Structure and Evolutionary History of the Solar System. (D.Reidel Publishing Co., Dordrecht, Holland, 1975).

[2] Barkin Yu.V. On Cassini's laws. Astronomichesky Jrnl. 55 (1978) 113-122.

[3] Beletsky V.V., The motion of the Earth's artificial satellite with respect to the center of mass , in: Artificial satellites of the Earth, vyp. 1 (Izd. AN SSSR, Moscow, 1958).

[4] Beletsky V.V., On libration of a satellite, in: Artificial satellites of the Earth, vyp. 3 (Izd. AN SSSR, Mosocw, 1959).

[5] Beletsky V.V., The Motion of an Artificial Satellite with Respect to the Center of Mass (Nauka, Moscow, 1965).

[6] Beletsky V.V., Resonance Rotations of Celestial Bodies and Cassini's Laws, Celestial Mechanics, 6 (1972) 356-378.

[7] Beletsky V.V., The Motion of a Satellite with Respect to the Center of Mass in the Gravitational Field (Izd. Moscovskogo Universiteta, Moscow, 1975).

[8] Beletsky V.V., Lavrovsky E.K., To the theory of Mercury's resonant rotation, Astronomichesky Jrnl. 52 (1975) 1299-1308.

[9] Beletsky V.V., Shlyakhtin A.N., Extremal properties of resonant motions, DAN SSSR 231 (1976) 829-832.

[10] Beletsky V.V., Essays on the Motion of Space Bodies (Izd. 2nd, Nauka, Moscow, 1981).

[11] Beletsky V.V., Kasatkin G.V., On extremal properties of resonant motions, DAN SSSR 251 (1980) 58-62.

[12] Beletsky V.V., Levin E.M., Pogorelov D.Yu., To the question of the resonant rotation of Venus. Astronomichesky Jrnl. 57(1980) 158-167.

[13] Beletsky V.V., Levin E.M., Pogorelov D.Yu., To the question of the resonant rotation of Venus. Astronomichesky Jrnl. 58 (1981) 198-207.

[14] Beletsky V.V., Levin E.M., Correctness of averaging in the plane problem on the resonant rotation of Venus, Astronomichesky Jrnl. 58 (1981) 416-421.

[15] Beletsky V.V., Tidal Evolution of Inclinations and Rotations of Celestial Bodies, Celestial Mechanics, 23 (1981) 371-382.

[16] Beletsky V.V., On slow rotations of celestial bodies, Cosmicheskie Issledovaniya, 19 (1981) 26-33.

[17] Beletsky V.V., Two-Legged Walking - Model Dynamics and Control Problems (Nauka, Moscow, 1984).

[18] Blekhman I.I., Sinchronization in Nature and Engineering (Nauka, Moscow, 1981).

[19] Chernous'ko F.L., Resonant phenomena in the satellite's motion with respect to the center of mass, Jrnl. Vychislitel'noi matematiki i matematicheskoi fiziki, 3 (1963) 528-538.

[20] Chernous'ko F.L., On the satellite's motion with respect to the center of mass under the action of gravitational torques, Prikl. matem. i mekh. 27 (1963) 473-483.

[21] Colombo G., Cassini's Second and Third Laws, Astrol. Jrnl. 71 (1966) 891-896.

[22] Ereev T.M., Kozlov N.N., The Problems of Simulation of Planetary Systems Accumulation processes. Adv. Space Res., 1 (1981) 201-215.

[23] Goldreich P. Explanation of frequent occurrence of commensurable mean motions in the Solar System, Mon. Not. Roy. Astron. Soc. 130 (1965) 159-181.

[24] Goldreich P., Peale S., The Dynamics of Planetary Rotations, Ann. Rev. Astron. and Astroph., Palo Alto, Calif., USA, 6 (1968).

[25] Grechko G.M., Sarychev V.A., Legostaev V.P., Sazonov V.V., Gansvind I.N., Gravity Gradient Stabilization of the Salyut 6/Soyuz Orbital Complex. Preprint, Inst. Appl. Math., the USSR Academy of Sciences, 18 (1983).

[26] Khentov A.A., Sinchronization of satellites, in: Dynamics of Systems vyp. 4 (Mezhvuzovskii sb., Gorkii, 1974).

[27] Khentov A.A., Dynamics of the formation of resonant rotations of natural celestial bodies, Astronomichesky Jrnl. 59 (1982).

[28] Kozlov V.V., Averaging in the vicinity of stable periodic motions, DAN SSSR, 264 (1982) 567-570.

[29] Lidov M.L., Neushtadt A.I., A method of canonic transformations in problems on the rotations of celestial bodies and Cassini's laws, in: Determination of Motion of Space Vehicles (Nauka, Moscow, 1975).

[30] Molchanov A.M., The Resonant Structure of the Solar System. Icarus 8 (1968) 203-215.

[31] Okhotsimsky D.E., Sarychev V.A., A gravity gradient stabilization system of artificial satellites, in: Artificial Satellites of the Earth,vyp. 16 (Izd. AN SSSR, 1963).

[32] Ovenden M.W., Feagin T., Graff O., On the Principle of least interaction action and the Laplacian satellites of Jupiter and Uranus, Celestial Mechanics, 8 (1974) 455-471.

[33] Peale S.I., Generalized Cassini 's Laws, Astron. Jrnl. 74 (1969) 483-489.

[34] Torzhevsky A.P., Motion of an artificial satellite with respect to the center of mass and resonances, Acta Astronautica, 14 (1969) 241-259.

Theoretical and Applied Mechanics
F.I. Niordson and N. Olhoff
Elsevier Science Publishers B.V. (North-Holland)

QUANTUM,CLASSICAL AND SEMICLASSICAL ADIABATICITY

M.V.Berry

H.H.Wills Physics Laboratory, Tyndall Avenue
Bristol BS8 1TL,U.K.

Adiabatic theorems are reviewed, and an asymptotic connection established between quantum and classical anholonomy properties of systems whose Hamiltonians contain parameters that are slowly varied round a cycle. The quantum property is a geometrical phase accompanying transport of an eigenstate round the cycle; the classical property is a shift in phase-space torus angle variables. Examples are given. Starting from the geometrical phase, different classes of quantum system are seen to have different degeneracy properties. Sometimes, quantum tunnelling causes discordance between the adiabatic and semiclassical limits; for these cases a uniform semiclassical adiabatic approximation can be found, encompassing both limits.

INTRODUCTION

Consider a dissipationless mechanical system with N freedoms executing bound motion under the action of forces that may be time-dependent. The evolution is governed by a Hamiltonian $H(q,p;X(t))$, where $q\equiv\{q_1...q_N\}$ and $p\equiv p_1...p_N\}$ are the coordinates and conjugate momenta [1,2] and $X(t)\equiv\{X_1(t),X_2(t),...\}$ are parameters controlling the forces. If the system is classical, q and p are ordinary variables; if it is quantum-mechanical, q and p are operators with the commutation law [3]

$$[q_i,p_j]=i\hbar\delta_{ij} \qquad (1)$$

where $\hbar$ is Planck's constant, leading to a Hermitian Hamiltonian operator H (after applying factor ordering conventions if necessary). This framework is very general and encompasses not only quantum and classical mechanics but also physical and geometrical optics and linear theories of acoustic, elastic and water waves and their ray approximations. Those less familiar with quantum mechanics might prefer to think in terms of wavevector k and frequency Ω, related to p and H by

$$\left.\begin{aligned} &k=p/\hbar \quad \text{(de Broglie relation)} \\ &\Omega(q,k;X(t)) = \frac{1}{\hbar}H(q,p;X(t) \quad \text{(dispersion relation)} \end{aligned}\right\}, \qquad (2)$$

with wave equations constructed by regarding q as an ordinary variable and k as the operator $-i\partial/\partial q$ (a choice which satisfies (1)).

The asymptotics of two limiting situations is particularly interesting and important. First, there is the semiclassical limit $\hbar\to 0$, corresponding (cf.(2)) to high frequency Ω or short wavelength $2\pi/k$. Second, there is the adiabatic limit $dX/dt\to 0$, corresponding to slow change of the parameters. These limits can be very subtle and complicated because quantities often turn out to depend nonanalytically on $\hbar$ and dX/dt. My purpose here is to give an informal account of the two limits and describe how they may be sometimes concordant, sometimes discordant.

ADIABATIC THEOREMS

For physicists, adiabatic theory began around 1910, with the embryonic quantum mechanics inspiring a result in classical mechanics: for a one-dimensional harmonic oscillator (e.g.a pendulum) whose frequency ω (now considered as the parameter X) is slowly changed (e.g. by altering the length of the string), what is conserved is not the energy E but the ratio E/ω. The argument was that for fixed ω the energies are restricted by quantum mechanics to values we would now write as

$$E_n = (n+\tfrac{1}{2})\hbar\omega \quad (n=1,2...), \tag{3}$$

and n, being integer, cannot vary continuously, so as ω is changed E_n must change proportionately, thus conserving E_n/ω which ought to be conserved classically too because it does not involve $\hbar$. Thus a classical adiabatic invariant (E/ω) has been derived from an (implied) quantal adiabatic theorem (persistence of state labelled n) via an (implied) semiclassical connection (the quantum state n is associated with the classical motion whose energy is E_n). In the succeeding years these ingredients of the adiabatic argument were made more precise and generalized as follows.

The quantal adiabatic theorem [4] concerns states $|\Psi(t)\rangle$ satisfying the time-dependent Schrödinger equation

$$H(q,p;X(t))|\Psi(t) = i\hbar\frac{\partial}{\partial t}|\Psi(t)\rangle, \tag{4}$$

and asserts that in the adiabatic limit of infinitely slow variation of the parameters X(t) there are states, labelled $|\Psi_n(t)\rangle$, which cling to the eigenstates $|n;X\rangle$ of the instantaneous Hamiltonians. These eigenstates, and their energies $E_n(X)$, are defined by

$$H(q,p;X)|n;X\rangle = E_n(X)|n;X\rangle, \tag{5}$$

and the adiabatic assertion is

$$|\Psi_n(t)\rangle = e^{i\phi_n(t)}|n;X(t)\rangle, \tag{6}$$

where $\phi_n(t)$ are phases to be discussed in the next section. Non-adiabatic behaviour, measured in terms of the amplitudes for transitions to states other than n, can be extremely small (of order $\exp\{-|dX/dt|^{-1}\}$ if X varies analytically with t [5,6]), even if the total change in the X is large.

The classical adiabatic theorem [1,2,7] does not have the generality of the quantal theorem, because it is restricted to Hamiltonians which for constant X support quasiperiodic motion, that is motion where orbits explore N-dimensional tori in the 2N-dimensional phase space of the variables q,p. Such tori fill phase space when N=1 and also when N>1 for integrable systems, that is systems with

N constants of motion. Generically, however, phase space has some regions where motion is chaotic [2,8,9] and orbits explore regions of dimensionality greater than N which in the extreme case of ergodicity consist of the entire energy surface H=constant, whose dimensionality is 2N-1. For nonchaotic motion, then, the theorem asserts that for slow variations of X the orbit clings to that torus which preserves the values of the N actions $I \equiv \{I_1 \ldots I_N\}$. These quantities, which are thus the adiabatic invariants, are defined as the line integrals

$$I_j \equiv \frac{1}{2\pi} \oint_{\Gamma_j} p_i(q;X)\,dq_i \tag{7}$$

around the irreducible cycles Γ_j of the torus. The succession of tori thus selected will, in general, have different energies whose instantaneous values are given by the Hamiltonian in 'action' representation, namely

$$E_I(t) = H(I;X(t)). \tag{8}$$

For N>1, application of the adiabatic theorem to nonintegrable systems depends on the continuing existence of the torus with actions I during the variation of the parameters X. Generically, however, the torus will be repeatedly disrupted near resonance zones where the ratios of its frequencies

$$\omega=\{\omega_j(X) \equiv \frac{\partial H(I;X)}{\partial I_j}\} \tag{9}$$

are close to rational. Nevertheless, there are some nontrivial cases where a torus can remain nonresonant - for example when an external force field is rigidly rotated or translated.

The semiclassical connection that the two adiabatic theorems strongly suggest, and which is confirmed by asymptotic analysis [10,11], is an association, valid as $\hbar \to 0$, between the quantum state labelled by quantum numbers $n \equiv \{n_1 \ldots n_N\}$ and the classical torus with actions I. The association is

$$I_j(n) = (n_j+\alpha_j)\hbar \tag{10}$$

where the indices α_j depend on how the torus is embedded in phase space [2,10]. The energy of the asymptotic state $|n;X\rangle$ is then given by

$$E_n(X) = H(I(n);X). \tag{11}$$

Preservation of the association between $|n\rangle$ and the torus I as X varies is guaranteed by the two adiabatic theorems. Although (11) does generalize the quantum-mechanical WKB approximation, it is of course restricted to systems whose classical motion lies on tori. For these systems, however, the method is very useful in practice for calculating energy eigenvalues, especially in quantum chemistry [12-18].

For classically chaotic motion, there seem to be no adiabatic invariants (except for systems very close to integrable ones [51]) and no asymptotic semiclassical relation giving (for example) the energies of the quantum states, except for the limiting case of ergodic motion. In this case [19] there is a single adiabatic invariant, namely the phase-space volume within the energy surface, defined by

$$\mathrm{Vol}(E) \equiv \int d^N q \int d^N p\, \Theta(E - H(q,p;X)) \tag{12}$$

where Θ denotes the unit step function. Thus, when X varies slowly, the energy changes in accordance with

$$\mathrm{Vol}(E(X(t))) = \mathrm{Vol}(E(X(0))). \tag{13}$$

This suggests that for classically ergodic systems the quantum states are labelled by a single quantum number n in terms of which the energy levels are given asymptotically by

$$\mathrm{Vol}(E_n) = (2\pi\hbar)^N (n+\tfrac{1}{2}). \tag{14}$$

As an expression of the old idea [11] that one quantum state 'occupies' a volume $(2\pi\hbar)^N$ of classical phase space, this approximation (augmented by asymptotic correction terms) gives surprisingly good agreement with the trend of exact quantal energy curves $E_n(X)$ for a range of parameters X (see fig.5 of ref [20]). (Superimposed on this trend are small-scale fluctuations whose statistics seem to be universal [21,22,40], and also larger-scale fluctuations connected with classical closed orbits [11,23,24,50].)

ADIABATIC ANHOLONOMY

For each of the two adiabatic theorems just described, a question is left unanswered. In the quantum case, the question is: what is the phase ϕ_n (eq.6) associated with the adiabatic convection of the eigenstate $|n\rangle$? In the classical case, the question is: how do the angle variables $\theta \equiv \{\theta_1 \ldots \theta_N\}$ (conjugate to the actions, and describing positions on the tori [2,8]) evolve as the orbit clings adiabatically to the torus I? Because of the arbitrariness in the definition of the phase of the eigenstate $|n\rangle$ and of the origin with respect to which angles are measured on each torus I, these questions only have invariant meaning if the parameter variation consists of a circuit C in the space X. We imagine this circuit to be traversed over a (long) time T, after which X(T)=X(0), the Hamiltonian has returned to its original form, the state $|\Psi_n(t)\rangle$ returns to the original eigenstate $|n(0)\rangle$ (up to a phase) and the classical orbit returns to the same torus I.

When thus posed, the two questions have immediate naive answers, both wrong. In the quantal case, the naive answer is that the phase change between t=0 and t=T is the 'dynamical' increase of the phase associated with the instantaneous eigenstate, namely

$$\frac{1}{\hbar}\int_0^T dt\, E_n(X(t)). \tag{15}$$

The correct answer [25] is that the states before and after the circuit are related by

$$|\Psi_n(T)\rangle = \exp\{\frac{i}{\hbar}\int_0^T dt\, E_n(X(t)) + i\gamma_n(C)\}\, |\Psi_n(0)\rangle \tag{16}$$

where $\gamma_n(C)$ is a geometrical phase depending, in a manner soon to be explained, on the circuit C in parameter space and the transported state $|n\rangle$, but not on the time T taken to traverse C. Thus $\gamma_n(C)$ is small in comparison with the dynamical phase but does not vanish in the adiabatic limit. In the classical case, the naive answer is that the increments in the angles θ are given by

$$\int_0^T dt\ \omega(X(t)) \tag{17}$$

where $\omega(X)$ are the frequencies (9) with which the orbit whirls around the cycles Γ of the instantaneous I-torus at the parameter-space point X. The correct answer [26,27] is that the angles before and after the circuit are related by

$$\theta(T) = \theta(0) + \int_0^T dt\ \omega(X(t)) + \Delta\theta(I;C), \tag{18}$$

where $\Delta\theta(I;C) \equiv \{\Delta\theta_i\}$ are classical adiabatic angles depending only on the circuit C and the action I of the torus being transported, but not on T.

Each of the quantities $\gamma_n(C)$ and $\Delta\theta(I;C)$ is naturally expressed as an integral in parameter space, over any surface A spanning C, that is: as the flux of a 2-form through C. Quantally, it can be shown [25] by analysis of the time-dependent Schrödinger equation (4) that

$$\gamma_n(C) = -\int_{\partial A=C} V(N;X) \tag{19}$$

where the phase 2-form V is

$$V(n;X) = \text{Im}\langle dn| \wedge |dn\rangle \tag{20}$$

and d refers to displacement (of the state $|n;X\rangle$) in parameter space. Classically, it can be shown [26,27] by analysis of Hamilton's equations that

$$\Delta\theta(I;C) = -\frac{\partial}{\partial I} \int_{\partial A=C} W(I;X) \tag{21}$$

where the angle 2-form is given in terms of

$$W(I;X) = \frac{1}{(2\pi)^N} \oint d^N\theta\ dp \wedge dq, \tag{22}$$

and the integral is of parameter-space-displacements of the p's and q's, and over the I-torus at X.

Both γ_n [28] and $\Delta\theta$ can be regarded as exemplifying anholonomy, that is closure failure under parallel transport, and the two-forms correspond to curvatures (analogous to those of spacetime in general relativity [29]). What turns mathematical connection phenomena into potentially observable physical phenomena is the fact that the Schrödinger and Hamilton equations induce precisely those connections which are mathematically natural.

The geometrical phase $\gamma_n(C)$ is a completely general feature of quantum mechanics. Its applicability is not restricted to the semiclassical limit or even to systems with classical analogues. But when $|n\rangle$ does have a classical analogue which moreover corresponds to quasiperiodic motion, the phase 2-form V, given by (20), can be shown [27] to be semiclassically related to the 2-form W, given by (22), by

$$V(n;X) = -\frac{1}{\hbar} W(I;X) \tag{23}$$

so that the classical adiabatic angles are given by

$$\Delta\theta_j(I;C) = -\hbar\frac{\partial\gamma_n}{\partial I_j}(C) = -\frac{\partial\gamma_n}{\partial n_j}(C) \tag{24}$$

where the association (10) enables the quantum numbers n to be treated as continuous variables.

Now we make two remarks which help in interpreting and evaluating the quantal formulae (19) and (20). Firstly, if there are three parameters, these can be regarded as components of a vector $\underset{\sim}{X}$, d becomes the gradient ∇ in $\underset{\sim}{X}$ space, and $\wedge$ becomes the cross product of ordinary vector analysis. If in addition the state $|n;X\rangle$ is expressed in position representation in terms of the wavefunction $\psi_n(q;\underset{\sim}{X}) \equiv \langle q|n;\underset{\sim}{X}\rangle$, then the geometrical phase can be written as

$$\gamma_n(C) = -\operatorname{Im}\int d^N q \iint_{\partial A=C} d\underset{\sim}{A}\cdot\nabla\psi_n^*(q;\underset{\sim}{X})\wedge\nabla\psi(q;\underset{\sim}{X}). \tag{25}$$

Secondly, if the Hamiltonian operator can be represented (non-classically) as a 2×2 Hermitian matrix, written in the general form

$$H(X) = \tfrac{1}{2}\begin{pmatrix} w(X) & u(x)-iv(X) \\ u(X)+iv(X) & -w(X)\end{pmatrix} \tag{26}$$

then the phase 2-form for the upper and lower eigenstates $|\pm\rangle$ can be calculated to be

$$V_\pm = \mp\frac{u\,dv\wedge dw + v\,dw\wedge du + w\,du\wedge dv}{2(u^2+v^2+w^2)^{3/2}}. \tag{27}$$

Physically, $\gamma_n(C)$ could be observed by the interference produced when the state that has been transported round C is superposed with a state which has evolved under a constant Hamiltonian (fixed X). A feasible realization of this idea employs a beam of polarized nuclei in a spin-n eigenstate of an external magnetic field B. The beam is split into two beams, one of which passes through a region of constant B while the other passes through a region where the vector $\underset{\sim}{B}$ is taken round a cycle C (in magnetic field space which is the $\underset{\sim}{X}$ space in this case); then the cycled beam can be shown [25] to have acquired a phase $\gamma_n(C)=n\Omega(C)$, where $\Omega(C)$ is the solid angle subtended by C at $\underset{\sim}{B}=0$, and this phase difference can be revealed as a fringe shift when the two beams are recombined.

The phase anholonomy embodied in $\gamma_n(C)$ is not confined to quantum mechanics but can be expected whenever Hermitian matrices arise in physics. One such area is the optics of anisotropic media [30], where the two polarization states of electromagnetic waves which propagate unchanged with wavevector k are the eigenvectors of the 2x2 transverse part of the inverse of the dielectric tensor. If the medium is both uniaxially birefringent and gyrotropic (i.e. the plane of polarization rotates) with both optic axes coinciding, and this common axis is made to turn once in a cone of semiangle θ about $\underset{\sim}{k}$, then [31] the two eigenpolarizations experience a phase shift

$$\gamma_\pm(\theta) = \pm 2\pi\left(1 - \frac{\sigma\cos\theta}{(\sin^4\theta+\sigma^2\cos^2\theta)^{\frac{1}{2}}}\right) \tag{28}$$

where σ is a constant involving the ratio of the strengths of the gyrotropy and birefringence. One way to produce this effect would be to employ a liquid with an external magnetic field (giving rise to gyrotropy via the Faraday effect) and, in the same direction, an electric field (giving rise to birefringence via the Kerr effect).

As an example of the classical anholonomy embodied in $\Delta\theta$, consider a particle sliding freely with speed v round a noncircular planar hoop with area $\mathcal{A}$ and perimeter $\mathcal{L}$. Now let the hoop be slowly rotated in its own plane and in the same sense as the particle so that it makes a complete turn in time T. The distance it has travelled round the hoop is not vT but vT + Δs, where [26,27]

$$\Delta s = \frac{\mathcal{L}\Delta\theta}{2\pi} = -\mathcal{L} + \mathcal{L}(1 - \frac{4\pi\mathcal{A}}{\mathcal{L}^2}). \tag{29}$$

The first term is the expected slippage through one turn; the second term is the anholonomy effect, and is always positive by the isoperimetric inequality. This is related to the Sagnac effect [32] which is the basis of the navigationally important ring gyroscopes [33].

DEGENERACY

Let us study the phase 2-form $V(n;X)$ more closely, by enquiring where it can have singularities. For this purpose a form more transparent than (20) is the sum [25]

$$V(n;X) = \text{Im} \sum_{m \neq n} \frac{\langle n|dH|m\rangle \wedge \langle m|dH|n\rangle}{[E_m(X) - E_n(X)]^2} \tag{30}$$

over all states $|m\rangle$ other than that being transported. The denominators indicate that V has singularities at those parameter values X* where $|n\rangle$ is <u>degenerate</u> with the state immediately above or below it in energy, and large phase changes $\gamma_n(C)$ can be expected for circuits near X*.

In the neighbourhood of X*, only the state $|m\rangle = |n+1\rangle$ or $|n-1\rangle$ with which $|n\rangle$ degenerates contributes (in the simplest case of an eigenvalue with multiplicity two at X*) to the sum (30). Thus H(X) can be represented by a 2×2 matrix, for which a convenient normal form, involving three parameters $\underset{\sim}{X}=(X,Y,Z)$, is

$$H(\underset{\sim}{X}) = \tfrac{1}{2} \begin{pmatrix} Z & X-iY \\ X+iY & -Z \end{pmatrix}, \tag{31}$$

where the degeneracy X* has been chosen as origin. The two eigenvalues are

$$E_\pm(\underset{\sim}{X}) = \pm\tfrac{1}{2}(X^2+Y^2+Z^2)^{\tfrac{1}{2}}. \tag{32}$$

This shows that the degeneracy is <u>isolated</u> at X*, illustrating the old result [34] that for families of Hermitian matrices without symmetry degeneracies have codimension 3.

By comparing (31) and (26), the result (27) can be used to show that near $\underset{\sim}{X}$* the phase 2-form is

$$\underset{\sim}{V}(\pm;\underset{\sim}{X}) = \pm \underset{\sim}{X}/2(X^2+Y^2+Z^2)^{3/2} \tag{33}$$

so the singularity is that of a monopole with strength $\pm\frac{1}{2}$. This leads to the result that with the normal coordinates as in (31) the geometrical phase is

$$\gamma_{\pm}(C) = \mp \Omega(C)/2 \tag{34}$$

where $\Omega(C)$ is the solid angle subtended by C at $\underset{\sim}{X}^*$.

Now let us concentrate on the important special case where H is real symmetric instead of Hermitian; this includes for example all of wave theory and quantum mechanics in the absence of external influences which are chiral or not invariant under time-reversal. In (31) this corresponds to restriction to the parameter plane Y=0, and (32) now illustrates the known result [34,2] that for families of real symmetric matrices degeneracies are of codimension 2.

A circuit enclosing such a degeneracy subtends the solid angle of a hemisphere, namely 2π, so (34) predicts $\gamma=\pm\pi$, implying eigenvectors of a real symmetric matrix will change sign when transported around a degeneracy. This has long been known in a special case: the index is $\pm$ 1/2 for generic singularities of tensor fields such as disclinations in liquid crystals [35] (around which molecules reverse) and umbilic points on surfaces [36] (around which lines of curvature reverse). In quantum mechanics it arises in the theory of molecular spectra, in the approximaton where the coordinates of nuclei can be considered as parameters on which the states of electrons depend [37-39]. The sign change has been seen in computations [20] of the spectra of families of vibrating membranes whose boundaries are triangular (with the two independent angles taken as the parameters X).

Equation (32) shows that in E,X space the connection between surfaces $E=E_n(X)$ and $E=E_{n\pm 1}(X)$ at a degeneracy X* has the form of a double cone (in the real symmetric case) or hypercone (in the Hermitian case); this association between diabolos and degeneracies suggests the term diabolical point [20] to describe the connection. Diabolical points are degeneracies occurring in generic cases, that is without symmetry, and are therefore what physicists have called 'accidental'. But, as with traffic accidents, an event that is accidental for an individual (i.e. a degeneracy between levels of a given Hamiltonian) becomes almost inevitable when a population is considered (i.e. two- or three-parameter families of Hamiltonians). Variation of a single parameter will not, in these generic cases, produce degeneracies, because a line in parameter space will miss the diabolical points; of course, close encounters with diabolical points may occur, causing the curves of E_n versus parameter to show hyperbolic avoided crossings. The existence of these avoided crossings in one-parameter families of spectra has now been established by calculations for a variety of quantal systems [41-44,20], whose classical counterparts have varying degrees of chaos but share the feature of nonintegrability.

This leads to the conjecture that those families of quantal systems with the generic property that their spectra display diabolical points will have classical limits whose motion is nonintegrable. Such behaviour differs strikingly from that of separable systems, or integrable systems whose levels are approximated by the torus quantization formula (11): in these cases, when N>1, degeneracies occur when a single parameter is varied (because, from (11), levels correspond to intersection of constant-energy surfaces with points n in the N-dimensional lattice of quantum numbers, and variation of a parameter will cause the surface through a given n to turn and repeatedly intersect additional points).

In the semiclassical limit, energy levels are very close together and it becomes meaningful to speak of the statistics of the levels near a given energy E and set of parameters X, with X regarded as parameterising an ensemble of spectra. A useful statistic is the probability distribution P(S) of the spacings between neighbouring levels E_{n+1} and E_n; it is convenient to measure S in terms of the average level spacing, which is of order $\hbar^N$ [24]. A simple geometrical argument [11,40,41] shows that the different codimensions K=3,2,1 of degeneracies of families of generic Hermitian, generic real symmetric, and torus-quantized Hamiltonians imply different behaviour of P(S) in the limit $S\to 0$ of small spacings (which is determined by behaviour close to degeneracies):

$$P(S) \propto S^{K-1} \text{ as } S\to 0. \tag{35}$$

For the nonintegrable cases (K=3 or K=2), P(S) is therefore predicted to vanish as $S\to 0$. This phenomenon, called level repulsion, has been observed in many calculations [20,41,44] for K=2 but to my knowledge no such calculatons have been performed for the Hermitian case K=3. For separable systems, or integrable systems with torus quantization, K=1 and P(S) is predicted to have a finite limit as $S\to 1$, indicating level clustering; a detailed study [45] of (11) shows that in fact $P(S)=\exp(-S)$ for generic integrable Hamiltonians H(I), and computational tests abundantly confirm this result. Therefor classically integrable and nonintegrable systems can show striking differences in the fine structures of their quantal spectra.

QUANTUM AND CLASSICAL ADIABATIC DISCORDANCE [46]

Sometimes, the semiclassical and adiabatic limits can come into conflict in the sense that they lead to opposite predictions for the evolution of a system, rather than being complementary as in the examples given so far. Such discordance can arise when there are regions of phase space that are classically separated but which may be quantally connected by tunnelling.

The simplest case in which this occurs is the one-dimensional Hamiltonian

$$H(q,p;X) = \frac{p^2}{2m} + V(q;X), \tag{36}$$

describing a particle of mass m moving along the q axis in a potential V consisting of two potential wells separated by a barrier. As the single parameter X increases, the left-hand well (L) gets shallower and the right-hand well (R) gets deeper.

Attention will be restricted to energies below the barrier top. Then classical orbits must always be confined to L or R, which correspont to disconnected sets of loops (tori) in the phase plane q,p. For fixed X, the simplest semiclassical theory, based on (11), predicts that the quantum energy eigenstates are similarly localized in L or R. And indeed for most values of X the exact eigenstates are indeed localized (with accuracy $\exp\{-1/\hbar\}$) in L or R. But close to 'semiclassical degeneracies', i.e. when X is such that an approximate eigenstate in L has energy differing from that of an approximate eigenstate in R by an amount of order $\exp\{-1/\hbar\}$ (small in comparison with the mean spacing which is of order $\hbar$) then simple semiclassical theory breaks down [47] and the eigenstates are shared by both wells (the extreme case is the symmetric double-well (L and R identical), for which the true states are, to a close approximation, even and odd superpositions of the simple semiclassical single-well states). As X varies, the energy level curves $E_n(X)$ do not cross as (11) would predict, but are smooth curves with exponentially small splittings instead of degeneracies; as X varies across a semiclassical degeneracy, the state corresponding to $E_n(X)$ transfers from one well to the other. All this is well known.

The adiabatic-semiclassical discordance arises as follows. Suppose H is made time-dependent by slowly increasing X across a semiclassical degeneracy at X*, starting at t_i and ending at t_f, and semiclassical conditions prevail so that the wells are deep (i.e. supporting many states) and the barrier high (i.e. with exponentially small penetration). If the initial quantum state (at t_i) is chosen to be an eigenstate localized in L, what is the state at t_f? Because $dX(t)/dt$ and $\hbar$ are small, it is natural to attempt predictions using adiabatic and semiclassical theory. Because classical particles remain in L throughout the change, the semiclassical theory leads to the expectation that at t_f the system will again be in an L-localized eigenstate. But because the exact quantum state which is initially localized in L transfers smoothly (albeit rapidly) to R as X passes X*, the quantum adiabatic theorem leads to the expectation that the system will cling to this eigenstate and so be localized in R at t_f.

To resolve this paradox it is necessary to make a detailed study [46] of the evolution as determined by the time-dependent Schrödinger equation, which governs the probability amplitudes $a_L(t), a_R(t)$ for finding the system in L,R, the initial condition being

$$a_L(t_i) = 1; \quad a_R(t_i) = 0 \tag{37}$$

(of course $|a_L(t)|^2 + |a_R(t)|^2 = 1$ for all t). The aim is to find $|a_L(t_f)|^2$, which is the probability of finding the system in L after passage through the semiclassical degeneracy.

The result is a <u>uniform asymptotic formula</u> for $|a_L(t_f|^2$, involving a single quantity Q, incorporating both the semiclassical and adiabatic limits. Semiclassical smallness is governed by the energy splitting Δ of the quantum states at the semiclassical degeneracy, given by standard WKB theory [47] as

$$\Delta = \frac{\hbar}{\pi}(\omega_L\omega_R)^{\frac{1}{2}} \exp\{-\frac{1}{\hbar}\int_{q_L^+}^{q_R^-} dq\, [2m(V(q)-E^*)]^{\frac{1}{2}}\} \tag{38}$$

where ω_L and ω_R are the frequencies of classical motion in the wells, E* is the energy of the semiclassical degeneracy, and q_L^+ and q_R^- are the inner limits (turning points) of classical motion in the wells. Adiabatic smallness is governed by the rate at which the difference between the wells changes, i.e. by

$$dX(t)/dt, \text{ where } X(t)=E_L(t)-E_R(t), \tag{39}$$

E_L and E_R being the semiclassical well energies given by (11). The uniform asymptotic formula is given [46] by

$$\left.\begin{aligned} &|a_L(t_f)|^2 = \exp\{-Q\} \\ &\text{where } Q = \frac{\pi\Delta^2}{2\hbar|dX(t^*)/dt|} = \frac{\hbar\omega_L\omega_R \exp\{-\frac{2}{\hbar}\int_{q_L^+}^{q_R^-} dq[2m(V(q)-E^*)]^{\frac{1}{2}}\}}{2\pi|\frac{d}{dt}(E_L(t^*)-E_R(t^*))|} \end{aligned}\right\} \quad (40)$$

The two discordant limits are easily recovered from this 'exponential of an exponential' : in the semiclassical limit, $\hbar\to 0$, so that $Q\to 0$ and $|a_L(t_f)|^2\to 1$, as predicted; in the adiabatic limit, $d(E_L-E_R)/dt\to 0$ so that $Q\to\infty$ and $|a_L(t_f)|^2 \to 0$ and the system switches to R, as predicted. There is no inconsistency in these limits being different, but it is important to notice that the quantum adiabatic limit is achieved not when the parameter speed dX/dt is small in comparison with the frequency of classical motion in each well, but when it becomes negligible in comparison with the much smaller frequency of tunnelling between the wells. Otherwise, the system jumps the energy gap and remains, classically, in L.

The simplicity of (40) for the final probability conceals considerable complexity in the evolution for times between t_i and t_f, involving parabolic cylinder functions of complex order and complex argument [48].

One feature of this simple system that is particularly worthy of emphasis, because it is likely to occur more widely, is the qualitatively different asymptotic behaviour associated with different degrees of adiabatic slowness. It might, for example, account for the difficulty in finding classical adiabatic invariants for quasi-integrable systems [52], where tori repeatedly dissolve into chaos as changes in parameters X take them through resonant zones: if dX/dt is not too slow, the orbit might hardly notice the chaos, and emerge from resonance onto a torus with the same actions as before; if dX/dt is very slow (and N>2), the orbit might 'forget' its adiabatic invariants whilst exploring the chaos, and emerge onto a torus with quite different actions (for systems whose closeness to integrability is measured by a small parameter ε, the time for such forgetting is exponentially long in ε[51]). In the analogous quantal problem, the system might, if dX/dt is not too small, jump the avoided level crossings [40-43] associated with classical resonances and so continue to be associated with semiclassical torus states as given by (11). These considerations might explain the accuracy of a calculation [49] of approximate quantum levels of non-integrable systems, based on a classical calculation of orbits of a system whose Hamiltonian is slowly changed from integrable to nonintegrable, together with the assumption that the association (11) holds, even though in these calculations no attention was paid to classical resonances or quantal avoided crossings.

REFERENCES:

[1] Landau, L.D. and Lifshitz,E.M. Mechanics (Vol.1of Course of Theoretical Physics) (3rd Ed.Pergamon:Oxford 1976)

[2] Arnol'd,V.I. Mathematical Methods of Classical Mechanics (Springer: New York 1978)

[3] Dirac,P.A.M.,The Principles of Quantum Mechanics (Oxford: Clarendon Press, 1947)

[4] Messiah,A.,Quantum Mechanics (North-Holland: Amsterdam 1962)

[5] Peierls, R.E.,Surprises in Theoretical Physics (New Jersey: Princeton University Press 1979)

[6] Hwang,J-T and Pechukas,P.,J.Chem.Phys. 67 (1977) 4640-4653

[7] Born,M.,The Mechanics of the Atom (Frederick Ungar:New York 1960)

[8] Berry,M.V., Regular and Irregular Motion,in:Topics in Nonlinear Mechanics (S.Jorna,ed.) Am.Inst.Phys.Conf.Proc. 46 (1978) 16-120.

[9] Lichtenberg,A.J. and Liberman,M.A., Regular and Stochastic Motion (Springer: New York 1983)

[10] Maslov,V.P. and Fedoriuk,M.V., Semiclassical Approximation in Quantum Mechanics (D.Reidel: Dordrecht 1981)

[11] Berry,M.V., Semiclassical Mechanics of Regular and Irregular Motion,in:Chaotic Behavior of Deterministic Systems (Les Houches Lectures XXXVI, eds.G.Iooss,R.H.G.Helleman and R. Stora (North-Holland: Amsterdam 1983) pp171-271

[12] Noid,D.W. and Marcus,R.A., J.Chem.Phys. 67 (1977)559-567

[13] Eastes,W and Marcus,R.A., J.Chem.Phys. 61 (1974) 4301-4306

[14] Noid,D.W. and Marcus,R.A., J.Chem.Phys. 62 (1975) 2119-2124

[15] Noid,D.W., Koszykowski,M.L., and Marcus,R.A., J.Chem.Phys. 71 (1979) 2864-2873

[16] Percival, I.C. and Pomphrey,N., Mol.Phys. 31 (1976) 97-114

[17] Jaffe,C. and Reinhardt,W.P., J.Chem.Phys. 71 (1979) 1862-1869

[18] Chapman,S., Garrett,B. and Miller,W.H., J.Chem.Phys. 64 (1976) 502-509

[19] Kahan,Th., Physique Theoreique,Vol.II (Paris: Presses Universitaires de France (1960))

[20] Berry,M.V. and Wilkinson,M., Proc.Roy.Soc.Lond. A392 (1984) 15-43

[21] Bohigas,O., Giannoni,M.J. and Schmit,C., Phys.Rev.Lett. 52 (1984) 1-4.

[22] Pechukas,P., Phys.Rev.Lett. 51 (1983) 943-946

[23] Balian,R. and Bloch,C., Ann.Phys. (N.Y) 69 (1972) 76-160

[24] Berry,M.V., Structures in Semiclassical Spectra: a Question of Scale,in:The Wave-Particle Dualism (ed.S.Diner, D.Fargue,

G.Lochak and F.Selleri) (D.Reidel: Dordrecht 1984) 231-252.

[25] Berry,M.V., Proc.Roy.Soc.Lond. A392 (1984) 45-57.

[26] Hannay, J.H., (1984) Submitted to J.Phys.A.

[27] Berry, M.V., (1984) Submitted to J.Phys.A.

[28] Simon, B., Phys.Rev.Lett. 51 (1983) 2167-2170.

[29] Misner, C.W., Thorne, K.S. and Wheeler,J.A., Gravitation (Freeman: San Francisco, 1973).

[30] Landau, L.D. and Lifshitz,E.M., Electrodynamics of continuous media (Vol.8 of course of Theoretical Physics) (Pergamon: Oxford, 1960).

[31] M.V.Berry. In preparation.

[32] Post,E.J., Revs.Mod.Phys. 39 (1967) 475-493.

[33] Forder,P.W., J.Phys.A. 17 (1984) 1343-1355.

[34] Von Neumann, J. and Wigner,E., Phys.Z. 30 (1929) 467-470.

[35] Frank,F.C., Disc.Far.Soc. 25 (1958) 19-28.

[36] Berry, M.V. and Hannay, J.H., J.Phys.A. 10 (1977) 1809-1821.

[37] Herzberg,G. and Longuet-Higgins,H.C., Disc.Far.Soc., 35 (1963) 77-82.

[38] Longuet-Higgins,H.C., Proc.Roy.Soc.Lond. A344 (1975) 147-156.

[39] Mead,C.A., Chem.Phys.(Netherlands) 49 (1980) 23-32, 33-38.

[40] Berry,M.V., Aspects of Degeneracy, in: Proc.Como Conference on Quantum Chaos (G.Casati,ed) (Plenum:London,1984). In press.

[41] Berry,M.V., Ann.Phys.(N.Y) 131 (1981) 163-216.

[42] Noid,D.W., Koszykowski,M.L., Tabor,M. and Marcus,R.A., J.Chem.Phys. 72 (1980) 6167-6175.

[43] Marcus,R.A., in:Nonlinear Dynamics (ed.R.H.G.Helleman) Ann N.Y.Acad.Sci. 357 (1980) 169-182.

[44] Richens,P.J. and Berry,M.V., Physica 1D (1981) 495-512.

[45] Berry,M.V. and Tabor,M., Proc.Roy.Soc.Lond. A356 (1977) 375-394.

[46] Berry,M.V., J.Phys.A. 17 (1984) 1225-1233.

[47] Fröman,N., Fröman,P.O., Myhrman,U. and Paulsson,R., Ann.Phys.(N.Y) 74 (1972) 314-323.

[48] Zener,C., Proc.Roy.Soc.Lond. A137 (1932) 696-702.

[49] Solovev,E.A., Sov.Phys.JETP 48 (1978) 635-639.

[51] Hannay,J.H. and Ozorio de AlmeidaA.M., J.Phys.A. (1984). In Press.

[52] Nekhoroshev,N.N., Russ.Math.Surv 32 (1977) 1-65.

[53] Arnold,V.I., Geometrical Methods in the Theory of Ordinary Differential Equations (Springer: New York, 1983).

Theoretical and Applied Mechanics
F.I. Niordson and N. Olhoff
Elsevier Science Publishers B.V. (North-Holland)

ASPECTS OF NONLINEAR ACOUSTICS

Leif Bjørnø

Technical University of Denmark
Building 352, DK-2800 Lyngby
Denmark

The historical background and bench marks of research in nonlinear acoustics are briefly discussed. A presentation is given of the theoretical basis of the propagation of plane, cylindrical and spherical finite-amplitude waves through lossless, thermo-viscous and relaxing fluids involving a generalized Burgers' equation, the theory of 'weak shocks' and a KdV-type equation. Characteristic features of the distortion course observed by finite-amplitude waves in various regions of propagation are emphasized in the time- and in the frequency domain. Recent changes from fundamental towards more applied research are reflected in detailed discussions of the applicability of nonlinear acoustics research results in fields like *underwater acoustics* and *biomechanics*.

INTRODUCTION

The non-linearity of Nature is a fact that has frequently been recognized by physicists, geologists, and engineers when considering motion in fluids and in solids. Nevertheless, a linearization of the governing equations has in a number of cases led to mathematically more simple expressions with solutions showing surprisingly good agreement with experimental results. Rather early it was recognized that the validity of the solutions was limited, especially for cases of strong nonlinearity of the material, represented for instance by its equation of state, and for cases of high amplitude of the disturbances propagating in the material.

The propagation of acoustic waves of finite, but moderate, amplitude in media of various degrees of nonlinearity will be discussed in this paper, thus excluding the consideration of strong shocks. The finite-amplitude wave propagation is comprised by the special field of acoustics termed '*Nonlinear Acoustics*', which includes waves with amplitudes ranging from infinitesimal - covered by the linear theory - to a magnitude leading to the formation of weak shocks.

Finite-amplitude wave phenomena influence a great number of acoustic topics of both theoretical and practical character. Steady and unsteady finite-amplitude waves in supersonic aerodynamics, weak-shock theory, finite-amplitude wave propagation in gases, liquids and solids, finite-amplitude bubble pulsations and cavitation in liquids, etc. Moreover, finite-amplitude wave propagation influences are found by high-power ultrasonic cleaning, filtering and welding processes, by high-intensity sound agglomeration of aerosols, trapping of fog and drying of powder. Furthermore, finite-amplitude waves are used for mixing of difficulty miscible materials, production of emulsions, dispersion of solids in liquids, coagulation processes, degassing of liquids and melts, etc.

Noise generation and propagation by high-powered jet engines and noise propagation in motor silencers are influenced by finite-amplitude wave phenomena leading to a change in the spectral composition of the noise.

Only wave propagation in unlimited fluids will be considered in this paper. Therefore, discussion of influence of free surfaces, solid walls, etc. on wave propagation in fluids will be omitted in what follows. This restriction further implies that reflection of finite-amplitude waves, standing waves and resonances, acoustical boundary layers, and finite-amplitude waves in tubes and horns, in general, will not be discussed, though fundamental solutions and conclusions derived from studies of these topics will be discussed to the extent that they may serve the purpose of explaining concepts of nonlinear acoustics used in this paper. The same limitations will apply to nonlinear impedance materials and nonlinear acoustic resonators.

The convective acceleration terms in the equations of motion and the nonlinear pressure-density relationship found by fluids and solids describe and contribute to the finite-amplitude wave distortion course where their cumulative, rather than merely their local, effect, eventually may produce steep wavefronts. These nonlinear processes are counteracted by linear processes like attenuation, dispersion, molecular relaxation and diffraction. It is this conflict between cumulative nonlinear wave distortion and the competing linear processes leading to a pronounced change in waveform which forms the key to the contents of this paper.

Like the development within about all scientific fields, the development within finite-amplitude wave propagation is based on works done by earlier generations. Although the greatest development in this field has taken place within the last quarter of a century - as evidenced by the still-increasing number of papers published on problems relating to the propagation of finite-amplitude waves through nonlinear media - the basic research and early developments in nonlinear acoustics can be traced back some hundred years.

HISTORICAL DEVELOPMENT

A brief survey of the history of finite-amplitude wave propagation might appropriately begin with Euler's formulation of the concervation equations for a fluid: The equation of continuity (1) and the equations of motion (2), respectively:

$$\rho_t + (\rho u)_x = 0 \; , \tag{1}$$

$$\rho u_t + uu_x = -p_x \; , \tag{2}$$

where ρ and p are the fluid density and the pressure, respectively. u is the particle velocity and t and x denote time and cartesian spatial coordinate, respectively. Eq. (2), where external body force influences are omitted, is known as the *Euler equation* for one-dimensional, frictionless flow. The character of Eq. (2) as a nonlinear partial differential equation appears from its convective term. Euler's interest in lossless flow led to his study of the propagation of both infinitesimal-amplitude and finite-amplitude waves in fluids [1].

In 1808 Poisson [2] worked on a one-dimensional, nonlinear wave equation in Eulerian coordinates of the form:

$$h^2\eta_{xx} - \eta_{tt} = 2\eta_x\eta_{xt} + (\eta_x)^2\eta_{xx} \; , \tag{3}$$

where η denotes a velocity potential and where h is the 'isothermal' (Newtonian) sound velocity. Poisson found an exact solution to this equation expressed by:

$$\eta_x = F\{x-(h + \eta_x)t\} \; , \tag{4}$$

for finite-amplitude wave propagation in the direction of increasing x in a gas governed by Boyle-Mariotte's equation of state. In order to be able to obtain a total solution to Eq. (3), Poisson coupled the solution (4) to an auxiliary equation:

$$\eta_t + h\eta_x + (\eta_x)^2/2 = 0 \; , \tag{5}$$

which is a reduced version of the wave equation for finite-amplitude waves propagating in the positive x-direction in a lossless fluid.

During the years around 1848, while revolution spread across Europe, there was an increase in nonlinear acoustics research. It is hard to say whether there was any connection between these incidents, but they were certainly both characterized by a breakthrough of new ideas. After a discussion between Airy and Challis [3] about the existence of plane sound waves, Stokes [4] published the first sketches showing how waveform distortion might occur due to the different phase velocity of different parts of a finite-amplitude wave. Stokes even treated the waveform in the limit of distortion - the shock wave - but because he considered a lossless fluid he omitted the thermodynamic consequence of shock formation. Little was known about thermodynamics in 1848, and it was to be another 20 to 30 years before the significance of energy loss for shock propagation was recognized and generally accepted.

The year 1860 could be considered as the end of the era of wave propagation in lossless fluids. The last essential contributions to the theory of lossless wave propagation, completing the period of development started by Euler, were made by Earnshaw [5] and Riemann [6]. A connection was established between the thermodynamics represented by the thermodynamical expression λ in Eq. (6) and the original mathematical basis for one-dimensional, finite-amplitude wave propagation represented by the equation of continuity (7), the equation of motion (8) and an equation of state (9):

$$\lambda = \int (c/\rho)d\rho \ , \tag{6}$$

$$\rho_t + u\rho_x + \rho u_x = 0 \ , \tag{7}$$

$$u_t + uu_x + \rho^{-1}p_x = 0 \ , \tag{8}$$

$$p = p(\rho) \ , \tag{9}$$

λ is the local thermodynamic state of the fluid, and c is the local 'isentropic' velocity of sound. Insertion of Eq. (6) into Eqs. (7) and (8) yields:

$$\lambda_t + u\lambda_x + cu_x = 0 \ , \tag{10}$$

$$u_t + uu_x + c\lambda_x = 0 \ , \tag{11}$$

which shows that for simple waves the phase velocity of a point of a wave having particle velocity u is given by $(u \pm c)$, + for waves propagating in the positive x-direction and - for waves propagating in the negative x-direction. It was, moreover, shown that the local velocity of sound might be written as:

$$c = c_o \pm \tfrac{1}{2}(\gamma-1)u \ , \tag{12}$$

where c_o is the velocity of sound for infinitesimal-amplitude waves, while the correction term $\frac{1}{2}(\gamma-1)u$ arises from the nonlinearity of the pressure-density relation of the gas. γ is the ratio of specific heats of the gas.

For wave propagation in the positive x-direction the phase velocity of a point of the wave having particle velocity u might therefore be written as:

$$(x_t)_{u=const} = u + c = c_o + \tfrac{1}{2}(\gamma+1)u \ , \tag{13}$$

which involves the two fundamental contributions to finite-amplitude wave distortion, the thermodynamic contribution $\frac{1}{2}(\gamma-1)u$ and the convective contribution arising from the fact that the local velocity of sound c is being convected along with the local particle velocity u [7].

Contributions to a better understanding of the nature of a shock were published by Rankine [8] and some years later by Hugoniot [9] who showed that an adiabatic, reversible transition in a shock would violate the principle of conservation of energy, and further that, in the absence of viscosity and heat conduction in the fluid outside the shock, the conservation of energy implied conservation of entropy across the shock. The conservation equations connecting the thermodynamic and kinematic quantities on the two sides of a shock, the so-called *Rankine-Hugoniot equations*, thus found their final form.

A deeper appreciation of the processes occurring in a shock was obtained through the work of Lord Rayleigh [10] and G.I. Taylor [11] published in 1910. Viscosity and heat conductivity had now been included in the equations governing finite-amplitude wave propagation. The problem was now to solve the nonlinear wave equations subjected to specific initial or boundary conditions considering the influence of the loss sources.

During the nineteen thirties development in nonlinear acoustics research was brought a great step forward by Fay's [12] publication of a solution for a periodic finite-amplitude wave in a viscous fluid and by Fubini's [13] explicit solution to the wave equation for finite-amplitude wave propagation in a lossless fluid.

Fay published a solution to the wave equation valid for the propagation of plane, finite-amplitude waves in a viscous fluid in the region showing a comparatively stable wave form:

$$(p-p_o)/(\rho_o c_o^2) = 2\alpha c_o/\beta\omega \sum_{n=1}^{\infty} \{\sin n(\omega t-kx)\}/\{\sinh n\alpha(x+x_o)\} , \tag{14}$$

where: $\alpha = (2/3)\omega^2\eta/\rho_o c_o^3$ and $\beta = \frac{1}{2}(\gamma+1)$,

η and ω are the shear viscosity and the angular frequency, respectively. n is the harmonics number. x_o is a constant related to the discontinuity distance, which is the distance from the wave source for the formation of a discontinuity in a plane sine wave of angular frequency ω propagating in a dissipationless medium.

While Fay's solution does not satisfy the boundary conditions of a sinusoidal source, the explicit solution to the wave equation obtained by Fubini some years later does. Fubini's solution shows how an original sinusoidal wave of finite-amplitude will distort by formation of higher harmonics during propagation:

$$u = 2u_o \sum_{n=1}^{\infty} (n\sigma)^{-1} J_n(n\sigma)\sin(n(\omega t-kx)) , \tag{15}$$

where: $\sigma = \frac{1}{2}(\gamma+1)u_o kx/c_o$.

The discontinuity is attained for a value of the dimensionless distance parameter σ=1. While Fubini's solution describes the finite-amplitude wave distortion close to the source and before shock formation, Fay's solution describes the waveform towards which the finite-amplitude wave tends if the original amplitude is high enough for shock formation.

The development of theoretical solutions for finite-amplitude wave propagation has since the Second World War mainly followed two tracks: (1) solutions based upon *Burger's equation*, and (2) solutions based upon the *theory of weak shocks*, which permit the solution of finite-amplitude wave propagation problems, when the waveform contains shocks of not too high amplitude. Both main approaches will be discussed in detail in the following.

FINITE-AMPLITUDE WAVE PROPAGATION OF TODAY

The fundamental equations governing the finite-amplitude wave propagation are usually: (a) the *equation of continuity* expressing the conservation of mass, (b) the three components of the *equation of motion* expressing the conservation of momentum, (c) an *energy equation* expressing the conservation of energy, and (d) an *equation of state*, a constitutive relation characterizing the fluid and its response to thermal or mechanical stress [7]. The equation of state may be written as:

$$p = p(\rho,s) , \tag{16}$$

where s is the specific entropy. Since sound propagation is a substantially isentropic process (16) can be written [7] in a Taylor series development as:

$$p-p_o = A\{(\rho-\rho_o)/\rho_o\} + B\{(\rho-\rho_o)/\rho_o\}^2/2 + C\{(\rho-\rho_o)/\rho_o\}^3/6 + \ldots\ldots , \tag{17}$$

$$\text{where: } A = \rho_o(\partial p/\partial\rho)_s = \rho_o c_o^2 , \; B = \rho_o^2(\partial^2 p/\partial\rho^2)_s , \; C = \rho_o^3(\partial^3 p/\partial\rho^3)_s . \tag{18}$$

The dimensionless ratios B/A and C/A are essential parameters for finite-amplitude wave propagation. These parameters may be written [14] as:

$$B/A = 2\rho_o c_o(\partial c/\partial p)_T + 2c_o T\beta(\partial c/\partial T)_p/c_p , \tag{19}$$

and

$$C/A = 3(B/A)^2/2 + 2\rho_o^2 c_o^3(\partial^2 c/\partial p^2)_s . \tag{20}$$

The third-order term in (17) is frequently some orders of magnitude smaller than the second-order term.

The perturbation analysis, first adapted to acoustics by Eckart [15], has proved to be an efficient tool in the evaluation of the significance of the terms of the fundamental equations. The modified perturbation analysis, leading to first-order, linearized, approximation and to higher-order approximations, may be found in [7].

Some of the fundamental features of finite-amplitude wave propagation may be illustrated by consideration of the adiabatic, one-dimensional wave propagation in a long straight tube containing a non-viscous fluid. The exact wave equation in Lagrangian coordinates thus writes:

$$\xi_{tt} = c^2(1+\xi_a)^{-(\gamma+1)}\xi_{aa} \; . \tag{21}$$

a is the space - or identification - coordinate for a particle, while the dependent variable ξ constitutes a convenient quantity for the description of displacement, velocity and acceleration of the fluid particle considered. For a simple harmonic piston motion at the tube end expressed by $f(t) = A\cos\omega t$, a solution to (21) may be written as:

$$\xi(a,t) = A\cos\omega(t-a/c_o)+(\gamma+1)\omega^2A^2a\{1-\cos 2\omega(t-a/c_o)\}/8c_o^2 \; . \tag{22}$$

Eq. (22) shows that the displacement of any particle in the wave motion is no longer simple-harmonic, but is made up of a term *independent of t* and of two simple harmonic terms, one with the piston frequency and one with *twice this frequency*. The occurrence of the factor a in the term of twice the piston frequency implies that the energy being supplied at the piston frequency will be transferred to the second and higher harmonic components, leading to a progressively steeper front of a condensation wave during its propagation. This transfer of energy, from the fundamental frequency to its higher harmonics, is one of the basic features of nonlinear acoustics and it will under lossless conditions result in the formation of a shock wave at a distance from the piston given by:

$$\ell = 2c_o^2\{(\gamma+1)\omega u_o\}^{-1} = 2\{(\gamma+1)kM\}^{-1} \; , \tag{23}$$

where u_o is the peak value of the particle velocity and where $M = u_o/c_o$ is the *acoustic Mach number*; $k(=\omega/c_o)$ is the propagation constant in the wave propagation with fundamental frequency ω. The distance ℓ is often called the *discontinuity distance*.

If dissipative mechanisms are taken into account, the increasing absorption at increasing frequencies - $\alpha \propto \omega^2$ for most fluids - implies that the wavefront will attain a maximum steepness when the distance of propagation is such that the rate of energy transfer to higher harmonics, due to the nonlinearities, is just equalized by the increase of absorption at the higher harmonics.

BURGERS' EQUATION

Let us consider the propagation of a *plane, finite-amplitude wave in a dissipative fluid*. It is assumed that heat conduction (constant heat conductivity) and viscous dissipation (constant viscosities) are the only sources contributing to the entropy production. Further, it is assumed that the acoustic Mach number $M \ll 1$, which is the most interesting case due to the interaction between nonlinearities and dissipation. Then the equation of motion without body forces may be written as:

$$\rho(u_t + uu_x) = -p_x + (4\eta/3 + \zeta)u_{xx} \; , \tag{24}$$

and the energy equation may through the use of the 2nd law of thermodynamics be written as:

$$\rho T(s_t + us_x) = \varkappa T_{xx} + \Phi \; . \tag{25}$$

By insertion of (17) into (25) linearizing the diffusion terms due to viscosity and heat conductivity, and by insertion of the result into the equation of motion (24), this equation may be written in the following approximate form:

$$\rho(u_t + uu_x) = -p_x + bu_{xx} \tag{26}$$

with $b = 4\eta/3 + \zeta + \varkappa(c_v^{-1} - c_p^{-1})$, (27)

where ζ and $\varkappa$ denote bulk viscosity and heat conductivity coefficients, respectively. p is here expressed by Eqs. (17) and (18). Introduction of a thermodynamic function $\Lambda = \int dp/\rho$ together with a velocity potential ϕ permits the equation of continuity (1) to be written as:

$$\Lambda_t + \phi_x \Lambda_x + c^2 \phi_{xx} = 0 \; . \tag{28}$$

Analogously, Eqs. (26) and (17) may be transformed using Λ and ϕ, and their results, when inserted into Eq. (28), lead to:

$$\phi_{xx} - c_o^{-2} \phi_{tt} + (b/\rho c_o^2) \phi_{xxt} + c_o^{-1} (B/A + 2) \phi_x \phi_{xx} = 0 \tag{29}$$

which is an approximation correct to the squared term in M, inclusive.

Transforming the variable t to the retarded time $t' = t - x/c_o$ and returning to the particle velocity $u = \phi_x$, Eq. (29) reduces to:

$$u_x - \tfrac{1}{2}(B/A + 2) c_o^{-2} u u_{t'} = b(2\rho_o c_o^3)^{-1} u_{t't'} \; . \tag{30}$$

Eq. (30) is of the same type as the *Burgers' equation*, and by omission of the nonlinear term (30) reduces to an equation of the diffusion type. Eq. (30) describes with sufficient accuracy, in spite of its approximate character, both the nonlinear and the dissipative processes during the propagation of finite-amplitude plane waves in a viscous, heat conducting fluid. Owing to the reduction of the retarded time t' to t for x = 0, Eq. (30) is well adapted for use with boundary-value problems, and it may be termed the *boundary-value form* of the Burgers' equation.

Using a characteristic distance x_c in the sound field together with the peak value of the particle velocity u_o (source value), the following dimensionless ratios can be established:

$$V = u/u_o \; ; \; \sigma = \tfrac{1}{2}(B/A + 2) M x/x_c \quad \text{and} \quad y = c_o t'/x_c \; . \tag{31}$$

which, when inserted into Eq. (30), yields the dimensionless form of the Burgers' equation suitable for boundary value problems.

$$V_\sigma - V V_y = \Gamma^{-1} V_{yy} \tag{32}$$

where $\Gamma = (B/A + 2) u_o \rho_o x_c / b = (B/A + 2) Re_a$ (33)

where $Re_a = \rho_o u_o x_c / b$ constitutes an *acoustic Reynolds number* analogous to the hydrodynamic one.

The coefficient Γ is an essential parameter in nonlinear acoustics. It describes the ratio of the influence of nonlinearity (including 'equation of state' and 'convective' nonlinearity contributions), represented by the nonlinear strength term (1+B/2A)M, to the influence of dissipation.

Γ was first introduced by Gol'dberg [16] as a criterion such that shock formation is not likely to take place if $\Gamma < 1$. For an originally sinusoidal wave, it will be appropriate to choose the characteristic distance x_c of the wave propagation as: $x_c = c_o/\omega = 1/k$, where k is the wave number. This leads to the expression:

$$\Gamma = (B/A + 2) u_o \rho_o / bk = (B/A + 2) p_o / b\omega \tag{34}$$

where p_o is the source peak pressure.

Eq. (32) has formed the basis for a number of theoretical analyses of finite-amplitude wave propagations in dissipative fluids. A complete steady-state solution to Eq. (32) can for instance be found in [17] and [18] for plane waves giving:

$$V = 2\Gamma^{-1} \, \partial(\ln \Xi)/\partial y \tag{35}$$

where $\Xi = \Sigma \varepsilon_n (-1)^n \, I_n(\Gamma/2) \exp(-n^2\sigma/\Gamma) \cos ny$, with ε_n being the Neumann factor ($\varepsilon_o = 1$, $\varepsilon_n = 2$ for $n \geq 1$) and I_n denoting the Bessel function of imaginary argument: $I_n(z) = i^{-n} \, J_n(iz)$.

The reduction of the hydrodynamical equations and the equation of state to one single equation for the approximate description of the propagation of cylindrical and spherical waves of finite amplitudes, analogous to the plane wave case expressed in Eq. (30) may be written as:

$$u_r + nu/r - \tfrac{1}{2}(B/A + 2) c_o^{-2} u u_{t'} = b(2\rho_o c_o^3)^{-1} \, u_{t't'} \tag{36}$$

where n = 1, ½ and 0 for spherical, cylindrical and plane waves, respectively. For n = 1 and ½, Eq. (36) is valid only in the domain $kr \gg 1$, and $t' = t \mp (r-r_o)c_o$ for divergent and convergent waves, respectively. r_o denotes a characteristic radius.

Using dimensionless quantities, Eq. (36) can be reduced to the dimensionless boundary value form of Burgers' equation given by:

$$W_f - WW_y = \Gamma^{-1} W_{yy} \qquad (37)$$

where: $W = (\sigma/\sigma_o)^n V = (r/r_o)^n V = (r/r_o)^n (u/u_o)$

$$\left.\begin{aligned} f &= 2\sqrt{\sigma_o}\,(\sqrt{\sigma} - \sqrt{\sigma_o}) = 2\sigma_o(\sqrt{r/r_o} - 1) \\ \Gamma &= \Gamma_o(1 + f/2\sigma_o)^{-1} \end{aligned}\right\} \text{ cylindrical waves}$$

$$\left.\begin{aligned} f &= \sigma_o \ln(\sigma/\sigma_o) = \sigma_o \ln(r/r_o) \\ \Gamma &= \Gamma_o \exp(-f/\sigma_o) \end{aligned}\right\} \text{ spherical waves}$$

$$y = \omega_o t' = \omega_o[t-(r-r_o)/c_o]$$

$$\sigma_o = \tfrac{1}{2}(B/A + 2)u_o k_o r_o/c_o$$

$$\Gamma_o = (B/A + 2)\rho_o u_o c_o/\omega b \ .$$

The equations (32) and (37) differ with respect to the spatial coordinate dependence of Γ in Eq. (37). This is a very interesting feature and owing to it the propagation of a finite-amplitude cylindrically or spherically divergent wave in a thermoviscous fluid is mathematically equivalent to the propagation of a plane finite-amplitude wave in a fluid which has a thermoviscous loss coefficient that grows linearly or exponentially with range, respectively.

Calculations based on the above equations and supported by experimental findings show that the propagation path of an original sinusoidal finite-amplitude wave in a dissipative fluid roughly may be divided into three regions. In the first region - in what follows termed *region I* - the dissipative effects are small compared to the prevailing nonlinear effects. Thus the nonlinear effects by accumulation during the wave propagation distort the sinusoidal wave and lead to an increasing steepness of the wave fronts, eventually resulting in the formation of a 'sawtooth' wave. But before the wave shape becomes a sawtooth the first discontinuity of the wave front (or maximum steepness) has formed at its zero-crossings at a source distance termed the *discontinuity length*. This distance must, owing to the dissipative effects in region I, be greater for wave propagation in a thermoviscous fluid than for wave propagation in a lossless fluid, Eq. (23), if the same initial wave amplitude is considered. The discontinuity length may appropriately form the boundary between region I and the subsequent *region II*, where the effects of nonlinear and dissipative processes cancel one another, leading to a stabilization of the wave shape. In region II a relatively stable sawtooth shape of the original sinusoidal wave may be obtained.

Owing to the relatively strong dissipation of energy in a sawtooth wave profile, the finite-amplitude wave gradually loses its wavefront steepness and the shock thickness of the weak periodic shock wave (the sawtooth wave) increases. Finally, the wave profile returns to its original sinusoidal shape at a distance from the source forming the beginning of region III, the so-called 'old-age-region' in which the amplitude of the wave will become of second order of smallness and in which further amplitude reduction is governed by small-amplitude (linear) absorption rules.

At a sufficiently high initial pressure, the later amplitude of the distorted originally sinusoidal wave becomes relatively independent of the initial amplitude, an effect termed *acoustical saturation*.

The description of finite-amplitude wave distortion in a thermoviscous fluid has been done using Burgers' equation (32) and its solution (35) for plane waves. For $\Gamma \gg 1$ Eq. (35) can be reduced to:

$$V = 2\Gamma^{-1} \sum_{n=o}^{\infty} \sin ny/\{\sinh n(1+\sigma)/\Gamma\} \tag{38}$$

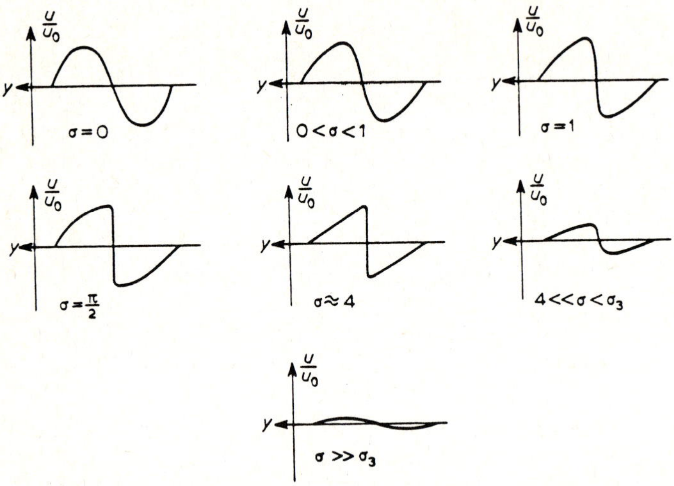

Figure 1. Distortion through regions I to III of one period of an originally plane, sinusoidal, finite-amplitude wave with Γ >> 1.

which is a form of Fay's frequency domain solution Eq. (14) for a nearly stable periodic waveform in a viscous perfect gas (a region II solution). A region I solution to Eq. (32) for $\Gamma >> 1$ has been given in [19] for an initially sinusoidal waveform expressed by $u = u_o \sin \omega t$ at $x = 0$. This solution for the distortion of the travelling plane wave has the form:

$$y = \arc\sin(u/u_o) - \sigma(u/u_o) \tag{39}$$

where $\sigma = \frac{1}{2}(\gamma+1)u_o\omega x/c_o^2$ and $y = \omega(t-x/c_o)$.

Eq. (39) may be used for a graphical construction of the waveform in region I since it consists of the sum of two functions, arc $\sin(u/u_o)$ and a straight line making an angle $\phi = \arctan \sigma$ with the u/u_o-axis.

An exact solution to Eq. (32) describing the structure of one period of a *sawtooth wave*, i.e. a periodic shock wave, in region II can be written as:

$$V = u/u_o = (1+\sigma)^{-1}\{-y + \pi \tanh y/\Delta\} \quad \text{for} \quad -\pi \leq y \leq \pi \tag{40}$$

where the dimensionles quantity Δ is a measure of the shock thickness given by:

$$\Delta = 2\pi^{-1}(1+\sigma)/\Gamma \, . \tag{41}$$

From (41) it appears that the shock thickness will decrease for increasing Γ, leading to $\Delta \to 0$ for $\Gamma \to \infty$. In this limit the solution becomes discontinuous at $y = 0$ and describes a real sawtooth expressed by:

$$V = (1 + \sigma)^{-1}\{-y \pm \pi\} \tag{42}$$

where the plus sign corresponds to $0 < y < \pi$ and the minus sign to $-\pi < y < 0$. If the particle velocity amplitude is designated u_p for $y = 0$, Eq. (42) gives:

$$u_p/u_o = |\pi/(1+\sigma)| \tag{43}$$

which determines the reduction of the peak amplitude of an originally sinusoidal wave, when the wave during its propagation has attained a sawtooth shape.

In order to bridge between the solutions discussed above for regions I and II, establishing an inner relation between the solutions, Blackstock [20] suggested a connection between the Fubini and the Fay solutions for plane sound waves of finite amplitudes. By use of the theory for weak non-uniform shock waves he developed a general solution that contains the Fubini and the sawtooth solutions as limiting cases, and which also covers the transition region between the two solutions. The solution was given as a Fourier series representation of the waves in the two regions:

$$B_n = (2/\pi n)V_p + (2/\pi\, n\, \sigma)\int \cos n\,(\Phi - \sigma \sin \Phi)d\Phi \tag{44}$$

where B_n is the Fourier coefficient for the nth harmonic, $V_p = u_p/u_o$ is the shock amplitude given by Eq. (43), and Φ is a function of $y = \omega t \mp kx$.

In the range $0 \leq \sigma < 1$, $V_p = 0$ and Eq. (44) reduces to the Fourier coefficient $B_n = (2/n\sigma)J_n(n\sigma)$ of the Fubini solution (15), while for $\sigma > 1$, V_p is given by Eq. (43) and Eq. (44) reduces to the Fourier coefficient of Fay's solution (14) in the lossless limit ($\Gamma \to \infty$) of this equation given by: $B_n = 2/\{n(1+\sigma)\}$.

THE WEAK-SHOCK THEORY

In Burgers' equation the nonlinear and the dissipative effects are attached to individual terms in the equation which then form the basis of analytical or approximate solutions in the regions I and II. An analytical method, which more quickly and more easily leads to results in region II, is the *weak-shock theory*. The basic features of the weak-shock theory are that the diffusion effects (viscosity and heat conduction effects) are disregarded except within the shock themselves, where they account for the entropy increase, and that the continuous parts of the wave in the regions between the shocks are governed by the Earnshaw solution for simple waves, thus assuming isentropic flow in these regions.

The Rankine-Hugoniot relations will be the bridging functions which connect the continuous parts of the wave by relating the values of the kinematic and the thermodynamic quantities on both sides of a surface of discontinuity.

Due to the fact that the discontinuity of entropy in a weak shock wave is of the third order of smallness relative to the discontinuity of pressure, even finite, but small, amplitude waves lead to approximately isentropic conditions. Friedrichs [21] showed that if changes up to only the second order of smallness in the shock strength $\sigma = (p_2 - p_1)/p_1$ are considered, no change of entropy will be observed. The subscripts 1 and 2 refer to the conditions just ahead of and just behind the shock, respectively.

As shown already by Earnshaw [5] the propagation of finite-amplitude waves in a perfect gas may subject to the boundary conditions $u(0,t) = G(t)$ be written in terms of a parameter ψ as:

$$u(x,t) = G(\psi), \text{ where } \psi = t - x\{c_o + [(\gamma+1)/2]G(\psi)\}^{-1} . \tag{45}$$

The parameter ψ may be interpreted as the time the 'wavelet' of particle velocity u left the origin x = 0.

If the continuous waveform solution represented by Eq. (45) leads to a shock formed at a distance $\underline{x}$ from the origin and at the time $\underline{t}$, then this shock will arrive at any subsequent point at a time t_s given by:

$$t_s = \underline{t} + \int_{\underline{x}}^{x} \partial\xi/U \tag{46}$$

where U is the velocity of propagation of the shock, which according to the Rankine-Hugoniot relations depends on the particle velocity values u_1 and u_2 on either side of the shock, and thus may be written as:

$$U = c_o + \tfrac{1}{2}(\gamma+1)(u_1 + u_2)/2 . \tag{47}$$

By use of the retarded time $t' = t - x/c_o$ and by insertion of Eq. (47) into Eq. (46) the following differential equation for the path and amplitude of the shocks may be found:

$$dt'_s/dx = -\tfrac{1}{2}(\gamma+1)(u_1+u_2)/2c_o^2 . \tag{48}$$

u_1 and u_2 are determined by the continuous solution given by Eq. (45) for the wave parts between the shocks, thus giving together with the Eq. (48) a system of equations to be solved giving u_1, u_2 and t'_s within the weak-shock theory.

The formulas given above for the weak-shock theory applied to the plane wave case may be generalized for $kr >> 1$ also to comprise cylindrical and spherical finite-amplitude waves [7].

PROPAGATION IN A RELAXING FLUID

The intensity of irreversible processes taking place during the restoration of the thermodynamic equilibrium and leading to a dissipation of energy depends on the relation between the *relaxation time* τ and the angular frequency ω of the propagating wave.

Owing to the lack of thermodynamic equilibrium during a relaxation process, it is necessary to specify one more thermodynamic variable in the equation of state for a fluid. Such a variable ξ, characterizing the state of non-equilibrium considered, may for small deviations from equilibrium be determined by the simple first-order reaction equation: $\partial\xi/\partial t = (\xi_o - \xi)/\tau$.

Thus the equation of state for a relaxing fluid may instead of Eq. (16) be written as: $p = p(\rho, s, \xi)$. On the basis of the equations of continuity and motion correct to second order a single equation similar to Burgers' equation and governing the propagation of finite-amplitude waves in a relaxing fluid with a single relaxation frequency may be written as:

$$(\tau\partial/\partial t'+1)\{\partial u/\partial x + c_o^{-2}(B/A+2)(u/2)\partial u/\partial t'\} = m\tau\partial^2 u/\partial t'^2/2c_o \qquad (49)$$

where $m = (c_\infty^2 - c_o^2)/c_o^2 \ll 1$ represents the dispersion.

On a par with the Gol'dberg number Γ for nonlinear acoustic effects in a thermoviscous fluid a characteristic quantity $\varkappa$ for finite-amplitude wave propagation in a relaxing fluid may be defined as: $\varkappa = mc_o/[(B/A+2)u_o]$, which expresses the ratio of the relaxing to the nonlinear effects [7].

Thermoviscous effects, relaxation effects, nonlinear effects and diffraction effects may all be included in a compound equation which for a quasi-plane wave case may be written in dimensionless form [22] as:

$$\partial/\partial y(V_\sigma - VV_y - GV_{yy} - D\partial/\partial y\int V_{y'}\exp\{-(y-y')/\omega\tau\}dy' = N(V_{\xi\xi}+\xi^{-1}V_\xi)/4 , \qquad (50)$$

where $V = u/u_o$ and where ξ is a diffraction variable.

Eq. (50) has the following properties:

(i) for $G \neq 0$ and $D=N=0$, it reduces to Burgers' Eq. (32) for a thermoviscous medium ($G = \Gamma^{-1}$);

(ii) for $D \neq 0$ and $G=N=0$, in the limit of $\omega\tau \ll 1$, it reduces to a symbiosis of the Korteweg-de-Vries equation involving nonlinearity and dispersion and Burgers' equation involving nonlinearity and dissipation (see also Eq. (49);

(iii) for $N \neq 0$ and $G=D=0$, it reduces to the equation governing the diffraction influence on the nonlinear wave propagation.

EXPERIMENTAL STUDIES

Experimental investigations in fundamental nonlinear acoustics may roughly be divided into three main groups. (1) The nonlinearity of the pressure-density relation of the fluids expressed by the linearity ratios B/A and C/A. (2) The growth and decay of the harmonic content of the finite-amplitude waves during propagation. (3) Absorption measurements.

Determination of the parameters of nonlinearity B/A has mainly been performed using *static*, *thermodynamic* or *finite-amplitude waveform distortion methods*, of which the thermodynamic method has been the one mostly used. Knowing from experiments how the sound velocity in a fluid depends on pressure and temperature the sound velocity derivatives in Eq. (19) may be determined. These derivatives form the most crucial factors of the equation for B/A. Some B/A-values are given for various fluids in table 1.

Investigation of the growth and decay of the harmonic contents of finite-amplitude waves during propagation is very much tied up with the determination of B/A and both optical and electro-mechanical devices have been used for the experimental study of finite-amplitude wave distortion in various fluids. A detailed account of these studies together with the studies of nonlinear absorption may be found in [7].

Subjects of particular interest to fundamental nonlinear acoustics research of today are for example Burgers' equation in two or three dimensions, wave propagation in tubes influenced by boundary layer absorption and dispersion, thermodynamic expressions for B/A and higher-order nonlinear acoustic constants of fluids and so-

Fluid (at atmospheric pressure)	T ($^{\circ}$C)	B/A	Fluid.	T ($^{\circ}$C)	B/A
Distilled water	0	4.16	Freon C51-12	20	12.41
Distilled water	10	4.63	Fluoro carbon 43	20	12.85
Distilled water	20	4.96	Fluoro carbon 75	20	12.19
Distilled water	30	5.22	Polystyrene	0	12.70
Distilled water	40	5.38	RTV 602	0	13.40
Distilled water	50	5.55	Syntactic foam	0	3.70
Distilled water	60	5.67			
Distilled water	80	5.96			
Distilled water	100	6.11			
Carbon tetrachloride	30	11.54	Liquid metals.		
Glycerine	20	8.80			
Mercury	40	8.33	Sodium	110	2.70
Methyl alcohol	30	9.62	Tin	240	4.39
Ethyl alcohol	30	10.57	Indium	160	4.55
Propyl alcohol	30	10.70	Bismuth	318	7.10
Liquid nitrogen	-199	9.69			
Liquid oxygen	-199	9.56			
Monatomic gas	20	0.67			
Diatomic gas	20	0.40			

Table I. Measured B/A-values for various liquids.

lids, etc. But the present well-developed mathematical-physical basis of nonlinear acoustics combined with the availability of fast computers have encouraged a widespread exploitation of the nonlinear acoustics research results, reflecting a change from fundamental towards more applied research. Fields of research in applications are now numerous, but only the *underwater acoustics* and the *biomechanical applications* shall be discussed in this paper.

UNDERWATER APPLICATIONS OF NONLINEAR ACOUSTICS

The generation of sum- and difference-frequencies by the interference between two finite-amplitude sound waves has been the subject of discussion for more than two hundred years. Helmholtz [23] and Lamb [24] credit the original observation of difference-frequency tones to Sorge (1745) and Tartini (1754). Since then the subject of difference-frequency wave generation has received the attention of several authors, but only the last 15 years have brought a strong development in the practical exploitation of the finite-amplitude wave interaction products.

The theoretical work on nonlinear interaction of two sound beams - the scattering of sound by sound - begins with the papers of M.J. Lighthill on sound produced by turbulence [25], [26]. He transformed the basic equations of fluid mechanics into a form being particularly suited for the study of sound generated aerodynamically. Based on Lighthill's analysis, Ingard & Pridmore-Brown [27] performed a theoretical and an experimental study of scattering of sound by sound by two perpendicularly intersecting sound waves. Their results, where a scattered signal level about 10 dB below the theoretically predicted level was found experimentally, started a discussion of assumptions and experimental procedures related to scattering of sound by sound, in particular whether interaction components propagate outside the region of interaction. This discussion is still going on [28].

The case of two collinear sound beams was treated by Westervelt [29] in a paper where he noted that the nonlinear terms made the beam act as a distribution of sources for the modulating frequency. In [29] and in particular in [30] Westervelt formulated his now classical theory for the *parametric acoustic array*, a name given to the collinear sound beam interaction region due to its resemblance to the corresponding sonar array. The following simplifying assumptions and approximations underlie Westervelt's work:

(a) The equation of motion for an ideal fluid is used and the attenuation effect is introduced in an 'ad hoc' way.
(b) The two superimposed, high-frequency, plane primary waves are assumed to form beams so narrow and so perfectly collimated that the volume distribution of sources may adequately be represented by a line distribution located along the axis of the primary waves. The cross-sectional dimensions of the primary wave interaction region are assumed to be small compared with the wavelength at the difference frequency.
(c) No attenuation of the difference-frequency wave is assumed to occur.
(d) The amplitude attenuation coefficients for each of the two primary waves are equal and assumed to be one or more orders of magnitude less than the wave number of the difference-frequency wave.
(e) Nonlinear attenuation is negligible.

By introduction of the assumptions (a)-(e) into Lighthill's equation for aerodynamic sound production and by using a perturbation analysis retaining terms to second order in the field variables only, Westervelt's quasi-linear approach led to the following inhomogeneous wave equation for the pressure amplitude p_s of the scattered, i.e. the difference-frequency, wave:

$$\square^2 p_s = -\rho_o \partial q/\partial t \quad \text{where} \quad q = \beta(\rho_o^2 c_o^4)^{-1} \, \partial p_i^2/\partial t \; . \tag{51}$$

q is the source strength density responsible for the generation of acoustic energy through the nonlinear interaction of the primary waves in which the instantaneous pressure at a source point is p_i. β is related to the second order nonlinearity ratio B/A of the fluid through: $\beta = 1 + B/A$.

The general solution to Eq. (51) may be written as a volume integral by:

$$p_s(\underline{R},t) = (\rho_o/4\pi) \int_V (\partial q/\partial t)\exp(jk_s|\underline{R}-\underline{r}|)/(|\underline{R}-\underline{r}|)dV \tag{52}$$

where $\underline{R}$ and $\underline{r}$ denote the position vectors from the origin to the location of the observer (i.e. the field point) and to the differential volume dV of the zone of volume integration V, respectively.

The integral (52) was used by Westervelt for a derivation of the difference-frequency sound field generated by the nonlinear interaction of the two perfectly collimated, plane, monochromatic, collinear, primary waves of equal source amplitude p_o. His expression for the pressure amplitude p_s as a function of the distance R from the projector emitting the primary waves to the observation point and as a function of the angle θ between the observation point and the acoustic axis of the projector may be written as:

$$p_s(R,\theta) = \omega_s^2 p_o^2 S\beta(8\pi\rho_o c_o^4 R\alpha_o)^{-1}(1+k_s^2/\alpha_o^2(\sin^4(\theta/2)))^{-\frac{1}{2}} \tag{53}$$

where the time and phase dependences have been omitted. ω_s is the angular frequency of the difference-frequency wave, i.e. $\omega_s = \omega_1-\omega_2$, S denotes the cross-sectional area of the collimated wave region and k_s denotes the wave number of the difference-frequency wave. α_o is the mean absorption coefficient of the primary waves for infinitesimal wave amplitudes. The solution Eq. (53) is restricted to the far field of the scattered, i.e. the difference-frequency, wave by the condition: $k_s R > (k_s/\alpha_o)^2$, thus permitting the Born approximation to be made on the Green's function of the exact nonhomogeneous Helmholtz equation forming the basis of the theory. The bracket in (53) leads to the half-power beamwidth θ_h of the difference-frequency wave given by:

$$\theta_h \simeq 2(\alpha_o/k_s)^{\frac{1}{2}} \tag{54}$$

which shows that a narrowing of the beam takes place for a decrease in the primary wave frequency, opposite to what is the case for a conventional linear projector. Further, a narrowing of the beam will follow an increase in the difference-frequency. It should be noted that the influence of the primary frequencies on (53) and (54) is only through the absorption coefficient α_o.

The difference-frequency signal amplitude p_s may be considered to be radiated from an array of sources distributed continuously throughout the interaction region, being bounded by the collimated beams and extending a distance along the acoustic axis determined by the small signal absorption of the carrier waves. The parametric array is shaded by virtue of the naturally smooth decay in the conversion of the carrier frequency waves to the difference-frequency wave with increasing distance from the signal source. The shading of the array, being due to the carrier wave absorption and diffraction within the interaction region, gives rise to a monotonically decaying angular response of the difference-frequency wave with increasing θ-values, thus avoiding the undesirable *minor lobes* that are common in conventional piston type transducers. Due to the small width of the interaction region compared to its length the parametric array produces a field of radiation much *narrower* than the one which would be produced by a conventional underwater sound source operating linearly at the difference frequency. Moreover, the *wide band* character of the parametric conversion process enables one to remedy some of the disadvantages of the rather *low efficiency* of the nonlinear conversion process by the use of wide-band signal processing techniques. In spite of the low source level efficiency - ranging from 10% down to 10^{-5}% - systems employing parametric arrays can be superior to conventional linear systems when the reduction of the beamwidth, the transducer size or the absorption - due to the low difference frequency - are taken into account.

Since Westervelt's publication of his quasilinear approach leading to his asymptotic solution being valid at long ranges from the interaction region, a great deal of theoretical and experimental works has been done in order to improve the understanding of the characteristics of the parametric acoustic arrays.

For spherically spreading primary waves confined to a cone of angular width $2\psi_1$ and by assuming a uniform intensity distribution across the cone, Berktay [31] found for the difference-frequency pressure amplitude along the axis of symmetry:

$$p_s(R,0) = \beta p_1 p_2 \omega_s^2 (\exp(-\alpha_s R))(2\rho_o c_o^4 R k_s)^{-1}\{(\tfrac{1}{2}\ln(1+\psi_h^4))^2 + (\tan^{-1}\psi_h^2)^2\}^{\frac{1}{2}} \tag{55}$$

where ψ_h is given by: $\psi_h^4 = (\psi_1/\theta_h)^4 \simeq (k_s/\alpha_T)(1-\cos(\psi_1))^2$, and where the half-power beamwidth ψ_h is given by: $\theta_h \simeq 2(\alpha_T/2k_s)^{\frac{1}{2}}$, for $\alpha_T/2k_s << 1$. α_T is for the spherical wave case given by: $\alpha_T = \alpha_1+\alpha_2-\alpha_s\cos\theta\cos\psi \simeq \alpha_1+\alpha_2-\alpha_s$.

In (55) p_1 and p_2 denote the initial pressure amplitude of the primary waves and α_1, α_2 and α_s denote the absorption coefficients of the primary waves and the difference-frequency wave, respectively.

The asymptotic solution (55) is like (53) valid for the field point at long ranges from the interaction region. The influence of the finite aperture in the parametric acoustic array which is of particular importance in relation to nearfield interaction has in particular been treated by Tjøtta, latest in reference [32]. When most of the interaction between the primary waves takes place in their nearfield, i.e. within the Rayleigh distance defined by $R_r = S/\lambda_o$, where λ_o is the average wavelength of the primary waves, the parametric array is assumed to be *absorption limited* in contrast to the *spreading-loss limited* array, where most interaction takes place in the farfield. While Westervelt's results characterize an absorption limited array, the results obtained in [33] is based upon all interaction taking place in the farfield of a piston projector, where the half-power beamwidth will increase with the source distance and asymptotically approach the half-power beamwidth of the product of the primary beam directivity patterns. This relation can be seen from the double integral in Eq. (56) where a two-dimensional convolution of the product of the primary wave directivity patterns $D_1(\ )$ and $D_2(\ )$ appears [33]:

$$p_s(R,\theta,\eta) \simeq \omega_s^2 p_1 p_2 \beta (4\pi\rho_o c_o^4 R)^{-1}(\exp(-(\alpha_s + j\,k_s)R))\ \times$$
$$\int_{-\pi/2}^{\pi/2}\int D_1(\gamma,\phi) D_2(\gamma,\phi)(\alpha_T + j\,k_s(1-\nu))^{-1}\cos\gamma\, d\gamma\, d\phi \tag{56}$$

where: $\nu \simeq 1(\gamma-\theta)^2 - \tfrac{1}{2}(\phi-\eta)^2$.

The difference-frequency pressure p_s at the field point (R,θ,η) in the geometry of Fig. 2 will reduce to the Weltervelt case for extremely narrow primary beams and for $D_1(\ ) = D_2(\ )$. A more detailed discussion of absorption and spreading-loss limited arrays can be found in [34].

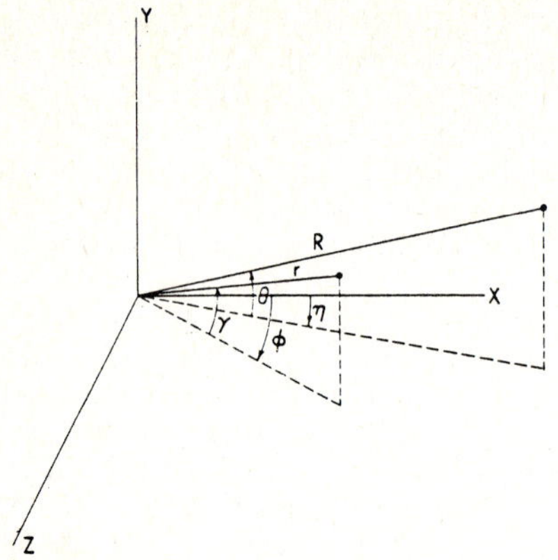

Figure 2. Geometry used in (54).

The full consequence of the asymptotic theory for parametric arrays cannot be met with the field point in the interaction region, which is frequently the case for laboratory tank experiments. Two different theoretical approaches have been used in order to calculate the difference-frequency signal level at points in the interaction region. The first approach is based on numerical integration of a volume integral and the second is based on the introduction of a correction factor to the asymptotic theory.

The numerical integration procedure was used in [35] where the generation in fresh water of the sum- and difference-frequency signal of two high-frequency primary waves (418 and 482 kHz) transmitted by a 3-in-diameter, circular piston projector is investigated. The following volume integral expression for the difference-frequency pressure amplitude, valid for R >> r, and approximately valid for the field point in the interaction region, is obtained:

$$p_s(R,\theta) = 2p_1p_2\omega_s^2\beta R_r^2(\pi\rho_o c_o^4 k_1 k_2 a^2)^{-1} \times$$
$$\int_{R_r'}^{R}\int_o^{\phi_e}\int_o^{\pi} J_1(k_1 a \sin\sigma)J_1(k_2 a \sin\sigma)/(\sin^2\sigma) \times \qquad (57)$$
$$(\exp-(((\alpha_1+\alpha_2)-jk_s)r-(jk_s-\alpha_s)r'))/r')\sin\phi \, d\psi \, d\phi \, dr$$

where $\sin\sigma = ((\sin(\theta-\phi)+\sin\phi\cos\theta(1-\cos\psi))^2 + \sin^2\phi\sin^2\psi)^{\frac{1}{2}}$ and $r' = (R^2+r^2-2rR\cos\phi)^{\frac{1}{2}}$. ϕ_e is an effective array angle and $R_r' = 3a^2/4\lambda_o$ with a being the projector radius. The results of a calculation using a modified version of (57) and comparing it with experimental results are given in [36] and are reproduced in Fig. 3.

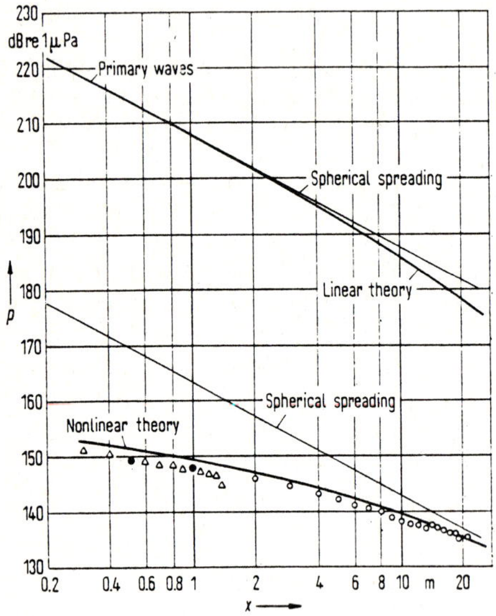

Figure 3. Pressure level p as a function of range x. [36]

A nearfield theory applicable to absorption as well as saturation limited arrays which can be used for a prediction of the farfield from nearfield measurements is given in [37].

Finite-amplitude effects, i.e. wave distortion and finite-amplitude absorption, in parametric transmitting arrays leading to *saturation limited* arrays have been dealt with by several authors [38] - [40]. The saturation effect manifests itself as an effective shortening of the array length leading to a broadening of the difference-frequency beam and a reduction of its source level [34].

The possibility of developing a *parametric receiving array* was first mentioned by Westervelt [30] and in spite of the fact that most efforts in developing parametric arrays have been laid down into the transmitting arrays in order to obtain highly directional, low frequency sources using relatively small transducers, the parametric receiving array has recently received considerable interest. In a parametric receiver, the nonlinear interaction process may take place between a low-frequency, plane signal wave of low intensity and a high-frequency pump wave of higher intensity being generated locally. The sum- or difference-frequency signal are then received by a transducer on the acoustic axis of the pump wave.

Farfield reception was considered by Barnard et al. [41] who studied a first-order sound field consisting of a 'low-amplitude', spherical, harmonic pump wave of frequency f_1 and a plane, harmonic signal wave of frequency f_2, ($f_1 > f_2$). No nearfield effects of the pump were assumed to be involved and only the pump wave was assumed undergoing absorption (viscous). In polar coordinates ((r,ψ,ϕ), where ϕ is the polar angle and ψ is the azimuthal angle) their first-order sound field can be written as:

$$p_f = p(r_o/r)(2J_1(k_1a\ \sin\phi))(k_1a\ \sin\phi)^{-1}\ x \\ (\exp(-\alpha_1 r))(\cos(\omega_1 t-k_1 r)) + p_2\cos(\omega_2 t-k_2 z) \qquad (58)$$

where p_1 is the peak sound pressure level of the pump wave at a distance r_o and p_2 is the sound pressure level of the signal wave at input to the parametric receiving array. a is the radius of the piston pump and k_2z is defined as the plane-wave phase at r, where: $z = r(\sin\nu\cos\phi - \sin\phi\cos\nu\cos\psi)$. The angle ν in the horizontal plane is the acute angle between the acoustic axis of the array and the plane wave front. Insertion of (58) into (52) and introduction of the viscous absorption in an 'ad hoc' way, lead to the following volume integral expression for the pressure amplitude of the sum- or difference-frequency signal:

$$p_s(R_o,\nu) = \beta\omega_s^2 p_1 p_2 r_o (2\pi\rho_o c_o^4 a k_1)^{-1}\ x \\ \int_o^R \int_o^{\phi_e} \int_o^{\pi} (J_1(k_1a\ \sin\phi))\exp(j(k_1 r \pm k_2 z + j\alpha_1 r))\ x \qquad (59) \\ (\exp(j(k_s+j\alpha_s)r'))(r')^{-1}\ r\ d\psi\ d\phi\ dr$$

where $\omega_s = \omega_1 \pm \omega_2$, and $r' = (r^2+R_o^2-2rR_o\cos\phi)^{\frac{1}{2}}$, while ϕ_e is the angle between the acoustic axis and the first null of the pump farfield radiation pattern. R_o is the distance along the acoustic axis between the pump and the receiving transducer. (59) has been solved numerically and the results have been tested against experiments for the sum-frequency wave [41] showing a good agreement, both concerning beam patterns and range (R_o) for various signal frequencies. Later works along the line introduced by (59) have been published in [42] and [43] and nearfield reception has been treated theoretically by Rogers et al. [44].

Only a small amount of work has been done in analyzing finite-amplitude effects in parametric receiving arrays. In general, some serious problems still have to be overcome before a practical system can be realized. For instance, motion of the pump and the transducer during reception [43], water noise at the signal frequency or at the sum- or difference-frequencies, electronic noise in the equipment, etc., can create serious problems for full-scale reception.

The *low conversion efficiency* of the parametric acoustic arrays previously mentioned is its capital weakness. The second-order effects on which it is based limit the energy transfer from the primary beams to the difference- and sum-frequency beams.

It was pointed out by Merklinger [45] that if the necessary bandwidth of the transmitter was available, a pulsed type of transmission would give an improvement in conversion efficiency of between 2 and 6 dB depending on the system constraints. Experimental evidence for this prediction has later been created [36]. 100% amplitude modulation of a primary wave was shown in [36] and in [46] to lead to a 2.5 dB increase in the difference-frequency sound pressure level compared to a two-component primary of the same total power.

From the parameters involved in the expressions (53), (55), and (56) for low-amplitude wave interactions in parametric transmitting arrays it may be concluded, that an increase in the virtual source strength and thus in p_s may be obtained primarily by the following procedures: a. An increase in the peak amplitudes of the carrier waves, and b. an increase in β and a decrease in ρ_o and in particular in c_o in the fluid. The influence of these parameters on the conversion efficiency of a parametric transmitting array has been studied in theory and through experiments in [36]. By putting the nearfield of the projector under pressure, cavitation effects could be avoided and the nearfield liquid was replaced by liquids showing more appropriate β, ρ_o and c_o values (methyl and ethyl alcohols), which led to an about 15 dB increase in the difference-frequency sound pressure level. Further, it was shown in [36] that some essential dB's at the difference-frequency sound pressure level could be gained through the use of an acoustic lens effect and through the use of a slow-waveguide antenna effect in the nearfield liquid cylinder, a subject which also has been studied by Ryder et al. [47] for a silicone rubber cylinder in contact with the projector. The increase in β by using air-bubbles in the interaction region of the primaries has been attempted by Lockwood et al. [48] and recently by Kosyaev et al. [49] and Kustov et al. [50] who introduced a bubble layer in the farfield of the primary waves. This bubble field will, due to the strongly increased nonlinearity of bubbly liquids [51] lead to an essential conversion efficiency improvement.

Fields of increasing importance in the study of parametric acoustic arrays are signal processing [52] and nonlinear tone/noise interaction [53] and [54] which for instance may explain anomalies found in jet noise reduction with propagation distance and in ambient noise attenuation in seawater. Recent studies of a focused parametric array [55] have shown that the difference-frequency beam also becomes very narrow in the focal plane and thus permits a higher resolution to be obtained which is a subject of particular interest to medical application of nonlinear acoustics, where a parametric echoscanning principle has been suggested used for medical diagnosis [56].

BIOMECHANICAL APPLICATIONS OF NONLINEAR ACOUSTICS

Ultrasound is new being used extensively for detection and display of interfaces between tissues in medical imaging systems and as a means of gaining information about tissue pathology. Various intensities of ultrasound are used for diagnostic purposes and the use of focal field intensities in the range of 100-1000 W/cm^2 is not uncommon. Nonlinear effects are therefore to be expected in relation to a number of applications of ultrasound in medicine [57].

In the diagnostic use of ultrasound, a deeper understanding of nonlinear acoustics phenomena in mammalian tissue should lead to more precise information on the state of the organ being examined. In the therapeutic use of ultrasound, the detailed understanding should lead to better control of, for example, the heat deposition of ultrasonic energy during hyperthermic treatment of cancer. For both applications it is important to determine whether the nonlinear acoustic phenomena may lead to harmful effects to patients. On this background a considerable amount of research in acoustic nonlinearity of biological media is now being done around the world in particular aiming at an evaluation of the applicability of the second-order acoustic nonlinearity parameter B/A as a *tissue characterizing parameter*. Development of reliable methods for *in vitro* and in particular for *in vivo* determination of B/A forms an essential part of this research.

Several methods have been used during the past for determination of B/A of liquids and gases and two of these, a *finite-amplitude method* and a *thermodynamic method*, have been of particular interest to the study of B/A in tissue. Most recently, a new promising method, *real-time nonlinear parameter tomography*, has been developed being able to give a B-scan like pattern of a function of B/A in tissue.

Using Fubini's solution (15) the pressure amplitude of the second harmonic as a function of source distance may be written as:

$$p_2 = p_o\{J_n(2\sigma)\}/\sigma \ . \qquad (60)$$

Expansion of the Bessel function in (60) in a power series, retaining only the first and the second term, and inserting (23) lead to:

$$\{p_2(x)/(xp_1^2(0))\}_{xp_1^2(0)\to 0} = (2+B/A)\pi f/(2\rho_o c_o^3) \ . \qquad (61)$$

From (61), B/A may be found by plotting the measured values of $\{p_2(x)/(xp_1^2(0))\}$ as a function of x and extrapolating back to x = 0. This approximation is accurate to within 2%, for $\sigma = x/\ell$ values up to 0.25 and has been used for determination of B/A in biological liquids like bovine serum albumin and porcine whole blood [58]. The most critical assumption in the derivation of (61) is that the medium shall be lossless [59]. Attenuation effects may be introduced using a procedure developed by Thuras et al. [60], which for the attenuation coefficients α_1 and α_2 for the fundamental frequency and for its second harmonic satisfying the relation $(\alpha_2 - 2\alpha_1)x < 0.5$, leads to the expression (62) for the second harmonic amplitude as a function of source distance x:

$$p_2(x) = (2+B/A)\pi f/(2\rho_o c_o^3) x p_1^2(0) \exp(-(\alpha_1+\alpha_2/2)x) . \qquad (62)$$

Expression (62) has been used for determination of B/A of homogenized and whole liver of 23°C [58].

Very recently, a more advanced finite-amplitude method has been developed [61] in which the influence of diffraction and phase cancellation effects over the receiver surface by nearfield measurements has been taken into account. Several of the B/A-values for biological tissues given in table II are based on this method.

B/A measurement techniques based upon (61), (62) and other finite-amplitude methods require very accurate transducer calibration and are prone to scattering effects, phase cancellation effects, etc. in particular by measurements in inhomogeneous media like biological tissue. It is therefore desirable to have a method completely independent of finite-amplitude waves. The *thermodynamical methods* posses this quality.

A frequently used thermodynamical method is based on expression (19) for B/A. The most crucial factors in (19) are the derivatives of the velocity of sound with respect to pressure and temperature for constant temperature and pressure, respectively. The other factors in (19) may be determined by standard laboratory procedures. (19) has been used for determination of B/A of inhomogeneous materials like water-saturated sediments [62] and most recently also biological tissue [63].

Another thermodynamic procedure [64] based on the expression $B/A = 2\rho_o c_o \{(\partial c/\partial p)_s\}$ has most recently been used for determination of B/A using phase measurements [65]. This procedure seems to be of interest to tissue measurements too.

The inherent disadvantages of using the finite-amplitude or the thermodynamic methods for clinical determination of B/A of tissue are obvious. These procedures are based on measurements on samples of tissue and form, therefore, more *in vitro* than *in vivo* methods. Moreover, these methods give information on the average B/A of an entire sample, i.e. they lead to B/A-values in discrete points where samples are taken. These disadvantages have to a certain extent been overcome in a new system recently investigated in Japan [66]. This system is based on *real-time nonlinear parameter tomography* in which a pulse of a relatively high power pump wave of low frequency interacts with a CW, low power probe beam of high frequency. This interaction leads to a sequential phase modulation of the probe beam by the product of B/A along the beam and the pressure of the pulsed pump wave. Detection and demodulation of the probe beam determine the distribution of B/A along the probe beam axis. The tomographic method has given some very promising results and work is now being done to develop clinical diagnostic equipment based on this method.

Biological material	T (°C)	B/A
Bovine serum albumin 17g/dl	25	6.0
Porcine blood	30	6.2
Dog blood	30	5.4
Pig liver	25	6.7
Dog liver	30	7.9
Human liver	30	7.6
Canine spleen		6.8
Dog spleen	30	6.8
Human spleen		7.8
Dog kidney	30	7.2

Table II. B/A-values of biological material.

CONCLUSIONS

The theoretical background of nonlinear acoustics has been developed over a couple of centuries, but most research activity in this field has taken place within the last two decades. The availability of fast and powerful computers permits us to solve nonlinear acoustic problems of very complicated character and the development in electronics permits us to do experiments studying fast events in real time. These facts form an essential part of the basis for the change from fundamental towards more applied nonlinear acoustic research, which has taken place within the last decade. This development will probably continue and it is to be expected that the nonlinear acoustics research results will lead to solutions to many yet unsolved mathematical-physical problems and will form the basis of several future inventions.

REFERENCES

1. Euler, L., Principes généraux du movement des fluides. Hist. de l'Acad. de Berlin, 11, 1757.
2. Poisson, S.D., Memoire sure la théorie du son. Journal de l'École Polytechnique, 7 (1808), 319.
3. Challis, J., On the velocity of sound. Phil. Mag., Ser. 3, 32 (1848), 494.
4. Stokes, G.G., On a difficulty in the theory of sound. Phil. Mag., Ser. 3, 33 (1848), 349.
5. Earnshaw, S., On the mathematical theory of sound. Phil. Trans. Roy. Soc. (London), 150 (1860), 133.
6. Riemann, B., Über die Fortpflanzung ebener Luftwellen von endlicher Schwingungsweite. Abh. Ges. der Wiss., Göttingen, 8 (1860), 43.
7. Bjørnø, L., Nonlinear Acoustics. In R.W.B. Stephens & H.G. Leventhall (Eds.) *Acoustics and Vibration Progress*, Vol. 2, pp. 101-198 (Chapman & Hall, Ltd., London, 1976).
8. Rankine, W.J.M., On the thermodynamic theory of waves of finite longitudinal disturbance. Trans. Roy. Soc. (London), 160 (1870), 277.
9. Hugoniot, H., Sur la propagation du movement dans les corps et specialement dans les gaz parfait. Journal de l'Ecole Polytechnique, 58 (1889), 1.
10. Lord Rayleigh, Aerial plane waves of finite amplitude. Proc. Roy. Soc., A84 (1910), 247.
11. Taylor, G.I., The conditions necessary for discontinuous motion in gases. Proc. Roy. Soc., A84 (1910), 371.
12. Fay, R.D., Plane sound waves of finite amplitude. J. Acoust. Soc. Amer., 3 (1931), 222.
13. Fubini, E., Anomalie nella propagazione de onde acustiche di grande ampiezza. Alta Frequenza, 4 (1935), 530.
14. Coppens, A.B. et al., Parameter of nonlinearity in fluids II, J. Acoust. Soc. Amer., 38 (1965), 797.
15. Eckart, C., Vortices and streams caused by sound waves. Phys. Rev., 73 (1948), 68.
16. Gol'dberg, Z.A., On the propagation of plane waves of finite amplitude. Soviet Phys. Acoust., 2 (1956), 346.
17. Mendousse, J.S., Nonlinear dissipative distortion of progressive sound waves of moderate amplitudes. J. Acoust. Soc. Amer., 25 (1953), 51.
18. Blackstock, D.T., Thermoviscous attenuation of plane, periodic, finite-amplitude sound waves. J. Acoust. Soc. Amer., 36 (1964), 534.
19. Khokhlov, R.V. & Soluyan, S.K., Propagation of acoustic waves of moderate amplitude through dissipative and relaxing media. Acustica, 14 (1964), 241.
20. Blackstock, D.T., Connection between the Fay and Fubini solutions for plane sound waves of finite amplitude. J. Acoust. Soc. Amer., 39 (1966), 1019.
21. Friedrichs, K.O., Formation and decay of shock waves. Comm. Pure. Appl. Math., 1 (1948), 211.
22. Rudenko, O.V. et al., Problems of the theory of nonlinear acoustics. In L. Bjørnø (Ed.), *Finite-amplitude wave effects in fluids*. (IPC Science and Technology Press, Ltd., London 1974).

23. Helmholtz, H., Die Lehre von den Tonempfindungen als physiologische Grundlage für die Theorie der Musik. (Braunschweig 1862).
24. Lamb, H., The dynamical theory of sound. (Dover, New York 1960).
25. Lighthill, M.J., On sound generated aerodynamically. I (General theory). Proc. Roy. Soc. (London), A211 (1952), 564.
26. Lighthill, M.J., On sound generated aerodynamically. II (Turbulence as a source of sound). Proc. Roy. Soc. (London), A222 (1954), 1.
27. Ingard, K.U. & Pridmore-Brown, D.C., Scattering of sound by sound. J. Acoust. Soc. Amer., 28 (1956), 367.
28. Westervelt, P.J., Recent advances in the theory of the nonscattering of sound by sound. Proc. 10 International Symposium on Nonlinear Acoustics, Kobe, Japan 1984.
29. Westervelt, P.J., Parametric end-fire array. J. Acoust. Soc. Amer., 32, (1960), 934(A).
30. Westervelt, P.J., Parametric acoustic array. J. Acoust. Soc. Amer., 35 (1963), 535.
31. Berktay, H.O., Possible exploitation of non-linear acoustics in underwater transmitting applications. J. Sound & Vib., 2 (1965), 435.
32. Tjøtta, J.N. & Tjøtta, S., Effects of finite aperture in a parametric acoustic array. J. Acoust. Soc. Amer., 68 (1980), 970.
33. Berktay, H.O. & Leahy, D.J., Farfield performance of parametric transmitters. J. Acoust. Soc. Amer., 55 (1974), 539.
34. Bjørnø, L., Parametric acoustic arrays. In G. Tacconi (Ed.), *Aspects of Signal Processing*, (D. Reidel Publishing Comp., Holland, 1977).
35. Muir, T.G. & Willette, J.G., Parametric acoustic transmitting arrays. J. Acoust. Soc. Amer., 52 (1972), 1481.
36. Bjørnø, L. et al., Some experimental investigations of the parametric acoustic array. Acustica, 35 (1976), 99.
37. Mellen, R.H. & Moffett, M.B., A numerical method for calculating the nearfield of a parametric acoustic source. J. Acoust. Soc. Amer., 63 (1978), 1622.
38. Bartram, J.F., A useful analytical model for the parametric acoustic array. J. Acoust. Soc. Amer., 52 (1972), 1042.
39. Fenlon, F.H., Approximate methods for predicting the performance of parametric sources at high acoustic Reynolds numbers. In L. Bjørnø (Ed.) *Finite-Amplitude Wave Effects in Fluids*, (IPC Science and Technology Press, Ltd., London 1974).
40. Fenlon, F.H., Parametric scaling laws, (Westinghouse Res. Labs., No. 0014-14-C.0214, 1974).
41. Barnard, G.R. et al., Parametric acoustic receiving array. J. Acoust. Soc. Amer., 52 (1972), 1437.
42. Truchard, J.J., Parametric receiving array and the scattering of sound by sound. J. Acoust. Soc. Amer., 64 (1978), 280.
43. Reeves, C.R. et al., Parametric acoustic receiving array response to transducer vibration. J. Acoust. Soc. Amer., 67 (1980), 1495.
44. Rogers, P.H. et al., Parametric detection of low-frequency waves in the nearfield of a directional pump source. In L. Bjørnø (Ed.), *Finite-Amplitude Wave Effects in Fluids*, (IPC Science and Technology Press, Ltd., London 1974).
45. Merklinger, H.K., Of finite amplitude plane waves and of endfire arrays. Proc. Symp. on Nonlinear Acoustics, University of Birmingham, 1971.
46. Eller, I.A., Application of the USRD type E8 transducer as an acoustic parametric source. J. Acoust. Soc. Amer., 56 (1974), 1735.
47. Ryder, J.D. et al. Radiation of difference-frequency sound generated by nonlinear interaction in a silicone rubber cylinder. J. Acoust. Soc. Amer., 59 (1976), 1077.
48. Lockwood, J.C. & Smith, D.P., Difference-frequency generation by forced-air bubbles. (AMETEK/STRAZA, Techn. Rep., 11-135E-74-1, 1974).
49. Kozyaev, E.F. & Naugol'nykh, K.A., Parametric sound radiation in a two-phase medium. Soviet Phys.-Acoust., 26 (1980), 48.

50. Kustov, L.M. et al., Parametric acoustic radiator with bubble layer. Acoustics Letters, 6 (1982), 15.
51. Bjørnø, L., Acoustic nonlinearity of bubbly liquids. Applied Scientific Research, 38 (1982), 291.
52. Muir, T.G. & Goldsberg, T.G., Signal processing aspects of nonlinear acoustics. In L. Bjørnø (Ed.) *Underwater Acoustics and Signal Processing*, (D. Reidel Publ. Comp., Holland 1981).
53. Stanton, T.K. & Beyer, R.T., Interaction of sound with noise in water II. J. Acoust. Soc. Amer., 69 (1981), 989.
54. Gurbatov, S. & Bjørnø, L., Interaction between a finite-amplitude tone and low-amplitude noise in a rigid-walled tube. Proceedings of 10 International Symposium on Nonlinear Acoustics, Kobe, Japan, 1984.
55. Lucas, B.G. et al., Field of a parametric focusing source. J. Acoust. Soc. Amer., 73 (1983), 1966.
56. Bjørnø, L. & Grinderslev, S., Parametric echoscanner for medical diagnosis. Journal de Physique, C8-40 (1979), 111.
57. Bjørnø, L. & Lewin, P.A., Measurement of nonlinear acoustic parameters in tissue. In J.F. Greenleaf (Ed.) *Tissue characterization with ultrasound* (CRC Press 1984).
58. Dunn, E. et al. Nonlinear ultrasonic properties in biological media. British Journal of Cancer, 45 (1982), 55.
59. Bjørnø, L., Comparison of procedures for determination of acoustic nonlinearity of some inhomogeneous materials. J. Acoust. Soc. Amer., 74 (1983), S27.
60. Thuras, A.L. et al., Extraneous frequencies generated in air carrying intense sound waves. J. Acoust. Soc. Amer., 6 (1935), 173.
61. Cobb, W.N., Measurement of the acoustic nonlinearity parameter for biological media. Ph.D. dissertation, Yale Univ., 1982.
62. Bjørnø, L., Finite-amplitude wave propagation through water-saturated marine sediments. Acustica, 38 (1977), 195.
63. Law, W.K. et al., Comparison of thermodynamic and finite-amplitude methods of B/A measurements in biological materials. J. Acoust. Soc. Amer., 74 (1983), 1295.
64. Bjørnø, L. & Black, K., Higher-order acoustic nonlinearity parameters of fluids. In U. Nigul and J. Engelbrecht (Eds.) *Nonlinear Deformation Waves*, (Springer-Verlag, 1983).
65. Zhu, Z. et al., Determination of the acoustic nonlinearity parameter B/A from phase measurements. J. Acoust. Soc. Amer., 74 (1983), 1518.
66. Ichida, N. et al., Nonlinear parameter tomography by using pump waves. Proc. 10 International Symposium on Nonlinear Acoustics, Kobe, Japan, 1984.

Theoretical and Applied Mechanics
F.I. Niordson and N. Olhoff
Elsevier Science Publishers B.V. (North-Holland)

HYDRODYNAMIC LOADS ON MARINE STRUCTURES

Odd M. Faltinsen

Division of Marine Hydrodynamics
Norwegian Institute of Technology
Trondheim, Norway

An overview over important wave load problems for ships and offshore structures are given. Three main topics are dealt with in details. These are a) ship motions at forward speed, b) slowly varying and mean wave loads and c) separated flow problems. A brief introduction to the topics are given. Unsolved theoretical problems of practical interest are discussed.

INTRODUCTION

Wave induced motions and loads on ships and offshore structures are of significance in several contexts. For ships it is for instance important to predict heave, pitch, roll, acceleration and relative vertical motions between the ship and the waves. The latter can be used to evaluate the possibility and damage due to slamming and green water on deck. For a ship it is important to avoid slamming as well as water on deck due to local damage of the structure.

Rolling may be a problem from an operational point of view of working vessels, passenger ships and naval vessels. Means to reduce the rolling of a ship is therefore of interest. For smaller ships, rolling in combination with for instance wind, water on deck and motion of the cargo can cause the ship to capsize. Other important reasons to capsizing are breaking waves and broaching. The latter is associated with directional instability of the ship in following waves when the frequency of encounter between the ship and the waves are small.

Liquid sloshing in tanks may be a problem for instance for LNG carriers. Sloshing is resonant liquid oscillations caused by ship motions. It is characterized by strong nonlinearities. Resulting high local pressures as well as total forces may be important in design.

For larger ships wave induced bending moments, shear forces and torsional moments are of interest. More special problems are whipping and springing. Whipping are transient elastic vibrations of the ship caused by for instance slamming or bowflare forces. Springing is steady state elastic vibrations caused by the waves and is of special importance for larger oceangoing ships. Springing is both due to linear and nonlinear excitation mechanisms. The linear exciting forces are associated with waves of small wave lengths relative to the ship length.

The added resistance of ships in waves and the change of propulsion coefficients in waves are generally of less importance than the still water resistance and propulsion of the ship. But it is necessary information in order to obtain realistic values for the shaft horse power of ships in a sea. Knowledge about the sensitivity of added resistance to ship form may be important information in the context of energy economization.

Knowledge about wave induced motions and loads for offshore structures indcluding offshore working vessels are generally more important than it is for ships used for transportation purposes. This is particularly true for offshore structures in hostile areas like the North Sea. Roughly 60% of the time the significant wave height is larger than 2 m. The most probable largest wave height in 100 years may be up to 40 m. For drilling operations heave motions is a limiting factor. It is important to design structures with low heave motion so that it is possible to drill in as high percentage of the year as possible. Semisubmersibles are examples on structures with very low heave motion in the actual frequency domain. Rolling may also be an important motion mode to evaluate. Examples are for operation of crane vessels or for transportation of jackets on barges.

In the design of mooring systems for offshore structures loads due to current, wind, wave drift forces and wave induced motion are generally of equal importance. Wave drift forces are mean forces due to waves. The wave induced motion can be divided into high-frequency motion and slowdrift oscillations. The high frequency motion are mainly linearly excited motion in the frequency domain of significant wave energy. The slowdrift oscillation are caused by nonlinear interaction mechanisms in the waves and the structural induced fluid motion. The slowdrift forces excite resonance oscillations of the moored system. Typical resonance periods are in the order of 1 or 2 minutes for a conventionally moored system. For a single-point mooring system with for instance a ship attached to a loading buoy instabilities in the horizontal plane may be an additional problem. This is similar to what may happen when one tow a ship.

Wind, current, mean wave drift forces and slowly varying wave drift forces are also important in the design of thrusters and in stationkeeping of crane vessels, diving vessels, supply ships, stand-by ships and pipelaying vessels. Interaction of thrusters with other thrusters and structures may also be important for dynamic positioning systems, towing and marine operations in waves.

Vortex shedding may cause resonant oscillations of some offshore structures or structural parts. One example is current and wave induced hydroelastic oscillations of risers where failure due to fatigue is a possible consequence. Similar problems may occur for pipelines. For long slender and vertical cylindrical moored structures we may also have resonance problems due to vortex shedding. An example on the latter type of structures is a moored loading buoy.

In some cases one is speculating that "Mathieu-equation instabilities" may cause problems. Mathieu-equation occur in problems when there is a time dependent restoring force. Examples are the analysis of horizontal motions of a tension leg platform or a tout-anchored buoy. Another example is rolling of ships with non vertical ship sides at the water line. To what extent the instabilities will occur is dependent on the damping level. But it is known that the Mathieu-equation phenomena may cause some ships to roll in head sea.

Both viscous effects and potential flow effects may be important in determining the wave induced motions and loads on marine structures. Included in the potential flow is the wave diffraction and scattering around the structure. In order to judge when viscous effects or different types of potential flow effects are important it is useful to refer to a simple picture like, Fig. 1. This drawing is based on results for horizontal wave forces on a vertical cylinder standing on the sea floor and penetrating the free surface. The incident waves are regular. The results are based on the use of Morison equation (see equation (13)) with a mass coefficient of 2 and a drag coefficient of 1. The linear MacCamy and Fuchs theory [1] has been used in the wave diffraction regime. Let us try to use this figure for offshore structures. We will consider a regular wave of wave height 30 m and wave length 300 m. This corresponds to an extreme wave condition. Let us consider wave loads on the caisson of a gravity platform where typical crossdimensions are 100 m. This implies equivalent H/D and λ/D-values of 0.3 and 3, respectively. This means that wave diffraction is most important. If we consider a semisubmer-

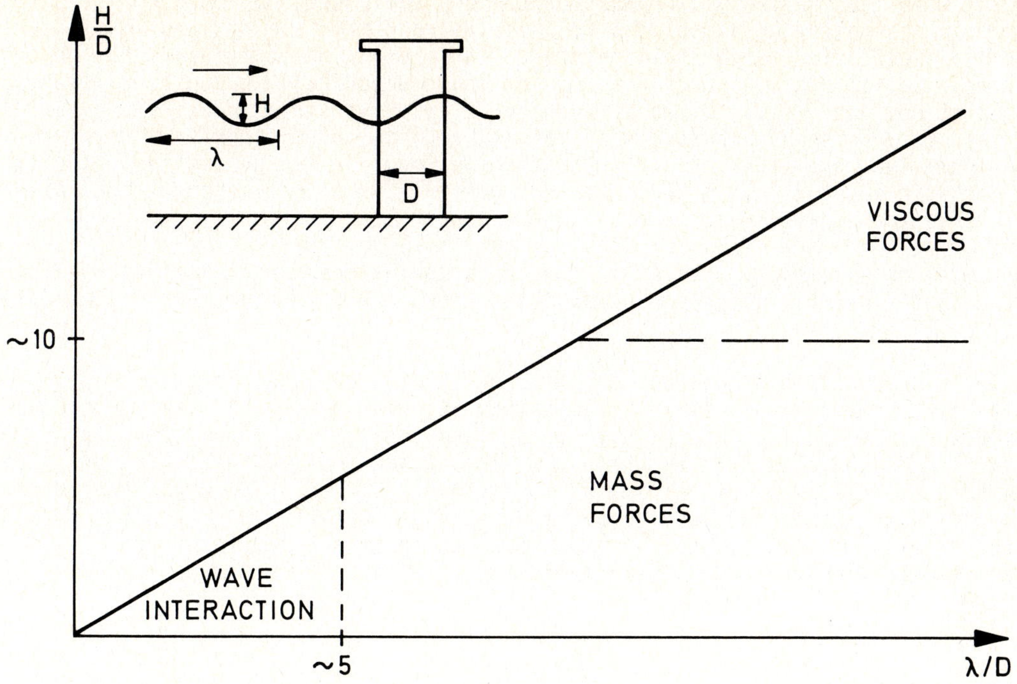

Figure 1
Relative importance of mass-, drag- and diffraction forces on marine structures

sible an equivalent diameter would be approximately 10 m. This implies λ/D = 30, H/D = 3, which means that the hydrodynamic forces are mainly potential flow forces in phase with the undisturbed local fluid acceleration. Wave diffraction and viscous forces are of less significance. For a jacket an equivalent diameter is approximately 1 m. This implies that viscous forces are most important. With viscous forces we do not mean shear forces, but pressure forces due to separated flow. The examples above are for an extreme wave condition. In an operational wave condition the relative importance of viscous and potential flow effects are different. We should have in mind that Fig. 1 only provide a very rough classification. For instance if a resonance oscillation in rolling occur, the damping may very well be significant influenced by viscous effects like eddymaking even if the wave diffraction is the most important according to Fig. 1.

In general, compressibility has no effect on the flow around marine structures. One possible exception is for impact between a structure and water. In that case air flow may also matter.

We will in the following text try to analyze some important marine hydrodynamical problems in more detail. We will try to examine the problem mostly from a numerical and theoretical viewpoint. The purpose is not to give a review, but try to give an indication of what tools are available for analysis and what are the unsolved problems. We have singled out three main items. These are a) ship motions at forward speed b) slowly varying and mean wave loads c) separated flow problems.

SHIP MOTIONS AT FORWARD SPEED

Generally speaking strip theories are good engineering tools to predict ship motions and sea loads. There are exceptions. We will in particular mention deficiencies when the frequency of encounter between the waves and the ship is low. This occurs in following waves. Strip theories have also limited applicability at high ship speeds, i.e. high Froude number. The prediction of relative motion is not always satisfactory, and one has to introduce empirical damping coefficients to predict realistic roll values at resonance. Even if strip theories are linear theories based on small amplitudes of motion, they may predict realistic values for extreme sea conditions. But linear theories have of course limited validity when the ship bow goes out of the water or green water on deck occur. Good predictions in these cases must be attributed to poor luck.

One example on a commonly used strip theory is the Salvesen-Tuck-Faltinsen [2] (STF) method. We will briefly discuss this method and try to explain its deficiencies.

Consider a ship advancing at constant mean forward speed with arbitrary heading in regular sinusoidal waves. It is assumed that the resulting oscillatory motions are linear and harmonic. Let (x,y,z) be a right-handed orthogonal coordinate system fixed with respect to the mean position of the ship, with z vertically upwards through the center of gravity, x in the direction of forward motion and the origin in the plane of the undisturbed free surface (see Fig. 2). We note that we assumed regular incident waves, but the combination of the results to an irregular sea is straightforward for a linear system.

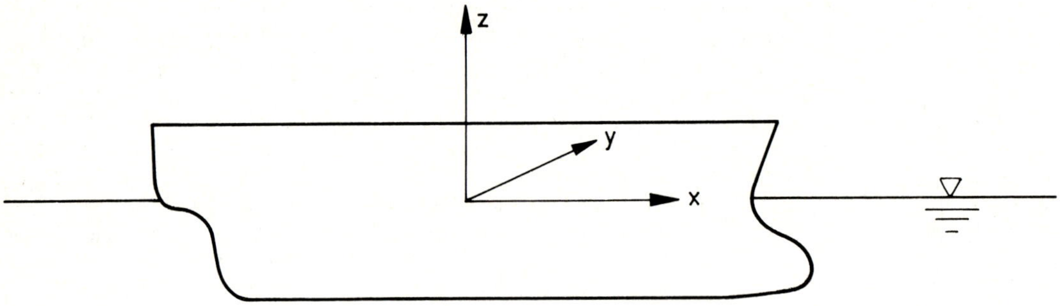

Figure 2
Coordinate system

If viscous effects are disregarded the fluid motion can be assumed to be irrotational, so that the problem can be formulated in terms of potential flow theory. The total velocity potential Φ must satisfy the three-dimensional Laplace equation. Separating the velocity potential $\Phi(x,y,z;t)$ into two parts, one the time-independent steady contribution due to the forward motion of the ship and the other the time-dependent part associated with the incident wave system and the unsteady body motion, we get

$$\Phi(x,y,z;t) = [-Ux + \phi_S(x,y,z)] + \phi_T(x,y,z)e^{i\omega t} \tag{1}$$

Here $-Ux + \phi_S$ is the steady contribution with U the forward speed of the ship, ϕ_T is the complex amplitude of the unsteady potential, and ω is the circular frequency of encounter in the moving reference frame. It is understood that real part is to be taken in expressions involving $e^{i\omega t}$.

By linearizing the free surface condition and neglecting the interaction with the steady velocity potential ϕ_S we can write the free surface condition as

$$\left[(i\omega - U\frac{\partial}{\partial x})^2 + g\frac{\partial}{\partial z}\right]\phi_T = 0 \qquad \text{on } z = 0 \tag{2}$$

In addition ϕ_T satisfies linearized body boundary conditions and radiation conditions.

The following two simplifications are made in the STF method

$$i\omega \gg U \frac{\partial}{\partial x} \tag{3}$$

$$\frac{\partial}{\partial x} \ll \frac{\partial}{\partial y}, \frac{\partial}{\partial z} \tag{4}$$

The first condition implies that the free surface condition can be written as

$$-\omega^2 \phi_T + g \frac{\partial \phi_T}{\partial z} = 0 \qquad \text{on } z = 0 \tag{5}$$

This simplification says that the theory is limited to high frequencies, i.e. that the strip method breaks down in following seas when the frequency of encounter is low. Further the theory is limited to small Froude numbers and is questionable in ship regions where the x-derivative of the velocity potential is high. The latter may be true in the bow region.

The argument for stating the second condition (4) is that the ship can be considered a slender body. This condition implies that the three-dimensional Laplace equation reduces to a two-dimensional Laplace equation.

By also taking into account the body boundary condition, we see that the STF-method is a strip theory for the velocity potential due to forced motion in the different six rigid body modes of motion. This implies that there is no hydrodynamic interaction between different cross-sections along the ship. In the STF-method the diffraction potential due to incident waves on the restrained ship is not calculated. Instead a generalized Haskind relation is used. The numerical solution of the two-dimensional problems for the forced motion potentials are not trivial, but well documented standard procedures exist. We will not deal with this topic here.

In order to get a better understanding of how we can improve the strip theory for low frequencies, high Froude number and in the bow region, it is useful to have a far-field point of view of the ship. In the far-field the ship shrinks down to a line and its physical effect can be represented by a line distribution of singularities. In the heave and pitch case and for the symmetric diffraction problem the singularities are sources that satisfy the linear free surface condition (2) and the radiation condition. The use of equation (2) is an approximation that can be questioned. But we will have this as a starting point.

Let us begin with zero forward speed and consider one source only. In the wave system generated by this source there is only one wave length. By combining the sources into a line distribution we see that the sources create a hydrodynamic interaction between the different cross-sections of the ship. This is not accounted for in strip theories like the STF-method. On the other hand it can be shown that in the case of high frequency heave and pitch motion that a first order approximation of the longitudinal interaction is zero close to the body. (Ogilvie and Tuck [3]). This is a validation of the strip theory. In the high frequency head sea diffraction case we cannot neglect the interaction. In this case the source density is likely to be rapidly oscillating like $\sigma(x)e^{ikx}$, where k is the wave number of the incident waves and σ is slowly varying with x. From the analysis of Faltinsen [4] the resulting flow field close to the body represents waves propagating along the ship in the same direction as the incident waves. To the lowest order it seems as if the disturbance created by the ship is carried along the ship. Looking on what is happening at some cross-section there is therefore an integrated effect of what is happening at sections from the forward perpendicular of the ship up to the cross-section.

The information that we get from the far-field point of view in the high frequency head sea diffraction case is telling us that condition (4) is questionable in this case. There will be a rapidly oscillating behaviour in the x-direction due to the term e^{ikx}. This means that the order of magnitude of $\frac{\partial}{\partial x}$, $\frac{\partial}{\partial y}$, $\frac{\partial}{\partial z}$ may be comparable and that a two-dimensional Helmholtz equation is in this case a more appropriate near-field equation that the two-dimensional Laplace equation. Further this example warns us that it is not only geometric considerations of the hull shape that determines the relationship between the order of magnitude of $\frac{\partial}{\partial x}$, $\frac{\partial}{\partial y}$ and $\frac{\partial}{\partial z}$ near the ship. We also have to consider characteristic length scales associated with the generated wave systems.

By using the information we get from the far-field picture it is possible to evaluate the hydrodynamic interaction effect in the near-field for any frequency. This has been done by Newman and Sclavounos in a series of papers [5], [6], [7] with satisfactory results.

It is of course also possible to solve the problem by a three-dimensional sink-source technique. This is a common procedure for offshore structures. Other boundary element methods and hybrid finite element methods have also been used.

Let us now turn to the forward speed problem. The wave system generated by a source at forward speed is more complicated than in the zero-speed case. There are many different wave systems. Let us show this with two examples for the deep water case. One case is for $\tau = \frac{\omega U}{g} < \frac{1}{4}$ and the other case is for $\tau > \frac{1}{4}$ (See Fig. 3 and 4). When $\tau > \frac{1}{4}$ there is no upstream far-field wave effect from a transla-

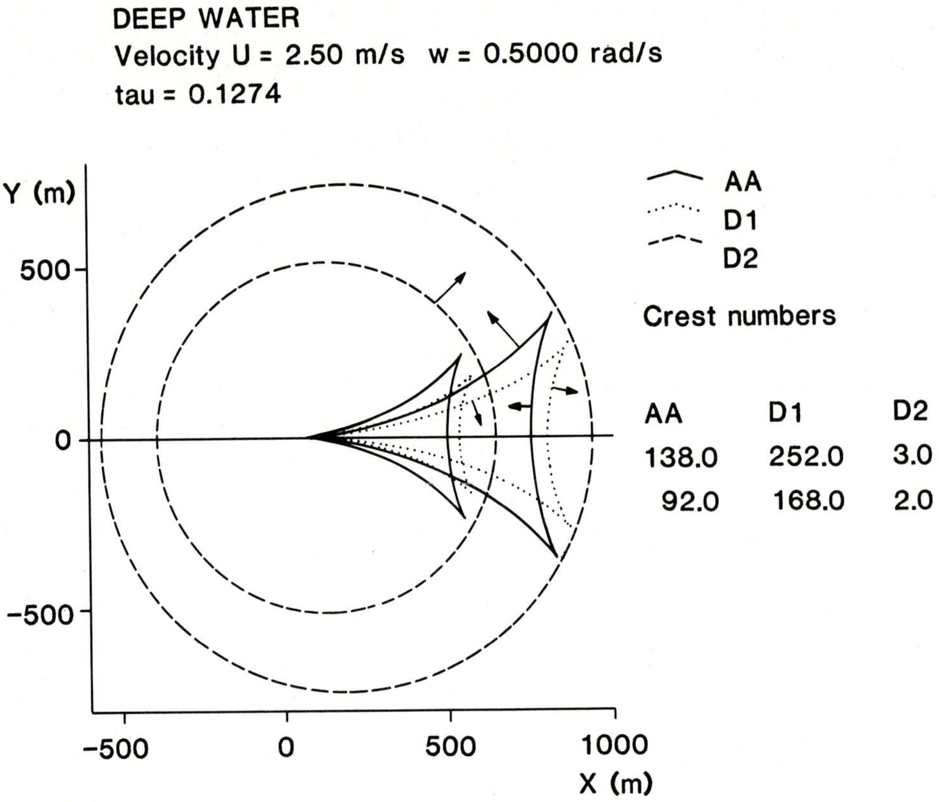

Figure 3
Wave system due to harmonically oscillating source, $\tau < 1/4$

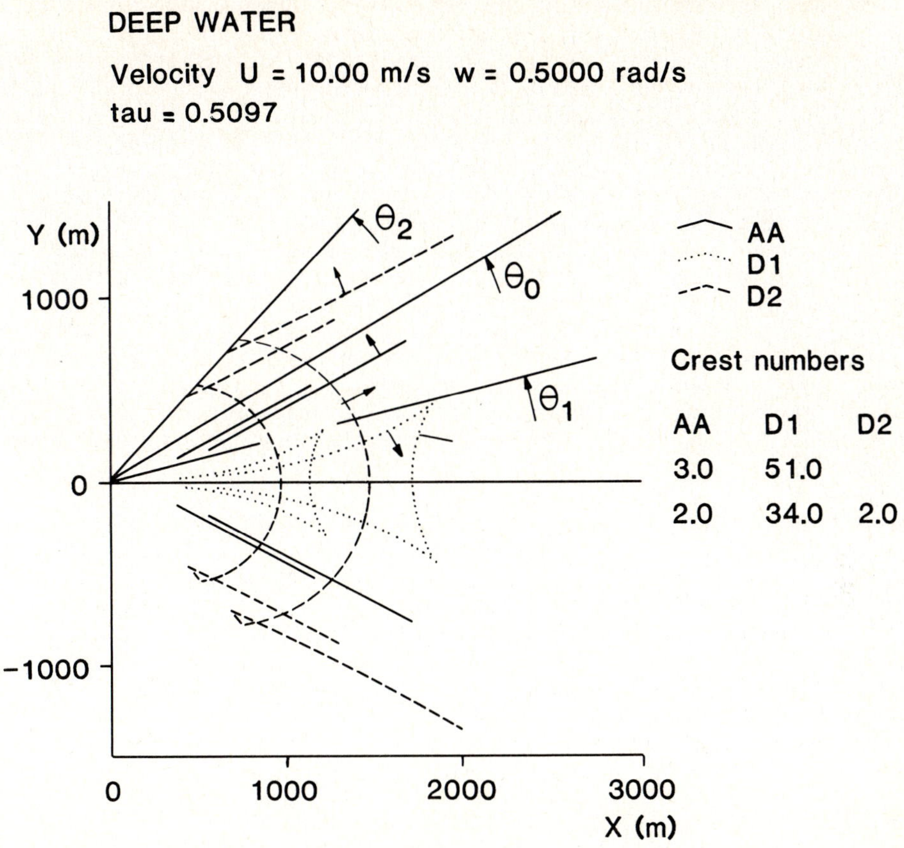

Figure 4
Wave system due to harmonically oscillating source, $\tau > 1/4$

tory harmonic oscillating source. In the figures there are written up crest numbers associated with the different wave systems AA, D1, D2. There are given two numbers and shown two "crests" for each wave system. What we call "crests" are of course only real crests for specific time instants with periodic intervals. The difference between the two wave crest numbers for each wave system says how many "crests" there are between the two "crests" which we have shown for each wave system AA, D1, D2. For each wave system we may have both divergent and transverse waves. The latter notation is commonly used for the two types of waves we see behind a ship in steady forward motion in calm sea.

By combining the sources into a line distribution representing the ship in the far-field it follows that there is in general a complicated hydrodynamic interaction between the different cross-sections of the ship. We may remember that strip theories show no hydrodynamic interaction. Ogilvie and Tuck [3] have studied the far-field solution near the ship in the case of forced high frequency heave and pitch motion. To the lowest order they end up with strip theory but the next order term account for the hydrodynamic interaction between cross-sections. In practical calculations this term has to be included. Faltinsen [8] has studied the bow-flow when $\tau > 1/4$. By representing the far-field effect by a single source like we presented in Fig. 4, he found that the dominant behaviour near the ship in the bow region is due to the divergent waves. In the near-field solution in the bow there is a strong upstream effect from the ship. Along the parallel midbody it is expected that the transverse waves are also of importance. Newman and Sclavounos have used an inner expansion of the far-field expansion near the ship which is in principle valid for all frequencies. This is used as information to account for the hydrodynamic interaction in the near-field. Newman and Sclavounos' unified theory represents an improved ship motion slender body theory from a rational point of view. But there are still possibilities for improvements. One can ask whether they take proper account of the forward speed effect. For instance they use the rigid free surface condition for the steady wave potential near the ship. With rigid free surface condition we mean that the free surface acts like a wall. How good is this for high Froude number? Another question is whether equation (2) is the right linear free surface condition to use. One has dropped out all interactions with the steady wave potential and assumed that U is the dominant steady velocity. This is for instance certainly not true near a stagnation point on the ship. This is of course a very special case. But it should warn us. To make further improvements in ship motion predictions at high Froude number it is felt that one first has to study the steady wave potential problem in more detail. This is the same as is referred to as the wave resistance problem in ship hydrodynamics and is known to be a difficult problem. It is fair to say that there exist no general practical numerical method for the wave resistance problem. The generalization of unified theory to finite water depth has been studied by Børresen [9].

There exists complete three-dimensional linear methods using singularities satisfying the classical linear free surface condition (2). (See for instance Inglis and Price [10]). In the future with increasing computer capabilities the three-dimensional methods may become practical tools to predict ship motions. But they will still have the deficiency that they are based on linear theory.

There exist nonlinear theories based on perturbation procedures. But it is hard to proceed longer than to second order in wave amplitude. It is felt that this is not the way to go in the case of extreme ship motions when the bow goes out of the water.

In order to be able to handle the finite amplitude potential flow problem of structures in waves it is felt that it is necessary to go back to studying two-dimensional flow problem without forward speed. There has been some success in numerically predicting breaking waves and its effect on submerged structures (See Vinje et al [11] for instance). Their numerical analysis is based on satisfying the exact nonlinear free surface condition at each timestep in a numerical time integration procedure. For a surface piercing body there are numerical difficulties in properly handling the intersection between the free surface and the body. There are also difficulties in formulating the proper matching condition far away. Can we for instance assume linearity and state that the waves generated by the body is only outgoing waves? These are interesting research problems.

SLOWLY VARYING AND MEAN WAVE LOADS

Mean wave loads and slowly varying drift loads are of importance in several contexts for marine structures. Examples are in the design of mooring and thruster systems, analysis of offshore loading systems, evaluation of towing of large gravity platforms from the contruction site to the operation site, added resistance of ships in waves, performance of submarines close to the free surface, investigation of tilting and capsizing of semisubmersibles, and analysis of slowly oscillating heave, pitch and roll of large volume structures with low waterplane area.

In a potential flow model driftforces are due to a structure's ability to create waves. The consequence of this is that driftforces are small in a potential flow model when massforces dominates. An example on the latter is semisubmersibles and tension leg platforms. But viscous effects may also contribute to driftforces.

Let us try to give a simple explanation of why we get a mean wave force on a structure in regular incident harmonically oscillating waves. For a surface piercing body a major contribution is due to the relative vertical motion between the structure and the waves. This causes some of the structural surface to be part of the time in the water and part of the time out of the water. Examining the pressure in one of the points in this surface zone, it is obvious from Fig. 5 that the result is a non-zero mean force even in regular harmonic oscillating waves. But there are also other causes. One of them we get by averaging the quadratic term in Bernoulli's equation.

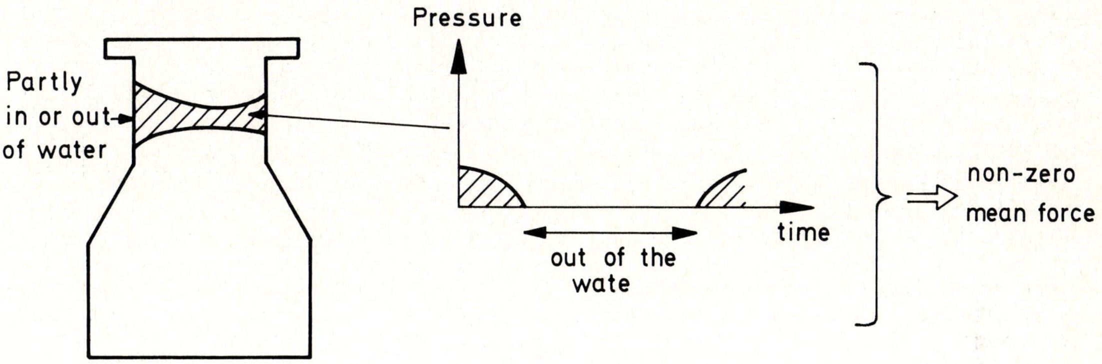

Figure 5
Contribution to mean wave force

There are several causes why viscous effects contribute to mean driftforces. By using a simple cross-flow principle for the flow around the pontoons and columns of a semisubmersible, decomposing the forces into components along an earth fixed coordinate system and averaging the forces over one period, we will find non-zero mean wave loads that are proportional to the third power of the wave amplitude. By evaluating the viscous dragforces on the surface zones of the platform that is part of the time in and part of the time out of the water we will also find non-zero mean forces. The Stokes drift in waves is another origin to viscous forces, that causes mean wave loads.

Hsu and Blenkarn [12] have given a simple and illustrative explanation of the slowly varying wave loads on marine structures in irregular sea. They imaging the irregular waves system divided into approximate regular wave parts. In each regular wave part the structure will experience a constant drift force (or moment). In this way slowly varying wave loads are obtained.

Let us now mainly concentrate on potential flow problems. At zero forward speed or zero current the mean wave forces in regular waves may be calculated by either a direct pressure integration method or by using the equations for conservation of momentum in the fluid. The latter approach may be combined with using the equations for conservation of energy in the fluid. There are standard procedures used for calculating mean forces on marine structures in regular waves and no current. Since the evaluated quantities are small, a high degree of accuracy is needed in the calculations. The results are sensitive to wave heading, structural form, structural motion, wave length and wave height. It is straightforward to combine the results for regular waves to obtain mean wave drift loads in irregular sea. It can be shown that mean wave loads are sensitive to the sea state.

The effect of current is very often neglected in the calculations. This is an area which require further research. Based on known results for added resistance of ships in waves, it is not unlikely that the interaction with current may represent a 20% increase in the mean wave loads. The reason why we refer to the added resistance results is that the forward motion of the ship may be considered to have similar physical effect as a current. But in the current case we must allow for any direction between the incident stream and the structure. Let us for instance consider a ship in a beam sea condition and let the current be in the same direction as the wave propagation direction. In order to find the mean wave drift loads one would assume small incident wave amplitudes and write the velocity potential as

$$\Phi = \sum_{i=0}^{\infty} \varepsilon^i \phi_i \tag{6}$$

where ε is proportional to the incident wave amplitude. The first term i = 0 is the steady flow and the second term i = 1 is the linear harmonically oscillating time dependent flow. In order to find the mean wave drift loads which are proportional to wave amplitude square it is only necessary to find the two first terms in equation (6). If we now use the same coordinate system as in Fig. 2, a first try at the linear free surface condition for ϕ_1 would be to use equation (2) with $\frac{\partial}{\partial x}$ replaced by $\frac{\partial}{\partial y}$. But as we discussed in connection with the ship motion chapter, this implies that U is the dominant steady velocity. This is certainly less true in a transverse steady flow condition like we are talking about here than in a longitudinal steady flow condition like we were talking about in the ship motion problem at forward speed. We would therefore have to solve for ϕ_S and correct the linear free surface condition (2) so that it also involves interaction terms with ϕ_S.

In solving for ϕ_S it is likely that we can use the rigid free surface condition. The reason for that is that current velocities normally are small and corresponding steady wave effects are not significant. But since we are talking about steady cross flow past a blunt body, we cannot neglect the fact that the flow separates. This will have a governing effect on the steady flow around the ship. The separation will also lead to unsteady effects. This leads into unsolved problem which we will discuss in the next chapter.

Let us now turn to the slowdrift oscillation problem. We will disregard the effect of the current. As long as the amplitude of oscillations are sufficiently small and there are no sharp corners on the body, it is reasonable to neglect separated flow effects. We have given a simplified explanation of why we get slowly varying wave loads on marine structures in irregular sea. We will now try to give a more rationally based explanation.

Let us first consider irregular longcrested incident waves in deep water. We can write the incident wave potential correct to first order in wave amplitude as

$$\phi^I = \sum_{i=1}^{N} \frac{g A_i}{\omega_i} e^{k_i z} \sin(k_i x \cos\beta + k_i y \sin\beta - \omega_i t + \varepsilon_i) \tag{7}$$

Here t is the time variable, g the acceleration of gravity, β the angle between the wave propagation direction and the x-axis, ω_i the circular frequency of oscillation and k_i the wave number of wave component number i. ω_i and k_i are connected through the dispersion relationship $\omega_i^2/g = k_i$. The phase angles ε_i may be considered as random phase angles and the amplitudes A_i may be determined by a wave spectrum $S(\omega)$ characterizing the sea state. If the important part of the wave energy is concentrated between the circular frequencies ω_{min} and ω_{max} we divide the frequency interval ω_{min} to ω_{max} into N equal subintervals and call the midpoints of the ith interval ω_i. A_i is then determined by

$$\frac{A_i^2}{2} = S(\omega_i) \frac{\omega_{max} - \omega_{min}}{N} \tag{8}$$

In principle we should let $N \to \infty$, $\omega_{min} \to 0$ and $\omega_{max} \to \infty$ so that the sum in equation (7) becomes an integral, but we will keep it as a finite sum.

The solution procedure will now be to solve the hydrodynamic boundary value problem correct to second order. We will not formulate this here, but as an example let us show the second order free surface condition, i.e.

$$\frac{\partial^2}{\partial t^2}\phi_2 + g\frac{\partial \phi_2}{\partial z} = -\frac{\partial}{\partial t}|\nabla\phi_1|^2 + \frac{1}{g}\frac{\partial\phi_1}{\partial t}\frac{\partial}{\partial z}\left[\frac{\partial^2\phi_1}{\partial t^2} + g\frac{\partial\phi_1}{\partial z}\right] \quad \text{on } z = 0 \tag{9}$$

On the right hand side we note there are products between first order quantities. There exist standard procedures to find the first order potential ϕ_1. This has been briefly discussed in the last chapter. The second order problem is more complicated due to the presence of the inhomogeneous free surface condition. Faltinsen and Löken [13] have analyzed this problem in the case of incident beam sea waves on an infinitely long horizontal cylinder of arbitrary shape floating in the free surface. In order to get a feeling of how the solution will look let us not include the whole first order potential. Let us only consider ϕ^I. We then see that the right hand side of equation (9) involves product of terms like

$$\sin(k_i y - \omega_i t + \varepsilon_i)\cos(k_j y - \omega_j t + \varepsilon_j) \tag{10}$$

By elementary trigonometric rules we know that this is equal to

$$\frac{1}{2}\sin[(k_i + k_j)y - (\omega_i + \omega_j)t + \varepsilon_i + \varepsilon_j)] + \tag{11}$$

$$\frac{1}{2}\sin((k_i - k_j)y - (\omega_i - \omega_j)t + \varepsilon_i - \varepsilon_j)$$

From this we see there is a sum frequency term and a difference frequency term. It is the last term which is of interest to us when we want to evaluate the slowly varying wave forces on structures. This last term will be one of many causes to the slowly varying forces. There will be other contributions which arise from nonlinearities in the body boundary condition and a consistent second order formulation of the forces. Let us study the slowly varying force term which arises from the inhomogenity of the second order free surface condition in some more detail. If the y-variation along $z = 0$ is $\sin((k_i - k_j)y)$ or $\cos((k_i - k_j)y)$, a particular solution of the ϕ_2 problem must have a z-dependence of the form $e^{|k_i-k_j|z}$. This implies that there is a small exponential decay with depth and that we in most practical cases should have included the effect of finite water depth in the ϕ_2-solution. This was not done by Faltinsen and Löken [13]. It would be interesting to study this finite depth effect. To my knowledge the ϕ_2-difference frequency problem has not been numerically solved for a general three-dimensional structure in irregular sea. It would also be of interest to study the sum frequency problem. In some problems "dangerous" natural periods may be lower than periods with significant wave energy. Then the sum-frequency terms may be a possible excitation mechanism of resonant oscillations.

We have discussed the slowly varying wave force problem for longcrested sea. But the procedure may be generalized to shortcrested sea by writing the first order incident wave potential as

$$\phi^I = \sum_{i=1}^{N}\sum_{j=1}^{M}\frac{g A_{ij}}{\omega_i} e^{k_i z}\sin(k_i x\cos\beta_j + k_i y\sin\beta_j - \omega_i t + \varepsilon_i) \tag{12}$$

To my knowledge nobody has studied this problem in a more general case.

For a recent comprehensive review of the slowly varying and mean wave load problems for marine structures it is recommended to read Ogilvie's article [14].

SEPARATED FLOW PROBLEMS

Flow around bluff bodies and associated vortex shedding is of importance in several marine hydrodynamic problems. Examples are wave and current induced loads on piles, jackets, risers and pipe lines, roll damping of ships and barges, slow drift oscillation damping of moored structures in irregular sea and wind, anchor line damping, large amplitude maneuvering forces on ships and still water resistance of blunt ship forms. All these cases are very high Reynolds number flows. For instance for a survival condition of a jacket the Reynolds number may be up to 10^7.

The state of the art in calculating the loads on bluff bodies at very high Reynolds numbers is not satisfactory. Traditionally the Morison's equation has been used in the offshore industry to calculate wave and current loads on cylindrical shapes applied in structural work. The formula is semiempirical and embarassing simple. For those who are not acquainted with the formula we will briefly discuss it. For simplicity let us consider a fixed vertical cylinder (see Fig. 6).

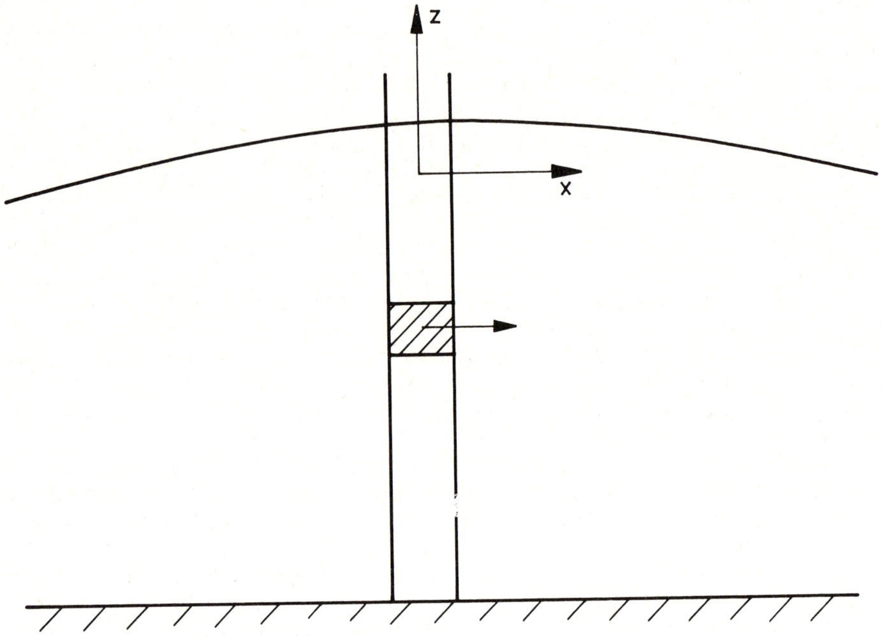

Figure 6
Wave forces on vertical piles

The horizontal force pr. unit length on a strip of the cylinder can according to Morison's equation be written as

$$dF = \rho\pi \frac{D^2}{4} C_M a_x + \frac{\rho}{2} C_D D \,|u|u \tag{13}$$

The force direction is in the wave propagation direction. Further ρ is the mass density of the water, D is the cylinder diameter, u and a_x are horizontal undisturbed fluid velocity and acceleration at the midpoint of the strip. The mass and dragcoefficients C_M and C_D have to be empirically determined and are dependent on many parameters like Reynolds number, Keulegan-Carpenter number, a relative current number and a roughness ratio. The application of Morison's equation in the free surface zone is particularly questionable. A limited number of full scale measurements indicate that one should be careful in applying equation (13) to that part of the pile surface that is part of the time in the water and part of the time out of the water. Furthermore Morison's equation cannot at all predict the oscillatory forces due to vortex shedding. The most important part of the oscillatory forces are the lift forces, which may have a stochastic behaviour even in regular incident flow. The correlation of the lift forces along the cylinder is an unsolved problem. We often see Morison's equation generalized to non-fixed structures. But important hydroelastic phenomena like "lock-in" cannot at all be explained in this way.

It is of course easy to critizise Morison's equation. But there is nothing better from a practical point of view. The reason for this is the very complicated flow picture that occur for separated flow around marine structures. No one has yet solved the Navier Stokes equations satisfactorily for so high Reynolds numbers that we are interested in. Instead one has tried to follow a different approach by using thin free shear layer models. The discrete vortex method is an example, on the latter (see for instance Sarpkaya and Shoaff [15]). We will not try to give a review of all the computational efforts here, but instead try to report some of our own work with thin shear layer models. We will use this as a mean to focus on some of the numerical and physical problems associated with separated flow around marine structures.

Our model differ from the discrete vortex methods. It has been described in detail by Faltinsen and Pettersen [16]. It is assumed that the vorticity is concentrated in thin boundary layers and free shear layers. At each time step one has to solve a potential flow problem outside the boundary layers and free shear layers. We decided to solve for the velocity potential and used a distribution of sources and dipoles to represent the potential flow. In order to update the position of the free shear layers at each time step we use the fact that the velocity potential jump across the free shear layer is convected with the mean velocity of the two sides of the free shear layer. In order to determine the separation point when the flow separates from a continuously curved surface one has to couple the potential flow calculation with a viscous flow calculation of the boundary layer. In developing the method we had in mind that the model should ultimately be able to handle general body configurations including several bodies in interaction as well as arbitrary motion of the bodies and incident flow.

In Fig. 7 is shown an example on numerical result for the free shear layer position at one particular time instant for steady incident flow past a circular cylinder. We will use this figure to focus on some of the problems with the method. One of them is to predict the separation points. In our analytical model the free shear layer has to leave the body surface tangentially at a separation point. In this way we avoid stagnation points at both sides of the free shear layer at the separation point. Stagnation points on both sides would mean no shed vorticity into the fluid. By a local potential flow solution near the separation point it is possible to show analytically that the pressure gradient ahead of the separation point has a square root singularity at the separation point. If we combine our potential flow model with a conventional boundary layer calculation procedure up to the separation point, the consequence may be that the separation point is continuously forced toward the forward stagnation point and end up to be unrealistic. There is evidently something wrong with the physical modelling around the separation point. The viscous flow has to interact with the potential flow so that the steep pressure gradient ahead of the separation point is modified. One way of doing this would be to generalize to unsteady problems the "triple-deck" solution studied by for instance Sychev [17] and Smith [18] for steady laminar flow. This has been done by Aarsnes [19].

In a realistic full scale condition for marine structure we will both have the effect of marine growth and turbulent boundary layers. The "triple-deck" solution cannot be used for this case. On the other hand calculations by Aarsnes [19] do not show the necessity of modifying the potential flow pressure gradient in the turbulent flow case. The reason to this is partly that a turbulent flow can sustain a larger adverse pressure gradient than a laminar flow before it separates.

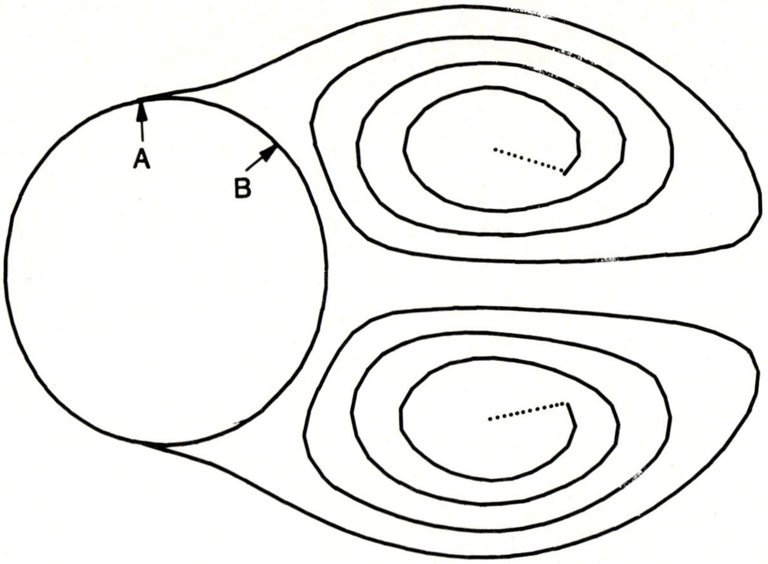

Figure 7
Example on numerical wake picture around a circular cylinder at one time instant

Another reason is due to numerical errors. Infinite or extremely large adverse pressure gradient will not occur in the numerical results.

All workers within discrete vortex modelling seem to agree on that it is necessary to reduce the circulation relative to what is predicted by potential theory. We have found that the amount of shed vorticity is sensitive to an accurate numerical modelling around the separated point. But there is obviously a need to reduce the circulation. One way of doing this would be to introduce secondary separation. Due to high backwards velocities induced by vortices close to the cylinder we may get separation points on the back of the cylinder. This is exemplified by point B in Fig. 7. The effect of secondary separation would be a reduction in vorticity in the wake.

An example on secondary separation is shown in Fig. 8. The shear layer emanating from the secondary separation point has wiggles. There is a possibility that is due to physical instabilities, but we have not examined this yet. In practical calculations it is necessary to "smooth" out the wiggles. It should also be pointed out that we have not been able to use a boundary layer calculation to predict the secondary separation point. Instead we relate the secondary separation point to the vicinity of peaks in the velocity outside the boundary layer. The introduction of secondary separation causes a more realistic pressure distribution around the cylinder. Without secondary separation unrealistic high pressure peaks may occur in the wake.

Another reason to reduction of vorticity may be three-dimensional effects associated with for instance instabilities. We could go on and mention other problems, but let us instead state some results. Aarsnes was able to find a mean C_d-value for a circular cylinder in subcritical flow on 1.4 and for transcritical flow on 0.7. The corresponding mean experimental values seem to be 1.2 and 0.6, respectively. The Strouhals numbers are correctly predicted, and the pressure distribution seemed reasonable both for subcritical and transcritical flow. The mean separation point was estimated to be 75^{0} and to 109^{0} for laminar and turbulent flow, respectively.

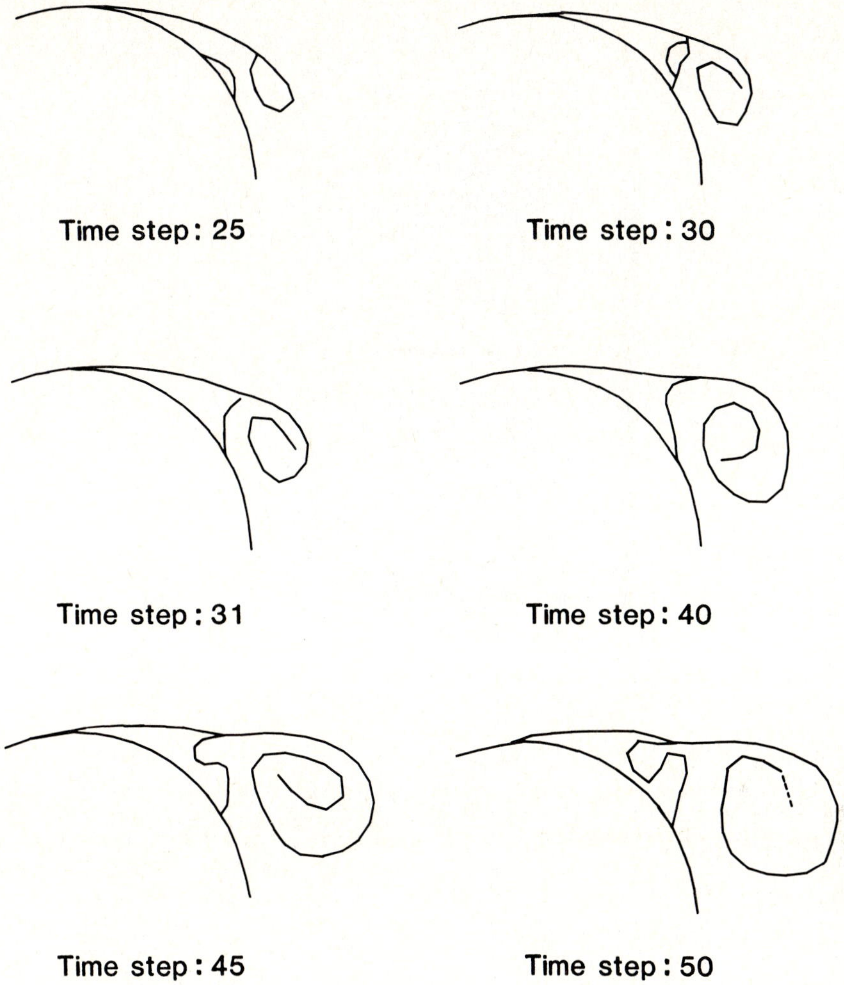

Figure 8
Development of secondary shear layer

The method was also applied to ship cross-sections in cross-flow. This is not without problems, but generally speaking the approach seems to work. For two-dimensional flow the method overpredicted the dragcoefficient by 10-20%. But if we used the two-dimensional results in a strip theory fashion to construct dragcoefficients for a ship in crossflow the theory overpredicted the dragcoefficient by 30%. One important reason to this three-dimensional effect is the closure of vortex lines in three dimensions. This results in strong vertical vortex lines (or surfaces) close to the ship ends. The consequence of this is a reduced effective inflow to the cross-sections along the ship.

The three dimensional effect was only evaluated and explained in a qualitative manner by Aarsnes. But we are presently generalizing our approach to three dimensions. In principle this is relative easy to do when we operate with the velocity potential and dipole sheets and stay out of the problem with predicting separation lines. This does not mean that it is easy to do in practice. Some preliminary three-dimensional results for cross-flow past a circular disk is shown in Fig. 9. These calculations have been done by N. Skomedal.

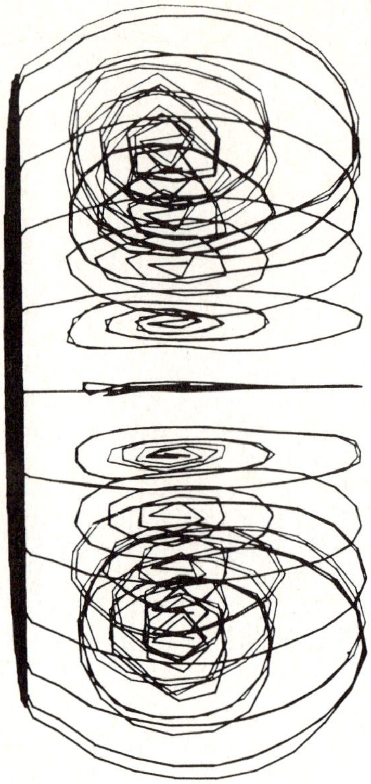

Figure 9
Example on numerical three-dimensional wake picture

CONCLUSIONS

An overview over important wave load problems for ships and offshore structures are given. Three main topics are dealt with in detail. These are a) Ship motions at forward speed b) Slowly varying and mean wave loads and c) Separated flow problems. The purpose was not to give a review, but try to give an indication of what tools are available and what are the unsolved problem. Particularly for viscous problems there exist many unsolved theoretical problems of practical importance.

REFERENCES

[1] Mac Camy, R.C. and Fuchs, R.A., Wave Forces on Piles: A Diffraction Theory, U.S. Army Corps of Engineering, Beach Erosion Board, Techn. Memo, No. 69 (Washington 1954).

[2] Salvesen, N., Tuck, E.O. and Faltinsen, O., Ship Motions and Sea Loads, Trans. SNAME, Vol. 78 (1970).

[3] Ogilvie, T.F. and Tuck, E.O., A Rational Strip Theory of Ship Motions: Part I, Report No. 013, The Department of Naval Architecture and Marine Engineering, The University of Michigan (March 1969).

[4] Faltinsen, O., Wave Forces on a Restrained Ship in Head-Sea Waves, Proc. 9th ONR, Paris (August 1972).

[5] Newman, J.N., The Theory of Ship Motions, Adv. Appl. Mech., 18 (1978).

[6] Newman, J.N. and Sclavounos, P., The Unified Theory of Ship Motions, Proc. 13th ONR/SNH (Tokyo 1980).

[7] Sclavounos, P.D., The Diffraction of Free-Surface Waves by a Slender Ship, Ph.D. Thesis, MIT (1981).

[8] Faltinsen, O., Bow Flow and Added Resistance of Slender Ships at High Froude Number and Low Wave Lengths, JSR, Vol. 27, No. 3 (September 1983).

[9] Børresen, R., The Unified Theory of Ship Motions in Water of Finite Depth, Dr.ing.thesis, Division of Marine Hydrodynamics, The Norwegian Institute of Technology (February 1984).

[10] Inglis, R.B. and Price, W.G., The Influence of Speed Dependent Boundary Conditions in Three-Dimensional Ship Motion Problems, ISP, Vol. 28, No. 318 (February 1981).

[11] Vinje, T. et al., A numerical Approach to Nonlinear Ship Motion, Proc. 14th ONR/SNH, Ann Arbor (1982).

[12] Hsu, F.H. and Blenkarn, K.A., Analysis of peak mooring forces caused by slow vessel drift oscillations in random seas, Offshore Technol. Conf. (Houston 1970) Paper 1159.

[13] Faltinsen, O.M. and Løken, A.E., Slow drift oscillations of a ship in irregular waves, Applied Ocean Research, Vol. 1, No. 1 (1979).

[14] Ogilvie, T.F., Ship and Platform Motions, International Workshop on University of California, Berkeley (October 1983).

[15] Sarpkaya, T. and Shoaff, R.L., A discrete-vortex analysis of flow about stationary and transversely oscillating circular cylinders, NPS-69SL79011, Naval Postgraduate School, Monterey (California 1979).

[16] Faltinsen, O. and Pettersen, B., Vortex shedding around two-dimensional bodies at high Reynolds number, 14th Symposium on Naval Hydrodynamics, The University of Michigan, Ann Arbor (August 23-27. 1982).

[17] Sychev, V.V., Laminar separation, English translation from Izveshya Akademii Nauk SSSR, Mekanika Zhidkosti, Gaza, No. 3. pp. 47-59 (May - June 1972).

[18] Smith, F.T., Laminar Flow of an Incompressible fluid past a bluff body, the separation, reattachment, eddy properties and drag, Journal of Fluid Mech., (1979 Vol. 92, part 1, pp. 171-205.

[19] Aarsnes, J.V., Current forces on Ships, Dr.ing.thesis, Division of Marine Hydrodynamics, The Norwegian Institute of Technology (1984).

Theoretical and Applied Mechanics
F.I. Niordson and N. Olhoff
Elsevier Science Publishers B.V. (North-Holland)

THE RECOVERY OF OIL FROM UNDERGROUND RESERVOIRS

E.J. Hinch

Department of Applied Mathematics and Theoretical Physics
University of Cambridge,
Silver Street, Cambridge CB3 9EW
England.

The first part of this paper reviews some of the processes involved in extracting oil from underground reservoirs. Typical magnitudes of the quantities relevant to the fluid mechanics are given. The second part of the paper describes recent studies which aim at generalising Darcy's law to the flow of two immiscible fluids through a porous medium.

DISCLAIMER

The reader must be warned that I am a fascinated newcomer and not an authoritative expert on the problems of fluid mechanics in the oil industry. I have dared to attempt this review because I myself was frustrated by the apparent absence of a brief introduction couched in terms readily digested by fluid dynamicists. I am happy to acknowledge my debt to the patience of many friends in the oil industry, particularly in the Schlumberger Corporation in the USA, England and France. I apologise to them and the reader for my inevitable failures to understand all that they have said.

OIL RESERVOIRS

Oil is found underground, trapped in the interstices of a porous rock like sandstone (and also limestone). An oil bearing stratum will have above it an impermeable rock such as clay which prevents the oil from migrating upwards. There will also be a horizontal variation of the depth of the stratum due to an upward dome or an upward slant terminated by an impermeable fault which prevents the oil from moving sideways. At lower levels in the oil bearing stratum, water, which is heavier than oil, will fill the interstices of the porous rock. Sometimes there can be a *gas cap* at the highest part of the stratum.

When the porous rock was formed as a sediment at the bottom of the sea, water filled all the interstices. Later in its geological evolution, hydrocarbon products from some rotten vegetation or marine sediments migrated up the stratum as far as they could go (dependent on the thermal and pressure history). As a result, some water usually remains trapped in amongst the oil, as *connate water*.

Oil reservoirs vary enormously in size. A new find can have a horizontal extent of 100 km, while an odd pocket in an old field might be only 500 m. Perhaps several kilometers is not untypical of the horizontal dimension of one working part of a field. The thickness of an oil-bearing stratum is also very variable. An economic deposit is likely to be 100 m thick, but with strong geological variations effectively dividing it into four or five independent production zones 10 m thick. These oil deposits were first discovered just under the surface of the Earth. Now it is becoming economic to drill as deep as 5 km.

At the depth of 3 km, which is typical of the North Sea, the geothermal gradient of 30°C/km means that the temperature has risen to 100°C, while the hydrostatic pressure in water would be 300 atmospheres. As the more dense rock is a deformed solid, its stress is not isotropic: the vertical normal stress is about 5/2 times the hydrostatic pressure of water with the horizontal components of the order of 400 atmospheres. It is this horizontal stress which pushes onto the side of an oil well when a hole is drilled (if it is drilled vertically), and which also gives the trapped fluids (gas, oil and water) their high pressure.

Oil is extracted from the reservoir by drilling wells, roughly 200 m apart. One drilling rig in a fixed location can sink many (10) wells using a special drilling bit which slowly veers the well from the vertical. Inclined (or *deviated*) wells obviously offer an advantage of greater surface area if they traverse an oil stratum acutely. The standard diameter of a bore hole is 22 cm, with 60 cm being used in some initial stretches. It takes about a month of continuous drilling to get 3 km down, but frequently serious problems cause delays.

Drilling muds

When the oil wells are drilled, the hole, which is somewhat wider than the rotating drill shaft (known as the drill stem), is kept filled with a drilling mud. Drilling muds consist of water and a variable amount of clay, which is expensive, together sometimes with a small amount of polymer and other additives, which are very expensive. The drilling mud performs three equally important tasks.

First, the density of the mud is controlled by adding varying quantities of clay. A density of 1.3 gm/cm^3 would be typical, and if need be 1.8 is possible. The purpose of controlling the density is to ensure that the hydrostatic pressure in the mud exceeds the horizontal stress in any soft rock which could flow plastically into the hole and cause a blockage. (This becomes a more serious problem with inclined holes which also feel some part of the larger, vertical component of stress.) The mud pressure at the drilling bit must similarly exceed the high pressure of any anticipated trapped fluids, if a *blow out* is to be avoided. If, on the other hand, there is no imminent danger from soft rock or trapped fluids, then the clay content can be reduced, because it is expensive and because a high mud pressure can produce a flow out into a porous stratum, thereby loosing mud and also blocking the pores against future flow of oil. (An excessively high mud pressure can also push the sand grains apart and so internally fracture the rock.) When a high mud pressure is needed, the flow out into a porous strata can be controlled by adding polymers which clog the surface of the bore hole.

A second purpose of the drilling mud is to lubricate the drill bit as it cuts the hole and also lubricate the rotating drill stem when it comes close to the side of the hole. Drilling muds have a viscosity of the order of 1 poise.

The final role of the drilling muds is to convey the rock cuttings to the top of the well. The cuttings vary in size from 1 cm chips to 10 μm splinters from a single sand grain. They are brought to the surface by circulating mud down the inside of the hollow drill stem with a return flow up the annular gap between the drill stem and the side of the bore hole. At a volume flux of 0.05 $m^3 s^{-1}$ the flow is nearly turbulent. The conflicting requirements of an efficient drag on the rock cuttings and low pumping costs are partially ameliorated by a shear thinning nature of the non-Newtonian drilling muds.

There are two possible further uses of the drilling mud. Low-frequency sound waves propagating through the mud can be used to communicate useful information such as the position and temperature of the drill from the drill bit to the surface. The circulation of the mud can also be harnessed to rotate the drill bit instead of rotating the drill stem from the surface.

Water from drilling muds can sometimes reduce disastrously the permeability of an oil-bearing strata. This occurs where oil wets the sand grains with no water present. Water invading from the drilling mud can then swell clays which coat the surface of the sand grains, thus reducing the permeability by a thousand fold. This problem can be overcome by switching to oil-based drilling muds, i.e. a stabilised emulsion of cheap water in oil, although there is then the problem of disposing of the rock chippings contaminated by oil.

Completion

After a well has been drilled, it is prepared for oil extraction in a process known as completion. The 22 cm hole is lined with a 15 cm steel pipe and the annular gap is filled with cement. This strengthening of the well is to prevent any soft rock caving in.

Wells are sometimes lined in stages (with a sequence of nesting sizes of pipes) in order to prevent soft rock flowing at one level without using a high mud pressure which would block some porous rock at another level.

When the well is lined, oil can be extracted from a selected section of a stratum by *perforating* the well. The perforations are 1 cm^2 holes which go about 30 cm deep into the rock, and so normally beyond the porous rock which has been badly invaded by the drilling mud. These perforations are produced at intervals of about 10 cm down the well with a special *gun*.

When the oil-bearing stratum is badly blocked by drilling mud invading the porous rock near to the well, two methods can be used to improve the oil flow, known as *stimulating* the well. An acid can be pumped down which dissolves the clays and rock: an unstable flow with the acid flowing preferentially where it has etched the rock. (As far as I am aware, this flow instability has not been analysed.) Alternatively water can be pumped down at sufficient high pressure to fracture internally the porous rock near to the well.

To facilitate a controlled development of a reservoir, different strata can be isolated by plugging the steel pipe each side of a selected section with *packers* and connecting each such section to the surface independently with small 5 cm pipes within the main 15 cm pipe.

Porous rock

The porosity, i.e. the void space between the sand grains of an oil-producing rock, is typically about 10%, but the value varies from 20% for a loose sandstone down to 2% for one compacted and with clays coating the surfaces of the sand grains. Even lower proposings of ½% would be acceptable in gas-producing rock, where the less viscous gas flows more easily.

The permeability of a rock is measured in units of the darcy. A pressure gradient of 1 atmosphere/cm applied to a rock with a permeability of 1 darcy containing a fluid with a viscosity of 1 centipoise would produce a flow of 1 cm^3/s across a surface of 1 cm^2. Oil-producing rocks have permeabilities in the range of 10^{-4} to 1 darcy, with 10^{-2} being typical. For loose open sandstones, the permeability scales simply with the square of the size of the sand grains, with 0.5 mm diameter sand grains producing a permeability of 1 darcy. In tighter rocks, the permeability is very sensitive to the degree of blocking within the constrictions which link neighbouring pores. It should also be noted that the effective vertical permeabilities can be between a tenth and a thousandth of the horizontal values given above, the reduction being caused by the thin layers of fine particles or clays within the geological strata.

The viscosity of both salt water and oil in a reservoir at 100°C is about 1 centipoise, although in pure water and in light oils it can drop to one tenth centipoise. The viscosity of a gas in a reservoir at 300 atmospheres is about 0.02 centipoise. The interfacial tension between oil and water is about 20 dynes/cm, although the very 'dirty' interface can have a complicated surface rheology.

Flow rates

The standard measure of volume in the oil industry is a barrel. One barrel is about $1/7\ m^3$ (it varies between countries). Production from wells varies enormously, from 3 barrels/day from some old U.S. wells to 3.10^4 barrels/day from a good Arabian well. Off-shore a well would have to produce about 300 barrels/day to be useful.

A production rate of 300 barrels/day would correspond to a flow of 25 cm/s up a 5 cm inner pipe, and this would just be turbulent. If the oil were produced from a 10 m stratum, then in the porous rock there would be a superficial flow rate (volume flux per unit area of rock) of 0.1 mm/s near to the bore hole, falling off inversely proportional to the distance from the hole, to 0.1 μ m/s at a distance of 100m. This latter superficial velocity corresponds to an average velocity in the fluids of 7 cm/day when the porosity is 10%. At this flow rate it would take the fluid two years to travel the 100 m to the bore hole.

The appropriate non-dimensional measure of these flow rates is given by the capillary number $\mu U/\gamma$, which measures the ratio of the viscous to surface tension forces. Near to the bore hole the capillary number based on the superficial velocity U would be 5.10^{-5}, dropping to 5.10^{-8} at 100 m out. These numbers should not be considered as negligibly small (see later). The Reynolds number UL/ν, however, measuring the ratio of the inertial to viscous forces, decreases from 10^{-1} at the bore hole to 10^{-4} at 100m away, when based on the superficial velocity U and a sand-grain diameter of $L = 0.1$ mm; and these Reynolds numbers can be considered small.

The flow rates given above would be produced in a porous rock of permeability 10^{-2} darcys by a pressure difference of 70 atmospheres between the bore hole and 100 m away. Because the pressure various logarithmically with distance from the bore hole, half this 70 atmosphere pressure drop occurs within 3 m of the bore hole.

STAGES OF OIL RECOVERY

There are three stages of oil recovery. In the first or *primary* stage, the oil (and/or gas) is driven out by the higher pressure of the oil trapped in the reservoir compared with the pressure of the oil (or gas) in the well at that depth. In the second or 'secondary' stage, the flow of oil is maintained by pumping water into the stratum through some *injection* wells. In the final or *tertiary* phase, expensive chemicals are added to or replace the injected water.

Secondary recovery may be embarked upon in order to maintain the pressure of the trapped fluids, in addition to the more obvious maintenance of an economic flow rate. If the pressure drops too far, particularly in reservoirs only 1 km deep, gas dissolved in the oil can come out of solution at the *bubble* pressure. There is a corresponding problem in gas reservoirs with heavy hydrocarbons condensing out when the pressure falls below the *condensation* pressure. (Note that the phase diagram of mixtures of many fractions of hydrocarbons is not simple.)

The simultaneous production of oil and gas is usually avoided, for many reasons. Within the reservoirs very small pockets of boiled-off gas (or condensed oil) can seriously impede the flow of oil (or gas). A mixed flow up the well can lead to large surges at the well head. Finally in a gas well, the high pressure at the bottom of the well can be harnessed to pump the gas to a distant storage if only its own low gravitational head is lost in coming up the well. (Only about one atmosphere is lost in friction.)

The loss of pressure in the reservoir during the primary stages depends on the permeability k of the rock and the compressibility c^2 of the rock and fluids (and is governed by a diffusion equation with a diffusivity $D = \frac{\rho k c^2}{\mu}$ which is of the order 10^2 cm^2/s). Thus the response of a well to being suddenly switched on or off can be used to measure the permeability of the rock at some distance from the contaminated region near the well, and also the horizontal extent of the whole reservoir. (Here it is useful to eliminate the large body of compressible material in the well by siting the valve near to the producing stratum.)

Problems in secondary recovery

Viscous fingering. When the viscosity of the injected water is less than that of the oil in the reservoir, there can be a viscous fingering instability in which the more mobile water bypasses the oil. It does take some time, however, for the water to first reach the producing well, although after the first *break through* of water the proportion of water produced (the *water cut*) increases as more directions and more levels yield water. Thus as well as the loss of the irrecoverably bypassed oil, there is a problem of separating the produced mixture of oil and water. (An oil cut of 20% can still be economic.)

Viscous fingering is not as universal as one might expect from laboratory experience. High levels of salt dissolved in the water increase the viscosity of the water, while the oils thin at the high temperatures in the reservoirs. The surface tension forces, which locally dominate the viscous forces, also complicate the viscous fingering. (See later.)

Inefficient areal sweep. The water injection wells used in secondary recovery are often positioned in the oil-bearing stratum with four or five surrounding and at roughly equal separation from an oil-producing well (the *four* and *five-spot* patterns). Now water passing nearby one of the stagnation points in the flow will take longer to travel from the injection well to the production well than that travelling in the straight line between the two wells. Thus the two-dimensional nature of the flow leads to only half the (recoverable) oil being produced before water on the most direct route first breaks through. This problem of the inefficient areal sweep can be tackled by varying in time the injection rates at the different wells so as to move around the stagnation point of the flow, by drilling an extra well in a bypassed pocket of oil, by employing a 'line drive' configuration instead of the four/five spot pattern, or by fracturing the strata in a major way.

In the line drive method, the injection and production rates are managed so as to keep the boundary between the oil and the water as a straight line which sweeps up all the (recoverable) oil as it migrates from one side of the reservoir to the other. While more oil is recovered, the rate of production from the reservoir can be low, because few wells are active at any one time.

In the *massive hydro cracking* treatment, a large quantity of water (10^5 barrels) is injected over a couple of days. If the pressure is sufficiently high, the sand grains of the porous sandstone can be driven apart, and a crack thus made to propagate as far as a kilometer (with a height of 10 m and a width of 1 cm, as far as one can guess -- little hard evidence is available). The least principal component of compressive stress in the porous rock will be in a horizontal direction which is determined by the local geological conditions, and it is in this direction that the sand grains are most easily parted; thus determining the direction of the crack, at least in theory. If the fluid pressure is much higher than this least principal component of stress, then the crack can also propagate in other directions. When calculating the appropriate pressure to use, it is necessary in deep, hot reservoirs to take account of the thermal contraction caused by pumping in cold sea water. Polymers are often added to the *cracking* fluid to reduce the leakage (90% or more) into the porous rock and so concentrate the flow towards the crack.

Once the crack is formed it must be held open when the high water pressure is turned off, and this is done by adding large particles (known as *propants*, usually coarse sand, but originally nut shells). To stop the large particles from sedimenting out too early along the crack, or even within the well, they can be held up in a polymer gel. An alternative treatment is to pump in acid which etches the surfaces of the crack so that they do not fit tightly together when the pressure is released.

The purpose of massive hydrofracking is to convert the line sink nature of the oil well into a plane wall sink; hence the need to ensure that the crack is much more permeable than the porous rock. The increased effective surface area of the well leads to an increased rate of production of oil, while the pattern of the flow lines for the plane wall sink leads to a greater efficiency of area sweep.

Gravity override. In a very thick oil-bearing stratum, the slightly higher density of the water can be exploited to take oil selectively from the top of the reservoir. There are other, less fortunate circumstances, however, in which a very permeable zone at the bottom of a thin stratum acts as a fast water channel, leaving the majority of the oil bypassed on the top. One solution to this problem is to block the water channel by injecting a polymer solution which is allowed to gel in the channel.

Chemical changes. As was mentioned in connection with drilling in in oil-wetted rocks, the introduction of water (now as part of the secondary recovery process) can cause clay deposits on the sand grains to swell, thereby reducing the permeability by several orders of magnitude. In other reservoirs, the introduction of salt water has caused mineral salts to precipitate and so block the pores.

Methods of tertiary recovery

Polymer additives. The viscous fingering instability can be controlled by increasing the effective viscosity of the injected water by adding some suitable polymers. The polymers (such as a 2% polyacrylimide solution with a shear viscosity of 6 times that of water and an effective viscosity in a porous medium of 40 times that of water) are chosen so that they do not clog the rock near to the injection well and also so that they do not degrade rapidly (they must remain active for a year). Some biological degradation can be avoided by sterilising the water. On the other hand, there are problems in finding a suitable polymer which can withstand the high temperature of deep wells (most degrade by 85°C).

Surfactants. The proportion of oil recovered from that initially in the reservoir can be increased by reducing the oil-water interfacial tension γ as measured by the capillary number $\mu U/\gamma$. To produce a significant effect, the capillary number must be increased to 5.10^{-3}, i.e. by more than four orders of magnitude. Certain very special chemicals can achieve this, although they are expensive, they absorb rapidly onto the rock surface, and they are so effective only within a narrow temperature range (10°C). Further, if the surfactant treatment is commenced long after the start of the secondary recovery, then an extra order-of-magnitude reduction in the surface tension is needed to mobilise disconnected pockets of oil (see later). It may be possible to improve the effective 'wettability' of the oil in the complex porous rock, without having to reduce the interfacial tension so much.

Carbon dioxide and nitrogen injection. Carbon dioxide from nearby natural reservoirs or industrial plant or nitrogen from the air can be compressed and injected instead of water. When they dissolve in oil, these gases make it less viscous and more mobile. Rather a large quantity of gas must be injected, about seven times the weight of oil recovered. Moreover in thick but shallow strata the large density difference produces a severe gravity override, with most of the oil bypassed.

Steam and air injection. Another alternative to water injection is to inject heat, either by means of steam (*steam soak*) or oxygen to burn with some of the oil in the reservoir (*in situ combustion*). The heat vapourises the lighter hydrocarbon fractions which then move more easily than the original oil (which might even have been bypassed after secondary recovery). Although the buoyant gases (steam, air and light hydrocarbons) will move along the top of the stratum leaving most of the oil bypassed at the bottom, an advantage of this form of tertiary treatment is that the heat will slowly diffuse downwards, so having an effect on the whole depth of the reservoir. Again steam or air of several times the weight of oil recovered must be injected. Predicting the progress of a heated reservoir is complicated by the need to keep track of the temperature and several different chemical products; thus, for example, some light hydrocarbons vapourised in the heated region, condense out in a cooler part and there, mixing with some trapped oil, mobilise it.

Platform management problem

The problem of managing an off-shore platform gives an illustration of the economic considerations which must be applied to the engineering possibilities, some of which have been outlined above. One must decide when and where to drill the next well. It is expensive to keep the drill (capital and manpower) idle. On the other hand, there is a limit to the production which can be handled in a confined space, particularly the separation of oil and water in secondary recovery. One has to optimise the company profits within the ever-changing market and, more imponderable, the vacillating tax laws. Often short-term profit considerations lead to fast but poor overall recovery. One part of the optimisation program will be a numerical model of how the oil flows in the reservoir. While the model is adaptive, i.e. it updates its parameters in the light of experience, its centre piece will be some equations, dubious in my opinion, for the flow of a mixture of oil and water through the porous rock. Hence the importance of a fundamental study of multiphase flow in a porous medium, to be described in the following second part of this review.

MULTIPHASE FLOW THROUGH A POROUS MEDIUM

Darcy's law for single-phase flow

The flow of a single fluid through a porous medium is governed by Darcy's law

$$v = -\frac{k}{\mu}\nabla p ,$$

where v is the superficial velocity (volume flux per unit area), k the permeability, μ the viscosity and p the pressure in the fluid. The law can be derived by various methods, such as *homogenisation*, if the flow has a low Reynolds number and if the length scale of the pressure variations greatly exceeds the size of the sand grains in the porous rock.

Attempts have been made recently by Koplik, Lin and Vermette (1984) to calculate the permeability of a porous, coarse sandstone from a series of 41 scanning electron micrograms of cross-sections of the rock taken at a separation of 10μm. The images were processed by a computer to identify the larger void spaces as *pores* and the connecting passages as *throats*. The flow resistance through each throat was approximated by applying Poiseuille's law for a tube with a slowly-varying elliptical cross section, and the overall resistance through the network of throats was calculated approximately using an effective medium theory. The calculated permeability was found to be ten times larger than the actual permeability, which is not bad for a first attempt. According to Poiseuille's law the resistance is very sensitive to the narrowest dimension of a constriction (inversely proportional to the cube for a slit-like constriction, and inversely proportional to the fourth power for a more circular constriction), and more care is probably needed in assigning this in their program.

The value of the permeability of a rock should be independent of the fluid flowing through it. As noted earlier, some fluids can however cause the rock to change, e.g. swelling of clays and precipitation of mineral salts when salt water is introduced, and this would make the permeability depend on the fluid flowing through it.

Extension of Darcy's law to multiphase flow

When two phases, say oil and water, are flowing mixed together through a porous medium, it is common to generalise Darcy's law by applying it to the two phases separately and simultaneously at each location, i.e. setting at each point both

$$v_o = -\frac{k_o}{\mu_o} \nabla p_o \quad \text{and} \quad v_w = -\frac{k_w}{\mu_w} \nabla p_w ,$$

where p_o and p_w are the pressures within the oil and water respectively. These two pressures are often taken to differ by a *capillary pressure*

$$p_o = p_w + p_c$$

which must have the magnitude of the surface tension divided by a pore diameter, i.e. about half an atmosphere. This capillary pressure and the separate permeabilities to the oil and water, k_o and k_w, are allowed to depend on the proportion of the void space occupied by the oil, which is known as the *oil saturation* s_o.

Before proceeding any further it must be declared that although this generalisation to multiphase flow is very natural and is universally employed, it has no theoretical justification, such as the derivation of the law for single-phase flow. There is one exception: this is when one phase does not progress through the rock (zero superficial velocity) and the fluid interface within each pore deforms reversibly and only a little.

Relative Permeabilities

Despite this uncertainty in the formulation of the multiphase flow, the two permeabilities k_o and k_w are routinely measured in laboratories by pumping a mixture of oil and water through a sample of rock. Quite small samples must be used in order that the flow reaches an equilibrium in a reasonable time, bearing in mind the very slow flow rates appropriate to reservoir conditions (cms/day).

The measured permeabilities are expressed as a fraction of the value that they have when there is only the one phase present, this fraction being known as the *relative permeability*. Quoting the figures for a typical sandstone, a sample of Berea, the relative permeability of the rock to the flow of oil, is found to decrease from 1 as the proportion of oil to water drops from 100% at $s_o = 1$.

Initially, while $s_o > 0.9$, the relative permeability drops gradually to 0.95. Then it drops rapidly to 0.1 by $s_o \approx 0.6$, and finally the permeability to oil drops to 0 at around $s_o = 0.3$. Thus oil does not move, whatever the pressure gradient (within some limits), once the oil saturation drops below a critical value, around 30%, known as the *residual oil saturation* s_{or}. This means 30% of the oil is irrecoverable.

In fact the permeability and the residual oil saturation do depend on the pressure gradient as measured by the capillary number $\mu U/\gamma$ for the total flow. Up to a capillary number of 5.10^{-5} (for this typical Berea) no dependence on the flow rate is detected. Then as the capillary number increases to 5.10^{-4}, the residual oil saturation drops to 15%, and at 2.10^{-3} only 3% of the oil is not recoverable. Note how small the surface tension must be made to increase the proportion of oil recovered; for the flow rate given earlier of 0.1 μ m/s at 100 m from the well, the surface tension must be reduced to 2.10^{-3} dynes/cm.

The picture presented above for the oil flow is mirrored in a similar dependence of the water flow on the *water saturation*, $s_w = 1 - s_o$.

Buckley-Leverett theory

The governing equations for the multiphase flow are completed by two conservation equations, one for the oil

$$\phi \frac{\partial s_o}{\partial t} + \nabla \cdot v_o = 0$$

where ϕ is the porosity of the rock, and a similar equation for the water saturation $1 - s_o$. Combining the two yields an equation for the pressure field

$$\nabla \cdot (k_o + k_w) \nabla p_o = \nabla \cdot k_w \nabla p_c \ .$$

where for this section only k/μ has been replaced by k. This elliptic equation for p_o is solved at each instant of time with the dependence on position of k_o, k_w and p_c known from the instantaneous distribution of oil saturation, $s_o(x,t)$. The solution p_o is then used to evaluate the oil flow v_o , which is substituted into the conservation equation to yield the oil saturation at the next instant. The hyperbolic nature of the conservation equation is the potential source of numerical instabilities, which are often avoided only with the introduction of artificial diffusion which far exceeds any real diffusion represented by $p_c(s_o)$.

In one dimension one can easily examine the behaviour of this system of equations. Let there initially be only oil ($s_o = 1$) in $x > 0$, let a flux Q of water be introduced at $x = 0$, and ignore the effects of capillary pressure ($p_c = 0$). The combined conservation of oil and water gives the pressure distribution $\nabla p_o = -Q/(k_o + k_w)$, which when substituted into the oil conservation yields

$$\frac{\partial s_o}{\partial t} + V \frac{\partial s_o}{\partial x} = 0 \quad \text{with} \quad V(s_o) = \frac{Q}{\phi} \frac{d}{ds_o} \left[\frac{k_o}{k_o + k_w} \right] .$$

Thus a particular value of oil saturation propagates at a constant velocity V (if Q is constant). Now because k_o vanishes in $s_o < s_{or}$ and k_w vanishes in $s_o > 1 - s_{wr}$, this propagation velocity vanishes outside the range of oil saturations $s_{or} < s_o < 1 - s_{wr}$, and is positive within the range. At $x = 0$ in the initial instant, effectively the whole range of oil saturation above the residual coexist. Hence a shock will inevitably be formed (with the propagating $s_{or} < s_o < 1 - s_{wr}$ overtaking the non-propagating $s_o = 1$). Ahead of the shock there will be a flow Q of pure oil (at $s_o = 1$). Just behind the shock there will be a lower value of oil saturation which is given by balancing its propagation velocity with the velocity of the water there (and for this water velocity one must use the superficial velocity divided by the product of the porosity and the water saturation), i.e.

$$V(s_o) = \frac{Q \quad k_w}{\phi \; s_w (k_o + k_w)} .$$

From the shock back to the water inlet at $x = 0$, the oil saturation drops from the value given by the solution to the above equation to the residual oil saturation. The details of this variation in oil saturation depend upon the two permeabilities and in particular the way in which k_o approaches zero at s_{or} (if $k_o \propto (s_o - s_{or})^\alpha$ as $s_o \to s_{or}$, then $s_o + c\,(x/t)^{1/\alpha - 1}$ near $x = 0$).

A non-zero capillary pressure adds a diffusion-like term to the oil conservation equation with a diffusivity

$$\frac{k_o \; k_w}{\phi (k_o + k_w)} \; \frac{dp_c}{ds_o}$$

The effect of this term is small because the capillary pressure of about one atmosphere is small compared with the pressure differences within the reservoir of 100 atmospheres.

MICRO MODELS

To understand the mechanics of multiphase flow through a porous medium a number of simple micro models have been studied. One is interested in such questions as whether relative permeabilities and capillary pressure represent a correct formulation for the flow and what particular features of the rock determine the residual oil saturation.

All the models replace the complex random geometry of the interstitial space between the sand grains with a network of *pores*, which represent the larger more cavernous parts, interconnected by *throats*, which represent the linking more constricted parts. This precise divide is never so clear when one is confronted with a die cast of a real rock!

For convenience the network of pores and throats is usually taken to be a regular lattice; in two or three dimensions, a simple cubic or perhaps, e.g. a hexagonal one. A serious topological deficiency of a strict two-dimensional lattice is if one phase connects the top to the bottom, then the other phase cannot connect one side to the other. This restriction on the coexistence of connected paths in the two phases can be overcome without the high cost of a fully three-dimensional network but using two interconnected layers or alternatively, a single layer with some diagonal links avoiding one another out of the plane.

The number of throats connected to a single pore is considered another important topological measure of a porous medium, and it is called the *coordination number*. Many lattices have rather larger coordination numbers than one might expect in a rock. Also in a rock the number of throats connected to a pore is random. This fact can be incorporated into the network by randomly turning off some possible throats.

Depending on the particular micro model, the diameter and volumes of the pores along with the diameter and length of the throats are assigned randomly, sometimes in accordance with some data on a real rock. Usually no correlation is made between the dimensions of a pore and the dimensions of the throats to which it connects. To minimise entry and exit effects the networks are usually taken to be at least 50% longer than they are wide, and periodic side-wall conditions are used.

Viscous flow networks

In the first group of micro models, typified by the work of Koplik & Lasseter (1982) and Payatakes & Dias (1984), the viscous flow through the network is calculated by applying Poiseuille's law to each random throat. If the meniscus is part way along a throat, then the total pressure drop for that throat is taken as the sum of Poiseuille's law for the two parts using the appropriate viscosities, together with the capillary pressure jump calculated for a static meniscus within the throat. This approximation, which avoids calculating the dynamical shape of the meniscus with the associated problems of describing the movement of the interface along the solid boundary (in conflict with the no-slip boundary condition), is known as the Washburn approximation and is probably quite reasonable.

The fluid dynamics within the pores tends to be fudged in the network micro models. Whilst there will be little viscous pressure drop within the wider regions of the pores, the precise motion of the oil-water interface will be important in determining whether throats become switched off with a meniscus closing off both ends or whether a droplet of oil is formed and trapped within the pore. Payatakes & Dias effectively collapse the junction of the throats to a single point, whereas Koplik & Lasseter use a variable saturation which changes according to some *upstream* influence.

The viscous flow network models are numerically expensive, with a large matrix for the pressure at the pores to be inverted at each instant, and then the motion of the meniscii to be followed carefully along all the throats. As a result, quite small networks are studied: Koplik & Lasseter choose a 10 x 10 two-dimensional square lattice with non-intersecting diagonals, whereas Payatakes & Dias used a 15 x 30 strictly two-dimensional square lattice. There is some advantage in Payatakes & Dias's choice of throats whose diameters change smoothly along their length, as opposed to Koplik & Lasseter's abrupt change at a pore which leads to some difficult nonlinear constraints. Both calculations could probably be made faster by exploiting the fact that the largest pressure changes occur near to the interface.

Bypassed oil

The most important result of these micro models is that oil is bypassed and left behind due to the randomness of the geometry. At low capillary numbers, surface tension pulls the interface along routes consisting of the narrower throats and pores. As these routes wander randomly sideways as well as towards the exit, they sometimes join, thereby possibly cutting off from the exit some oil in wider throats and pores. By this mechanism roughly 50% of the oil was found to be left behind in the two studies. Koplik & Lasseter found this figure was insensitive to the coordination number of the lattice (number of throats connected to a pore), a low coordination number leading to more numerous and more fragmented drops with a similar total volume. Payatakes & Dias found that the residual oil saturation was not very sensitive to the ratio of the viscosities of the oil and water (they used 50, 1 and 0.2) at the low capillary numbers relevant to reservoirs, where the local dynamics is dominated by the surface tension forces.

There are some other interesting details in the numerical simulations. First, the water flood proceeds with a finite transition zone, with only oil moving ahead of the zone and only water moving (passed stationary trapped oil) behind the zone. The thickness of this transition zone decreases with increasing capillary number, although the size of this zone is probably influenced considerably by the width of the network in the current simulations. Second, as soon as some oil is bypassed and becomes disconnected from the exit, it rarely moves again (at the low capillary numbers relevant to reservoirs). The size of a region of disconnected oil which can be mobilised and the thickness of the transition regions can both be determined for a given imposed pressure gradient by balancing the viscous pressure drop over that distance with an appropriate capillary pressure; for the disconnected oil it is the capillary pressure which must be overcome in a tight pore on the exit route, while for the transition zone it is the capillary pressure in the widest trailing pore which is available to push along the oil. Finally the viscous flow network models show that at low capillary numbers the interface spends most of the time making minor adjustments and then suddenly a single meniscus will dart across a pore. The pore selected tends to be smallest one along the current interface, i.e. the one with the highest capillary pressure. This observation leads to a simplified *capillary jump* model to be discussed shortly.

The smallness of the capillary numbers

The capillary numbers of the flows in reservoirs initially look extremely small with values in the range 10^{-8} to 10^{-4}. Measurements on rock samples and also the viscous flow network models show, however, that it is the comparison with about 10^{-4} and not 1 that is important, because at a capillary number of 10^{-4} the residual oil saturation begins to decrease.

The problem here is that the non-dimensional capillary number $\mu U/\gamma$ based on the superficial velocity U is not the most relevant dynamical measure of the viscous to surface tension forces. The correct measure is to compare the capillary pressure with the viscous pressure drop between two pores. Now the lowest capillary pressure occurs when the interface is in the pore, and so is γ/R where R is the radius of the pore. On the other hand, the viscous pressure drop between the two pores occurs mainly in the constriction of the throats and is given by Poiseuille's law as $\mu R v/r^2$, where r is the radius of a throat which is assumed to be R long and v is the typical velocity in the throat which can be related to the superficial velocity by equating fluxes $r^2 v = R^2 U$. Hence the correct measure of viscous to surface tension forces is a modified capillary number $C^* = R^4 \mu U / r^4 \gamma$.

We thus see that, if the geometry of the rock has throats which are one-tenth as wide as the pores, then the viscous forces will become important at a capillary number of 10^{-4}. For a rock with such a geometry and for a slower flow with a standard capillary number of 10^{-8}, my modified capillary number is 10^{-4} and so the viscous forces would still dominate over the length of 10^4 pores, i.e. one would expect the transition zone where both oil and water are mobile to be about 1 m thick (for 0.1 mm sand grains).

Capillary jump model

It was observed in the viscous flow networks that at low capillary numbers the interface progresses mainly through isolated and sudden movements of the meniscus through the smallest pore on the instantaneous interface, i.e. the one with the highest capillary pressure. This observation leads logically (if not historically) to a *capillary jump* micro model in which the interface moves only according to this criterion. Abandoning the calculation of the viscous pressure field greatly reduces the computational cost, and has allowed Wilkinson & Willemsen (1983) to study 100 x 200 two-dimensional and 30 x 30 x 60 three-dimensional lattices.

An alternative jump criterion based on the largest throat on the instantaneous interface rather than the smallest pore was studied earlier by Chandler, Koplik, Lerman & Willemsen (1982). This alternative criterion corresponds to the so-called *drainage* process of oil advancing into a water-saturated rock, with the advance being restricted least by the surface tension forces in the largest throat. This process should be contrasted with the usual *imbibition* process in which water displaces oil with the maximum assistance from the surface tension forces occurring in the smallest pore.

As the interface advances, some oil is bypassed and becomes trapped. Now because the oil is incompressible, the interface must be prevented from advancing into trapped oil. Thus in the numerical simulations, the trapped oil is removed from the list of the active interface, and just how this should be done efficiently is an amusing programming problem. [Treating the oil as perfectly compressible may correspond to the *porosimetry* test in which mercury is forced into an evacuated porous rock. This test is supposed to yield the distribution of the sizes of the throats from the dependence of the volume of mercury absorbed on the pressure required to absorb it.]

The advance of the water into the oil is found to be different in the capillary jump model to that in the viscous flow network models: there is no advancing front, but instead the water moves through tortuous random fingers which are nearly isotropic in their wanderings. This micro model for vanishingly small capillary numbers effectively has an infinitely thick transition zone, and so given the isotropic nature of the fingering it is the width of simulation lattice which limits the large-scale features.

When the water first reaches the exit, the *break through* point, only a small proportion of the oil has been removed. This proportion is found to decrease as the lattice size increases: it is only necessary for one random walk to break through and on a larger lattice a longer less probable random walk can be expected. The dependence on the lattice size N of this small proportion of recovered oil at the water break through follows a strange fractal power law, proportional to $N^{-\alpha}$ where $\alpha = 0.48$ or 0.18 for three or two-dimensional lattices respectively. Note that this index α is found to be independent of the type of lattice (except for a two-dimensional triangular lattice which has $\alpha = 0.12$).

If the numerical simulation is continued after the break through of water at the exit, oil starts to become trapped by the bypass mechanism. When the active interface all finally reaches the exit, about 35% of the oil remains trapped in three dimensions, the precise figure depending on the particular type of three-dimensional lattice. (Note that the present simulations assign the same volume to all the pores, although they have different diameters!) In strictly two-dimensional lattices, the impossibility of both phases being simultaneously connected (with the isotropic wanderings) leads to a residual oil saturation approaching 100% as the lattice size is made larger, although at $N = 100$, $S_{or} \approx 0.55$ and the approach to 1 is like $N^{-0.18}$, which is very slow.

The capillary jump micro model yields some information about the sizes of the regions of trapped oil, called oil *blobs*. the proportion of trapped blobs involving n pores decreases like $n^{-\tau}$, where $\tau = 2.07$ in three dimensions and 1.80 in two dimensions, again these indices are independent of the type of lattice. Note that, because τ is only just greater than 2 in three dimensions, quite large oil blobs make a significant contribution to the total residual oil (and hence assigning the same volume to each pore may not lead to a large error - it all depends on the number of pores on the surface of the irregular blobs compared with the number in the interior). In strictly two dimensions the same sum for the total residual oil over the sizes of trapped blobs is divergent.

Classical percolation theory

Many aspects of the capillary jump model can be understood in terms of percolation theory. This branch of statistical theory was invented by Broadbent & Hammersley in 1957 to describe, amongst other things, fluid moving through a porous medium. It has since found many applications, particularly in critical phenomena in solid-state physics.

According to classical percolation theory we suppose that on a network a certain proportion p of the pores are suddenly *switched on*, the pores being selected randomly and independently. If the proportion p is small, then there will be only finite clusters of connected pores which are switched on. Above a critical proportion, there will always exist an infinite cluster which may be said to allow percolation from one side of the network to the other. The critical proportion is called the *percolation threshold*, p_c, and it depends on the type of lattice. For a simple cubic lattice

$p_c = 0.312$.

Further study of this percolation process shows that just below the percolation threshold the size of the clusters increases like $(p_c - p)^{-\nu}$, where the exponent $\nu = 0.88$ in three dimensions and 1.33 in two dimensions, and that just above the threshold the proportion of pores on the percolating cluster increases like $(p - p_c)^{\beta}$ where $\beta = 0.45$ and 0.15 in three and two dimensions respectively. Note that these exponents are independent of the type of lattice, and are unchanged if it is the throats rather than the pores which are switched on. This universality is connected with the large-scale interactions near to the threshold which do not depend on the small-scale details.

Invasion percolation theory

The process in the capillary jump micro model is not quite identical to that of the classical percolation, hence the qualification *invasion*. In the classical process all the pores, say those smaller than a certain size, are switched on at the same instant. In the invasion process the pores are switched one at a time as the interface progresses from the entry to the exit, and only those on the current interface are considered for switching on. Thus there can be very tight pores within some trapped oil which never come under consideration. Also in order to propagate at every instant, sometimes moderately large pores must occasionally be selected (so long as they are smallest on the interface at that instant).

At the break through of water, it is argued that there must be a percolating water path. The proportion of water on this percolating path can be calculated by first equating the size of the network with the size of the largest cluster $N = (\Delta p)^{-\nu}$, and then using this deviation from the threshold Δp to find the proportion on the percolating cluster, Δp^{β} . Thus one can find the water saturation at break through varies like $N^{-\alpha}$ with $\alpha = \beta/\nu$. The classical theory predicts here $\alpha =$ 0.54 and 0.11 compared with the observed invasion 0.48 and 0.18.

At the end of the water flood when the active interface all reaches the exit, the oil has just stopped percolating, and so one would expect the residual oil saturation to equal the percolation threshold (if the pores can be treated as having equal volumes). Again the classical theory gives too low a prediction, for a simple cubic lattice $p_c = 0.312$ while in the invasion process $S_{or} = 0.341$. This low estimate results from some small pores which would have been switched on in the classical process being enclosed within the trapped oil and so are never visited by the invading interface.

A jump model with a finite transition zone

A major deficiency of the capillary jump model is that working at vanishing small capillary numbers it has an infinitely thick transition zone between where only oil and where only water is mobile. The full viscous flow network model, on the other hand, is very expensive in the calculation of the changing pressure field. If however the approximation is made of replacing the fluctuating pressure field by the average viscous pressure gradient, then one has a fast model with a finite non-zero capillary number. Thus in considering which pore on the current interface is to move, one adds to the capillary pressure for the randomly chosen size of each pore a viscous pressure which decreases linearly with the distance from the entrance and which is proportional to the capillary number; and here one should use the C^* introduced earlier rather than the usual $\mu U/\gamma$. It immediately follows that this model has a transition zone of thickness $1/C^*$.

This modified capillary jump model was introduced by Wilkinson (1984) in the context of buoyancy forces producing pressure differences along the interface. Unlike the viscous forces, no approximation is involved in the pressure field which will be proportional (at slow speeds) to the density difference and vertical distance from the start of the network.

The size of oil blobs

An important contribution of Wilkinson is his observation that the size of the largest blobs of trapped oil is considerably smaller than the thickness of the transition zone. He argues that near to the percolation threshold the largest blob has a size $L_{max} \propto (\Delta p)^{-\nu}$ (non-dimensionalised by the size of a single pore). Now at a distance L_{max} from the the back of the transition zone where the oil is percolating, the oil will be off the percolation threshold by the ratio of L_{max} to the thickness of the transition zone $1/C^*$, i.e. $\Delta p = C^* L$. Hence

$$L_{max} \propto C^{*-\mu} \quad \text{with} \quad \mu = \frac{\nu}{1+\nu} = 0.48 \quad .$$

Wilkinson's numerical simulations support this argument. There is also some support from the observations of Chatzis & Morrow (1981) of hysteresis in the dependence of the residual oil saturation on flow rate: they found that a high residual oil saturation from a low capillary number flood was not necessarily mobilised by a faster flow, even though the faster flood would have produced a lower residual oil saturation had it been applied from the beginning, before the oil had become disconnected. One interpretation of this hysteresis is that the first flood produces blobs of size $C^{*-\mu}_1$ which are only mobilised if $C^{*-1}_2 < C^{*-\mu}_1$. Note that if $C^* = 10^{-4}$ and the size of the sand grains is 0.1 mm, then the size of the largest blob of oil is 1 cm, which would only be mobilised by a flood with $C^* = 10^{-2}$. Thus if surfactants are introduced in tertiary recovery after the oil has become disconnected in secondary recovery, then much higher capillary numbers must be used.

Flow dependence of the residual oil

Wilkinson also gives an argument, due to Halperin, for the dependence of the residual oil saturation on flow rate. The proportion of oil on the infinite cluster near to the percolating threshold is proportional to $(p - p_c)^\beta$. Now if the infinite percolation process is truncated by the spatial variation across the transition zone at $\Delta p = C^{*1/1+\nu}$, then the amount of oil truncated from the infinite cluster would be $\Delta p^{1+\beta}$. This oil is replaced by water, thus decreasing the residual oil saturating

$$\Delta S_{or} \propto C^{*\lambda} \quad \text{with} \quad \lambda = \frac{1+\beta}{1+\nu} = 0.77 \ .$$

Note that the oil trapped in the finite clusters varies more slowly, like Δp^β, and so makes a smaller contribution to the change in the residual oil. The numerical simulations give a good agreement, with an exponent $\lambda = 0.76$.

The above argument is perhaps less obvious than Larson, Davis & Scriven's (1981) appealing suggestion that the residual oil can be calculated from the sum over the different sizes of blobs $\sum n^{1-\tau}$. If this sum is truncated when the number of pores within a blob n_{max} corresponds to the maximum linear dimension L_{max}, the two being related by the fractual dimension $d - \beta/\nu$ (d = dimension of space), $n_{max} = L_{max}^{d-\beta/\nu}$, one estimates a reduction in the residual

oil like $C^{*\kappa}$ with $\kappa = (\tau - 2)(d - \beta/\nu)\nu/(1+\nu) = 0.23$, which is not supported by the numerical simulations. It appears that the oil on the truncated part of the infinite cluster would have produced trapped oil blobs of all sizes, not only those larger than n_{max}.

Speculation on relative permeablities

I hope that my personal selection of micro models and emphasis on particular features has left an impression that the mechanics of the microscale do not naturally lead to relative permeabilities and the associated standard formulation of multiphase flow in a porous medium. It seems to me that, based on the investigations of the micro models, an adequate description of the motion on the large scale of the reservoir would be a transition zone moving as a negligibly thin shock wave, with pure oil moving ahead and only water moving behind. One would have to use the water permeability of the rock containing the stationary trapped oil. This permeability would, however, be fixed by the local capillary number when the shock wave passes through that position, and being constant in time the difficult hyperbolic equation for the varying oil saturation is avoided. (Where the water permeability is also constant in space as would be the case on the low capillary number asymptote, the Poisson problem for the pressure field can further be solved efficiently by a boundary integral technique to yield the velocity of the shock wave.)

The model suggested above (not an original proposal) has in common with the Buckley-Leverett theory a moving shock wave, but does not share that theory's evolving oil saturation behind the shock. This structure behind the shock is a consequence of using relative permeabilities, which implicitly assumes that the two phases can flow together over large distances. I know of no experimental evidence to support this hypothesis. The experiments which do measure relative permeabilities are firstly steady state whereas the propagating shock is a transient phenomenon, and secondly performed on samples which are much smaller (in at least two directions) than the expected transition zone (and often smaller than the expected blob size).

The nature of the viscous fingering instability at low capillary numbers can be discussed easily in terms of the proposed model. The water will be more mobile only if the water viscosity divided by the water permeability (reduced by the trapped oil) is less than the oil viscosity divided by oil permeability (not reduced, with only oil present). As the oil will be trapped preferentially in the larger pores and throats the reduction in the water permeability might lead to a useful stabilising factor.

Finally I must remark that this section is very speculative and I could be totally wrong. Certainly the very simple model is inadequate to discuss details of the transition zone, which for example are essential in predicting thermal fronts in *in situ combustion*.

THE MOTION OF THE MENISCUS

The micro models presented so far do not examine in any detail the motion of the oil-water interface; in fact, the viscous network models simply assume that the miniscus jumps across a junction. It is now time to reduce the scale of consideration to that of a single pore in order to understand how the network models should be refined. At the scale of a single pore, more ideas have come from experiments than from theoretical analyses.

Haines jumps

In the simplest experiments a meniscus (or a pair, one either end of a finite drop) moves down a pipe which has varying diameter; see for example the recent work of Legait, Sourieau & Combarnous (1983). At low flow rates, the meniscus adopts a succession of quasi-static configurations, which depend on the local diameter and the contact angle to the local tangent to the wall. The associated capillary pressure difference across the meniscus will thus vary along the pipe. At locations where the upstream pressure decreases as the meniscus advances, the meniscus is unstable. If one is applying boundary conditions on the ends of the pipe of prescribed pressure, the meniscus will advance rapidly through all the unstable locations until it reaches a stable location. This motion is called a *Haines (1930) jump*.

Except for experiments with very viscous oil, the Haines jumps in pipes are limited by inertia, because the appropriate Reynolds number $\gamma R/\nu\mu$ will be large. In a porous medium, however, the effect of inertia will be reduced by the geometry factor r^4/R^4, introduced earlier, if fluid has to be moved through the constricted throats during the jump.

Mobilisation of oil blobs

The motion of an isolated blob of oil through a transparent pack of glass beads has been studied by Ng, Davis and Scriven (1978) and Payatakes (1982). In some films they show that blobs can be mobilised if the viscous pressure drop over the length of the blob exceeds the capillary pressure resisting the motion into the widest throat on the exit route. Once mobilised, a blob sometimes advances through a couple of pores. It can also branch out and try two different routes simultaneously. Often this branching leads to parts breaking off, resulting in smaller blobs which are less mobile. It is not clear how the motion of an isolated blob can be related to the motion of many blobs, which are quite crowded at an oil saturation of 50%. There then would be the possibility of blobs fusing together, which takes longer than breaking one blob into two (a problem worth further study). The investigation of mobilising blobs may, however, be less important to secondary recovery (but not tertiary recovery) if the blobs which are formed are much smaller than the transition zone, as explained earlier.

Choke off

An important observation of Mohanty, Davis & Scriven (1980) is that the shape adopted by an oil blob in a constriction is an unstable shape. If water can be supplied to replace the oil in the throat, then the neck of the oil linking two pores will break. They call this process of disconnecting oil *choke off*.

<u>Supply of water</u>. In the experiments with packs of glass beads water starts by wetting all the glass surfaces. As an oil blob moves through, it may wet some of the glass, but in the throats it will not penetrate the cusp-like crevices, due to the strong surface tension forces. Thus there exists a *privileged route for water* to supply the choke-off instability.

In a rock, the surfaces of the sand grains are rough, particularly when there are clay deposits on the sand grains, and the oil surface will be kept away from the solid surface by the peaks of the roughness. Thus the *roughness* provides a further route for the water, in addition to crevices between grains. The existence of such connected connate water in what is apparently an oil-saturated rock can be detected by measuring its electrical conductance (through the salt water).

Non-choke off. The choke-off instability does not act in every throat, however, because static oil can be found connected through many pores. Some constrictions will not have an unstable shape. For trapped blobs of oil, the pressure on the oil to leave the unstable throats may be less than that required to push the oil into any of the throats on the outside of the blob. Finally during a water flood, some instabilities may take too long (the flux of water through the thin films and crevices decreases with the cube of their thickness), if the instability is not possible until the transition zone arrives to provide water in a nearby pore, and if the instability is no longer possible when the oil becomes bypassed due to the inhibition described above for trapped oil.

Theoretical studies. In some theoretical studies in Cambridge by Hammond (1982) using lubrication theory for thin films of water and by Duffy using a numerical technique based on the boundary integral technique, it has been found that a drop of oil straddling a constriction in a water-wetted pipe does not always suffer from the choke-off instability. While a consideration of the shape of the oil-water interface in the constriction and at the ends of the drop might lead one to expect a choke off, the shape of the interface at all locations has a bearing on the fluid dynamics of the evolution.

In an unstable throat the curvature of the surface of the oil drop is highest in the throat and low on the spherical cap ends of the drop, but it is lowest on the shoulder of the drop where the spherical cap gives way to a cylinder of the same diameter. This minimum in the surface curvature produces a maximum of the pressure in the water just outside the oil drop, and this maximum in the pressure can act as a barrier to water flowing from the high pressure at the spherical cap ends to the minimum pressure in the water at the neck. For thin films of water, the maximum pressure pushes the water away from it and so draws the drop surface towards the pipe wall. This leaves a disconnected, static annular ring of water trapped in the constriction, which will not have sufficient volume to bridge across the width of the pipe if the initial volume of water in the constriction was small.

As a note of caution, it must be pointed out that this theoretical result for an axisymmetric geometry, may be misleading about the behaviour in a non-symmetric throat. The study does demonstrate, however, that the entire oil-water interface must be considered before the evolution can be predicted.

Etched networks

A two dimensional pattern of pores and throats can be etched chemically onto a flat glass slide. When covered by a second slide, the etched channels form a network similar to the theoretical ones described earlier. Separately Chandler and Dawe have produced some fascinating films of oil and water flowing through such networks.

Surface roughness. One immediate feature is that the meniscii do not always progress as one might guess from their position and curvature. Part of the reason is that the unseen curvature between the top and bottom slides can dominate the curvature which is seen in the plane of the slides, but also surface roughnesses do hold up the motion of the meniscus. This effect of surface roughness must be considered an essential ingredient in a rock.

Water channelling. Another feature seen vividly in the films is the effects of water moving invisibly along its privileged route in the corners of the etched channels, which surface tension prevents the oil from penetrating. This movement of water enables the choke-off instability to occur: oil becomes disconnected in a throat with water seeming to appear from nowhere. A similar movement of water is also seen in a partial contraction of the oil in the throats and pores some way ahead of the advancing front of the transition zone. It must however be pointed out that the water channels are particularly well connected in the etched networks and this may be untypical of a rock.

Meniscii at junctions. Lenormand, Zarcone & Sarr (1983) have examined the quasi-steady shapes that the oil-water interface can adopt at junctions in a symmetric square network. They find that the capillary pressure on the oil to leave is less if two adjacent exits are available compared with just one exit. Such priorities for the movement of the meniscii is illustrated in some clear experiments. This line of study needs to be developed to include irregularly shaped pores of a rock.

FUTURE WORK

The following is a brief shopping list of areas which I see as worthy of further study, starting from the small scale.

1. The moving contact line (involving the contradiction of the no-slip boundary condition and the observed progress of a two fluid interface over a solid surface) has become a classical unsolved problem. Perhaps the pragmatic solution of *slip at height* is adequate, or even correct. More attention should probably be given to dirty interfaces and

rough surfaces.

2. More revealing experiments are needed to examine the motion of meniscii at the junctions inside a pore. To provide some theoretical support, numerical methods such as the boundary integral technique need to be developed for the non-axisymmetric geometry and for the moving contact line.

3. It is time to incorporate into the capillary jump model (the one with the finite transition zone) some of the ideas about choke off and the motion of the meniscus at junctions. The privileged water channels can be modelled by a separate network with a high resistance to flow. The capillary pressure assigned to a pore on the instantaneous interface has, so far, been chosen to reflect the random size of the pore and the mean viscous pressure gradient. This random capillary pressure could perhaps be chosen not independently but taking account of the status of neighbouring pores, and reviewing that value during the progress of the flood.

4. This paper has concentrated on the simplest water flood. There are more complicated floods in tertiary recovery involving fluids with varying viscosities and surface tension, which will be affected by dispersion; and then there are floods involving heat, phase changes at the bubble or dew point and also chemical reactions. The porous strata are not necessarily homogeneous. Some reservoirs have porosity on different length scales, e.g. a system of cracks on the scale of meters or 1 mm particles which are themselves porous on a sub-micron scale. In addition, near to the bore hole the conditions are special with higher flow rates, lower pressures and contaminated rock.

REFERENCES

Broadbent, S.R. & Hammersley, J.M. 1957, Proc.Camb.Phil.Soc. 53, 629.

Chandler, R. Koplik, J., Lerman, K. & Willemsen, J.F. 1982, J.Fluid.Mech. 119, 249.

Chatzis, I. & Morrow, N.R. 1981, SPE paper, 10114.

Haines, W.B. 1930, J.Agricult.Sci. 20, 97.

Hammond, P.S. 1982, Ph.D. thesis, Cambridge.

Koplik,J. & Lasseter 1982, SPE paper, 11014.

Koplik,J., Lin,C. & Vermotte,M. 1984, submitted to J.Appl.Phy.

Larson,R.G., Davis,H.T. & Scriven,L.E. 1981, Chem.Eng.Sci. 53, 57.

Legait,B., Sourieau,P. & Cambarnous,M.J. 1983, J.Colloid Interface Sci. 91, 400.

Lenormand,R., Zarcone,C. & Sarr,A. 1983, J.Fluid Mech. 135, 337.

Mohnaty,K.K.,Davis,H.T. & Scriven,L.E. 1980, SPE paper, 9406.

Ng,K., Davis,H.T. & Scriven,L.E. 1978, Chem.Eng.Sci. 33, 1009.

Payatakes,A.C. & Dais,M.M. 1984, paper at Euromech 179.

Wilkinson,D. & Willemsen,J.F. 1983, J.Phys.A. 16, 3365.

Wilkinson,D. 1984, Phys.Rev.A. 30, 520.

Theoretical and Applied Mechanics
F.I. Niordson and N. Olhoff
Elsevier Science Publishers B.V. (North-Holland)

ICE AND SNOW MECHANICS
A CHALLENGE TO THEORETICAL AND APPLIED MECHANICS

Kolumban Hutter and Thorsten Alts

Laboratory of Hydraulics, Hydrology and Glaciology
ETH, Gloriastrasse 37-39, CH-8092 Zürich

Snow and Ice Mechanics has an engineering and a geophysical component, and both embrace virtually all aspects of theoretical and experimental mechanics. We shall present expositions on (i) the mathematical prediction of flacier and ice sheet geometry, (ii) the thermodynamics of temperate ice, (iii) convection in thawing subsea permafrost and, (iv) dynamics of snow and ice avalanches, all topics of geophysical fluid mechanics.

(i) Determination of the velocity and temperature distribution in cold glaciers and ice sheets is strongly coupled because the stress tensor depends on temperature and dissipative heat cannot, in general, be ignored. Further complications are the unknown free surface and highly nonlinear basal boundary conditions. Stretching of coordinates permits simplification under the restriction of slowly varying basal and free surface geometry. The emerging initial boundary value problem is ill posed and singular along the unknown grounding line unless basal friction behaves accordingly. Numerical difficulties of the steady problem are discussed.

(ii) Temperate ice is a mixture of an ice matrix at melting with trapped or perculating water (and salt). The water content is chiefly governed by viscous heat produced by the creeping motion. The continuum behavior of the two-phase model is dominated by the phase change processes taking place at the phase boundaries between the liquid inclusions and the ice matrix. Impurities (salt) may have significant quantitative effects. A thermodynamic model of surfaces is presented which are subject to phase changes and implications are drawn for the structure of a theoretical model of temperate ice.

(iii) When permafrost becomes submerged because of shore line erosion the covering ocean acts as a thermal insulator and the submerged permafrost starts to melt. The thawed layer is bounded above by the ocean bed through which salt may intrude and by the phase boundary which for a fixed offshore position is known to progress with the square root of time. This situation gives rise to non-steady double diffusion of the Darcy-Oberbeck-Boussinesq equations, coupling Bénard convection and liquefaction. The boundary value problem is formulated and scalings are introduced which indicate three different ranges of physical behavior depending on the relative magnitudes of phase boundary and convective bulk velocities. The perturbation analysis yielding observed thaw rates and the corresponding Bénard convection problem is presented in more detail.

(iv) Ice and snow avalanches are gravity flow phenomena in which collision of ice particles or eddies is the dominant mechanism of momentum transfer among flow entities. Ice and flow avalanches can be described by models appropriate for chute flow of granular materials but existing statistical theories are insufficient for their description. Powder snow avalanches are turbulent two phase flows akin to turbidity currents, but because of the importance of snow settlement phase separation is important. The two models are incompatible; transitions from flow avalanches to powder snow avalanches need be modelled by a granular material which incorporates the dynamics of the interstitial air. We present a description of the phenomena and review some aspects of the existing modelling, but concentrate on newer developments.

I. INTRODUCTION

Snow and ice mechanics has an engineering and a geophysical component, as both embrace virtually all aspects of theoretical and experimental mechanics. In glaciology observational and experimental techniques are better developed than the mathematical aspects, however, during the last decade a small community of applied mathematicians and mechanicians has contributed to the basic understanding of the science of snow and ice and promises to contribute more in the near future. The field is relatively unexplored in all those aspects which require a combination of several topics of classical physics and methods of analytical and computational solid and fluid mechanics. So, there is room for many more theoretically inclined people. They are likely to push glaciology ahead, but to be understood, they must bridge the communication gap that exists within the current glaciological community between those who favor physical and mathematical approaches, respectively.

It is hardly possible to do justice to all facets of snow and ice mechanics; we must necessarily be selective. Because of our own interest we shall emphasize mathematical approaches selected from four different branches:

(i) Mathematical prediction of ice sheet geometry,

(ii) Thermodynamics of temperate ice,

(iii) Convection in thawing subsea permafrost,

(iv) Dynamics of snow and ice avalanches.

Three of these deal with snow or ice in the geophysical environment and one is a challenging phase change problem that is not touched upon in usual treatizes on engineering thermodynamics. Physically, topic (i) deals with the response of large ice masses to variations in the climatological input, which is directly related to the solar irridation (Milankovitch hypothesis, see C. Covey, 1984) which, in turn varies according to the fluctuations in the earth's orbit. Ultimate aim is the forecasting of the extent of glaciers and ice sheets hundred, thousands or hundred thousend years ahead (Figures 1 & 2). The ice is modelled as a creeping non-Newtonian fluid on a possibly deformable visco-elastic bed. Mathematically, highly nonlinear partial differential equations must be solved in a domain which is also determined by the solution. Small parameters suggest simplifications by using perturbations, but the equations may be ill-posed and difficult to solve.

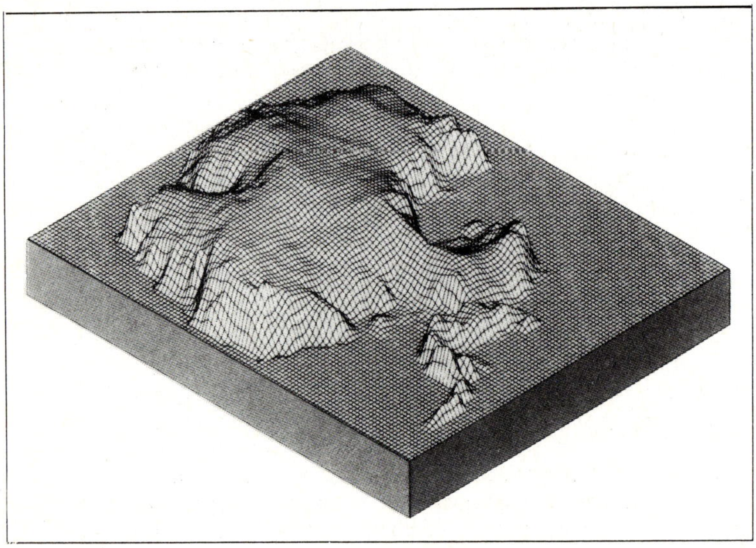

Figure 1

Isometric view of the surface of the Antarctic Ice sheet. The figure is oriented with the Antarctic Peninsula at the bottom. Fringing mountains are not reproduced. *(From Drewry, 1983).*

Temperate ice is ice at melting; thus dissipation due to stretching cannot, in general, be absorbed into heat but must rather be used up in melting. It follows that large temperate ice masses consist of ice and trapped or percolating water, subjected to persistent phase changes according to the amount of dissipation, and advected and conducted heat (Figure 3). A continuum model of temperate ice must

Figure 2

A view of the *Fieschergletscher* in Switzerland, position of snout as of 14 July 1968. This photograph shows the typical channelized character of Alpine glaciers and it illustrates that the snout position may fluctuate with time. At the time of the shot the entire length was 16 km, but as can be seen from the photograph the foreground was once (40 years earlier) covered with ice.

(Courtesy M. Aellen, Laboratory of Hydraulics, Hydrology and Glaciology, ETH, Zürich)

somehow account for these inner surfaces of phase changes and incorporate the water content into the dynamical equations and the constitutive rleations. Simple binary mixture and two phase models have been deduced by Hutter (1982) and Fowler (1984) but amendments are needed, if glacier flow and glacier hydraulics is to be coupled.

Glaciology has as challenging stability problems as any other branch of solid or fluid mechanics. The following is one, which offers complicated mathematics to both

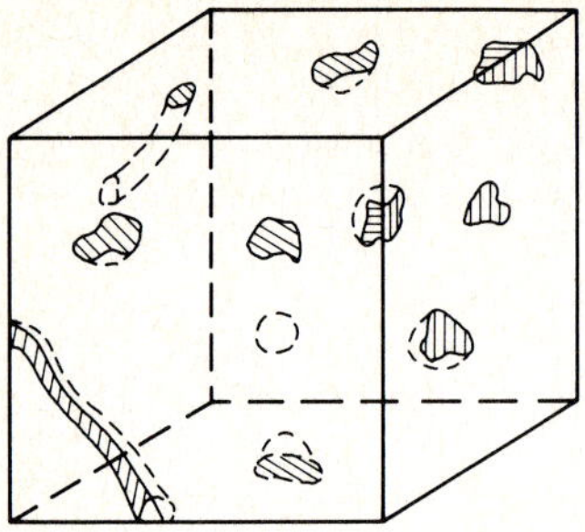

Figure 3

A cube of temperate ice consists of ice and liquid inclusions. These inclusions are isolated or interconnected and contribute to the behavior of wet ice. In real ice salt impurities are an important factor in the description of the proportion of such brine inclusions.

linear instability and non-linear energy stability analysis. It arose in a study just off-shore the coast of Alaska (Figure 4). When permafrost becomes submerged because of shore line erosion the covering ocean acts as a thermal insulation and the submerged permafrost starts to melt. The thawed layer is bounded above by the ocean bed through which salt may intrude and below by the phase boundary (the melting surface of the permafrost) which for a fixed offshore position is known to progress with the square root of time. Of interest is the flow of water and salt through the thawed layer of soil and its stability. Specialists will recognize that this problem gives rise to non-steady double diffusion of the Darcy-Oberbeck-Boussinesq (DOB) equations, coupling Bénard convection and liquefaction. In approach and methods it is very similar to problems of crystal growth from saturated melts.

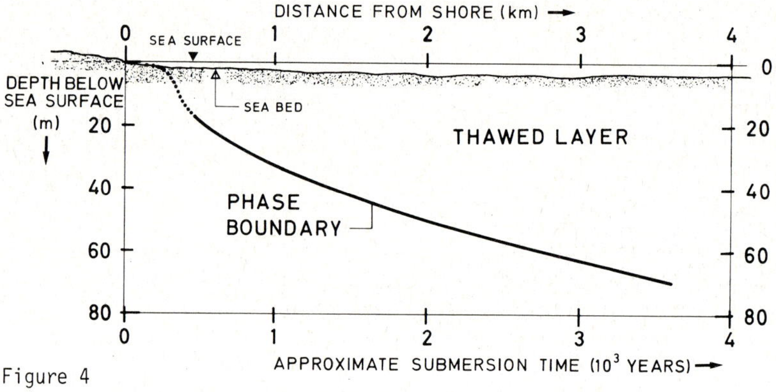

Figure 4

Approximate configuration of subsea permafrost at Prudhoe Bay, Alaska, near the West Dock. The phase boundary is shown dashed near the shore where its details are not well known. The submersion time scale is based on an assumed shoreline retreat rate of 1 m a^{-1}.
(From Harrison, 1982)

Snow under creeping motion and subject to varying atmospheric conditions suffers in general a multitude of phase and crystallographic changes. It represents the most difficult material known to us, and by complexity could absorbe a large number of thermodynamicists, rheologists etc. By contrast, snow under avalanching conditions is much easier to understand even though present knowledge is also limited in this regard. Snow avalanches, basically occur in two limiting forms, *flow avalanches* and *powder snow avalanches* (Figures 5 & 6). The first manifests itself as a rather compact shear flow down a mountain flank of tucky snow balls bouncing against each others. In this motion the interstitial air plays a negligible role. Powder snow avalanches are air-borne gravity driven turbulent wall jets of snow particles. They are faster and more damaging than flow avalanches, but the dynamics of both is important to know, because endangered zones can for instance be inferred from snow settlement estimates under given topographic and avalanching conditions, knowledge that would be valuable for intermauntane communities.

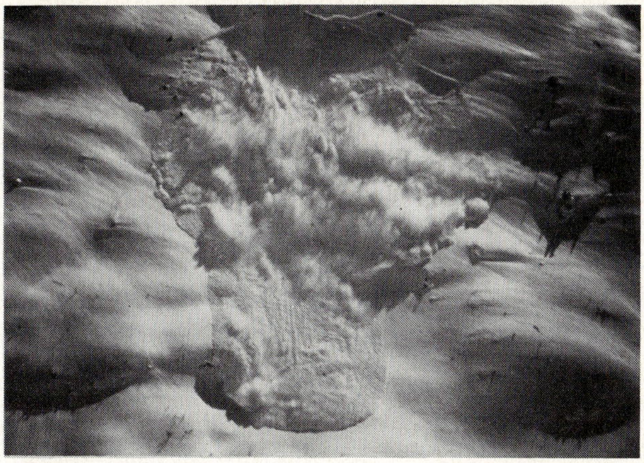

Figure 5

Artificially released flow avalanche "Marianne" at Bernina, 8. April 1975, Switzerland, under motion. Note the clouds of air borne snow in the upper parts. For wet spring avalanches these clouds are not formed. The proportions of air borne to flowing snow depends largely on wetness. *Photo: Swiss Federal Institute of Snow and Avalanche Research, Weissfluhjoch-Davos.*

Figure 6

Artificially released powder snow avalanche (Lavin, Engadin, Switzerland, 25. February 1970). *Photo: W. Porton, Federal Institute of Snow and Avalanche Research, Weissfluhjoch-Davos, Switzerland.*

2. MATHEMATICAL PREDICTION OF ICE SHEET GEOMETRY

2.1 State of the art

Morland (1984) in his recent article on thermomechanical balances of ice sheet flows, quoting Ahlman (1947) and Budd & Radok (1971), states "... that glaciers [and ice sheets] are our climatographical register, providing the most reliable evidence of the history of our climate once we understand how they move and vary under changing conditions. There is [indeed] a need for thermomechanical treatments to describe the effects of long time climate variation; ... systematic approximations (of very complex problems) which can be evaluated are required, in contrast to physically inferred simplifications in which neglected features are not assessed," (Paterson, 1980a, 1980b; Thomas, 1979; Robin, 1979). We fully concur with this statement.

The first consistent treatment of cold glacier flow is by Fowler & Larson (1978, 1980). They analyse plane ice flow through scaled mass-, momentum- and energy balances and apply their analysis to steady and unsteady plane flow including (kinematic) surface waves (Fowler, 1982) and seasonal waves or surges (Fowler, 1982). A more general treatment of plane flow of cold, but isothermal glaciers or ice sheets was given by Morland & Johnson (1980, 1982) and Johnson (1981). They demonstrate that in a scaling process of the governing equations essentially two dimensionless parameters arise whose order of magnitudes differ depending upon whether steep glaciers or flat ice sheets are analysed. One of these parameters occurs in the constitutive relations of stress and measures the amount of stress that is needed to produce a stretching of given order, the other is simply an aspect ratio expressing the fact that glaciers and ice sheets are long but shallow. Almost simultaneously, Hutter (1981, 1982a) deduced a similar model; he also used scaling arguments and rediscovered that asymptotic analysis of steep and flat ice sheets had to be differentiated. He later proved (unpublished, 1982) that his own and the Morland-Johnson schemes are identical. He further indicated (Hutter, 1983, Chapter 5) how cold ice masses had to be treated, however, it was Morland (1984) who presented a full scale-analysis of plane flow of cold ice sheets. He and Smith (1984), using prescribed temperature fields demonstrated that the motion and temperature distribution in cold ice masses are strongly coupled, so that neither the flow field, nor the temperature distribution, nor the geometry of the ice sheet can be determined without the simultaneous determination of the other. Yakowitz, Hutter & Szidarovsky (1984), finally, presented the first coupled calculation of the plane flow problem.

All the above deals with plane or pure radial flow, was demonstrated to be useful also in ice shelf balances (Morland & Shoemaker, 1982) and proved capable of reproducing certain plane flow features of the Greenland ice cap (Boulton, Smith & Morland, 1984). An asymptotic model for shallow cold (but isothermal) ice sheets in three dimensions is given by Hutter (1983, Chapter 7). Unlike one would expect, this three-dimensional extension yields properties of the flow field not seen by the plane flow assumption and subjectable to observational corroboration. It also paves the route for possible amendments and demonstrates that the most serious complication which is somewhat hidden in the two-dimensional plane flow situation is the fact that the ice sheet flow must be determined along with its geometry so that the domain of the solution for which the mathematical boundary value problem must be solved is itself unknown. This places serious difficulties in any endeavour to find computational solutions to the ice sheet equations.

To do justice, a word should also be included on the Russian literature. Grigoryan, Krass & Shumsky (1976) and Shumsky & Krass (1976) presented models of non-isothermal glaciers and ice shelves similar to those of Fowler, Hutter and Morland. In their more recent work (most of which is in Russian) they essentially "prescribe" temperature from surface temperature records and geothermal conditions, similar in spirit to Morland & Smith (1984), but they attack the time-dependent problem ab initio. (Buyanov, 1983; Krass, 1981, 1983; Verbitsky & Guvornkha, 1979; Verbitsky & Chalikov, 1980; Verbitsky, 1980, 1981; Grigoryan et al., 1984). A full thermomechanical problem is not treated.

2.2 Model equations of cold ice sheets

Consider a three-dimensional ice sheet occupying the domain $D \in \mathbb{R}^3$ with bounding surfaces ∂D_S (free surface) and ∂D_B (base). Let x, y, z be Cartesian coordinates (see Figure 7); x, y are horizontal while z is vertical. The top free surface and the basal surface will be denoted by $z = z_S(x,y,t)$ and $z = z_B(x,y,t)$, respectively and the domain D is assumed to be continuously filled with polycrystalline in situ-

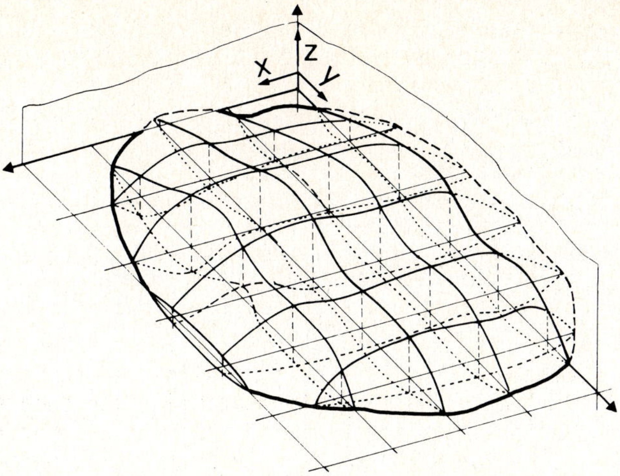

Figure 7

Sketch of a threedimensional ice sheet with chosen Cartesian co-ordinate system.

x and y are horizontal, z is vertical. (From Hutter, 1983)

ice, of which the constitutive properties will be defined in a moment. The effect of the Earth's mantle and the climate on the ice sheet is supposed to affect the ice sheet through thermal and kinematic boundary conditions on ∂D_B and ∂D_S, respectively. In other words, we assume the geosphere and the atmosphere to act as prescribed driving mechanisms and ignore interactions.

All continuum mechanical models of creeping flow of in-situ ice assume Stokes flow (negligible acceleration) and postulate ice as a *non-Newtonian viscous heat conducting fluid* which obeys Fourier's law of heat conduction. Deviator stress $\underset{\sim}{t}'$, stretching $\underset{\sim}{D}$ = sym grad $\underset{\sim}{u}$ and temperature T are related by

$$\underset{\sim}{D} = A(T)\, f(t'_{II})\, \underset{\sim}{t}', \qquad t'_{II} = \frac{1}{2} t'_{ij}\, t'_{ij} \tag{2.1}$$

in which A is a temperature dependent rate factor and f is a creep response function depending on the second stress deviator invariant. Traditionally, A is an exponential, but this Arrhenius relation is known to be inaccurate close to melting; f is a power law and then (2.1) is known as Glen's law (1955). In newer developments A and f are represented as polynomials. They give better fits to the scattered data (Hutter, 1983, Chapter 2; Smith & Morland, 1981) and avoid in certain circumstances unwanted boundary singularities (Johnson & McMeeking, 1982).

The boundary value problem for the flow of cold ice through D is given by the balances of mass, momentum and energy,

$$\begin{aligned} \operatorname{div} \underset{\sim}{u} &= 0, \\ -\operatorname{grad} p + \operatorname{div} \underset{\sim}{t}' + \rho \underset{\sim}{g} &= 0, \qquad (x,y,z)\, \varepsilon D \\ \rho c_p \dot{T} &= -\operatorname{div}(\kappa \operatorname{grad} T) + \operatorname{tr}(\underset{\sim}{t}', \underset{\sim}{D}) \end{aligned} \tag{2.2}$$

and the boundary conditions at the free surface

$$\begin{aligned} \frac{\partial z_S}{\partial t} + \frac{\partial z_S}{\partial x} u + \frac{\partial z_S}{\partial y} v - w &= a(x,y,z_S,t) \\ T &= T_S(x,y,z_S,t) \qquad (x,y,z)\, \varepsilon\, \partial D_S \\ \underset{\sim}{t} \cdot \underset{\sim}{n}_S = \underset{\sim}{0}, \quad \underset{\sim}{t} &= -p\underset{\sim}{1} + \underset{\sim}{t}' \end{aligned} \tag{2.3}$$

and the bed
$$\frac{\partial z_B}{\partial t} + \frac{\partial z_B}{\partial x} u + \frac{\partial z_B}{\partial y} v - w = 0,$$

$$\kappa \operatorname{grad} T \cdot \underset{\sim}{n}_B = -Q(x,y,z_B,t), \qquad (x,y,z) \in \partial D_B. \tag{2.4}$$

$$\underset{\sim}{u} = -F(\underset{\sim}{t}_\perp, \underset{\sim}{t}_\parallel, T)\, \underset{\sim}{t}_\parallel ,$$

Here, $\underset{\sim}{n}_S$ and $\underset{\sim}{n}_B$ are exterior unit vectors and $\underset{\sim}{u} = (u,v,w)$, p, ρ, g, c_p, κ are the velocity field, pressure, ice density, gravity constant, specific heat and heat conductivity of ice and grad, div, tr denote the common 3D vector operations. $a(x,y,z_S(x,y,t),t)$, $T_S(x,y,z_S(x,y,t),t)$ are the prescribed functions of surface accumulation rate and surface temperature; they constitute the driving forces modeling the climate. Moreover, $Q(x,y,z_B(x,y,t),t)$ is the geothermal heat flow, assumed small enough that basal melting cannot occur. The last of equations (2.4) expresses a sliding law, relating the basal velocity with the tangenial traction $\underset{\sim}{t}_\parallel = (\underset{\sim}{t}' - \underset{\sim}{n}_B \cdot \underset{\sim}{t}' \cdot \underset{\sim}{n}_B) \cdot \underset{\sim}{n}_B$ through a non-linear coefficient function F, which itself may depend on the normal stress $\underset{\sim}{t}_\perp = \underset{\sim}{t} \cdot \underset{\sim}{n}_B - \underset{\sim}{t}_\parallel$ and on temperature. Functional relationships for F are difficult to deduce, because F emerges from a boundary analysis within a layer dominated by a roughness scale; for a brief description see Hutter (1982c), for summaries Lliboutry (1979) and Weertman (1979), for new developments Fowler (1981, 1984b). The boundary conditions involve kinematic, temperature and stress statements. Because of a non-vanishing accumulation rate the top surface is non-material; equation $(2.3)_1$ serves as evolution equation for the ice sheet geometry and $(2.4)_1$ describes the motion of the base, ignoring abrasion but incorporating its deformation. In the simplest relaxation-type response model one could relate $\dot{z}_B$ to the basal pressure p_B according to

$$\dot{z}_B = \int_{t=-\infty}^{t} G(t-\tau) \frac{dp_B(\tau)}{d\tau} d\tau \tag{2.5}$$

with a monotonically decreasing relaxation function G. Better models would include non local effects. Henceforth the deformation of the base will be ignored.

2.3 Scaling arguments

Nondimensionalizations of the above equations are motivated by the fact that the sheets are long and wide, but shallow. We thus scale horizontal and vertical lengths differently by employing the *shallow ice approximation:* The transformations are

$$\begin{aligned}
(x,y) &= [L]\,(\xi,\eta) & [L] &= \text{typical length} \\
z &= [H]\,\zeta & [H] &= \text{typical depth} \\
(u,v) &= [U]\,(\bar{u},\bar{v}) & [U] &= \text{typical horizontal velocity} \\
w &= \varepsilon[U]\,\bar{w} & [\mathfrak{T}] &= \text{characteristic time} \\
a &= \varepsilon[U]\,\bar{a} & T_f &= \text{melting temperature at atmospheric pressure} = 273.15\ ^\circ\text{C} \\
t &= \frac{1}{\varepsilon}[\mathfrak{T}]\,\bar{t} & & \\
T &= T_f + [\Delta T]\,\theta & [\Delta T] &= \text{temperature range within the ice} \sim 20\ ^\circ\text{C} \\
(\underset{\sim}{t}', p) &= \rho g [H]\,(\underset{\sim}{\sigma}', \bar{p}) & & \\
& & \varepsilon &= H/L \ll 1
\end{aligned} \tag{2.6}$$

in which Greek, or overbarred quantities are dimensionless. Substituting (2.6) into (2.1)-(2.4) yields the dimensionless form of the field equation (Hutter, 1983, Chapters 3, 7)

$$\frac{\partial \bar{u}}{\partial \xi} + \frac{\partial \bar{v}}{\partial \eta} + \frac{\partial \bar{w}}{\partial \zeta} = 0,$$

$$-\varepsilon \frac{\partial \bar{p}}{\partial \xi} + \varepsilon \frac{\partial \sigma_x}{\partial \xi} + \varepsilon \frac{\partial \tau_{xy}}{\partial \eta} + \frac{\partial \tau_{xz}}{\partial \zeta} + \varepsilon g_1 = 0,$$

$$\varepsilon \frac{\partial \tau_{xy}}{\partial \xi} - \varepsilon \frac{\partial \bar{p}}{\partial \eta} + \varepsilon \frac{\partial \sigma_y}{\partial \eta} + \frac{\partial \tau_{yz}}{\partial \zeta} + \varepsilon g_2 = 0,$$

$$\varepsilon \frac{\partial \tau_{xz}}{\partial \xi} + \varepsilon \frac{\partial \tau_{yz}}{\partial \eta} - \frac{\partial \bar{p}}{\partial \zeta} + \frac{\partial \sigma_z}{\partial \zeta} - g_3 = 0,$$

$$\frac{\partial \theta}{\partial t} + \operatorname{grad}_{\underset{\sim}{\xi}} \theta \cdot \underset{\sim}{\bar{u}} = \mathbb{D}\left[\frac{\partial^2 \theta}{\partial \zeta^2} + \varepsilon^2 \left(\frac{\partial^2 \theta}{\partial \xi^2} + \frac{\partial^2 \theta}{\partial \eta^2}\right)\right] + 2\,\mathbb{E}\,\frac{\mathbb{G}}{\varepsilon}\,\bar{A}(\theta)\,\mathbb{f}(\tau_{II})\,\tau_{II},$$

$$\varepsilon \frac{\partial \bar{u}}{\partial \xi} = \mathbb{G}\,\bar{A}(\theta)\,\mathbb{f}(\tau_{II})\,\sigma_x, \tag{2.7}$$

$$\varepsilon \frac{\partial \bar{u}}{\partial \eta} + \varepsilon \frac{\partial v}{\partial \xi} = 2\,\mathbb{G}\,\bar{A}(\theta)\,\mathbb{f}(\tau_{II})\,\tau_{xy}$$

$$\frac{\partial \bar{u}}{\partial \zeta} + \varepsilon^2 \frac{\partial \bar{w}}{\partial \xi} = 2\,\mathbb{G}\,\bar{A}(\theta)\,\mathbb{f}(\tau_{II})\,\tau_{xz}$$

$$\varepsilon^2 \frac{\partial \bar{w}}{\partial \eta} + \frac{\partial \bar{v}}{\partial \zeta} = 2\,\mathbb{G}\,\bar{A}(\theta)\,\mathbb{f}(\tau_{II})\,\tau_{yz}$$

$$\varepsilon \frac{\partial \bar{v}}{\partial \zeta} = \mathbb{G}\,\bar{A}(\theta)\,\mathbb{f}(\tau_{II})\,\sigma_z,$$

$$\varepsilon \frac{\partial \bar{w}}{\partial \eta} = \mathbb{G}\,\bar{A}(\theta)\,\mathbb{f}(\tau_{II})\,\sigma_y,$$

in which $(g_x, g_y) = \varepsilon(g_1, g_2)$ are the horizontal components of the gravity vector and $g_3 = -g_z$ is its vertical component, g_1, g_2, g_3 being order unity or smaller. Further, $\tau_{II} = 1/2\,\sigma_{ij}\,\sigma_{ij}$ is the dimensionless second stress deviator invariant and $\bar{A}(\theta)$ and $\mathbb{f}(\tau_{II})$ are dimensionless order unity rate factors and creep response functions, respectively, and $\mathbb{G}$, $\mathbb{D}$ and $\mathbb{E}$ are dimensionless characteristic parameters defined in Table 1. From it we conclude with Hutter (1983) and Morland (1984) that thermal advection, both horizontal and vertical plays a dominant role in the balance of heat and cannot be neglected. The diffusion coefficient $\mathbb{D}$ is order unity for thin ice sheets but not small enough for thick ice sheets in order that a sharp thermal boundary layer could develop at the base. Alternatively, dissipation is larger for thick than for thin sheets and non-negligible in general. These deductions assume that $\mathbb{G}^{-1}$ and ε are of the same order, which is the case. For completeness we also give expressions for $\bar{A}$ and $\mathbb{f}$:

$$\begin{aligned} \bar{A} &= \mathbb{A}(\theta) = 0.7242 \exp(11.597\,\theta) + 0.3438 \exp(2.9494\,\theta) \\ \mathbb{f} &= \mathbb{f}(x^2) = (x^{n-1} + \mathbb{k})/(1+\mathbb{k}), \quad n = 2\text{-}3, \quad \mathbb{k} = 0.1\text{-}0.01 \end{aligned} \tag{2.8}$$

but these are not the only possible ones (Hutter, 1983, Chapter 2).

$[H] =$	100 m	1000 m	4000 m
$\mathbb{G} = [\mathfrak{r}]\, A_0\, \rho g [H]\, f((\rho g [H])^2) =$	10^2	10^2	10^2
$\mathbb{D} = \dfrac{\kappa\, [\mathfrak{T}]}{\rho c_p\, [H]^2\, \varepsilon} =$	0.4	0.04	0.01
$\mathbb{E} = \dfrac{g\, [H]}{c_p\, [\Delta T]} =$	0.025	0.25	1.0

Table 1

Approximate values of $\mathbb{G}$, $\mathbb{D}$, and $\mathbb{E}$ for different values of the scaling height $[H]$ using $[\mathfrak{T}] = [U]/[H]$, $[U] = 100\ \mathrm{m\, a^{-1}}$ (longitudinal velocity scale and $[HU/L] = 1\ \mathrm{m a^{-1}}$ acceleration scale, corresponding to $\varepsilon = 10^{-2}$). Numerical values for physical constants are as listed in Hutter (1983, p. 151).

Dimensionless forms of the boundary conditions are: At the free surface

$$
\begin{aligned}
&\frac{\partial \bar z_S}{\partial \bar t} + \frac{\partial \bar z_S}{\partial \xi}\bar u + \frac{\partial \bar z_S}{\partial \eta}\bar v - \bar w = \bar a(\xi, \eta, \bar z_S, \bar t) \\
&\theta = \theta_S(\xi, \eta, z_S, t) \\
&-\varepsilon(-\bar p + \sigma_x)\frac{\partial \bar z_S}{\partial \xi} + \tau_{xz} - \varepsilon\,\tau_{xy}\frac{\partial \bar z_S}{\partial \eta} = \varepsilon\,\bar p^{atm}\frac{\partial \bar z_S}{\partial \xi}, \\
&-\varepsilon\,\tau_{xz}\frac{\partial \bar z_S}{\partial \xi} + (-\bar p + \sigma_z) - \varepsilon\,\tau_{yz}\frac{\partial \bar z_S}{\partial \eta} = -\bar p^{atm}, \\
&-\varepsilon\,\tau_{xy}\frac{\partial \bar z_S}{\partial \xi} + \tau_{yz} - (-\bar p + \sigma_y)\frac{\partial \bar z_S}{\partial \zeta} = \varepsilon\,\bar p^{atm}\frac{\partial \bar z_S}{\partial \eta},
\end{aligned}
\qquad (2.9)
$$

and at the rigid base

$$
\begin{aligned}
&\frac{\partial \theta}{\partial \zeta} - \varepsilon\left(\frac{\partial \theta}{\partial \xi}\frac{\partial \bar z_B}{\partial \xi} + \frac{\partial \theta}{\partial \eta}\frac{\partial \bar z_B}{\partial \eta}\right) = -\mathfrak{Q}^{geoth}\, Y, \\
&\bar u = -\frac{\mathbb{F}}{Y}\left[\varepsilon\left(\sigma^\star_x\frac{\partial \bar z_B}{\partial \xi} + \tau^\star_{xy}\frac{\partial \bar z_B}{\partial \eta}\right) - \tau^\star_{xz}\right], \\
&\bar v = -\frac{\mathbb{F}}{Y}\left[\varepsilon\left(\tau^\star_{xy}\frac{\partial \bar z_B}{\partial \xi} + \sigma^\star_y\frac{\partial \bar z_B}{\partial \eta}\right) - \tau^\star_{yz}\right], \\
&\varepsilon\bar w = -\frac{\mathbb{F}}{Y}\left[\varepsilon\left(\tau^\star_{xz}\frac{\partial \bar z_B}{\partial \xi} + \tau^\star_{yz}\frac{\partial \bar z_B}{\partial \eta}\right) - \sigma^\star_z\right], \\
&Y = \left[1 + \varepsilon^2\left(\frac{\partial \bar z_B}{\partial \xi}\right)^2 + \varepsilon^2\left(\frac{\partial \bar z_B}{\partial \eta}\right)^2\right]^{1/2},
\end{aligned}
\qquad (2.10)
$$

in which $\sigma^\star_x$ etc. are the dimensionless counter parts of t_x etc. and

$$\mathbb{Q}^{geoth} = \frac{[H]\,Q^{geoth}}{\kappa\,[\Delta T]} \lessapprox 0.01, \qquad \mathbb{F} = \frac{\rho g\,[H]}{[U]}\,F. \tag{2.11}$$

Equations (2.7)-(2.11) comprise the formulation of the boundary value problem in dimensionless form. Of the dimensionless parameters in the field equations $\mathbb{D}$, $\mathbb{E}$ and $\mathbb{G}$ are fixed but ε which is expected to be small, is still free. In the subsequent asymptotic analysis we shall choose $\varepsilon = o(\mathbb{G}^{-1})$ or

$$\varepsilon = \mathbb{G}^{-1} \ll 1. \tag{2.12}$$

This choice is compelling. Indeed for $g_1 = g_2 = 0$ the lowest order force balances $(2.7)_{2,3}$ require that $\partial\tau_{xz}/\partial\zeta$ and $\partial\tau_{yz}/\partial\zeta$ vanish unless $\tau_{xz} = \mathcal{O}(\varepsilon)$ and $\tau_{yz} = \mathcal{O}(\varepsilon)$. In this latter case the transverse (vertical) shear stress gradients can be balanced by the horizontal pressure gradients $\partial\bar{p}/\partial\xi$ and $\partial\bar{p}/\partial\eta$, respectively. We assume such order of magnitude relationships to hold. It then follows from $(2.7)_{8,9}$ that

$$\mathbb{G}\,\bar{A}(\bar{\theta})\,\mathbb{f}(\tau_{II}) \begin{bmatrix} \tau_{xz} \\ \tau_{yz} \end{bmatrix} = \mathcal{O}(1) \implies \mathbb{G} = \mathcal{O}(\varepsilon^{-1}) \tag{2.13}$$

if $\bar{A}(\bar{\theta})$ and $\mathbb{f}(\tau_{II})$ are both order unity. With this choice of $\mathbb{G}$ $(2.7)_{6,7,10,11}$ imply that $\sigma_x, \sigma_z, \sigma_y$ and τ_{xy} are all $\mathcal{O}(\varepsilon^2)$ or smaller, so that τ_{II} is of $\mathcal{O}(\varepsilon^2)$; with the functional relationships (2.8), $\bar{A}$ and $\mathbb{f}$ are thus indeed $\mathcal{O}(1)$.

The above analysis suggests construction of the following perturbation solutions

$$\begin{aligned} (\bar{u}, \bar{v}, \bar{w}, \theta, \bar{p}) &= \sum_{\nu=0}^{\infty} \varepsilon^{\nu} (\bar{u}_\nu, \bar{v}_\nu, \bar{w}_\nu, \theta_\nu, \bar{p}_\nu), \\ (\tau_{xz}, \tau_{yz}) &= \sum_{\nu=1}^{\infty} \varepsilon^{\nu} (\tau_{xz_\nu}, \tau_{yz_\nu}), \\ (\sigma_x, \sigma_y, \sigma_z, \tau_{xy}) &= \sum_{\nu=2}^{\infty} \varepsilon^{\nu} (\sigma_{x_\nu}, \sigma_{y_\nu}, \sigma_{z_\nu}, \tau_{xy_\nu}). \end{aligned} \tag{2.14}$$

No systematic analysis has been performed so far. We therefore focus attention to the lowest order model.

2.3 Lowest order model equations

Introducing $(\bar{u}_o, \bar{v}_o, \bar{w}_o, \theta_o, \bar{p}_o) := (U, V, W, \Theta, P)$, $(\tau^1_{xz}, \tau^1_{yz}) =: \varepsilon(T_x, T_y)$ and incorporating the scaling (2.13) yields the lowest order field equations

$$\left.\begin{aligned} &-\frac{\partial P}{\partial \xi} + \frac{\partial T_x}{\partial \zeta} + g_1 = 0, \qquad \frac{\partial U}{\partial \xi} + \frac{\partial V}{\partial \eta} + \frac{\partial W}{\partial \zeta} = 0, \\ &-\frac{\partial P}{\partial \eta} + \frac{\partial T_y}{\partial \zeta} + g_2 = 0, \qquad \frac{\partial U}{\partial \zeta} = 2\,\bar{A}(\Theta)\,\mathbb{f}(\varepsilon^2 T_{II})\,T_x, \\ &\frac{\partial P}{\partial \zeta} + g_3 = 0, \qquad \frac{\partial V}{\partial \zeta} = 2\,\bar{A}(\Theta)\,\mathbb{f}(\varepsilon^2 T_{II})\,T_y, \\ &\frac{\partial \Theta}{\partial t} + \frac{\partial \Theta}{\partial \xi} U + \frac{\partial \Theta}{\partial \eta} V + \frac{\partial \Theta}{\partial \zeta} W = \mathbb{D}\,\frac{\partial^2 \Theta}{\partial \zeta^2} + 2\,\mathbb{E}\,\bar{A}(\Theta)\,\mathbb{f}(\varepsilon^2 T_{II})\,T_{II}, \\ &T_{II} := T_x^2 + T_y^2. \end{aligned}\right\} \text{ in D} \tag{2.15}$$

A similar reduction is also possible for the boundary conditions. Deleting momentarily the kinematic surface condition the boundary conditions (2.9) and (2.10) reduce to

$$\left.\begin{aligned} &\Theta = \theta_S(\xi, \eta, z_S(\xi, \eta, t), t), \\ &T_x = p^{atm}\frac{\partial z_S}{\partial \xi}, \quad P = \bar{p}^{atm}, \quad T_z = \bar{p}^{atm}\frac{\partial z_S}{\partial \eta} \end{aligned}\right\} \text{ on } \partial D_S \tag{2.16}$$

and

$$\left.\begin{aligned} &\frac{\partial \Theta}{\partial \zeta} = -\mathbb{Q}^{geoth}, \quad W = \frac{\partial z_B}{\partial \xi}U + \frac{\partial z_B}{\partial \eta}V, \\ &U = \tilde{\mathbb{F}}\, T_x, \quad V = \tilde{\mathbb{F}}\, T_y, \end{aligned}\right\} \text{ on } \partial D_B \tag{2.17}$$

where $\mathbb{F} = \varepsilon\, \tilde{\mathbb{F}}$ must be of order unity if basal sliding is significant. $\tilde{\mathbb{F}} = 0$ corresponds to no slip and $\tilde{\mathbb{F}} \to \infty$ to perfect sliding.

Consider the special case with $g_1 = g_2 = 0$, $g_3 = 1$. Straightforward manipulations with (2.15)-(2.17) then show that

$$\begin{aligned} &T_x = -\frac{\partial z_S}{\partial \xi}(z_S - \zeta), \quad T_y = -\frac{\partial z_S}{\partial \eta}(z_S - \zeta), \quad P = (z_S - \zeta), \\ &U = I_\xi - \tilde{\mathbb{F}}\frac{\partial z_S}{\partial \xi}(z_S - z_B), \quad V = I_\eta - \tilde{\mathbb{F}}\frac{\partial z_S}{\partial \eta}(z_S - z_B), \end{aligned} \tag{2.18}$$

in which

$$\begin{aligned} &I_\xi(\zeta, z_S, z_B, \frac{\partial z_S}{\partial \xi}, \frac{\partial z_S}{\partial \eta}) = -\int_{z_B}^{\zeta} 2\,\bar{A}(\Theta)\, \mathbb{f}(K^2(\bar{\zeta}))\frac{\partial z_S}{\partial \xi}(z_S - \bar{\zeta})\, d\bar{\zeta}, \\ &I_\eta(\zeta, z_S, z_B, \frac{\partial z_S}{\partial \xi}, \frac{\partial z_S}{\partial \eta}) = -\int_{z_B}^{\zeta} 2\,\bar{A}(\Theta)\, \mathbb{f}(K^2(\bar{\zeta}))\frac{\partial z_S}{\partial \eta}(z_S - \bar{\zeta})\, d\bar{\zeta}, \\ &K^2 = \varepsilon^2(z_S - \zeta)^2\left[\left(\frac{\partial z_S}{\partial \xi}\right)^2 + \left(\frac{\partial z_S}{\partial \eta}\right)^2\right]. \end{aligned} \tag{2.19}$$

These results provide informative implications: Clearly, if the geometry of the ice sheet and the temperature distribution within it were known, the lowest order stresses and the velocities could be calculated. Even without this knowledge, the following practically relevant inferences can be drawn (Hutter, 1983, Chapter 7):

(i) To lowest order, the shear stress components are proportional to both the depth above the point of consideration and the surface gradient in the same direction. Thus, maximum shear is in the direction of steepest descent and zero shear is in the direction of the lines of equal height.

(ii) At any given (ξ, η)-position and any depth the flow is always in the direction of the steepest descent of the top surface. A dome or a trough is, consequently, a position of vanishing velocity for all depths and serves as an ice divide. Flow lines can be constructed as the orthogonal trajectories of the lines of equal height from topographic maps, see Figures 8 & 9.

These results can be tested by field observations. As far as this has been done, they are reasonably fulfilled; improvements of the theoretical model could be constructed by higher order expansion but are probably premature.

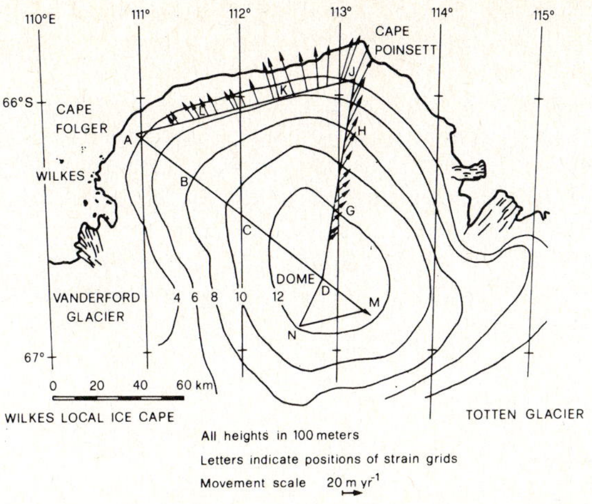

Figure 8

The Wilkes local ice cap showing movement vectors calculated from the differences between tellurometer traverses carried out in 1965 and 1966. *(From Hutter, 1983)*

Figure 9

Selected ice flowlines of the Antarctic ice sheet. Ice divides are dashed, and surface levels are indicated by varying intensities of the gray background. (From Drewry, 1983)

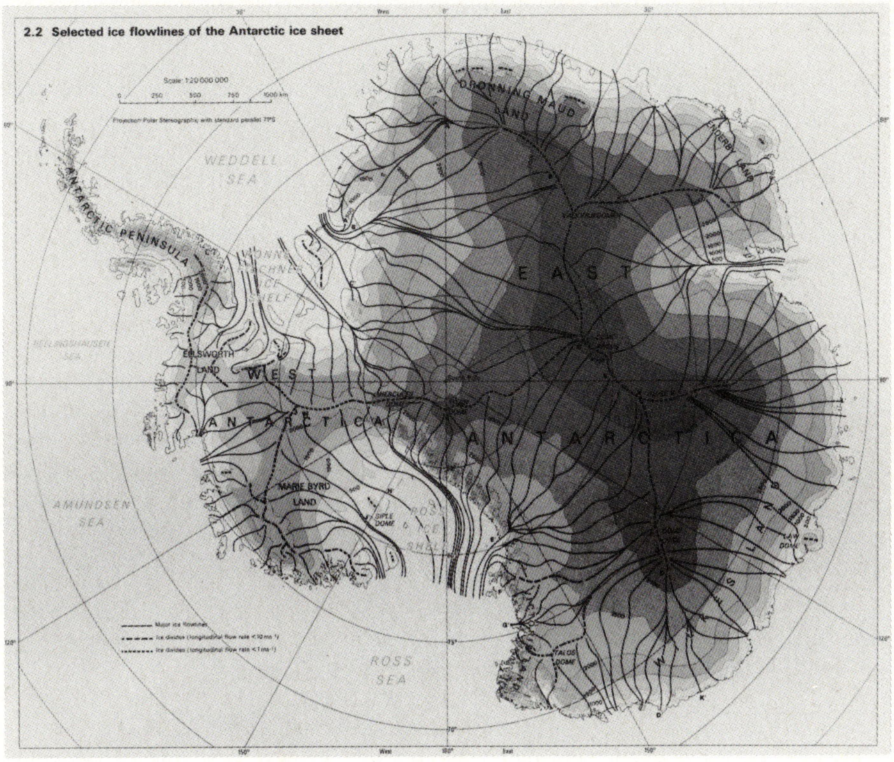

The evolution equation describing the surface geometry follows from the kinematic surface condition $(2.9)_1$ by integration over depth and satisfaction of appropriate boundary conditions at the base and the free surface (Hutter, 1983, Chapter 7). The result is

$$\frac{\partial z_S}{\partial \bar{t}} + \frac{\partial \mathcal{Q}_\xi}{\partial \xi} + \frac{\partial \mathcal{Q}_\eta}{\partial \eta} = a(\xi, \eta, z_S(\xi, \eta, \bar{t}), \bar{t}),$$
$$\mathcal{Q}_\xi = \int_{z_B}^{z_S} U(\cdot, \zeta)\, d\zeta, \quad \mathcal{Q}_\eta = \int_{z_B}^{z_S} V(\cdot, \zeta)\, d\zeta, \tag{2.20}$$

which is a forced *kinematic wave equation*. Because $\mathcal{Q}_\xi$ and $\mathcal{Q}_\eta$ are nonlinear functions of z_S, z_B, $\partial z_S/\partial\xi$, $\partial z_S/\partial\eta$, (2.20) is a nonlinear advection diffusion equation in two dimensions.

The lowest order ice flow problem is now complete. It is governed by the unsteady advection-diffusion equation for temperature Θ, $(2.15)_7$ which must be integrated subject to the thermal boundary conditions $(2.16)_1$ and $(2.17)_1$. They must be satisfied at the free surface and the base, respectively. The latter is known and the former can be obtained from the kinematic wave equation (2.20). In the process of its evaluation the velocity and stress fields can be determined as side results. The domain of integration $D(\xi,\eta)$ is given by the condition $z_S > z_B$ and the boundary $\partial D(\xi,\eta)$ evolves from the equation $z_S = z_B$.

2.4 Remarks on integration procedures

In steady state equation (2.20) has the form

$$D_\xi \frac{\partial^2 z_S}{\partial \xi^2} + D_\eta \frac{\partial^2 z_S}{\partial \eta^2} + C_\xi \frac{\partial z_S}{\partial \xi} + C_\eta \frac{\partial z_S}{\partial \eta} + A(z_S) = a(\zeta,\eta,z_S,\bar{t}) \qquad (2.21)$$

where D_ξ, D_η, C_ξ, C_η are diffusion coefficients and wave speeds in the ξ- and η-directions, which are in general functions of z_S, z_B, $\partial z_S/\partial\xi$, $\partial z_S/\partial\eta$, $\partial z_B/\partial\xi$, $\partial z_B/\partial\eta$ and the temperature $\Theta(\xi, \eta, z_S)$. Therefore, (2.21) is a *quasilinear* second order partial differential equation, but it is singular because D_ξ and D_η vanish at the grounding line $z_S = z_B$. If the equation (2.21) would be regular the condition $z_S = z_B$ along the grounding line $\partial D(\xi,\eta)$ would suffice as boundary condition to integrate (2.21), see Figure 10.

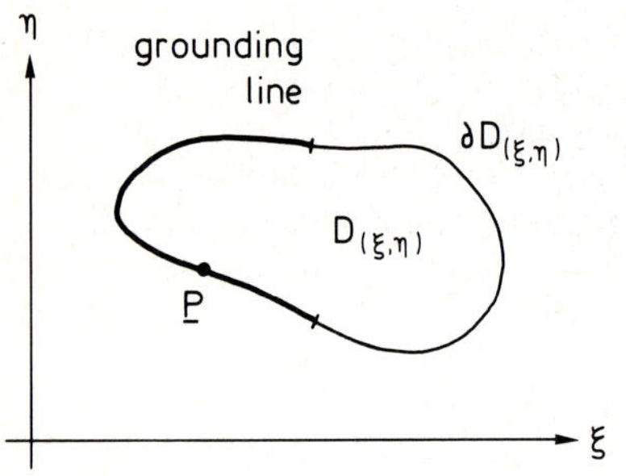

Figure 10

Domain $D(\xi,\eta)$ of the horizontal extent of the ice sheet in the stretched coordinate system (ξ,η).

For singular equations it is, however, known that Frobenius type expansions in the neighborhood of a point on the grounding line may determine the local solution. In particular, from this Frobenius expansion the normal derivative $\partial z_S/\partial n$ is known. We conclude that the proper boundary conditions for (2.21) are $z_S = z_B$ and $\partial z_S/\partial n$ = known along one portion of the grounding line, the extent of which must also follow from the properties of (2.21). The other portion of the grounding line is then a deduced result. The details of this behavior are not yet understood.

One other complexity lies in the form of the steady state heat equation $(2.15)_7$. To understand the difficulty, assume that the ξ-coordinate is locally in the direction of steepest descent while the η-coordinate is tangential to the line of constant height. Then, $(2.15)_7$ reduces to

$$U \frac{\partial\Theta}{\partial\xi} = \mathbb{D} \frac{\partial^2\Theta}{\partial\zeta^2} + \ldots\ldots, \qquad (2.22)$$

which for $U > 0$ (< 0) is a "forward" ("backward") heat equation. Numerical schemes are known to be stable only if integration is in the direction of U but not against it. Therefore, because $U > 0$ when $\partial z_S/\partial\xi < 0$, see (2.18), (2.19), integration must always follow the downhill direction, because the problem is ill-posed. Thus integration must start at a dome (whose location and ice depth may both be unknown). However, difficulties even persist when the location of the dome is known (i.e. from symmetry arguments). The height must then be estimated, but even with a prescribed height the heat equation remains *sensitive*, because $u = v = 0$ at the dome, and u, v are small close to the ice divide. Integration out of the divide is possible but computational proficiency required to numerically integrate the formulated boundary value problem is substantial and quick solutions are not in sight. This is why Yakowitz, Hutter ans Szidarovszky (1984) solve the plane flow problem first. But a clear analysis of the three dimensional problem is urgently needed.

On the other hand, if the time dependent initial value problem is solved, the geometry of the ice sheet is known at each time step. The kinematic wave equation determines the new geometry and the heat equation the new temperature field one time step ahead, but in order to start the integration an initial state must be known. Perhaps searching for the large time solution of a transient problem is easier.

3. ON THERMODYNAMICS OF TEMPERATE ICE

3.1. Review and motivation

No glaciologist would nowadays deny the role that water plays in temperate glaciers. Indeed, an abundance of evidence exists, which demonstrates that water affects the behavior of a glacier in its geophysical environment in more than a simple and peripherical fashion. We mention sliding with and without cavity formation (Weertman, 1979; Lliboutry, 1979; Fowler, 1981, 1984; Hallet, 1981, to name a few), surges (Meier and Post, 1969; Robin and Weertman, 1973), intraglacial channel flow (Mathews, 1969; Shreve, 1973; Roethlisberger, 1972; Nye & Frank, 1973; Nye, 1976; Spring & Hutter, 1982), and outbursts of ice dammed lakes and jökulhlaups (Björnsson, 1974, 1975; Spring & Hutter, 1981). A theoretical basis which would embrace all or some of these phenomena is still missing.

Water movement in temperate glacier ice operates essentially on two different scales, Figure 11. Over tens of meters or more a material part D of the glacier may consist of a substantial amount of "wet ice" and a number of relatively large cracks, crevasses and channels. The latter may be part of the heterogeneous intraglacial channel system and connect the surface and the bottom, or end within the ice. On much smaller length scales, in the range from 1 to 10^{-6} m, say (detail in Figure 11b) glacier ice appears as a polycrystalline assemblage of ice crystals and water, which primarily melts in tripple and quadruple connections and perculates through the vein system of connecting grain boundaries. The water flow through the large-scale channel system is fed by surface water and, to a lesser extent, by the melt water in the veins of the sheared ice. Water content in the large scale system is chiefly responsible for basal sliding with cavity formation (Iken, 1981; Iken et

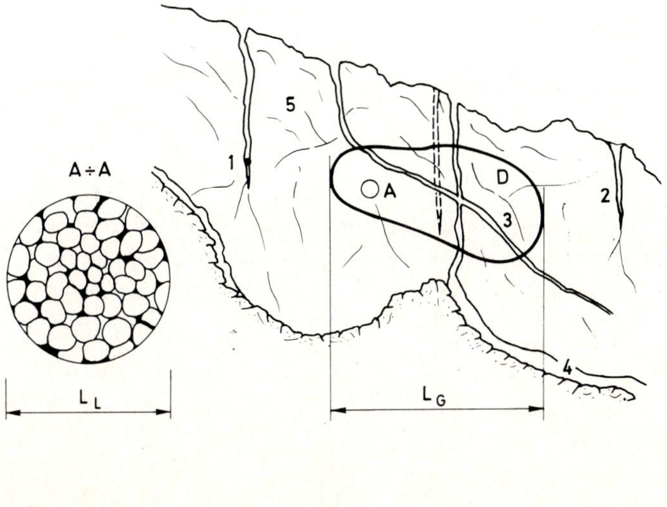

Figure 11

a) Schematic view of a temperate glacier:
   1 moulins,
   2 crevasses,
   3 intraglacial channels,
   4 subglacial channels,
   5 "wet ice",
   L_G is a global length scale.

b) Magnification of portion A of "wet ice". This is a mixture of polycrystalline ice with water containing inclusions and veins along grain boundaries. L_L is a local length scale.

al., 1983); it sets the water table to which this sliding is connected. However, the shearing motion of a glacier under its weight (called gliding) equally depends on the moisture content along the grain boundaries, because the local stress tensor for wet ice depends on this moisture content (Lliboutry, 1976).

These considerations should give sufficient indication that a continuum description of processes, which are typical of the large length scale variations, is obtained from averaging physical properties over both length scales. A first step will describe "wet ice" as an average over many grains of polycrystalline ice and water along grain boundaries. A theoretical model for the continuum description of wholly temperate glaciers will then, in a second step, involve an averaging process of this wet ice with the intraglacial channels.

We emphasize that the idea of a double averaging is typical of wholly temperate glaciers. For *polythermal* glaciers with limited geothermal heat flow a simpler model with only one averaging process may be satisfactory. Such simpler mixture models have been given by Hutter (1982) and Fowler (1984). But these authors ignore the surface effects of phase boundaries (along the grains). It is felt that the amount of curved internal surfaces may be considerable and, that energy and mass attributed to these surfaces may not be neglected. Support for this follows from the fact that freezing and melting at the phase boundaries are non-equilibrium processes, which continuously redistribute mass and energy between the bulk phases and change the geometries and the amount of the interfaces within the ice-water mixture. To derive a sound thermodynamical mixture theory of water bearing ice it is therefore necessary to study the dynamics and themodynamics of phase boundaries between ice and water.

Here, we present the basic ideas of a thermodynamic treatment of curved phase boundaries; new results regarding melting and freezing at grain boundaries of polycrystalline ice are deduced and implications for a mixture theory of "wet ice" are drawn.

To begin with and by way of motivation, we review the literature on phase-boundaries. A phase boundary is not a surface of sudden change of mass, energy and entropy density; it rather constitutes a thin layer across which mass, energy etc. change smoothly but rapidly between the densities of the adjacent bulk materials. The reason for this is the necessity of molecular adjustment between different molecular arrays of the adjacent bulk phases. Granted that such a boundary layer is stable it possesses a thickness of a few molecular distances and therefore carries mass, energy and entropy which in a physically reasonable description should not be neglected. Moreover, due to its different molecular ordering, the boundary layer has different constitutive properties than the adjacent bulk materials. One manifestation of this is the appearance of surface tension. Another one is the existence of the boundary layer thickness. Indirect experimental evidence for the existence of liquid-like films below 0°C is given by Nakaya & Matsumoto (1954) and Gubler (1982). Direct measurements of the boundary layer thickness by proton channeling experiments are due to Golecki & Jaccard (1977).

The smooth boundary layers between the bulk phases, ice and water, have a thickness of only a few atomic distances. Compared to the dimensions of the adjacent bulk materials, they are almost infinitely thin and can therefore be described as two-dimensional continua, representing mathematically singular surfaces with own thermomechanical properties. We follow the work of Scriven (1960), Gurtin & Murdoch (1975), Moeckel (1974), Lindsay & Straughan (1979) and extend it to thermodynamics and statics of phase boundaries. The theory is founded on the conservation laws for surface mass, momentum and energy and a balance equation for surface entropy and on constitutive assumptions for surface fields and jump-relations for the bulk fields across the phase boundary. The approach furnishes a consistent set of field equations for the motion of the phase boundary and its density and temperature. However, generalized boundary conditions for the dynamics and thermodynamics of freezing-melting processes at grain boundaries are also obtained.

These generalized boundary conditions are important relations in the deduction of a mixture theory for "wet ice", obtained by averaging over many grain diameters. The boundary conditions in such a mixture theory appear as "surface-weighted" production densities of the volume balance laws, and thus introduce a measure for the amount of internal phase boundaries in addition to the volume content of water. Since an ice-water mixture with a given water content behaves quite differently, when the water is concentrated in a few big inclusions rather than dispersed along many grain boundaries, it is likely that the amount of internal phase boundaries is a more important measure of the macroscopic behavior of temperate ice than water content. Two-phase-flow theories of i.e. Ishii (1975) and Drew (1983) have incorporated flux contributions from the phase boundaries, however, they contain no direct measure of the amount of internal interfactial area. We shall proceed in the spirit of these two-phase theories, but modify the constitutive assumptions and other details.

One comment on the mathematical treatment of phase boundaries should be added. It is possible to correlate the fields of surface mass and momentum densities etc. of the singular surface with mean values of excess fields over the smooth three-dimensional boundary layer, representing the phase boundary. Gibbs (1928, 1948), Buff & Saltsburg (1957a,b), Slattery (1967) and Deemer & Slattery (1978) follow this approach. It allows a physical interpretation of the surface fields and provides a basis for the constitutive theory for phase transitions. Here we do not go into details, but refer to Alts & Hutter (1984).

The mentioned second averaging over the larger length scale is not treated here. For a continuum description of the behavior of temperate zones inside the whole glacier, the second averaging may be likewise important for the development of thermo-mechanical field equations. This, however, is a matter for future research.

3.2. Thermodynamics on phase boundaries

The phase boundaries between ice and water are considered as non-material curved singular surfaces carrying mass, momentum, energy and entropy. A dynamic theory of freezing/melting can thus be founded on surface-balance equations for these fields and on constitutive assumptions for a part of them and for the transport properties across the phase boundary. Thermodynamics consists in the formulation of restrictions for the constitutive equations due to invariance requirements and the entropy principle. As a mathematical tool differential geometry is needed, Aris (1962) or Bowen & Wang (1976).

a) Geometry and motion of curved surfaces

The singular surface $\mathcal{S}(t)$, which divides the three-dimensional regions R^+ and R^- occupied by water and ice, respectively is given here in the parameter representation, Figure 12,

$$x^k = {}_a\chi^k(\xi^\alpha, t) . \tag{3.1}$$

x^k $(k = 1,2,3)$ are Cartesian co-ordinates in some inertial frame and ξ^α $(\alpha = 1,2)$ are surface co-ordinates in the local tangent planes of $\mathcal{S}(t)$. The tangent vectors $\tau_\alpha{}^k$ and the unit normal vector e_k

$$\tau^k_\alpha := \frac{\partial_a\chi^k(\xi^\alpha, t)}{\partial\xi^\alpha} , \quad e_k := \frac{\varepsilon_{k\ell m}\, \tau^\ell_1\, \tau^m_2}{|\underset{\sim}{\tau}_1 \times \underset{\sim}{\tau}_2|} \tag{3.2}$$

form a right-handed system with $\underset{\sim}{e}$ pointing into R^+. $\varepsilon_{k\ell m} = \varepsilon^k{}_{\cdot\ell m} = \ldots$ is the complete antisymmetric unit tensor with which the exterior vector or cross product is formed. Cartesian components of the unit tensor are denoted by $\delta_{k\ell} = \delta^{k\ell} = \delta_\ell{}^k$. The geometry of the surface is specified by its metric $g_{\alpha\beta}$ and its curvature $b_{\alpha\beta}$:

$$g_{\alpha\beta} := \delta_{k\ell}\, \tau^k_\alpha\, \tau^\ell_\beta = g_{\beta\alpha}, \qquad b_{\alpha\beta} := e_k \frac{\partial \tau^k_\alpha}{\partial \xi^\beta} = b_{\beta\alpha}. \tag{3.3}$$

The inverse metric $g^{\alpha\beta}$ with $g^{\alpha\beta} g_{\beta\gamma} = \delta^\alpha_\gamma$ is introduced as usual to raise surface indices. Mean and Gaussian curvature are defined by

$$K_M := \frac{1}{2} b^\alpha_\alpha = \frac{1}{2}(K_1 + K_2), \qquad K_G := \det \| b^\alpha_\beta \| = K_1 \cdot K_2, \tag{3.4}$$

where K_1 and K_2 are the eigenvalues of $b^\alpha_\beta = g^{\alpha\gamma} b_{\gamma\beta}$. Finally, summation convention with respect to twice diagonally repeated Latin and Greek indices is applied in this section.

To describe the motion of the singular surface information about the development of the actual surface co-ordinates ξ^α from corresponding surface co-ordinates Ξ^Γ $(\Gamma = 1,2)$ in some reference configuration $\mathcal{R}$ at time $t_{\mathcal{R}} (\leq t)$ is needed. This motion may be given by the transformation

$$\xi^\alpha = \xi^\alpha_s(\Xi^\Gamma, t) \quad \text{with} \quad \xi^\alpha_s\Big|_{t=t_{\mathcal{R}}} = \delta^\alpha_\Gamma\, \Xi^\Gamma. \tag{3.5}$$

The velocity of the non-material singular surface $s(t)$ is then the time derivative of (3.1) at fixed Ξ^Γ. In view of (3.5) and (3.2) we thus obtain

$$w^k = \frac{\partial_a \chi(\xi^\alpha_s(\Xi^\Gamma, t), t)}{\partial t} = w^\alpha \tau^k_\alpha + w_n e^k, \tag{3.6}$$

where the tangential (w^α) and normal (w_n) velocity components are

$$w^\alpha = \frac{\partial \xi^\alpha_s(\Xi^\Gamma, t)}{\partial t}, \qquad w_n = e_k \frac{\partial_a \chi^k(\xi^\alpha, t)}{\partial t}. \tag{3.7}$$

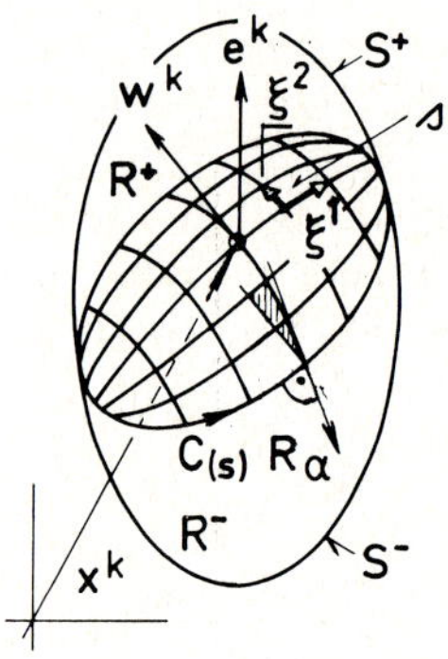

Figure 12

Geometry and motion of a curved singular surface representing a non-material phase boundary.

b) Balance laws on phase boundaries

Any additive quantity Ψ on a phase boundary $s(t)$ can be balanced, see Figure 13. Its change within an area, that is bounded by a curve $C(s)$ moving with velocity w^α, is caused by influx into the phase boundary through $C(s)$ and from the adjacent bulk materials and by surface production and supply. Let

$\psi_s(\xi^\beta, t)$ = surface density of Ψ per unit area;

$\phi^\alpha_s(\xi^\beta, t)$ = influx of Ψ along the tangent planes per unit time and unit length through co-ordinate lines moving with tangent velocity w^α;

π_s, σ_s = surface production and supply densities of Ψ per unit time and unit area, respectively;

$\phi^k(\underset{\sim}{x}, t)$ = bulk influx of Ψ per unit time and unit area through a material surface in the bulk moving with material velocity v^k;

$\psi_v(\underset{\sim}{x}, t)$ = bulk density of Ψ per unit volume of the bulk material;

$[\![\phi^k + \psi_v(v^k - w^k)]\!]\, e_k =$

$= [\phi^k_+ + \psi^+_v\,(v^k_+ - w^k)]\, e_k - [\phi^k_- + \psi^-_v\,(v^k_- - w^k)]\, e_k =$

= normal jump contribution of the bulk flux of Ψ per unit time and unit area through $s(t)$, which represents the influx to $s(t)$ from the adjacent bulk materials ($\phi^k_\pm$, $\psi^\pm_v$, $v^k_\pm$ are the limiting values of the bulk materials as the singular surface $s(t)$ is approached from R_+ and R_-, respectively).

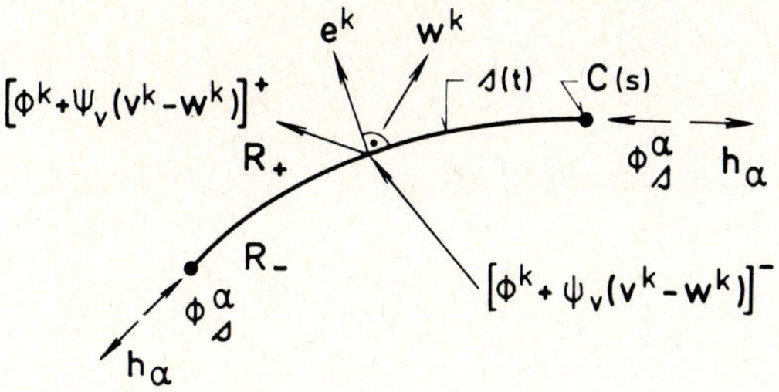

Figure 13

Flow contributions for the general balance law on moving non-material phase boundaries.

Then the local form of a balance equation for Ψ on a moving non-material singular surface takes the form (see Moeckel, 1974, or Alts & Hutter, 1984):

$$\begin{aligned}\frac{\partial\psi_s}{\partial t} - 2\psi_s K_M w_n + (\phi^\alpha_s + \psi_s w^\alpha)_{;\alpha} &= \\ = - [\![\phi^m + \psi_v (v^m - w^m)]\!] e_m + \pi_s + \sigma_s .&\end{aligned} \tag{3.8}$$

$\partial/\partial t$ denotes the partial time derivative at constant ξ^α, and the semicolon indicates covariant differentiation with respect to the actual surface co-ordinates ξ^α.

The special balance equations for mass, momentum, energy and entropy follow with the identifications of Table 2.

quantity	(1) ψ_s	(2) ϕ^α_s	(3) π_s	(4) σ_s	(5) ψ_v	(6) ϕ^m
mass	ρ_s	0	0	0	ρ	0
momentum	$\rho_s w^k$	$-t^{k\alpha}_s$	0	$\rho_s g^k$	ρv^k	$-t^{km}$
energy	$\rho_s (u_s + \frac{1}{2} \underset{\sim}{w}^2)$	$q^\alpha_s - w_k t^{k\alpha}_s$	0	$\rho_s w_k g^k + \pi_s r_s$	$\rho (u + \frac{1}{2} \underset{\sim}{v}^2)$	$q^m - v_k t^{km}$
entropy	$\rho_s s_s$	ϕ^α_s	$\pi_s \geq 0$	σ_s	ρ_s	ϕ^m

Table 2

Identification of fields for surface-balance equations.

The new quantities have the following meaning: ρ_s = surface mass density, u_s and s_s = surface internal energy and entropy per unit mass, $t^{k\alpha}_s$ = surface stress, q^α_s and ϕ^α_s = surface heat and entropy flux, $\rho_s r_s$ = density of surface heat supply due to radiation, π_s and σ_s = entropy production and supply densities. The quantities without subscript s denote fields in the adjacent bulk materials: ρ = mass density, u and s = internal energy and entropy per unit mass, t^{km} = Cauchy's stress, q^m and ϕ^m = heat and entropy flux. Finally, g^k is the Earth's acceleration. The production densities in the third column for mass, momentum and energy vanish, since these quantities are conserved. We emphasize, that the non-convective surface fluxes in the second column are referred to moving co-ordinate lines on $s(t)$. These are chosen as "semi"- material lines formed by surface particles with velocity w^α. Hence the mass flux vanishes and the other surface fluxes have the given interpretations.

Introducing the total time derivative

$$\frac{d\psi_s}{dt} = \frac{\partial\psi_s}{\partial t} + w^\alpha \frac{\partial\psi_s}{\partial\xi^\alpha} = \frac{\partial\psi_s}{\partial t} + w^\alpha \psi_{s,\alpha} \tag{3.9}$$

the following balance laws for mass, momentum and internal energy can be deduced from Table 2 and (3.8):

$$\begin{aligned}
&\frac{d\rho_s}{dt} - 2\,\rho_s K_M w_n + \rho_s w^{\alpha}_{;\alpha} = -\llbracket \rho(v^m - w^m)\rrbracket\, e_m ,\\
&\rho_s \frac{dw^k}{dt} - t^{k\alpha}_{s;\alpha} = -\llbracket -t^{km} + \rho(v^k - w^k)(v^m - w^m)\rrbracket\, e_m + \rho_s g^k ,\\
&\rho_s \frac{du_s}{dt} + q^{\alpha}_{s;\alpha} = -\llbracket q^m - (v_k - w_k)t^{km} + \rho\left[(u-u_s) + \tfrac{1}{2}(\underset{\sim}{v}-\underset{\sim}{w})^2\right](\underset{\sim}{v}^m - \underset{\sim}{w}^m)\rrbracket\, e_m\\
&\qquad + t^{k\alpha}_s w_{k;\alpha} + \rho_s r_s .
\end{aligned} \tag{3.10}$$

Furthermore, the balance equation for the surface entropy takes the form

$$\rho_s \frac{ds_s}{dt} + \phi^{\alpha}_{s;\alpha} = -\llbracket \phi^m + \rho(s-s_s)(v^m - w^m)\rrbracket\, e_m + \pi_s + \sigma_s . \tag{3.11}$$

The surface stress $t^{k\alpha}_s$ can be decomposed into tangential and normal components

$$t^{k\alpha}_s = \tau^k_\beta\, S^{\beta\alpha} + e^k\, S^{\alpha} . \tag{3.12}$$

$S^{\beta\alpha}$ are the contravariant components of the tensor of surface tension, traction being positive; β indicating the direction of force and α the co-ordinate line the unit length of which the force is referred to. S^{α} is a surface shear stress in the direction normal to the surface and referred to the unit length along the co-ordinate ξ^{α}. By inserting (3.6) and (3.12) into $(3.10)_2$ one yields the decomposition of the momentum balance into tangential and normal components and the identity

$$t^{k\alpha}_s w_{k;\alpha} = S^{\alpha\beta}(w_{\alpha;\beta} - b_{\alpha\beta} w_n) + S^{\alpha}(w_{n;\alpha} + b_{\alpha\beta} w^{\beta}) . \tag{3.13}$$

If moment of momentum of the phase boundary is also conserved, it can be shown that

$$S^{\alpha\beta} = S^{\beta\alpha}, \qquad S^{\alpha} = 0 . \tag{3.14}$$

For the description of simple melting freezing processes at the ice-water interface and not too small phase boundary surfaces (3.14) seem to be permissible assumptions. However, (3.14) may be inadmissable for such small geometries as appear i.e. in nucleation and gliding on a solid support. Then a surface couple stress and surface spin may develop inside the fluid-like boundary layer due to electrical polarization; for details see Alts & Hutter (1984).

c) *Constitutive assumptions*

The fields to be calculated in a thermodynamical theory of phase boundaries are the surface density $\rho_s(\xi^{\alpha}, t)$, surface motion $_a\chi^k(\xi^{\alpha}, t)$ and the surface temperature $T_s(\xi^{\alpha}, t)$. For given supplies g^k and r_s the differential equations (3.10) can be solved for these provided the surface fields $t^{k\alpha}_s(\xi^{\beta}, t)$, $q^{\alpha}_s(\xi^{\beta}, t)$, $u_s(\xi^{\beta}, t)$ and the jump contributions in (3.10) are related in a materially dependent manner to the fields of interest. Such relations are called *constitutive equations*. In a thermodynamic treatment we need also constitutive equations for the fields $s_s(\xi^{\beta}, t)$, $\phi^{\alpha}_s(\xi^{\beta}, t)$ and the jump contribution in the entropy balance (3.11).

Consequently, constitutive equations have to be formulated for the *surface fields*

$$\mathfrak{L}_s(\xi^{\alpha}, t): \quad S^{\alpha\beta}(\xi^{\gamma}, t),\ S^{\alpha}(\xi^{\gamma}, t),\ q^{\alpha}_s(\xi^{\gamma}, t),\ u_s(\xi^{\gamma}, t);\ \ \phi^{\alpha}_s(\xi^{\gamma}, t),\ s_s(\xi^{\gamma}, t); \tag{3.15}$$

and the *jump contributions*

$$\mathfrak{L}_s(\xi^\beta,t): \begin{cases} \mathbb{R}(\xi^\beta,t) := [\![\rho(v^m-w^m)]\!] e_m , \\ \Pi^\alpha(\xi^\beta,t) := \tau_k^\alpha [\![-t^{km} + \rho(v^k-w^k)(v^m-w^m)]\!] e_m , \\ \Pi_n(\xi^\beta,t) := e_n [\![-t^{km} + \rho(v^k-w^k)(v^m-w^m)]\!] e_m , \\ \mathfrak{U}(\xi^\beta,t) := [\![q^m-(v_k-w_k)t^{km} + \rho\left[(u-u_s)+\frac{1}{2}(\underset{\sim}{v}-\underset{\sim}{w})^2\right](\underset{\sim}{v}^m-\underset{\sim}{w}^m)]\!] e_m , \\ \mathfrak{S}(\xi^\beta,t) := [\![\frac{1}{T} q^m + \rho(s-s_s)(v^m-w^m)]\!] e_m . \end{cases} \qquad (3.16)$$

The entropy flux ϕ^m in the bulk materials has been replaced by $(1/T)\cdot q^m$, since ice and water are both simple materials; T is the absolute temperature.

The constitutive behavior of the *fluid-like interface* between ice and water is manifest in a compromise between the material properties of ice and water. The simplest possibility is to regard it as as viscous, heat conducting membran, defined by the constitutive assumptions for the surface fields:

$$\mathfrak{L}_s(\xi^\alpha,t) = \mathfrak{L}_s(\rho_s, T_s, w^\alpha, w_n, T_{s,\alpha}, w^\alpha_{;\beta}, w_{n,\beta};\ \tau^k_\alpha, e^k, \tau^k_{\alpha;\beta}, e^k_{,\beta}). \qquad (3.17)$$

$\psi_{,\alpha} = \partial\psi/\partial\xi^\alpha$ denotes the partial derivative with respect to ξ^α. $T_{s,\alpha}$ and $w^\alpha_{;\beta}$, $w_{n,\beta}$ account for heat conduction along and internal friction within the phase boundary, respectively. The tangent vectors and the normal vector and their derivatives allow for a dependence on the geometry of the surface.

For vanishing surface fields in (3.10) and (3.11) the jump contributions (3.16) must vanish, and thus degenerate to the classical boundary conditions at immaterial phase interfaces. Mass, momentum etc. bearing real phase boundaries, however, require constitutive assumptions for these jump quantities. Inspection shows that they are explicit functions of the limiting values

$$\begin{aligned} W^\alpha_+ &:= v^\alpha_+ - w^\alpha = (v^k_+ - w^k)\tau^\alpha_k , & t^\alpha_{n+} &:= \tau^\alpha_k t^{km}_+ e_m , \\ W^+_n &:= v^+_n - w_n = (v^k_+ - w^k)e_k , & t^+_n &:= e_k t^{km} e_m , \\ \theta_+ &:= T_+ - T_s , & q^+_n &:= q^m_+ e_m , \end{aligned} \qquad (3.18)$$

of relative tangential velocity W^α_+ *and* shear stress t^α_{n+}, relative normal velocity W^+_n *and* normal stress t^+_n, and temperature difference θ_+ *and* normal heat flux q^+_n. Morevoer, the jump contributions (3.16) contain explicitely $[\![\rho]\!]$, $[\![u]\!]$, $[\![s]\!]$ and the following jumps

$$K_s(\xi^\alpha,t): \begin{cases} \mathfrak{v}^\alpha := \tau^\alpha_k [\![v^k]\!], & \mathfrak{G}^\alpha := \tau^\alpha_k [\![t^{km}]\!] e_m , \\ \mathfrak{v}_n := e_k [\![v^k]\!], & \mathfrak{G}_n := e_k [\![t^{km}]\!] e_m , \\ K := [\![\frac{1}{T}]\!], & \mathfrak{Q} := e_k [\![q^k]\!] . \end{cases} \qquad (3.19)$$

The fields (3.18) can be independently prescribed on one side of the phase boundary or are given by the behavior of the bulk on this side. The outcome on the other side depends then on this input *and* on the constitutive properties of the phase interphase. Hence, the jump contributions (3.19) may depend on the set of variables in (3.17) and (3.18):

$$K_s(\xi^\alpha, t) = K_s(\rho_s, T_s, w^\alpha, w_n, T_{s,\alpha}, w^\alpha_{;\beta}, w_{n;\beta}; \tau^k_\alpha, e^k, \tau_{\alpha;\beta}, e^k_{,\beta};$$
$$w^\alpha_+, w^+_n, t^\alpha_{n+}, t^+_n, \theta_+, q^+_n). \tag{3.20}$$

The same holds for the jump contributions in (3.16), since they are combinations of (3.19) and (3.18).

We emphasize, that insertion of the constitutive equations (3.17) and (3.20) together with (3.16) into (3.10) yields a set of field equations for the determination of $\rho_s(\xi^\alpha, t)$, ${}_a\chi^k(\xi^\alpha, t)$ and $T_s(\xi^\alpha, t)$. Every solution for given supplies $g^k(\xi^\alpha, t)$ and $r_s(\xi^\alpha, t)$ and given bulk fields $\rho^+(\xi^\beta, t)$, $T_+(\xi^\beta, t)$, $v^\alpha_+(\xi^\beta, t)$, $v^+_n(\xi^\beta, t)$, $t^\alpha_{n+}(\xi^\beta, t)$ $t^+_n(\xi^\beta, t)$ and $q^+_n(\xi^\beta, t)$ on one side of the phase boundary is called a *thermodynamic process* on phase interfaces separating ice and water.

d) Thermodynamic restrictions

The constitutive assumptions must be restricted by the following general rules.

Principle of material frame indifference:

It states independence of the constitutive equations on observer frames in Euklidean space. The following restrictions can be derived:

$$\mathcal{L}_s(\xi^\alpha, t) = \mathcal{L}_s(\rho_s, T_s, T_{s,\alpha}, D_{\alpha\beta}; \; g_{\alpha\beta}, b_{\alpha\beta}),$$
$$K_s(\xi^\alpha, t) = K_s(\rho_s, T_s, T_{s,\alpha}, D_{\alpha\beta}; \; g_{\alpha\beta}, b_{\alpha\beta}; \; w^\alpha_+, w^+_n, t^\alpha_n, t^+_n, \theta_+, q^+_n), \tag{3.21}$$

where

$$D_{\alpha\beta} := w_{(\alpha;\beta)} - w_n b_{\alpha\beta} = \frac{1}{2}(w_{\alpha;\beta} + w_{\beta;\alpha}) - w_n b_{\alpha\beta}. \tag{3.22}$$

Details may be found in Lindsay & Straughan (1972) and Alts & Hutter (1984).

Principle of material symmetry:

It states that the constitutive equations must be independent of symmetry transformations of the material. Application to phase boundaries faces the difficulty, that the boundary layer has to match the crystalline ice and the isotropic water, which requires continuous loss of symmetry across the thickness. The representation of the boundary layer as a singular surface requires a symmetry group which is a compromise between those of ice and water. This would require a detailed study of the boundary layer, which has not been done so far. Hence we assume, less accurately, transversal isotropy of the phase boundary around its local normal directions. This then requires isotropy of all bulk and surface constitutive equations with respect to arbitrary two-dimensional rotations around the surface normal. The irreducible representations for the constitutive equations $(3.21)_1$ of the surface fields, which are linear in $T_{s,\alpha}$ and $D_{\alpha\beta}$, are then (cp. Smith (1965), Alts & Hutter (1984)):

$$\begin{aligned}
u_s &= u_s(\rho_s, T_s, K_M, K_G; \; I_D, II_D),\\
s_s &= s_s(\rho_s, T_s, K_M, K_G; \; I_D, II_D),\\
q^\alpha_s &= -\left[\kappa\, g^{\alpha\beta} + \kappa_2 (b^{\alpha\beta} - K_M g^{\alpha\beta})\right] T_{s,\beta},\\
\phi^\alpha_s &= -\left[\phi\, g^{\alpha\beta} + \phi_2 (b^{\alpha\beta} - K_M g^{\alpha\beta})\right] T_{s,\beta};\\
S^\alpha &= 0,\\
S^{\alpha\beta} &= \sigma g^{\alpha\beta} + \sigma_2 (b^{\alpha\beta} - K_M g^{\alpha\beta}) + (\mu_1 I_D + \mu_2 II_D) g^{\alpha\beta} +\\
&\quad + (\mu_3 I_D + \mu_4 II_D)(b^{\alpha\beta} - K_M g^{\alpha\beta}) + \mu_5 D^{\alpha\beta} + \mu_6 (\underset{\approx}{b}\underset{\approx}{D} - \underset{\approx}{D}\underset{\approx}{b})^{(\alpha\beta)};
\end{aligned} \tag{3.23}$$

where

$$I_D := \mathrm{tr}\,\underset{\approx}{D}, \qquad II_D := \mathrm{tr}\,(\underset{\approx}{b}\,\underset{\approx}{D}) \tag{3.24}$$

and the scalar coefficients $\kappa, \ldots, \mu_6$ may be functions of ρ_s, T_s, K_M, K_G.

The ice-water mixture cannot sustain shear stresses in thermostatic equilibrium: $t^{\alpha}_{n\pm}\big|_E = 0$ on both sides of the phase boundary; the normal stresses $t^{\pm}_n\big|_E = -p^E_{\pm}$ are merely hydrostatic. (The index E stands for equilibrium). Instead of the normal stress t^+_n in $(3.21)_2$ we may thus introduce the difference to its equilibrium value

$$\pi^+_n := t^+_n - t^+_n\big|_E = t^+_n + p^E_+, \tag{3.25}$$

and then obtain the linear irreducible representations for the jump contributions $(3.21)_2$

$$\begin{aligned}
\mathfrak{D}_n &= \cdot + \mathfrak{D}_1 I_D + \mathfrak{D}_2 II_D + \mathfrak{D}_3 w^+_n + \mathfrak{D}_4 \pi^+_n + \mathfrak{D}_5 \theta_+ + \mathfrak{D}_6 q^+_n ,\\
\mathfrak{G}_n &= \mathfrak{G}_0 + \mathfrak{G}_1 I_D + \mathfrak{G}_2 II_D + \mathfrak{G}_3 w^+_n + \mathfrak{G}_4 \pi^+_n + \mathfrak{G}_5 \theta_+ + \mathfrak{G}_6 q^+_n ,\\
\mathfrak{K} &= \cdot + \mathfrak{K}_1 I_D + \mathfrak{K}_2 II_D + \mathfrak{K}_3 w^+_n + \mathfrak{K}_4 \pi^+_n + \mathfrak{K}_5 \theta_+ + \mathfrak{K}_6 q^+_n ,\\
\mathfrak{Q} &= \cdot + \mathfrak{Q}_1 I_D + \mathfrak{Q}_2 II_D + \mathfrak{Q}_3 w^+_n + \mathfrak{Q}_4 \pi^+_n + \mathfrak{Q}_5 \theta_+ + \mathfrak{Q}_6 q^+_n ;
\end{aligned} \tag{3.26}$$

$$\begin{aligned}
\mathfrak{D}^{\alpha} &= \left[\mathfrak{D}_1 g^{\alpha\beta} + \mathfrak{D}_2 (b^{\alpha\beta} - K_M g^{\alpha\beta})\right] T_{s,\beta} + \left[\mathfrak{D}_3 g^{\alpha\beta} + \mathfrak{D}_4 (b^{\alpha\beta} - K_M g^{\alpha\beta})\right] w^+_{\beta} + \\
&\quad + \left[\mathfrak{D}_5 g^{\alpha\beta} + \mathfrak{D}_6 (b^{\alpha\beta} - K_M g^{\alpha\beta})\right] t^n_{\beta+} ,\\
\mathfrak{G}^{\alpha} &= \left[\mathfrak{C}_1 g^{\alpha\beta} + \mathfrak{C}_2 (b^{\alpha\beta} - K_M g^{\alpha\beta})\right] T_{s,\beta} + \left[\mathfrak{C}_3 g^{\alpha\beta} + \mathfrak{C}_4 (b^{\alpha\beta} - K_M g^{\alpha\beta})\right] w^+_{\beta} + \\
&\quad + \left[\mathfrak{C}_5 g^{\alpha\beta} + \mathfrak{C}_6 (b^{\alpha\beta} - K_M g^{\alpha\beta})\right] t^n_{\beta+} ,
\end{aligned}$$

where

$$w^+_{\alpha} := g_{\alpha\beta}\, w^{\beta}_+ , \qquad t^n_{\alpha+} := g_{\alpha\beta}\, t^{\beta}_{n+} , \tag{3.27}$$

in which the scalar coefficients $\mathfrak{D}_1, \ldots, \mathfrak{C}_6$ in (3.26) may again depend on ρ_s, T_s, K_M, K_G. Since $\mathfrak{D}_n$, $\mathfrak{K}$ and $\mathfrak{Q}$ vanish in equilibrium they contain no constant contributions; this is expressed by the dots in $(3.26)_{1-4}$. The general linear representations (3.26) contain 37 unknown coefficients; too many for a reasonable theory of the dynamics of phase transitions. We shall reduce their number later by additional requirements.

Entropy principle:

We use the following generalization of Müller's (1973) formulation:

(i) There exists an additive surface entropy. It satisfies the balance law (3.11) for every surface point.

(ii) The specific surface entropy s_s, its non-convective flux ϕ^{α}_s and its input density $\mathfrak{S}$ from the adjacent bulk materials are given by constitutive equations, which satisfy the principles of material frame indifference and material symmetry. For the ice-water interface they are given by (3.21), $(3.23)_{2,4}$ and (3.26) together with $(3.16)_5$.

(iii) The entropy-supply density σ_s is proportional to the densities of the supplies of momentum and internal energy:

$$\rho_s = \lambda^s \cdot \rho_s r_s + \lambda^s \cdot \rho_s g^k , \tag{3.28}$$

where λ^s and λ^s_k are independent of r_s and g^k.

(iv) The entropy-production density π_s is non-negative for every thermodynamic process:

$$\pi_s \geq 0 \quad \forall \text{ thermodynamic processes.} \tag{3.29}$$

The entropy inequality (3.29) is the key for the evaluation of further restrictions on the constitutive equations. It is valid for *all physically admissible* processes. These are those fields which are solutions of the conservation laws for mass, momentum and energy incorporating the constitutive relations. The balance laws (3.10) may therefore be regarded as constraints on all processes that satisfy (3.29). Using Lagrangian parameters (3.29) may be replaced by the equivalent inequality, Liu (1972),

$$\begin{aligned}
\pi_s = {} & \rho_s \frac{ds_s}{dt} + \phi^\alpha_{s,\alpha} - \sigma_s + \mathfrak{S} \\
& - \Lambda^s_\rho \left[\frac{d\rho_s}{dt} + \rho_s (w^\alpha_{;\alpha} - 2 K_M w_n) + \mathbf{R} \right] \\
& - \Lambda^s_k \left[\rho_s \frac{dw^k}{dt} - t^{k\alpha}_{s;\alpha} - \rho_s g^k + \tau^k_\alpha \Pi^\alpha + e^k \Pi_n \right] \\
& - \Lambda^s_\varepsilon \left[\rho_s \frac{du_s}{dt} + q^\alpha_{s;\alpha} - S^{\alpha\beta} D_{\beta\alpha} - \rho_s r_s + \mathfrak{U} \right] \geq 0
\end{aligned} \tag{3.30}$$

which holds for arbitrary surface fields $\rho_s(\xi^\alpha, t)$, ${}_a\chi^k(\xi^\alpha, t)$, $T_s(\xi^\alpha, t)$ and for unrestricted bulk fields $w^\alpha_+(\xi^\beta, t)$, $w_n(\xi^\beta, t)$, $t^\alpha_{n+}(\xi^\beta, t)$, $\pi^+_n(\xi^\beta, t)$, $\theta_+(\xi^\beta, t)$ and $q^+_n(\xi^\beta, t)$ on one side of the phase boundary.

Substituting the constitutive equations, performing all indicated differentiations and free variations of the surface fields and their derivatives yields a set of equations from which the Lagrange-parameters and the connection between the entropic and energetic constitutive equations can be determined. The results derived in detail in Alts & Hutter (1984) are

1) The Lagrange-parameters are

$$\Lambda^s_\varepsilon = \frac{1}{T_s}, \quad \Lambda^s_k = 0 , \quad \Lambda^s_\rho = \frac{\sigma}{T_s \rho_s} . \tag{3.31}$$

Λ^s_ε is the inverse (absolute) surface temperature, Λ^s_ρ is given by the scalar surface tension, $\sigma = 1/2\, S^\alpha_\alpha\big|_E$, which is the first coefficient of $(3.23)_6$ in equilibrium.

2) The parameters λ^s and λ^s_k in (3.28) are

$$\lambda^s = \frac{1}{T_s}, \quad \lambda^s_k = 0 : \sigma_s = \frac{1}{T_s} \cdot \rho_s r_s . \tag{3.32}$$

Hence the entropy-supply density is given by the heat supply.

3) The surface entropy flux is proportinal to the surface heat flux, the factor of proportionality being $1/T_s$:

$$\phi^\alpha_s = \frac{1}{T_s} q^\alpha_s . \tag{3.33}$$

This connection for the surface fluxes is akin to the corresponding one $\phi^k = 1/T \cdot q^k$

in simple bulk materials.

4) Surface entropy, internal energy and surface tension

$$s_s = s_s(T_s, \rho_s), \quad u_s = u_s(T_s, \rho_s), \quad \sigma = \sigma(T_s, \rho_s \tag{3.34}$$

depend only on surface temperature and density. They are connected by the differential form

$$ds_s = \frac{1}{T_s}(du_s + \frac{\sigma}{\rho_s^2}\, d\rho_s), \tag{3.35}$$

which is Gibbs' relation for the surface entropy. From this relation further restrictions on the constitutive equation of entropy and internal energy can be deduced. Together with the definition of the specific heat c_{ρ_s} at constant density the integrability condition for entropy yields the relations

$$\frac{\partial u_s}{\partial \rho_s} = -\frac{1}{\rho_s^2}(\sigma - T\frac{\partial \sigma}{\partial T_s}), \quad \frac{\partial u_s}{\partial T_s} := c_{\rho_s}, \tag{3.36}$$

implying

$$\frac{\partial s_s}{\partial \rho_s} = \frac{1}{\rho_s^2}\frac{\partial \sigma}{\partial T_s}, \quad \frac{\partial s_s}{\partial T_s} = \frac{c_{\rho_s}}{T_s}, \quad \frac{\partial c_{\rho_s}}{\partial \rho_s} = \frac{T_s}{\rho_s^2}\frac{\partial^2 \sigma}{\partial T_s^2}. \tag{3.37}$$

Indirect measurements of surface tension between ice and water indicate, that it is a linear function of temperature, Hobbs (1974). Consequently, the specific heat capacity $c_{\rho_s} = c_{\rho_s}(T_s)$ is independent of density, and internal energy and entropy follow from (3.36) and (3.37) by integration

$$u_s(T_s, \rho_s) = u_s^o + \int_{T_s^o}^{T_s} c_{\rho_s}(T_2')\,dT_s' - (1 - T_s\frac{\partial}{\partial T_s})\int_{\rho_s^o}^{\rho_s}\frac{\sigma(T_s,\rho_s')}{\rho_s'^2}\,d\rho_s',$$

$$s_s(T_s, \rho_s) = s_s^o + \int_{T_s^o}^{T_s}\frac{c_{\rho_s}(T_s')}{T_s'}\,dT_s' + \frac{\partial}{\partial T_s}\int_{\rho_s^o}^{\rho_s}\frac{\sigma(T_s,\rho_s')}{\rho_s'^2}\,d\rho_s'. \tag{3.38}$$

Hence, internal energy and entropy of the phase boundary are known (except for integration constants) whenever $c_{\rho_s}(T_s)$ and $\sigma(T,\sigma_s)$ are known.

Remark: Consider the special case in which temperature T_s and density $\rho_s = m_s/A_s$ are uniform over a phase boundary with mass m_s and area A_s. Total entropy $S_s = m_s\, s_s$, internal energy $U_s = \rho_s\, u_s$ and surface tension are then connected by

$$dS_s = \frac{1}{T_s}(dU_s - \sigma\, dA_s) - \frac{1}{T_s}\, G_s \frac{1}{m_s}\, dm_s,$$

where $G_s := U_s - T_s S_s - \sigma A_s$ is the total free enthalpy. The above differential is an immediate consequence of (3.35). For surfaces with constant mass ($dm_s = 0$) it contains the well known result for the surface entropy. But, growing phase boundaries do not preserve their mass, and since $G_s \neq 0$ (as we shall prove later) the application of the above differential without the last term to problems of nucleation growth (as it is done) is questionable.

5) With the foregoing results, the entropy-production density can be brought into the form

$$\pi_s = -\frac{1}{T_s^2}\, q_s^{\alpha}\, T_{s,\alpha} + \frac{1}{T_s}(S^{\alpha\beta} - \sigma g^{\alpha\beta})\, D_{\beta\alpha} + \mathbf{P} \geq 0, \tag{3.39}$$

where the jump contribution $\mathfrak{P}$ is given by

$$\mathfrak{P} := \mathfrak{S} - \frac{1}{T_s}(\mathfrak{U} + \frac{\sigma}{\rho_s} \cdot \mathfrak{R}) . \qquad (3.40)$$

This inequality is valid for unrestricted surface fields and bulk fields on one side of the phase boundary.

All the results, (3.31)-(3.40), are valid for general admissible thermodynamic processes.

e) Thermostatic equilibrium

The entropy-production density is a non-linear function of ρ_s and T_s and the set

$$x_A := \left\{ T_{s,\alpha},\ D_{\alpha\beta};\ w_n^+,\ \pi_n^+,\ \theta_+,\ q_n^+;\ w_+^\alpha,\ t_{n+}^\alpha \right\}; \qquad (3.41)$$

it vanishes when $x_A = 0$. We call such a state thermostatic equilibrium and denote it with an index $|_E$. π_s takes its minimum value zero in thermostatic equilibrium. Necessary conditions for equilibrium are therefore

$$\left.\frac{\partial \pi_s}{\partial x_A}\right|_E = 0, \qquad \left\| \left.\frac{\partial^2 \pi_s}{\partial x_A\, \partial x_B}\right|_E \right\| = \text{non-negative definit} \qquad (3.42)$$

Evaluation of $(3.42)_1$ yields essentially:

1) The second coefficient of the tensor of surface tension in $(3.23)_6$ vanishes

$$\sigma_2 = 0. \qquad (3.43)$$

2) The temperature of the adjacent bulk materials and the surface temperature are equal $T_+ = T_- = T_s$ as are the free equilibrium enthalpies of the bulks and the interface

$$g^+(T_s, p^+)\Big|_E = g^-(T_s, p^-)\Big|_E = g_s(T_s, \sigma)\Big|_E . \qquad (3.44)$$

The specific free enthalpies are defined as follows $g^\pm(T_\pm, p^\pm) := u^\pm - T_\pm s^\pm + p^\pm/\rho^\pm$ and $g_s(T_s, \sigma) := u_s - T_s s_s - \sigma/\rho_s$ and satisfy the differential forms

$$dg^\pm = - s^\pm dT_\pm + \frac{1}{\rho^\pm} dp^\pm, \qquad dg_s = - s_s\, dT_s - \frac{1}{\rho_s} d\sigma . \qquad (3.45)$$

3) The conditions for mechanical equilibrium result from the momentum balance $(3.10)_2$ together with (3.12), $(3.23)_{5,6}$ and (3.43) and are

$$\begin{aligned} (p^+ - p^-)\Big|_E &= (\sigma \cdot 2\, K_M + \rho_s\, e_k\, g^k)\Big|_E , \\ \left(\frac{\partial \sigma}{\partial \rho_s} \rho_{s,\alpha}\right)\Big|_E &= - (\rho_s\, \tau_{\alpha k}\, g^k)\Big|_E . \end{aligned} \qquad (3.46)$$

$(3.46)_1$ holds for every surface point. Differentiation with respect to ξ^α, use of $(3.46)_2$ and the tangential components $p^+_{,\alpha}\big|_E = (\rho^+ \tau_{\alpha k}\, g^k)\big|_E$ and $p^-_{,\alpha}\big|_E = (\rho^- \tau_{\alpha k}\, g^k)\big|_E$ of the barometric formulas for the adjacent bulk materials yields then

$$(\sigma \cdot 2\, K_{M,\alpha})\Big|_E = \left[\left\{ [\![\rho]\!]\, \delta_\alpha^\beta + \rho_s (b_\alpha^\beta + 2\, K_M\, \delta_\alpha^\beta) + \rho_s \frac{\partial \rho_s}{\partial \sigma} (e_m\, g^m)\, \delta_\alpha^\beta \right\} \tau_{\beta k}\, g^k \right]\Big|_E , \qquad (3.47)$$

where the acceleration of the Earth g^k is constant. This shows, that the curvature possesses a gradient along the phase boundary. With $g^k \neq 0$ its geometry in equilibrium is *not* spherical.

A second set of conclusions for phase equilibrium can be inferred from (3.44) and $(3.46)_1$. According to our definition of equilibrium, the phase boundary is fixed in space. Hence, κ_M and $e_k g^k$ are fixed and (3.44), $(3.46)_1$ are three equations for the determination of four fields, namely $p^+|_E$, $p^-|_E$, $\sigma|_E$ and T_s. Thus,

$$
\begin{aligned}
p^+|_E &= p_+^E(T_s;\ \kappa_M,\ e_k g^k),\\
p^-|_E &= p_-^E(T_s;\ \kappa_M,\ e_k g^k),\\
\sigma|_E &= \sigma^E(T_s;\ \kappa_M,\ e_k g^k).
\end{aligned}
\tag{3.48}
$$

In freezing/melting equilibrium ice and water pressures are not the same. The densities, entropies and internal energies of the bulk materials and the phase boundary are only functions of the temperature (and geometry), viz.

$$
\begin{aligned}
\rho_\pm^E(T_s;\ \kappa_M,\ e_k g^k) &= \rho_\pm(T_s,\ p_\pm^E(T_s;\ \kappa_M,\ e_k g^k))\\
\rho_s^E(T\ ;\ \kappa_M,\ e_k g^k) &= \rho_s(T_s,\ \sigma^E(T_s;\ \kappa_M,\ e_k g^k))\ \text{etc.}
\end{aligned}
\tag{3.49}
$$

A change in temperature shifts the phase equilibrium. We may ask how the equilibrium bulk pressures are changed at fixed geometry of the phase interface. Differentiation of (2.44) at fixed geometry yields with (3.45) and $(3.64)_1$ two Clausius-Clapeyron-equations

$$
\frac{dp_+^E}{dT_s} = \frac{\mathcal{L}_+}{T_s\left(\frac{1}{\rho_+^E} - \frac{1}{\rho_-^E}\right)}, \qquad \frac{dp_-^E}{dT_s} = \frac{\mathcal{L}_-}{T_s\left(\frac{1}{\rho_+^E} - \frac{1}{\rho_-^E}\right)}
\tag{3.50}
$$

with different latent heats for freezing and melting

$$
\begin{aligned}
\mathcal{L}_+ :&= T_s(s_+^E - s_-^E) - \frac{T_s}{\rho_-^E}\left(\frac{d\sigma^E}{dT_s}\,2\kappa_M + \frac{d\rho_s^E}{dT_s} e_k g^k\right),\\
\mathcal{L}_- :&= T_s(s_+^E - s_-^E) - \frac{T_s}{\rho_+^E}\left(\frac{d\sigma^E}{dT_s}\,2\kappa_M + \frac{d\rho_s^E}{dT_s} e_k g^k\right),\\
\mathcal{L}_+ - \mathcal{L}_- &= \left(\frac{1}{\rho_+^E} - \frac{1}{\rho_+^E}\right) T_s\left(\frac{d\sigma^E}{dT_s}\,2\kappa_M + \frac{d\rho_s^E}{dT_s}\, e_k g^k\right) \neq 0.
\end{aligned}
\tag{2.51}
$$

An order of magnitude for this difference is $|\mathcal{L}_+ - \mathcal{L}_-| = \frac{1.82\cdot 10^{-5}}{r}\,\frac{m\cdot J}{kg}$ (Alts & Hutter, 1984); it can not be neglected for radii $r \leq 10^{-4}$ m. This demonstrates the importance of the curvature dependent corrections to the classical theory of phase transitions. In a mixture of polycrystalline ice with water inclusions and veins along grain boundaries, these corrections cannot be neglected.

δ) Linear transport equations

The linear transport equations $(3.23)_{2,6}$ and (3.26) are further restricted by the entropy production inequality (3.39). This is tediuos work and not very enlightening.

In view of the thickness of the phase-boundary layer of only several atomic distances we make therefore the following simplifying *assumptions*:

(i) The temperature $T_+ = T_- = T_s$ is continuous across the phase boundary and equal to the surface temperature.

(ii) The relative tangential velocities $w_+^\alpha = v_+^\alpha = w^\alpha = 0$ and $w_-^\alpha = v_-^\alpha - w_-^\alpha = 0$ vanish on both sides of the phase boundary; that is, we assume adherence.

(iii) The melting/freezing process at the phase boundary is reversible. This requires vanishing entropy-production density $\pi_s = 0$ *for all* thermodynamic processes.

(iv) We take the following simplified constitutive equations for the jump contributions

$$
\begin{aligned}
\mathfrak{E}_n &= \mathfrak{E}_0 + \quad \cdot \quad + \mathfrak{E}_3\, w_n^+ + \mathfrak{E}_4\, \pi_n^+ ,\\
\mathfrak{P}_n &= \quad \cdot \quad + \quad \cdot \quad + \mathfrak{P}_3\, w_n^+ + \mathfrak{P}_4\, \pi_n^+ ,\\
\mathfrak{E} &= \quad \cdot \quad + \mathfrak{Q}_1\, I_D + \mathfrak{Q}_3\, w_n^+ + \mathfrak{Q}_4\, \pi_n^+ ;\\
\mathfrak{E}^\alpha &= 0 .
\end{aligned}
\tag{3.52}
$$

From assumptions (i) and (ii) we have

$$
\mathfrak{K} = 0, \qquad \mathfrak{P}^\alpha = 0. \tag{3.53}
$$

The constitutive equations (3.52) are the simplest set that meets all above requirements.

With (3.52), (3.53) and (3.23) the requirements (i) to (iii) yield

$$
\begin{aligned}
&\kappa = 0, \quad \kappa_2 = 0; \quad \mu_1 - K_M\,\mu_3 = 0, \quad \mu_2 + \mu_3 = 0, \quad \mu_4 = \mu_5 = 0;\\
&\mathfrak{P}_3 = 1, \quad \mathfrak{E}_3 = \frac{1}{\mathfrak{P}_4}, \quad \mathfrak{E}_4 = 1.
\end{aligned}
\tag{3.54}
$$

The transport equations, which guarantee reversibility at phase transitions are thus

$$
\begin{aligned}
&q_s^\alpha = 0, \qquad \phi_s^\alpha = 0 ;\\
&s^{\alpha\beta} = \rho\, g^{\alpha\beta} + \mu_3 (I_D\, b^{\alpha\beta} - II_D\, g^{\alpha\beta}) + \mu_6 (\underset{\approx}{b}\,\underset{\approx}{D} - \underset{\approx}{D}\,\underset{\approx}{b})^{(\alpha\beta)}
\end{aligned}
\tag{3.55}
$$

$$
\pi_n^- = -\frac{1}{\mathfrak{P}_4}\, w_n^+ , \qquad w_n^- = -\,\mathfrak{P}_4\, \pi_n^+ , \tag{3.56}
$$

which are just simpler writings for $(3.52)_{1,2}$. The tensor of surface tension may have viscous contributions in reversible boundaries ($\mu_3 \neq 0$)! Departure of the normal stress at one side of the interface from its equilibrium value produces relative motion on the other side and vice versa, and hence melting or freezing.

The coefficients μ_6 and $\mathfrak{Q}_1$, $\mathfrak{Q}_2$, $\mathfrak{Q}_3$ are not restricted by the requirements of reversibility. However, the assumption of continuity of temperature makes the surface balance equation for internal energy a constraint. Neglecting small contributions proportional to ρ_s ($\approx 10^{-8}\ g/cm^2$!), this determines the coefficients in the jump contribution $\mathfrak{Q}$:

$$
\mathfrak{Q}_1 = T_s \frac{d\sigma^E}{dT_s}, \qquad \mathfrak{Q}_3 = -\rho_+^E (h_+^E - u_s^E), \qquad \mathfrak{Q}_4 = -\rho_-^E (h_-^E - u^E)\, \mathfrak{P}_4 ; \tag{3.57}
$$

where $h_\pm^E = u_\pm^E + p_\pm^E/\rho_\pm^E$ is the specific enthalpy of the adjacent bulk materials in phase equilibrium. With $(3.56)_2$ the jump for the heat flux may then be written

$$\mathcal{Q} = T_s \frac{d\sigma^E}{dT_s} \cdot I_D - (h_+^E - u_s^E)\, \rho_+^E\, w_n^+ + (h_-^E - u_s^E)\, \rho_-^E\, w_n^- \tag{3.58}$$

Neglecting small contributions proportional to ρ_s in the mass balance equation $(3.10)_1$ and in (3.51) for the latent heats and using (3.44) relation (3.58) can be rewritten in either of the two following forms

$$\begin{aligned} \mathcal{Q} &= T_s \frac{d\sigma^E}{dt_s} (w^\alpha_{;\alpha} - 2 K_M v_n^+) - \mathcal{L}_- \cdot \rho_+^E\, w_n^+ \\ &= T_s \frac{d\sigma^E}{dT_s} (w^\alpha_{;\alpha} - 2 K_M v_n^-) - \mathcal{L}_+ \cdot \rho_-^E\, w_n^+ \, . \end{aligned} \tag{3.59}$$

The *classical result* is obtained, when the phase boundary is assumed to be "incompressible", $d\rho_s/dT = 0$, $I_D = 0$, $\mathcal{R} = 0$, so that (3.58) simplifies to

$$\mathcal{Q} = - T_s [\![\rho]\!]^E \rho_+^E\, w_n^+ = - \mathcal{L}\, \rho_+^E\, w_n^+ \, . \tag{3.60}$$

However, the boundary layer-theory shows, that the assumption of incompressibility for phase boundaries between ice and water is inacceptable. Details may be found in Alts & Hutter (1984).

3.3 Thermodynamics of ice-water mixture

The foregoing results on phase boundaries are needed in a mixture theory of ice and water with large amounts of internal phase boundaries, as appear in "wet ice". Under these conditions mass, energy and entropy contributions of the phase boundyries cannot be neglected.

We shall illustrate orders of magnitude by the following *example:* Consider a closed fcc - package of ice-spheres of radius r whose void volume is filled with water. The volume filling factors for ice and water are $\alpha_I = V_I/V = \sqrt{2} \cdot \pi/6 = 0.74$ and $\alpha_w = (V-V_I)/V = 0.26$, respectively. The volume of a cube surrounding one sphere is thus $V = (4\pi/3 \cdot r^3)/(\sqrt{2} \cdot \pi/6) = \sqrt{2} \cdot 4\, r^3$. Hence, the area of the internal phase boundaries per unit volume is accordingly $\beta_s = (4\pi\, r^2)/(\sqrt{2} \cdot 4\, r^3) = (2.22)/\gamma$, and this can take considerable values for small radii. The mixture densities of ice and water are $\bar{\rho}_I = \rho_I\, \alpha_I = 0.67\ g/cm^3$, $\bar{\rho}_w = \rho_w\, \alpha_w = 0.26\ g/cm^3$ at 0^oC. The density of the interface is $\rho_s \approx 10^{-8}\ g/cm^2$. In order that the mass of the internal phase boundaries is more than x % of the mass of water in the mixture, viz. $(\rho_s\, \beta_s)/\bar{\rho}_w = (\rho_s \cdot 2.22)/(r \cdot 0.26) \geq x\,\%$, the radius r must be smaller than the values given in Table 3.

x[%]	x[1]	r[cm]	d = 2r[Å]	$\beta_s \left[\frac{cm^2}{cm^3}\right]$
1	0.01	$8.54 \cdot 10^{-6}$	1708	$2.60 \cdot 10^5$
5	0.05	$1.71 \cdot 10^{-6}$	342	$1.30 \cdot 10^6$
10	0.10	$8.54 \cdot 10^{-7}$	171	$2.60 \cdot 10^6$

Table 3

Radii and diameters of ice grains in the mixture for which the mass of the internal phase boundaries is equal to x% of the mass of the water. Area of internal phase boundaries per unit volume.

For "wet ice" this indicates, that the interfacial mass may be neglected, when the diameters of typical ice grains, water inclusions and veins are larger than 1500 Å, say. For smaller internal geometries, however, the interfacial mass must be taken into account.

In any case the water content $\alpha_W (= 1-\alpha_I)$ *and* the amount β_s of internal phase boundaries are important measures for the description of the constitutive behavior of "wet ice". It is also clear, that both measures are necessary; for ice with larger water inclusions, α_W behaves differently from ice with the same content α_W but distributed water over many grain boundaries, veins and small inclusions. Evolution equations for "wet ice" must include both measures. In classical approaches of thermodynamics they would enter as hidden or internal variables. Averaging the thermodynamic statements over many grains and the water along the grain boundaries provides (i) evolution equations for them, (ii) balance laws for the averaged fields and (iii) constitutive relations for some of the averaged fields in terms of volume integrals of constitutive relations of local fields, deduced above. The role that the averaging process plays thereby is similar to that of statistical mechanics in the kinetic theory.

This is a long and tedious scheme that cannot be presented in this short review. Instead, we may explain the averaging procedure just with the mass balance equations and provide some comments on the structure of the equations.

Mean values for the mass densities over a fixed test volume Δv_x as indicated in Figure 2 for the different phases $p = I, W$ and the phase boundary s are defined by

$$\overset{p}{\bar{\rho}} := \frac{1}{\Delta v_x} \int_{\Delta v_p} \rho_p \, dv = \frac{\Delta v_p}{\Delta v_x} \cdot \frac{1}{\Delta v_p} \int_{\Delta v_p} \rho_p \, dv = \alpha_p \bar{\rho}_p ,$$
$$\overset{s}{\bar{\rho}} := \frac{1}{\Delta v_x} \int_{\Delta A_s} \rho_s \, da = \frac{\Delta A_s}{\Delta v_x} \cdot \frac{1}{\Delta A_s} \int_{\Delta A_s} \rho_s \, da = \beta_s \bar{\rho}_s , \qquad (3.61)$$

where Δv_p is all the volume inside Δv_x that is occupied by phase p, and ΔA_s is the area of all phase boundaries inside Δv_x. The volume fraction $\alpha_p := \Delta v_p / \Delta v_x$ of phase p and the amount $\beta_s := \Delta A_s / \Delta v_x$ of internal phase boundaries are functions of $\underset{\sim}{x}$ and $\underset{\sim}{t}$, where $\underset{\sim}{x}$ is some point within Δv_x.

The balance laws for the bulk masses and the interfacial mass can be written in the forms

$$\frac{\partial \rho_p}{\partial t} + \operatorname{div}(\rho_p \underset{\sim}{v}_p) = 0,$$
$$\frac{\partial \rho_s}{\partial t} + \operatorname{div}(\rho_s \underset{\sim}{w}) = - [\![\rho (\underset{\sim}{v} - \underset{\sim}{w})]\!] \cdot e, \qquad (3.62)$$

where ρ_s and $\underset{\sim}{w}$ are considered as functions of $\underset{\sim}{x}$ and t and $\partial/\partial t$ is the time derivative of constant $\underset{\sim}{x}$. Integration over the bulk phases Δv_p and the internal surfaces ΔA_s, respectively, yields

$$\frac{\partial \overset{p}{\bar{\rho}}}{\partial t} + \operatorname{div}(\overset{p}{\bar{\rho}} \, \overset{p}{\underset{\sim}{\bar{v}}}) = \overset{p}{\bar{c}} = \pm \beta_s \, \overset{s}{\overline{\rho_p (\underset{\sim}{v}_p - \underset{\sim}{w}) \cdot \underset{\sim}{e}}} \quad \binom{I}{W}$$
$$\frac{\partial \overset{s}{\bar{\rho}}}{\partial t} + \operatorname{div}(\overset{s}{\bar{\rho}} \, \overset{s}{\underset{\sim}{\bar{w}}}) = \overset{s}{\bar{c}} = - \beta_s \, \overset{s}{\overline{[\![\rho (\underset{\sim}{v} - \underset{\sim}{w})]\!] \cdot \underset{\sim}{e}}} \qquad (3.63)$$

where $\overset{p}{\underset{\sim}{\bar{v}}} = \underset{\sim}{\bar{v}}_p := 1/\Delta v_x \int_{\Delta v_x} \underset{\sim}{v}_p \, dv$ and $\overset{p}{\bar{c}}$, $\overset{s}{\bar{c}}$ have the meaning of production densities that arise from a redistribution of mass between the bulk phased and the interfaces during phase transition.

The phase mixture, however, satisfies a conservation law, since mass cannot be produced in phase transitions. Hence

$$\frac{\partial\bar{\rho}}{\partial t} + \operatorname{div}(\bar{\rho}\,\bar{\underset{\sim}{v}}) = 0, \tag{3.64}$$

where the mixture density and the mean mixture velocity are defined by

$$\begin{aligned}
\bar{\rho} &:= \sum_p \bar{\overset{p}{\rho}} + \bar{\overset{s}{\rho}} = \sum_p \alpha_p \bar{\rho}_p + \beta\,\bar{\rho}_s,\\
\bar{\rho}\,\bar{v} &:= \sum_p \bar{\overset{p}{\rho}}\,\bar{\overset{p}{v}} + \bar{\overset{s}{\rho}}\,\bar{\overset{s}{\underset{\sim}{w}}} = \sum_p \alpha_p \bar{\rho}_p \bar{v}_p + \beta\,\bar{\rho}_s \bar{\overset{s}{\underset{\sim}{w}}},\\
0 &= \sum_p \bar{\overset{p}{c}} + \bar{\overset{s}{c}}\,.
\end{aligned} \tag{3.65}$$

The phase boundary, when considered as a singular surface, has no volume. Hence

$$\sum_p \alpha_p = 1\,. \tag{3.66}$$

In order to further reduce (3.63) additional assumptions are needed. In slowly creeping temperate glacier ice all the bulk and surface fields do not change much over the extension of the test volume Δv_x. Hence they are approximately equal to the local fields, viz $\bar{\rho}_p \approx \rho_p$, $\bar{v}_p \approx v_p$, $\bar{\rho}_s \approx \rho_s$, $\bar{\overset{s}{\underset{\sim}{w}}} \approx \underset{\sim}{w}$. Insertion into (3.63) and use of (3.62) and (3.66) then yields

$$\begin{aligned}
(\rho_W-\rho_I)\,\frac{\partial\alpha_W}{\partial t} + (\underset{\sim}{v}_W-\underset{\sim}{v}_I)\cdot\operatorname{grad}\alpha_W &= \beta_s\,\overline{\overset{s}{[\![\rho(\underset{\sim}{v}-\underset{\sim}{w})]\!]\cdot\underset{\sim}{e}}}\\
\frac{\partial\beta_s}{\partial t} + \underset{\sim}{w}\cdot\operatorname{grad}\beta_s &= 0\,,
\end{aligned} \tag{3.67}$$

which are the desired evolution equations for α_W and β_s.

Similar, though more complicated balance laws can be derived for average momentum, energy and entropy of the mixture. These, together with assumptions about the mean geometry of the interfaces and constitutive assumptions for the pure bulk phases form a closed set of equations. The derivation is based on the preceding results about phase boundaries and partly on two-phase theories of Drew (1983) and Ishii (1975).

The structure of the equations is that of a binary mixture of a slowly moving viscous matrix containing a fluid which is dispersed in it and capable of perculating through the pores. Among other things, the stress tensor of the matrix contains also α_W and β_s as independent variables and thus substantiates statements made by Lliboutry (1976), according to which the stress tensor of the ice must depend on the moisture content. The procedure gives also a set of evolution equations for α_W and β_s and thus provides the essentials for the description of the hydrology of the ice mass. It is to be seen whether these field equations will do better in modelling temperate ice of a polythermal ice mass than the presently existing formulations of Hutter (1982) and Fowler (1984). One critical, and often neglected point is hereby the role that boundary conditions play. The present formulation must necessarily cope with them and include statements on α_W and β_s along the free surface, the cold-temperate transition surface and the base. The result must mathematically be a well set boundary value problem. If some of the boundary conditions cannot be physically motivated, or if measurements do not permit to provide the data of a well set mathematical problem sophistication in thermocynamic formulations may be of limited help. This remark applies to many situations in geophysics where data aquisition is difficult.

4. CONVECTION IN THAWING SUBSEA PERMAFROST

4.1 Statement of the problem

Glaciology is also fraught with challenging instability problems. Here we present a general formulation of a double-diffusion problem that offers difficult analyses in coupled Bénard convection and solidification problems, usually arising in other branches of fluid dynamics.

Consider Figure 4, which shows the approximate configuration of subsea permafrost just offshore Prudhoe Bay, Alaska. The permafrost is separated from the thawed layer by a phase boundary whose depth below the ocean bed grows approximately with distance from shore. The permafrost exists because the ocean shelves were not always submerged as the shoreline retreats. At earlier times, submerged portions were exposed to subaerial temperatures long enough for substantial permafrost growth to occur. Upon submergence, the overlying ocean acts as an insulator to the atmospheric cold: The submerged portion of the permafrost melts from above with a phase boundary separating the thawed layer from the frozen underground. Because this subsea thawing occurs at negative temperatures salt must play a significant role in the mechanism and perculate from the ocean bed through the thawed layer. This gives rise to non-buoyant conditions due to salt variations. Temperature induced density variations in the pore water of the thawed layer may also occur, but are likely to be less significant because the water temperature is close to the point of anomaly. Nonetheless, since relatively fresh water is generated at the lower phase boundary, melting must be regarded as the driving mechanism of the convective regime and associated stabilities. The problem has been analysed by Sellmann (1980), Smith et al. (1980), Harrison & Osterkamp (1978, 1984), Harrison (1982), Osterkamp & Harrison (1982), Swift et al. (1983) and Swift & Harrison (1984). We follow Hutter & Straughan (1984). From these works we infer the following observational facts:

(i) Except very close to the shore the phase boundary is parabolic and has a constant temperature.

(ii) Within the thawed layer, the pore water salinity is about 25 % higher than that of normal sea water. It is fairly constant, except in the bottom fraction of a meter near the phase boundary.

(iii) Temperature variations through the layer are fairly linear.

(iv) The present shoreline retreat is fairly steady and approximately 1 m a^{-1}.

A mathematical description of the convection problem must bear on these experimental facts.

4.2 Double diffusion equations in a porous solid with liquefaction at a boundary

The theory of flows through porous media with incompressible pores is governed by the Darcy-Overbeck-Boussinesq equations. They have the form (see Joseph, 1976, Vol. 2, p. 55 for a general derivation and Hutter & Straughan for the subsea permafrost applications).

$$
\begin{aligned}
&\rho_0 \dot{\underset{\sim}{u}} = -\nabla p' - \rho_0 \underset{\sim}{g}\, \alpha (T-T_0)^2 + \rho_0 \underset{\sim}{g}\, \beta (S-S_0) - \frac{\mu}{k'} \hat{\phi}\, \underset{\sim}{u}, \\
&\nabla \cdot (\hat{\phi}\, \underset{\sim}{u}) = 0, \\
&\frac{\partial S}{\partial t} + \underset{\sim}{u} \cdot \nabla S = \kappa_S \nabla^2 S, \qquad (4.1) \\
&\frac{\partial T}{\partial t} + \gamma\, \underset{\sim}{u} \cdot \nabla T = \kappa_T \nabla^2 T .
\end{aligned}
$$

In these equations $\underset{\sim}{u}$ is the pore water velocity, T the temperature and S the salinity; $\underset{\sim}{g}$ is the gravity vector, ρ_0 a reference density of water, μ the dynamic viscosity, k' the permeability, $\hat{\phi}$ the porosity, α the quadratic coefficient of thermal expansion, β the thermohaline coefficient, s_0 the reference salinity s_0. More-ocean, T_0 the temperature for which the density is a maximum at salinity s_0. Moreover, κ_S and κ_T are a salt and thermal diffusivity and $\gamma = \hat{\phi}(\rho c)_f/(\rho c)_m$ is the product of the porosity with the ratio of the heat capacity of the fluid to that of the bulk material. Finally, $p' = p - \rho_0 gz$ is the pressure reduced by the mean hydrostatic pressure below the ocean surface. The real density ρ is related to ρ_0, T and S through

$$\rho = \rho_0\left\{1 - \alpha(T - T_0)^2 + \beta(S - S_0)\right\} . \tag{4.2}$$

In ensuing developments it will be assumed that ρ_0, α, β, μ, k', $\hat{\phi}$, κ_S, κ_T, γ are constant and that the buoyancy forces due to temperature variations can be ignored.

Boundary conditions must be satisfied at the ocean bed $z = Z_B(x,y)$ and the phase boundary $z = Z(x,y,t)$. At the former we assume that the pore water velocity has no tangential component and the temperature and salinity are prescribed,

$$\underset{\sim}{u} - (\underset{\sim}{u}\cdot\underset{\sim}{n})\underset{\sim}{n} = \underset{\sim}{0}, \quad T = T_B(x,y,t), \quad S = S_B(x,y,t), \quad \text{at} \quad z = Z_B(x,y). \tag{4.3}$$

The interface between the rigid permafrost and the mobile thawed layer is a discontinuity surface at which a kinematic condition, a mass balance statement, Stefan conditions for heat and salt and a phase equilibrium condition must hold:

$$\begin{aligned}
&\text{(i)} && \frac{\partial Z}{\partial t} - N\,\underset{\sim}{u}\cdot\underset{\sim}{n} = -N a_\perp , && \text{kinematic condition} \\
&\text{(ii)} && \underset{\sim}{u}\cdot\underset{\sim}{n} = -\left(\frac{\rho_L}{\rho_S} - 1\right) a_\perp , && \text{mass balance} \\
&\text{(iii)} && k_S\, \nabla T\cdot\underset{\sim}{n}\Big|_S - k_L\, \nabla T\cdot\underset{\sim}{n}\Big|_L = -\rho_L L a_\perp , && \text{Stefan condition for heat} \qquad z = Z(x,y,t) \qquad (4.4) \\
&\text{(iv)} && \kappa_S\, \nabla S\cdot\underset{\sim}{n}\Big|_L = -S a_\perp , && \text{Stefan condition for salt} \\
&\text{(v)} && T = T_f{}^0 - c_S(S - S_0) , && \text{phase equilibrium}
\end{aligned}$$

where

$$N = \left[1 + \left(\frac{\partial Z}{\partial x}\right)^2 + \left(\frac{\partial Z}{\partial y}\right)^2\right]^{1/2} .$$

In (4.3) and (4.4), $\underset{\sim}{n}$ is the unit normal vector exterior to the thawed layer. Further, $a_\perp$ is the freezing ($a_\perp > 0$) or melting ($a_\perp < 0$) rate at the phase boundary having the dimension of a velocity; it is one of the unknown surface fields. k_S and k_L are the thermal conductivity of the permafrost (solid) and the thawed layer (liquid) and L is the latent heat of fusion per unit mass. Also, depending on the index (L or S), the normal derivatives at $z = Z$ on the left in the Stefan conditions for heat and salt must be evaluated as the surface is approached from below and above, respectively. Finally, $T_f{}^0$ is the freezing temperature of salt water with salinity S_0 and $c_S = -dT/dS$ describes the change of the melting temperature with salinity. Physical constants are listed in Table 4.

A deduction of (4.4) from first principles is given by Hutter & Straughan (1984). Here, it may suffice to state that the Stefan conditions emerge from an entropy balance and a mass balance of salt across the discontinuity surface. On the other hand, the condition of phase equilibrium ignores a pressure dependent term, since ocean depths are small, in general.

$\kappa_S = 4\times10^{-3}\ m^2\ a^{-1}$	$\dfrac{k'g}{(\mu/\rho_0)} = 20\ m\ a^{-1}$	$[\Delta T] = 2\ ^oC$
$\kappa_T = 21\ m^2\ a^{-1}$	$\hat{\phi} = 0.5$	$T_f = 273\ ^oK$
$\rho_0 = \rho_L = 2\times10^3\ kg\ m^{-3}$	$\beta = 0.773$	$S_0 = 0.034$
$k_S = 7\times10^7\ J\ k^{-1}\ m^{-1}\ a^{-1}$	$c_S = 54.11\ ^oK/\text{parts}$	$[\Delta S] = 0.034$
$k_L = 7\times10^7\ J\ K^{-1}\ m^{-1}\ a^{-1}$	$[L] = 10 - 50\ m$	$T_f^o = 271\ ^oK$
$L = 3.3\times10^5\ kg^{-1}\ J$		$\gamma \cong 0.4$

Table 4

Physical constants appropriate for the subsea permafrost problem.

4.3 Scaling arguments

Non-dimensionalizations of equations (4.1)-(4.4) must be motivated by the fact that the process of convection in thawing subsea permafrost is characterized by two sub-processes which have distinctly different scales but interact. On the one hand, melting occurs at the phase change boundary of which the speed is in the order of 2 to 5 cm a^{-1}, (Harrison, 1982). Appreciable changes in the position of the phase boundary are therefore recognized on time scales of years. On the other hand, convection in the thawed layer above the phase boundary is likely to occur with much larger velocities, say cm/day; corresponding time scales are therefore considerably shorter. In other situations ratios of melting rate velocities to bulk velocities may be order unity or large. In short, two velocity scales must be introduced; [U] is representative for the bulk motion and [V] for the moving interface. With

$$\begin{aligned}
(x,y,z,Z_B,Z) &= [L](\bar{x},\bar{y},\bar{z},\bar{Z}_B,\bar{Z}), & [L] &= \text{typical length},\\
(u,v,w) &= [U](\bar{u},\bar{v},\bar{w}), & [U] &= \text{velocity scale for bulk},\\
a_\perp &= [V]\,\bar{a}_\perp, & [V] &= \text{velocity scale for moving interface},\\
t &= [\mathfrak{T}]\,\bar{t}, & [\mathfrak{T}] &= \text{time scale},\\
T &= T_f + [\Delta T]\,\theta, & [\Delta T] &= \text{temperature range},\\
S &= S_0 + [\Delta S]\,s, & [\Delta S] &= \text{range for salinity},\\
& & [P] &= \frac{\mu}{k'}[U] = \text{scale for pressure}
\end{aligned} \tag{4.5}$$

the field equations and boundary conditions assume the form (overbars are omitted)

$$\left.\begin{aligned}
&\mathbb{B}\left\{\frac{\partial \underset{\sim}{u}}{\partial t} + \Omega\, \underset{\sim}{u}\cdot\nabla\underset{\sim}{u}\right\} = -\nabla p' - \mathbb{R}_T^2\,\theta^2\,\underset{\sim}{k} + \mathbb{R}_S^2\,s\,\underset{\sim}{k} - \hat{\phi}\,\underset{\sim}{u}\\
&\nabla\cdot\underset{\sim}{u} = 0,\\
&\frac{1}{\Omega}\frac{\partial\theta}{\partial t} + \gamma\,\underset{\sim}{u}\cdot\nabla\theta = \frac{\kappa_S}{[UL]}\nabla^2\theta,\\
&\frac{1}{\Omega}\frac{\partial s}{\partial t} + \underset{\sim}{u}\cdot\nabla_s = \frac{\kappa_S}{[UL]}\nabla^2 s,
\end{aligned}\right\} Z_B < z < Z \tag{4.6}$$

where $\underset{\sim}{k}$ is the unit vector in the z-direction.

$$\underset{\sim}{u}-(\underset{\sim}{u}\cdot\underset{\sim}{n})\underset{\sim}{n} = 0, \qquad \theta = \theta_B(\underset{\sim}{x},t), \qquad s = S_B(\underset{\sim}{x},t), \qquad z = Z_B(\underset{\sim}{x}) \tag{4.7}$$

and

$$\begin{aligned}
&\frac{1}{\Omega}\frac{\partial z}{\partial t} - N\,\underset{\sim}{u}\cdot\underset{\sim}{n} = -\frac{[V]}{[U]}\,N\,a_\perp, \\
&\underset{\sim}{u}\cdot\underset{\sim}{n} = -\left(\frac{\rho_L}{\rho_S}-1\right)\frac{[V]}{[U]}\,a_\perp, \\
&\frac{k_S}{k_L}\nabla\theta\cdot\underset{\sim}{n}\Big|_S - \nabla\theta\cdot\underset{\sim}{n}\Big|_L = -\frac{\rho_L\,L\,[VL]}{k_L\,[\Delta T]}\,a_\perp, \qquad z = Z(x,y,t) \\
&\nabla s\cdot\underset{\sim}{n}\Big|_L = (1+s)\frac{[VL]}{\kappa_S}\,a_\perp, \\
&\theta = \theta_0 + \mathscr{C}\,s,
\end{aligned} \tag{4.8}$$

in which

$$\mathcal{B} = \frac{k'}{\frac{\mu}{\rho}[\mathfrak{T}]}, \quad \Omega = \frac{[U\mathfrak{T}]}{[L]}, \quad \mathcal{R}_T^2 = \frac{\alpha g[\Delta T]^2}{\nu[U]}k', \quad \mathcal{R}_S^2 = \frac{\beta g[\Delta S]}{\nu[U]}k',$$

$$\mathscr{C} = c_s\frac{[\Delta S]}{[\Delta T]}, \quad \theta_0 = \frac{T_f{}^0 - T_0}{[\Delta T]}$$

are four dimensionless characteristic parameters, a dimensionless phase equilibrium constant and a dimensionless reference temperature. $\mathcal{B}$ measures the significance of the acceleration terms in Darcy's law and Ω that of the time dependent terms in the energy equation and the kinematic boundary condition; $\mathcal{R}_T{}^2$ and $\mathcal{R}_S{}^2$ are Rayleigh-Darcy numbers for temperature and salt, respectively. Ensuing developments will be based on the assumptions that (i) $\mathcal{B}\to 0$, $\mathcal{B}\Omega\to 0$ and (ii) buoyancy effects due to temperature variations are ignored. The first of these implies that acceleration terms in Darcy's law are negligible, the second sais that the Rayleigh-Darcy number for temperature drops out of the problem. Temperature still affects the Rayleigh-Darcy number for salt, because the salt and temperature problems are coupled through the boundary conditions. The field equations governing pressure and flow now read

$$\nabla p' = \mathcal{R}_S{}^2\, s\underset{\sim}{k} - \hat{\phi}\underset{\sim}{u}, \qquad \nabla\cdot\underset{\sim}{u} = 0. \tag{4.9}$$

Because thawing is an essential process, it is reasonable to suppose that the right hand side of $(4.8)_3$ is order unity, implying $[V] = k_L[\Delta T]/(\rho_L\,L\,[L]) \lesssim 0.05\ \mathrm{m\,a^{-1}}$ (Table 4), in good qualitative agreement with measured thaw rates ($[V] \sim 0.03\ \mathrm{m\,a^{-1}}$, Harrison, 1982). If we also choose [L] as a typical depth of the thawed layer there still remain the scales [U] and $[\mathfrak{T}]$ to be selected in order to fix the numerical values of the coefficients in (4.6)-(4.8). Depending upon which values of [UL] are selected, different regimes can be distinguished. The results are summarized in Table 5, in which, besides the important scales and dimensionless parameters, we have also listed the important dimensionless parameters, the transport equations for heat and salt as well as all boundary equations at the surface of phase change. In the *thermal diffusive regime* the melting rate is large in comparison to a typical bulk velocity ($[V]/[U] = O(10^2)$). Heat transport is primarily conductive (η_T is small) and temperature profiles probably close to linear. By contrast salt transport is both diffusive and convective. In the phase boundary condition a second small parameter, ε_T arises, but the limit $\varepsilon_T \to 0$ needs careful consideration (see Hutter & Straughan, 1984).

	Thermal diffusive regime	Salt convective regime	Intermediate regime
$[UL] =$	κ_S	κ_T	$\sqrt{\kappa_T \kappa_S}$
$[U] =$	$\frac{\kappa_S}{[L]} \sim 10^{-4}\ \mathrm{m\,a^{-1}}$	$\frac{\kappa_T}{[L]} \sim 0.4 - 2\ \mathrm{m\,a^{-1}}$	$\frac{\sqrt{\kappa_T \kappa_S}}{[L]} \sim 10^{-2}\ \mathrm{m\,a^{-1}}$
$\frac{[V]}{[U]} =$	$O(10^2) = \frac{1}{\varepsilon_T}$ (large)	$O(10^{-2}) = \varepsilon_S$ (small)	$O(1) = \chi$
	$\eta_T = \frac{\kappa_S}{\kappa_T} = 0.2 \times 10^{-3}$	$\eta_S = \frac{\kappa_S}{\kappa_T} = 0.2 \times 10^{-3}$	$\psi = \sqrt{\frac{\kappa_T}{\kappa_S}} = 72$
heat and salt transport equations	$\frac{1}{\Omega}\frac{\partial\theta}{\partial t} + \gamma\, \underset{\sim}{u}\cdot\nabla\theta = \frac{1}{\eta_T}\nabla^2\theta$	$\frac{1}{\Omega}\frac{\partial\theta}{\partial t} + \gamma\, \underset{\sim}{u}\cdot\nabla\theta = \nabla^2\theta$	$\frac{1}{\Omega}\frac{\partial\theta}{\partial t} + \gamma\, \underset{\sim}{u}\cdot\nabla\theta = \psi\,\nabla^2\theta$
	$\frac{1}{\Omega}\frac{\partial s}{\partial t} + \underset{\sim}{u}\cdot\nabla s = \nabla^2 s$	$\frac{1}{\Omega}\frac{\partial s}{\partial t} + \underset{\sim}{u}\cdot\nabla s = \eta_S\,\nabla^2 s$	$\frac{1}{\Omega}\frac{\partial s}{\partial t} + \underset{\sim}{u}\cdot\nabla s = \frac{1}{\psi}\nabla^2 s$
conditions at phase boundary	$\frac{1}{\Omega}\frac{\partial Z}{\partial t} - N\,\underset{\sim}{u}\cdot\underset{\sim}{n} = -\frac{1}{\varepsilon_T} N\, a_\perp$	$\frac{1}{\Omega}\frac{\partial Z}{\partial t} - N\,\underset{\sim}{u}\cdot\underset{\sim}{n} = -\varepsilon_S N\, a_\perp$	$\frac{1}{\Omega}\frac{\partial Z}{\partial t} - N\,\underset{\sim}{u}\cdot\underset{\sim}{n} = -\chi N a_\perp$
	$\underset{\sim}{u}\cdot\underset{\sim}{n} = -(\frac{\rho_L}{\rho_S} - 1)\frac{1}{\varepsilon_T} a_\perp = -\frac{\xi}{\varepsilon_T} a_\perp$	$\underset{\sim}{u}\cdot\underset{\sim}{n} = -(\frac{\rho_L}{\rho_S} - 1)\,\varepsilon_S\, a_\perp = -\varepsilon_S\, \xi\, a_\perp$	$\underset{\sim}{u}\cdot\underset{\sim}{n} = -(\frac{\rho_L}{\rho_S} - 1)\,\chi\, a_\perp = -\xi\,\chi\, a_\perp$
	$\frac{k_S}{k_L}\nabla\theta\cdot\underset{\sim}{n}\Big\vert_S - \nabla\theta\cdot\underset{\sim}{n}\Big\vert_L = -a_\perp$	$\frac{k_S}{k_L}\nabla\theta\cdot\underset{\sim}{n}\Big\vert_S - \nabla\theta\cdot\underset{\sim}{n}\Big\vert_L = -a_\perp$	$\frac{k_S}{k_L}\nabla\theta\cdot\underset{\sim}{n}\Big\vert_S - \nabla\theta\cdot\underset{\sim}{n}\Big\vert_L = -a_\perp$
	$\nabla s\cdot\underset{\sim}{n}\Big\vert_L = (1+s)\frac{1}{\varepsilon_T} a_\perp$	$\nabla s\cdot\underset{\sim}{n}\Big\vert_L = (1+s)\frac{\varepsilon_S}{\eta_S} a_\perp$	$\nabla s\cdot\underset{\sim}{n}\Big\vert_L = (1+s)\,\psi\,\chi\, a_\perp$
	$\theta = \theta_0 + \mathcal{L}\, s$	$\theta = \theta_0 + \mathcal{L}\, s$	$\theta = \theta_0 + \mathcal{L}\, s$

Table 5

Scales, important parameters and form of the heat and salt transport equations and the conditions of the phase boundary in three different regimes that are governed by the choice of the scale for the bulk velocity.

In the *salt convective regime* melting rates are small in comparison to characteristic bulk velocities, heat transport is conductive and convective, but salt transport is primarily conductive except in a thin boundary layer near the phase boundary. Balancing tangential advective with normal diffusive terms discloses a boundary layer thickness $\sqrt{\eta_S}$, corresponding to 0.1 - 0.5 m in physical space, in good agreement with observations. Conditions at the phase boundary contain a second small parameter, but again limits $\varepsilon_S \to 0$ need careful consideration. In the *intermediate regime*, characteristic melting rates and bulk velocities are of the same order. Heat flow is primarily diffusive and salt transport primarily advective, except in a bottom boundary layer.

Because processes in the thawed layer and at the phase boundary take place with different speeds in the three different regimes, it seems natural that this difference must manifest itself also in corresponding time scales. Thus we expect two time scales to be effective in the thermal diffusive and the salt convective regimes but only one in the intermediate regime. A careful analysis of this is given by Hutter & Straughan (1984); here we focus attention only on the salt convective regime.

4.4 Asymptotic analysis of the salt convective regime

So far, no time scale was selected. Physically, since processes at the phase boundary are the dominant feature in the salt convective regime, the appropriate time scale must most likely be based on melting speeds [V] and local depths [L]: so $[\mathfrak{T}] = [LV^{-1}]$, implying $\Omega^{-1} = \varepsilon_S \sim 10^{-2}$. In the analysis which follows a simplified situation of the permafrost problem will be attacked. The sea bed will be assumed horizontal, while the phase boundary for convectionsless pore water will be assumed mobile but always horizontal. Undulations in the phase boundary are supposed to be introduced by convection, see Figure 14. This imposes restrictions on the boundary conditions; for instance, the ocean salinity and temperature and the "geothermal" heat flux from below must not, to first order, be functions of position.

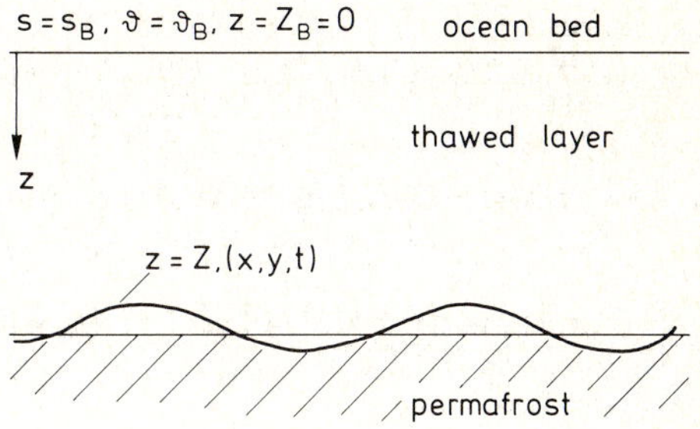

Figure 14

Simplified geometry of the thawed layer of permafrost. The ocean bed is horizontal and rigid, the basic state phase boundary is horizontal but mobile. Undulations are introduced by convective instabilities.

With $\Omega^{-1} = \varepsilon_S$, the governing field equations and boundary conditions (see Table 5) in the limit $\eta_S = 0$ contain a small parameter, ε_S, suggesting perturbation procedures to construct approximate solutions. In view of the two time scales on which melting and convection operate we use two timing with $\tau_0 = t$ typical for melting and $\tau_1 = \Omega t \gg \tau_0$ characteristic for convection. Order unity τ_1 then implies a small real time (in units of the time τ_0). Thus we use

$$\frac{\partial}{\partial \bar{t}} = \frac{\partial}{\partial \tau_0} + \varepsilon_S^{-1} \frac{\partial}{\partial \tau_1}, \tag{4.10}$$

and shall also employ the expansion (for justifications see Hutter & Straughan, 1984)

$$\Theta = \tilde{\Theta}(\underset{\sim}{x},T_0) + \sum_{j=1}^{\infty} \varepsilon_S^{j/2} \Theta_j(\underset{\sim}{x},T_1), \qquad a_\perp = \tilde{a}_\perp(\underset{\sim}{x},T_0) + \sum_{j=1}^{\infty} \varepsilon_S^{j/2} a_{\perp j}(\underset{\sim}{x},T_1),$$

$$s = \tilde{s}(\underset{\sim}{x},T_0) + \sum_{j=1}^{\infty} \varepsilon_S^{j/2} s_j(\underset{\sim}{x},T_1), \qquad Z = \tilde{Z}(\underset{\sim}{x},T_0) + \varepsilon_S^{\beta} \sum_{j=1}^{\infty} \varepsilon_S^{j/2} Z_j(\underset{\sim}{x},T_1). \tag{4.11}$$

$$u = \sum_{j=1}^{\infty} \varepsilon_S^{j/2} u_j(\underset{\sim}{x},T_1),$$

Accordingly, there is a *motionless basic state* (denoted by tildes) which depends on space and melting time, and there is a family of perturbed fields whose time dependence, however, is governed by the convective time T_1. The perturbation terms progress in half powers of ε_S and the series expansion for Z involves an additional factor ε_S^β, features that will be explained in a moment. In short, the expansions (4.10), (4.11) emerge from a delicate balance with the desire that (i) the solution of the basic state would reproduce the observational facts described at the end of section 4.1, (ii) the first order $(j=1)$ boundary value problem would yield the eigenvalue problem for the Rayleigh-Darcy number and (iii) higher order equations, and the expansion

$$\mathbb{R}_S^2 = R_0 + \varepsilon_S^{1/2} R_{1/2} + \varepsilon_S R_1 + \dots, \tag{4.12}$$

would permit determination of the weakly nonlinear convective flow. This is indeed so: Substitution of (4.10) and (4.11) into the transport equations of heat and salt yields

$$O(\varepsilon_S^0):\quad \Delta\tilde{\theta} = 0, \qquad \Delta\tilde{s} = 0,$$

$$O(\varepsilon_S^{1/2}):\quad \frac{\partial\theta_1}{\partial T_1} + \gamma\, \underset{\sim}{u}_1\cdot\nabla\tilde{\theta} - \Delta\theta_1 = 0, \qquad \frac{\partial s_1}{\partial T_1} + \underset{\sim}{u}_1\cdot\Delta\tilde{s} - \Delta s_1 = 0, \tag{4.13}$$

$$O(\varepsilon_S^3):\quad \frac{\partial\theta_2}{\partial T_1} + \gamma\, \underset{\sim}{u}_2\cdot\nabla\tilde{\theta} - \Delta\theta = -\frac{\partial\tilde{\theta}}{\partial T_0} - \gamma\, \underset{\sim}{u}_1\cdot\nabla\theta_1, \qquad \frac{\partial s_2}{\partial T_1} + \underset{\sim}{u}_2\cdot\Delta\tilde{s} - \Delta s_2 = -\frac{\partial\tilde{s}}{\partial T_0} - \gamma\, \underset{\sim}{u}_2\cdot\nabla s_1,$$

which are linear equations, if they are solved successively. The basic states $\tilde{\theta}$ and $\tilde{s}$ are governed by the Laplace equations which will imply linear temperature and salinity profiles (as observed!). The $O(\varepsilon_S^{1/2})$-equations are homogeneous; together with the momentum and the continuity equations and the boundary conditions at the sea bed and the phase boundary, these equations give rise to the eigenvalue problem for the Rayleigh-Darcy number*). In this eigenvalue problem one amplitude is still an undetermined function of T_0. Equations at the $O(\varepsilon_S^1)$-level are inhomogeneous and the right hand sides are functions of T_0 and T_1, but the T_0-dependence of the amplitude functions will follow from an orthogonality condition, see Joseph (1976).

Boundary conditions should similarly be treated. Table 5 shows that the crucial equation is the kinematic equation,

$$\frac{\partial\tilde{Z}}{\partial T_0} + \varepsilon_S^{\beta-1/2} \frac{\partial Z_1}{\partial T_1} + \varepsilon_S^{\beta} \frac{\partial Z_2}{\partial T_1} + O(\varepsilon^{\beta+1/2}) = -\frac{\rho_L}{\rho_S} N_0\, \tilde{a}_\perp - \frac{\rho_L}{\rho_S} N_0\, \varepsilon_S^{1/2} a_{\perp_1} + O(\varepsilon_S), \qquad N_0 = 1 \tag{4.14}$$

*) *Had the expressions (4.11) be chosen in integer powers of* ε_S *and not in fractional powers, no eigenvalue problem for the Rayleigh-Darcy number would have resulted at the first level.*

in which all quantities are evaluated at the deformed surface $z = \tilde{z} + \sum_{j=1}^{\infty} \varepsilon^{\beta+j/2} z_j$. To lowest order a dependence on basic state variables alone is sought, implying $\beta > 1/2$. We shall choose $\beta = 1$. Thus

$$\frac{\partial \tilde{z}}{\partial T_0} = -\frac{\rho_L}{\rho_S}\tilde{a}_\perp, \qquad \frac{\partial z_1}{\partial T_1} = -\frac{\rho_L}{\rho_S} a_{\perp 1}, \qquad \text{at} \quad z = \tilde{z}(T_0) \tag{4.15}$$

are the lowest and first order kinematic boundary conditions at the surface of phase change.

Next, the mass balance statement across this discontinuity surface requires $\xi\, \varepsilon_S\, \tilde{a}_\perp = 0$, where $\xi = (\rho_L/\rho_S - 1)$ is small. Thus, necessarily $\xi = 0$ or $\varepsilon_S = 0$ or $\tilde{a}_\perp = 0$. $\varepsilon_S = 0$ is not permissible because the perturbation scheme would fail and $\tilde{a}_\perp = 0$ does not admit a basic state solution (Hutter & Straughan, 1984). Hence for consistency we are forced to request that $\xi = 0$, appropriate for a material with no density change under phase changes. In view of the slow thawing rates this approximation is not damaging, however. The boundary value problem for the basic state is now

$$\begin{aligned} &D^2\tilde{\theta} = 0, \quad D^2 \tilde{s} = 0, && 0 < z < \tilde{z}(T_0),\\ &\tilde{\theta} = \theta_0, \quad \tilde{s} = s_0, && z = 0,\\ &\frac{\partial \tilde{z}}{\partial T_0} = -\tilde{a}_\perp, \quad D\tilde{\theta} = \tilde{a}_\perp, && \\ &\tilde{\theta} = \theta_0 + \mathscr{C}\tilde{s}, \quad D\tilde{s} = (1+\tilde{s})\,\delta_S\,\tilde{a}_\perp, && \Big\} \; z = \tilde{z}(T_0), \end{aligned} \tag{4.16}$$

in which $D = d/dz$. This boundary value problem admits the solution

$$\begin{aligned} \tilde{\theta} &= \theta_0 + \frac{\tilde{a}_\perp^0}{\tilde{z}(T_0)} z,\\ \tilde{s} &= s_0 + \frac{(1+s_0)\,\delta_S\,\tilde{a}_\perp^0/\tilde{z}(T_0)}{1-\delta_S\,\tilde{a}_\perp^0} z,\\ \tilde{z} &= \sqrt{-2\,\tilde{a}_\perp^0\, T_0 + \text{const}} \end{aligned} \tag{4.17}$$

in which

$$\tilde{a}_\perp^0 = \frac{1-\mathscr{C}\delta_S}{2\delta_S} \,(\pm)\, \sqrt{\left(\frac{1-\mathscr{C}\delta_S}{2\delta_S}\right)^2 - \mathscr{C}\frac{s_0}{\delta_S}}, \qquad \delta_S = \frac{\varepsilon_S}{\eta_S} \simeq 0.5. \tag{4.18}$$

Only the solution with the positive square root is physically meaningful. Both, the temperature and salinity profiles are linear functions of z, and the time evolution of the interface is a square root function of the slow thaw rate time. This same functional dependence arises also in the coefficients (4.17). The behavior, expressed in $(4.17)_{1,2,3}$ is in agreement with the features deduced from observations. In particular, the depth of the thawed layer grows with the square root of the time T_0, a result which would not have been obtained without the two-timing perturbation and without the expansions (4.11).

4.5 Remarks on stability analyses

Lineare stability analyses in fluid problems with solidification or liquefaction along a phase change boundary are usually performed with liquid solid interfaces in which the liquid obeys the Navier-Stokes equations, rather than the Darcy-Oberbeck-Boussinesq equations. Such analyses are by Mullins & Sekerka (1964), Coriell et al. (1980), Davis (1980), Davis et al. (1984) and Hurl et al. (1982, 1983), to name a few. A particular difficulty that is encountered in these problems is that the basic state (which may be subject to a bifurcation) is either not motionless, or its domain is time-dependent. This implies that a linear eigenvalue problem has time-dependent coefficients and would be difficult to solve. All authors known to us formulate problems for which the basic state solution is time independent. In Davis (1980) and Davis et al. (1984) melting arises because of convection and can be ignored or is null when the basic state is motionless. Hurl et al. (1983) in their crystal growth problem assume that the phase boundary in the basic state is plane and moves with constant speed and that the solid and liquid have their other boundaries infinitely far away. In a coordinate system moving with the phase boundary, the basic state is then obviously stationary. The scaling analysis of section 4.3 and the multiple time scale-perturbation procedure of section 4.4 was motivated by the desire to obtain a basic state which is motionsless, has a moving boundary, yet yields linear heat and salt profiles, for which the emerging double diffusion Bénard problem coupled with liquefaction is a linear eigenvalue problem with constant coefficients. This required the introduction of bulk and phase boundary velocity scales and the use of multiple variable perturbation expansions. The salt convective regime led to such simplified formulations and permits determination of instability conditions; for the other regimes situations are more difficult, Hutter & Straughan (1984).

Linear instability analyses or non-linear energy stability analyses along these lines are still missing, or those which are quoted (Harrison, 1982) are doubtful because they do not properly explore the role of the phase boundary. There is sufficient room for the specialists to move in!

5. DYNAMICS OF SNOW AND ICE AVALANCHES

5.1 An almanac of phenomena

Snow mechanics has many facets, and, despite intensive research, many of them are relatively unknown, both experimentally and theoretically. Conceptually rather unexplored areas within the global theme "snow at rest" are the hydrology of snow under varying climatic conditions (multi-component mixture thermodynamics with simultaneous solid-liquid, solid-gas, liquid-gas phase transitions) and rheological properties of snow under slow, creeping conditions. The study of the latter is important, because avalanche initiation depends on it. In the fracture initiation and fracture propagation mechanisms substratum weakening (depth hoar) due to thermodynamic metamorphosis is important, but existing constitutive models of dry snow, which are based on non-linear viscoelastic models, are still insufficient for prognostic purposes of failure (see Perla & La Chapelle, 1970; Brown et al., 1973); Desrues et al., 1980; Salm, 1982). This is one reason why the study of the motion of avalanches is important, if only in order to properly perform the avalanche-zoning in endangered areas.

The behavior of avalanches which form from glaciers, snow, rocks or soil has very little in common with the material from which they form. *Ice avalanches* develop from large ice masses (10^5 - 10^7 m^3) breaking off from glaciers. Upon hitting the ground, the few compact ice blocks generally disintegrate into a mass of smaller particles (Figure 15), which may be dispersed with a few larger blocks and move as a dense granular gravity flow down the mountain flank. Further particle fracturing during the motion is unknown, but probably much smaller than at the initial impact. Debris deposits indicate that grain diameter distribution is broad with a relatively large portion of particles in the millimeter range and a reasonable portion with the size of a fist (Figure 16). The majority of ice avalanches is small and moves as a compact mass of ice grains down the slide and spreads as it moves downhill. Only the biggest ones are believed to move as long approximately steady granular gravity currents. In areas of rapid flow their resisting forces are due to

a) Shot, just after breaking-off.

b) Shot, after hitting the ground. Note the substantial fracturing into particles.

Figure 15

Ice mass breaking-off from the tongue of the Festigletscher, Switzerland. *(From Alean, 1984)*

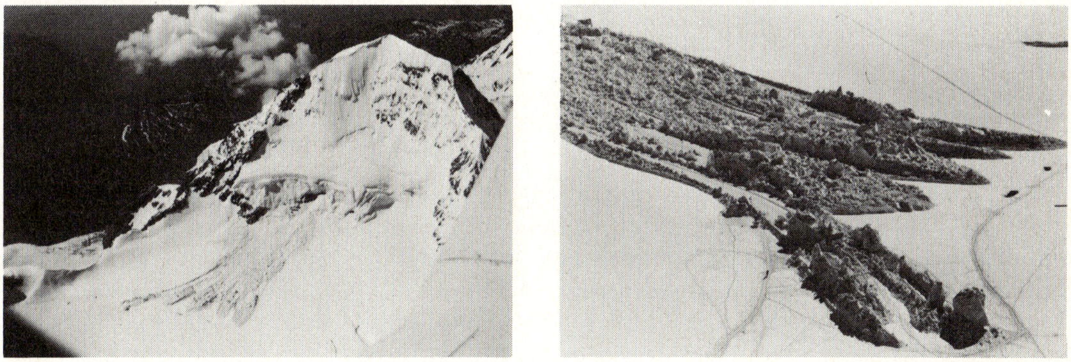

Figure 16a Ice avalanche deposit from a glacier breaking-off at the mountain Mönch, Switzerland, on 6 July 1984.

Figure 16b Close-up of the ice debris structure in the avalanche tongue.

(Photos taken 7 July by J. Alean, Laboratory of Hydraulics, Hydrology and Glaciology, ETH, Zürich)

two sources, basal sliding and particle collisions (called turbulence by avalanche specialists). Most likely, interstitial air plays an insignificant role. In the decelerating, settling zone the collision mechanism is replaced by internal friction. Larger portions of the avalanche may then move as nearly rigid bodies and slide as individual entities as i.e. in the frequent fingers of the tongue (Figure 17). Most of this is indirect evidence because only few ice avalanches have been observed (Alean, 1984; Heim, 1886; Roethlisberger, 1978).

As mentioned in the introduction, *snow avalanches* exist in two limiting forms; The motion close to the ground and following the ground contours of a snow mass of a large bulk density (0.1 - 0.4 g cm^{-3}) is called a *flow avalanche* (Figs.5,17). It de-

Figure 17

Deposit of a flow avalanche (Mühlebachlawine, 10 March 1970, Fieschtal, Switzerland). Note the granular structure of the snow and the many individual "fingers" in the zone where the avalanche came to rest. *(Photo: E. Wengi, Swiss Federal Institute of Snow and Avalanche Research, Weissfluhjoch-Davos, Switzerland)*

velops from relatively wet and settled snow after initial fracturing of gliding snow slabs into snow balls. The motion is a gravity driven shear flow of usually several layers of balls similar to that of ice avalanches described above; this can be inferred from avalanche deposits (Figure 18). The other limit is the motion of a snow cloud, consisting of a snow particle suspension of low bulk density ($\leq 10^{-2}$ g cm^{-3}) called air-borne powder snow avalanche, in short *powder snow avalanche* (Figure 6). In reality mixed forms do also exist, with all proportions of flowing and air borne parts. Powder snow avalanches always develop from flow avalanches but require dry snow for the transition to take place. For a small relatively compact snow mass the avalanche forms a buoyant cloud convecting down the incline akin to a thermal, for a continuous release source it behaves rather like a steady turbulent buoyant wall jet. Both cases and all intermediate stages occur in nature.

The above indicates that mathematical models of ice avalanches and flowing snow avalanches could be described by very similar models, while powder snow avalanches must be treated differently. This is indeed so, but the proposed models are still in the conjectural stage and must further be explored and be made available to the avalanche community. A large body of information on snow avalanches is contained in reviews of Mellor (1968, 1978), La Chapelle (1977), Perla (1980), Scheiwiller & Hutter (1982) and Hopfinger (1983). Of these, the latter two and Mellor (1978) are probably best suited for mechanicians. Ice avalanches have found little attention. A first systematic collection of observations and an attempt of interpretation is given by Alean (1984). Of general interest may also be a general text on rockslides and avalanches (Voight, 1978).

Figure 18 b

Deposit of a wet flow avalanche. *(Photo: Swiss Federal Institute of Snow and Avalanche Research, Weissfluhjoch-Davos, Switzerland)*

Figure 18 a

Snow deposit of a dry flow avalanche. *(Photo: E. Wengi, Swiss Federal Institute of Snow and Avalanche Research, Weissfluhjoch-Davos, Switzerland)*

5.2 Flow avalanches

Most of what will be said about flow avalanches also applies qualitatively to ice avalanches, but has so far not been tested in nature.

First theoretical models of avalanche motion are based on rigid body and open-channel hydraulic analogies. Voellmy (1955), Salm (1979), Perla et al. (1980) and Koerner (1976,1980,1983) deduce equations of motion for a snow mass sliding down an inclined plane surface. Frictional resistance is incorporated through (i) a basal drag, assumed proportional to the basal pressure and (ii) an internal, turbulent, drag proportional to the velocity squared. Differences in the various models lie in the handling of volume flux (mass balance) and changes of slope angles, but no clue as to which model would be superior to the other is available, mainly because running distances are almost the only observational facts against which the models can be tested. Scatter of data is naturally broad and agreement between model and observations depend on reliable knowledge of the drag coefficients.

Kulikowskii & Eglit (1973), Grigoryan & Ostroumov (1977) and Brugnot & Pochat (1981) use non-dimensional unsteady equations of mass and momentum balances by incorporating the hydrostatic pressure assumption, and thus arrive at model equations similar to the de Saint Venant equations

$$\frac{\partial \rho S}{\partial t} + \frac{\partial Q}{\partial x} = 0$$

$$\frac{\partial Q}{\partial t} + \frac{\partial (Q^2/\rho S)}{\partial x} + \rho g S \frac{\partial h}{\partial x} \cos\alpha = \rho g S\left(\sin\alpha - \mu\cos\alpha - \frac{f Q^2}{\rho^2 A^2 R}\right) \qquad (4.1)$$

$$S = \tilde{S}(h),$$

where ρ is density ($\sim 0.1 - 0.4$ g cm^{-3}), S the cross sectional area, $Q = \rho S U$ the mass flux, h the avalanche depth (measured from the talweg), α the (constant) talweg inclination, μ a dry friction and f a turbulent friction coefficient and R the hydraulic radius. $\tilde{S}(h)$ functionally relates S with h. The model is essentially the same as that of Voellmy-Salm-Perla, but mass balance is now properly treated. A difficult problem in these unsteady models is, however, the formulation of the jump condition at the front. Brugnot & Pochat (1981) use the classical hydraulic-jump frontal conditions by balancing mass and momentum across the frontal discontinuity in a frame moving with the front. They obtain reasonable variations in front velocity as long as topography changes smoothly, sudden changes in geometry, however, are not adequately predicted.

The major difficulty in flow avalanche forecast modelling lies in the lack of precise measurements of flow depths and velocities (both in plan view and at various depths). Gubler (1981, 1984) uses micro-wave radar systems capable of measuring speed (profiles) and flow heights. The system, able to perform several measuring cycles in one avalanche event, has been tested for dry snow avalanches, but data exploration and comparison with existing models is still under way.

Meanwhile, it was suggested that flow avalanches, and even more so ice avalanches, could well be modelled as a rapid shear flow of a cohesionless granular material (Scheiwiller & Hutter, 1982; Hopfinger, 1983). This analogy has led to a new and more detailed understanding (Bagnold, 1966; Savage, 1979; Savage & Jeffrey, 1981; Jenkins & Savage, 1983; Lun et al., 1984, and others) and offers the possibility of laboratory experiments. Steady rapid shear flow of glass beads down an inclined chute shows mean particle velocity and density profiles, qualitatively as shown in Figure 19a, untypical for Newtonian behavior. The variation of particle density with depth (< 0.7), in particular the relatively low density near the boundary corresponds to a variation of the effective angle of internal friction which will strongly affect the sliding and, as a result, absolute surface velocities (10-50 m s^{-1}). Savage measures a strong dependence of absolute velocities on the angle of inclination, which casts doubts on the correctness of the Manning-Gauckler-Strickler friction formula used in (4.1) and the Voellmy-Salm-Perla equations.

The theoretical models mentioned above are deduced from considerations of statistical mechanics in which shear deformation leads to nonisotropy of the particle distribution function. Equations of balance of mass, momentum and particle fluctuation energy for the mean motion are obtained, including expressions for the stress tensor, the flux of fluctuation energy and its annihilation by the non-elasticity of the binary collisons. In these equations, the balance law of fluctuation energy may be said to play the role of a mediator from the microscale to the mesoscale of mean behavior; it is similar to the higher order closure equations in the theory of turbulence and has the form of a heat equation. Independent field variables in a boundary value problem are therefore the particle density ν, the velocity field and the fluctuation energy Θ. The structure of the equations requires also a boundary condition for the fluctuation energy at each boundary, but these must insofar be compatible with the field equations as emerging boundary value problems should admit a mathematical solution. For instance, the stress tensor of the Jenkins-Savage (1983) model,

$$\underset{\sim}{t} = \underset{\sim}{t}^{P} = A(\nu,\sigma,e)\sqrt{\Theta}\left\{(\mathrm{tr}\,\underset{\sim}{D})\,\underset{\sim}{1} + 2\underset{\sim}{D}\right\}, \tag{4.2}$$

in which ν is the particle density, σ the particle diameter, e the coefficient of restitution and $\underset{\sim}{D}$ the stretching tensor, does not permit a solution of the steady plane chute flow when aerodynamic resistance at the upper surface is included, or when the no-slip condition at the base is applied, see Hutter et al., (1984). For vanishing aerodynamic drag and viscous sliding condition at the base these authors show, however, that for nearly elastic collisions the particle distribution func-

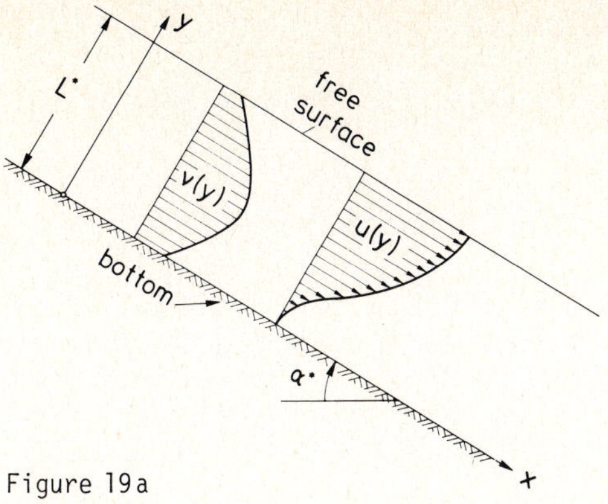

Figure 19a

Steady, gravity driven flow of a layer of granular material. Typical velocity profile and density distribution as surmised by the literature.

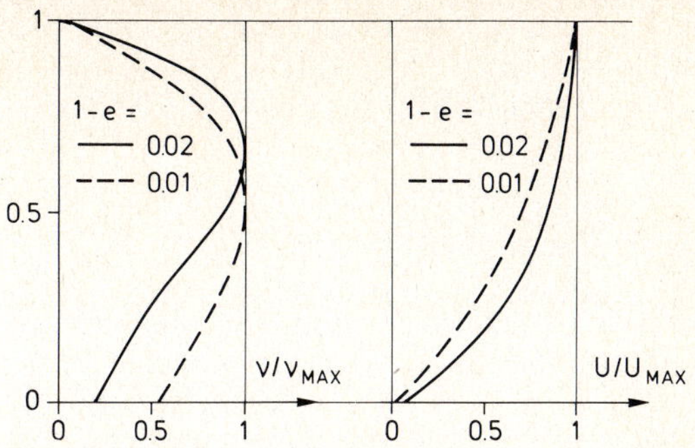

Figure 19b

Selective results of density distribution and velocity profiles for a steady granular chute flow using the Jenkins - Savage model. The two different results are for two different values of the coefficient of restitution.

tion and the velocity distribution is as shown in Figure 19b, in qualitative areement with the measured findings. On the other hand, the application of the model to the situation when there is a non-vanishing aerodynamic drag or when the no slip condition is applied requires an alteration of the stress tensor in the form

$$\underset{\sim}{t} = \underset{\sim}{t}^P + \chi \, \nabla\nu \otimes \nabla\nu , \qquad \chi = \text{constant} \tag{4.3}$$

The dyadic product of the gradient of the particle distribution function has already appeared in earlier theories (Goodman & Cowin, 1972) and it implies that second order spatial derivatives of the particle distribution function enter the field equations (through $\operatorname{div} \underset{\sim}{t}$). Because such higher derivatives of ν had not entered the equations of the Jenkins-Savage model, a stress tensor (4.3) requires one additional boundary condition in any mathematically well set boundary value problem, but such a condition is physically not transparent. An amendment of the Jenkins-Savage model is also needed because a stress-free upper boundary of a chute flow always yields $\nu = kz$, where z is the distance from the free surface (see Figure 19b). By contrast, a stress tensor (4.3) permits a finite nonzero particle density at the free surface, which is physically more realistic. Preliminary results for a steady chute flow are given by Hutter et al. (1984). No similar exploitation of the other extended statistical theories is known to us. Rather, approximate "solutions" to the governing equations have been constructed for situations for which the exact equations admit no solutions (see i.e. Jenkins & Savage, 1983; Savage, 1982, 1983).

This discussion transpires the following conclusions:

(i) The analogy between gravity driven granular flow and flow of dense flow avalanches must be experimentally substantiated.

(ii) The beauty of the statistical models describing rapid shear of granular materials is partly destroyed by unceraintes in the formulation of boundary conditions.

(iii) The dynamic models of rapid flow of granular materials must be explored with regard to various chute flow problems and the emerging results for flow velocities and particle distribution must be compared with findings from laboratory experiments.

(iv) The existing models should also be tested under conditions of variable topography and with varying total mass of the avalanche

(v) Experiments in the field and the laboratory are as important as theoretical considerations.

5.3 Powder snow avalanches

Powder snow avalanches are much larger and faster than flow avalanches with heights up to 50 m and velocities in the order of 100 to 150 m s^{-1}. Even though their density (ρ = 0.05 - 0.1 g cm^3) is small as compared to that of flow avalanches, powder snow avalanches are more damaging. The reason is the substantially enhanced kinetic energy that is involved and can develop because basal friction is negligible.

A powder snow avalanche (Figure 6) is essentially a turbidity current flowing down the mountain flank. It develops from flow avalanches by an abrupt transition in which particles suddenly become air-borne. A dramatic volume and density change is associated with the transition, which is achieved by entrainment of air through the free upper surface. The mechanism of cloud formation for powder snow avalanches is still not clearly understood. Hopfinger (1983) conjectures a kinematic-dynamic shock that forms; Hutter & Scheiwiller (1984) regard the formation as an instability mechanism that is triggered by transverse density gradients and transverse shear akin to the Kelvin-Helmholtz type instabilities. Because air content is crucial, a proper granular model for the prediction of the instability must be a mixture theory of granules and air. Given the difficulties that exist with granular materials excluding the role of the interstitial fluid it is no surprise that no stability analysis has been performed so far. The cause for powder snow avalanche formation is, therefore, still doubtful.

Theoretical models for powder snow avalanches are by Voellmy (1955), Tochon-Danguy (1977), Béghin (1979) and Béghin et al. (1981). Voellmy employs the same channel hydraulic equations as for flow avalanches by adjusting the friction coefficients to the larger turbulence. Tochon-Danguy (see also Hopfinger, 1983) applies statements of integrated mass and momentum balance for the air-snow mixture. He incorporates an entrainment hypothesis for air suggested by Ellison & Turner (1959) (entrainment equals longitudinal depth gradient) and so accounts for the substantial growth of air mass along the avalanche. He distinguishes between two limiting forms of avalanches. One is the motion of a wall-near jet from a steady source of mass. The corresponding avalanche consists of a head at the front and a steady tail or body (Figure 20). The other is the motion of a single buoyant mass (only the head in Figure 20) and emerges from an impulsive release of a concentrated snow mass. He deduces a relationship for the asymptotic body velocity. When basal friction can be ignored (steep slopes) one has

$$U^2 = c' \frac{g' h \sin \alpha}{E_a}, \quad g' = \frac{\rho - \rho_a}{\rho_a} g, \tag{4.4}$$

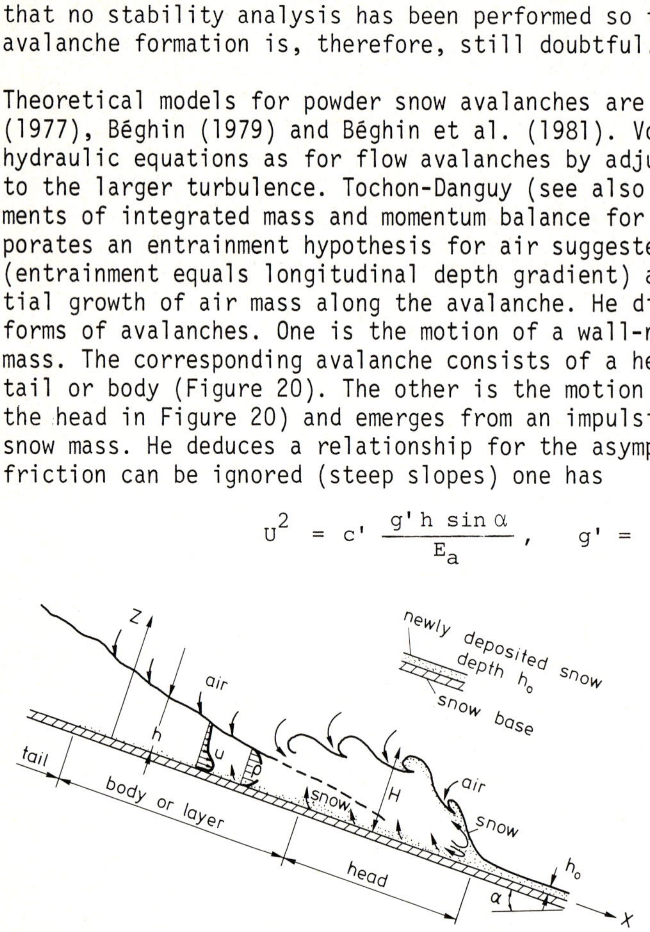

Figure 20

Sketch of a powder snow avalanche, indicating the body and head flow and the snow and air entrainment mechanisms. *(After Hopfinger, 1983)*

in which ρ is the mean bulk density, ρ_a the density of air, $E_a = dh/dx$ = constant and c' a constant, and the remaining quantities are defined in Figure 20. Thus, the mean avalanche body velocity is proportional to $[g' h \sin\alpha]^{1/2}$, typical for dense gravity flow, and inversely proportional to $E_a^{1/2}$. Entrainment effectively serves as a resistance. Tochon-Danguy uses arguments of dimensional analysis to obtain relationships for the frontal velocity for both long and short gravity flow. For Boussinesq conditions Britter & Linden (1980) obtain

$$U_f = (1.5 \pm 0.2)(g_0' Q_0)^{1/3}, \quad \text{long gravity flow} \tag{4.5}$$

in which the index "0" denotes a reference position and Q_0 is the mass flux.

A more detailed analysis of short gravity flows is given by Béghin (1979). He applies mass and momentum balances to a concentrated buoyant mass and combines these with an entrainment hypothesis for air and various ad-hoc assumptions which allow him to find expressions for the asymtotic velocity of the center of gravity of the concentrated mass, that of the front in terms of varying conditions of air entrainment, density differences, slope angle, snow entrainment from below etc.

The most important mechanism a model of powder snow avalanches must incorporate is *snow settlement*, because it determines running distances that affect avalanche zoning. Neither Tochon-Danguy nor Béghin can deal with it. Savage (in a presentation at EUROMECH 172: Mechanis of glaciers) extended Tochon-Danguy's "inclined thermal model" by looking at a cloud moving down a slope, in which the sedimentation of the particles was accounted for by a mass balance of the settling particles. Using this, an entrainment hypothesis for the ambient fluid and the integrated momentum equation, predicted flows compared favorably with laboratory experiments, but we have not seen any published results.

A more complete model that incorporates snow settlements is a two phase turbulence flow model in which air and snow particles are separate entities carrying their own mass and momentum, see Scheiwiller & Hutter (1982) and Scheiwiller (1984). Because snow and air are capable of having different velocities snow settlement is incorporated ab initio. On the basis of incompressible grains, balance laws of mass and momentum for each constituent and assumptions on drag resistance and turbulence (k-ε model for the air phase), field equations are obtained for the particle concentration and the peculiar velocities. Boundary conditions at the interface avalanche-air involve a kinematic condition and jump conditions of mass and momentum for each component and reveal an evolution equation of the avalanche height in which air entrainment enters as a source term. In a similar way the bottom boundary condition can be handled and if desired, snow entrainment be incorporated. A boundary layer approximation for plane flows down an inclined plane allows simplification of the governing equations. Scheiwiller uses two separate numerical methods (FD versus Kantorovich) by which the boundary layer equations are integrated. Results (density profiles, velocity distributions) look promising but must await adjustments of free phenomenological coefficients by experiments.

Experimental study of the flow of powder snow avalanches is impossible in nature, so laboratory experiments are a necessity. This was already recognised by Tochon-Danguy and Béghin, who calibrated their density current models by performing such experiments. Scheiwiller (1984) uses plastic particles (diameter ~ 200 - 400 μ) suspended in still water and succeeds in producing a turbidity current along an inclined chute which strongly resembles the typical features of powder snow avalanches, Figure 21. His experiments are performed in a 4 m deep tank which allows variation of chute inclination ($0 \le \alpha \le 90^o$). Ultrasonic techniques are used to measure mean profiles of particle density and particle velocities at individual cross sections and water velocities are estimated from dye insertion. Experiments of this kind are vital for the calibration of the turbulent two-phase model, but

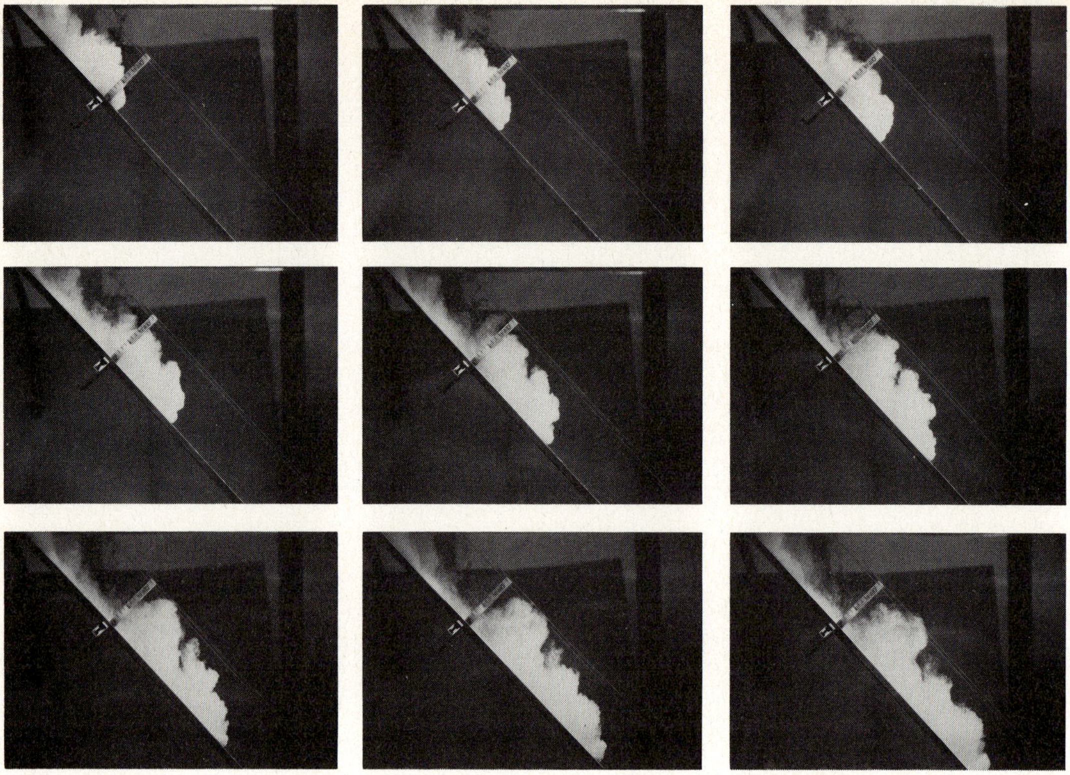

Figure 21

A sequence of shots of a laboratory powder snow avalanche consisting of water borne plastic particles moving down a submerged chute. *(Photo and experiment: T. Scheiwiller, Laboratory of Hydraulics, Hydrology and Glaciology, ETH, Zürich)*

this calibration, provided it is successful, will yield a model of a laboratory powder snow avalanche. There are unknown problems of scale effects (i.e. non-Boussinesq behavior) but this deficiency must be accepted in favor of at least a more detailed understanding of the physics of the processes.

We may summarize:

(i) Existing models based on turbidity current diffusion are inadequate for powder snow avalanche prediction because they ignore snow settlement.

(ii) Two-phase flow models which treat the particle and air phase separately are promising but still require calibration by laboratory experiments.

(iii) Experimental techniques, adequate for determining profiles of particle density and peculiar velocities, are presently being developed but need more emphasis to achieve advancement. Successful understanding of powder snow avalanches hinges on the fact how much support is given to investigations such as those mentioned in items (ii) and (iii).

ACKNOWLEDGEMENTS

The writing of this review has profited from many individuals and agencies, both directly and indirectly. We thank M. Aellen, J. Alean, T. Scheiwiller and B. Salm for providing us with photographs on glaciers and snow and ice avalanches.
B. Straughan permitted us to include results, he and K. Hutter deduced elsewhere,

and F. Szidarovszky and S. Yakowitz influenced our writings on numerical procedures in the glacier and ice sheet and the avalanche sections.

The work of K. Hutter was partly supported by the U.S. National Science Foundation through Contract: DPP 8219439. "Numerical Solutions of Dynamical Problems Concerning Cold and Temperate Ice Sheets" and by the British Science and Engineering Council (via a fellowship to Glasgow to study Thawing of Subsea Permafrost). The work of T. Alts was wholly supported by the Swiss National Science Foundation through Contract: 2.526-0.82. "Mixture Theory of Polythermal Ice". Finally we thank Prof. D. Vischer, Director of the Laboratory of Hydraulics, Hydrology and Glaciology for the continuing support of this work.

References

Ahlmann, H.W., 1947. Foreward to J. Glaciology, 1.

Alean, J.C., 1984. Untersuchungen über Entstehungsbedingungen und Reichweiten von Eislawinen. Abhandlung zur Erlangung des Titels eines Doktors der Naturwissenschaften der Eidg. Technischen Hochschule, Zürich

Alts, T. and Hutter, K., 1984: Towards a Theory of Temperate Glaciers. Part I: Dynamics and Thermodynamics of Phase Boundaries between Ice and Water. To be published.

Aris, R., 1962: Vectors, Tensors and Basic Equations of Fluid Mechanics, Prentice-Hall, Englewood Cliffs, N.J.

Bagnold, R.A., 1966. The shearing and dilatation of dry sand and the "singing" mechanisms. Proc. R. Soc. London Ser. A 295: 219-32

Béghin, P., 1979. Etude des bouffées bidimensionelles de densité en écoulement sur pente avec application aux avalanches de neige poudreuse. Thèse, Grenoble

Béghin, P., Hopfinger E.J. and Britter R.E., 1981. Gravitational convection from instantaneous sources on inclined boundaries. J. Fluid Mech. 107: 407-422

Björnsson, H., 1974. Explanation of Jökulhlamps from Grimsvötn, Vatnjaökull, Iceland. Jökull 24: 1-26

Björnsson, H., 1975. Subglacial water reservoirs, Jökulhlamps and volcanic eruptions. Jökull 25: 1-14

Boulton, G.S., Smith, G.D. and Morland, L.W., 1983. The reconstruction of former ice sheets and their mass balance characteristics using a nonlinearly viscous flow model. J. Glaciol. (in the press).

Bowen, R.M. and Wang C.-C., 1976. Introduction to Vectors and Tensors. Vol. 1: Linear and Multilinear Algebra. Vol. 2: Vector and Tensor Analysis. Plenum Press, New York, London

Britter, R.E. and Linden, P.F., 1980. The motion of the front of a gravity current travelling down an incline. J. Fluid Mech. 99: 531-43

Brown, R.L., Lang, T.E., St. Lawrence W.F. and Bradley, C.C., 1973. A failure criterion for snow. J. Geophys. Res. 78: 4950-58

Brugnot, G. and Pochat, R., 1981. Numerical simulation study of avalanches. J. Glaciol. 27: 77-88

Budd, W.F. and Radok, U., 1971. Glaciers and other large ice massers. Rep. Prog. Phys., 34: 1-70

Buff, F.P., 1956. Curved fluid interfaces. I. The generalized Gibbs-Kelvin equation. J. of Chem. Phys. 25: 146-153

Buff, F.P. and Saltsburg, H., 1957a. Curved fluid interfaces II. The generalized Neumann formula. J. of Chem. Phys. 26: 23-31

Buff, F.P. and Saltsburg, H., 1957b. Curved fluid interfaces III. The dependence of the free energy on paramters of external force. J. Chem. Phys. 26: 1526-1533

Buyanov S.A., 1983. Numerical modelling of ice sheet-evolution specifics. In: In Problems of mechanics of natural processes. M., Moscow University Publishers: 139-154 (in Russian)

Coriell, S.R., Cordes, M.R., Boottinger, W.J. and Sekerka, R.F., 1980. Convective and interfacial instabilities during unidirectional: solidification of a binary alloy. J. Crystal Growth, 49 (1982), 13-28

Covey, C., 1984. The earth's orbit and the ice ages. Scientific American, 84: 42-50

Davis, S.H., 1980. Energy stability theory for free-surface problems, buoyancy-thermo-capillary layers. J. Fluid. Mech. 98: 527-553

Davis, S.H., Müller, V. and Dietschi, C., 1983. Pattern selection in single component systems coupling Bernard convection and solidification. Report KfK 3641 Kernforschungszentrum Karlsruhe.

Deemer, A.D. and Slattery, J.C., 1978. Balance equations and structural models for phase interfaces. Int. J. Multiphase Flow 4: 171-197

Drew, D.A., 1983. Mathematical modeling of two-phase flow. Ann. Rev. Fluid Mech. 15: 261-291

Drewry, D.J., (editor) 1983. Antarctica. Glaciological and geophysical folio. Scott Polar Research Institut, University of Cambridge

Ellison, T.H. and Turner, J.S., 1959. Turbulent entrainment in stratified flow. J. Fluid Mech. 6: 423-48

Fowler, A.C., 1979. The use of a rational model in the mathematical analysis of polythermal glacier. J. Glaciology, 24: 443-456

Fowler, A.C., 1981. A theoretical treatment of the sliding of glaciers in the absence of cavitation. Phil. Trans. R. Soc. London. Ser A 298 (1445): 637-685

Fowler, A.C., 1982. Waves on glaciers. J. Fluid Mech., 120: 283-321

Fowler, A.C., 1984a. On the transport of moisture in polythermal glaciers. Geophys. Astrophys. Fluid Dynamics, 28: 99-140

Fowler, A.C., 1984b. A sliding law for temperate glaciers in the presence of subglacial cavitation. Proc. Royal Soc. London (to appear)

Fowler, A.C. and Larson, D.A., 1978. On the flow of polythermal glaciers. I. Model and preliminary analysis. Proc. Roy. Soc. A, 363: 217-242

Fowler, A.C. and Larson, D.A., 1980. The uniqueness of steady state flows of glaciers and ice sheets. Gephys. J. Roy. Astr. Soc., 63: 333-345

Gibbs, J.W., 1928. The collected work of J. Willard Gibbs. Vol. 1, p. 219, Yale University Press, New Haven

Gibbs, J.W., 1948. Collected works of J. Williard Gibbs. Voll I, pp. 55-353, Yale University Press, New Haven

Glen, J.W., 1955. The creep of polycrystalline ice. Proc. R. Soc. Lond. A 228: 515-538

Golecki, I. and Jaccard, C., 1977. The surface of ice near 0 Grad C Studied by 100 keV proton channeling. Physics Letters 63A: 374-376

Goodmann, M.A. and Cowin S.C., 1972. A continuum theory for granular materials. Arch. Rational Mech. Anal. 44: (4), 249-266

Grigoryan, S.S., Ostroumov, A.V., 1977. Mathematical simulation of the process of motion of a snow avalanche (summary only). J. Glaciol. 19: 664-65

Grigoryan, S.S., Buyanov, S.A., Krass, M.S. and Shumsky, P.A., 1984. The mathematical model of ice sheets and the calculation of the evolution of the Greenland ice sheet. J. Glaciology (submitted)

Gubler, H.U., 1981. Messungen an Fliesslawinen, EISLF Rep. No. 600, Davos, Switz.

Gubler, H.U., 1982. Strength of Bonds between ice grains after short contact times. J. of Glaciology 28: 457-473

Gubler, H.U., 1984. The use of microwave FMCW radar in snow and avalanche research. Cold Reg. Sci. Tech. (to appear)

Gurtin, M.E. and Murdoch, A.I., 1975. A continuum theory of elastic material surfaces. Arch. Rat. Mech. Anal. 57: 291-323

Hallet, B., 1981. Glacier abrasion and sliding: their dependence on the debris concentration in basal ice. Annals of Glaciology, 2: 23-28

Harrison, W.D., 1982. Formulation of a model for pore water convection in thawing subsea permafrost. Mitteilung No. 57 der Versuchsanstalt für Wasserbau, Hydrologie und Glaziologie an der ETH Zürich

Harrison, W.D. and Osterkamp, T.E., 1978. Heat and mass transport processes in subsea permafrost: 1 An analysis of molecular diffusion and its consequences. J. Geophys. Res. 83: 4707-4712

Harrison, W.D. and Osterkamp, T.E., 1982. Measurements of electrical conductivity of interstitial water in subsea permafrost. H. French, ed. Porceedings of the Fourth Canadian Permafrost Conference: Ottawa, National Research Council of Canada: 229-237

Heim, A., 1896. Die Gletscherlawine an der Altels am 11. September 1895. 98. Neujahrsblatt der Zürcherischen Naturforschenden Gesellschaft auf das Jahr 1896

Hobbs, P.V., 1974. Ice Phycics. Clarendon Press, Oxford

Hopfinger, E.J., 1983. Snow avalanche motion and related phenomena. Ann. Rev. Fluid. Mech. 15: 47-76

Hurle, D.T.J., Jakeman, E. and Wheeler, A.A., 1982. Effect of solutal convection on the morphological stability of a binary alloy. J. Crystal Growth, 58: 163-179

Hurle, D.T.J., Jakeman, E. and Wheeler, A.A., 1983. Hydrodynamics stability of the melt during solidification of a binary alloy. Phys. Fluids, 26 (3): 624-626

Hutter, K., 1981. The effect of longitudinal strain on the shear stress of an ice sheet. In defense of using stretched coordinates. J. Glaciology, 27: 39-56

Hutter, K., 1982a. Dynamics of glaciers and large ice masses. Ann. Rev. Fluid Mech. 14: 87-130

Hutter, K., 1982b. A mathematical model of polythermal glaciers and ice sheets. Geophys. Astrophys. Fluid Dynamics. 21: 201-224

Hutter, K., 1982c. Glacier flow. Am. Sci. 70, 1: 26-34

Hutter, K., 1983. Theoretical Glaciology. Reidel, Dordrecht

Hutter, K. and Straughan B., 1984. Convection in thawing subsea permafrost. Manuscript in preparation.

Hutter, K. and Scheiwiller, T., 1984. Flow avalanches treated as a binary mixture of granules and air. (Pending publication)

Hutter, K., Szidarovszky, F. and Yakowitz, S., 1984. Wet-snow-flow avalanches treated as a plane steady shear flow of a cohesionless granular material down an inclined plane (in preparation)

Iken, A., 1981. The effect of the subglacial water pressuere on the sliding velocity of a glacier in an idealized numerical model. J. Glaciology, 27 (97): 407-421

Iken, A., Röthlisberger, H., Flotran, A. and Haeberli, W., 1983. The uplift of Unteraargletscher at the beginning of the melt season - a consequence of water storage at the bed. J. Glaciology, 29 (101): 28-47

Ishii, M., 1975. Thermo-fluid dynamic theory of two-phase flow. Direction des Etudes et Recherches d'Electricité de France, Eyrolles, Paris

Jenkins, J.T. and Savage, S.B., 1983. A theory for the rapid flow of identical smooth, nearly elastic spherical particles. J. Fluid Mech. 130: 187-202

Johnsons, R.E. and McMeeking, R.M., 1982. Near-surface flow in glaciers obeying Glen's law. Univ. Illinois T & AM Rep. 454

Joseph, D.D., 1976. Stability of Fluid Motions. Vol I, II, Springer Verlag, Berlin, New York, Heidelberg

Körner, H.J., 1976. Reichweite und Geschwindigkeit von Bergstürzen und Fliessschneelawinen. Rock Mechanics, 8: 225-256.

Körner, H.J., 1980. Modelle zur Berechnung der Bergsturz- und Lawinenbewegung. Interpraevent, Bad Ischl, Vol. II: 15-55

Körner, H.J., 1983. Zur Mechanik der Bergsturzströme vom Huascaran, Peru. In Patzelt, G. (Herausgeber): Die Berg- und Gletscherstürze von Huascaran, Cardittera Blanca, Peru. Hochgebirgsforschung Heft 6, Universitätsverlag Wagner, Innsbruck: 71-100

Krass, M.S., 1981. Mathematical models and numerical modelling in glaciology. M., Moscow State University Publishers. (Russian)

Krass, M.S., 1983. Mathematical theory of glaciomechanics. Glaciology, vol. 3 (Results of sciences and technology. VINITI. AN SSSR) M. (in Russian)

Kulikovskii, A.G. and Eglit, M.E., 1973. Two-dimensional problem of the motion along a slope with smoothly changing properties. PMM J. Appl. Math. Mech. 37: 792-803. Transl. from Prikl. Mat. Mekh. 37: 837-48

La Chapelle, E.R., 1977. Snow avalanches: a review of current research and applications. J. Glaciol. 19: 313-24

Lindsay, K.A. and Straughan, B., 1979. A thermodynamic viscons interface theory and associated stability problems. Arch. Rat. Mech. Anal. 71, 307-326

Liu, I-Shih, 1972: Method of Lagrange multipliers for exploitation of the entropy principle. Arch. Rat. Mech. Anal. 46: 131-148

Lliboutry, L.A., 1976. Physical processes in temperate glaciers. J. Glaciology, 16: 151-158

Lliboutry, L.A., 1979. Local friction laws for glaciers: A critical review and new openings. J. Glaciology, 23: 67-96

Lun, C.K.K., Savage, S.B., Jeffrey, D.J. and Chepurniy, N., 1984. Kinetic theories for granular flow: Inelastic particles in Couette flow and sligthly inelastic particles in a general flow field. J. Fluid Mech. 140: 223-256

Lun, C.K.K. and Savage, S.B., 1984. A simple binary theory for granular flow of rough, inelastic spherical particles. J. Fluid Mech. (in press)

Mathews, W.H., 1973. Record of two Jökulhlamps. Symposium on the Hydrology of Glaciers, Cambridge, 1969. Publ. No. 95, International Association of Scientific Hydrology: 99-110

Meier, M.F. and Post, A., 1969. What are surges? Can. J. Earth Sci. 6: 807-816

Mellor, M., 1968. Avalanches Cold Regions Sci. Eng. Monogr. III-A3d. Hannover, N. H.: US Army Cold Reg. Res. Eng. Lab. 215 pp.

Mellor, M., 1977. Engineering properties of snow. J. Glaciol. 19: 15-65

Mellor, M., 1978. Dynamics of snow avalanches. In Rockslides and Avalanches. I. Natural Phenomena, ed. B. Voight, Amsterdam, Elsevier: 753-92.

Moeckel, G.P., 1974. Thermodynamics of an Interface. Arch. Rat. Mech. Anal. 57, 255-280

Morland, L.W., 1984. Thermomechanical balanches of ice sheet flows. Geophys. Artrophys. Fluid Dynamics: (to appear)

Morland, L.W. and Johnson, I.R., 1980. Steady motion of ice sheets. J. Glaciology, 25: 229-246

Morland, L.W. and Johnson, I.R., 1982. Effects of bed inclination and topography on steady isothermal ice sheets. J. Glaciology, 28: 71-90

Morland, L.W. and Shoemaker, E.M., 1982. Ice shelf balances. Cold Reg. Sci. Tech. 5: 235-251

Morland, L.W. and Smith, G.D., 1984. Influence of non-uniform temperature distribution on the steady motion of ice sheets. J. Fluid Mech. 140: 113-930

Morland, L.W., Smith, G.D. and Boulton, G.S., 1984. Basal sliding relations deduced from ice sheet data. J. Glaciology (in press)

Müller, I., 1973: Thermodynamik-Grundlagen der Materialtheorie. Bertelsmann Universitätsverlag, Düsseldorf, pp 232

Mullins, W.W. and Sekerka, R.F., 1964. Stability of a planar interface during solidification of a dilute binary alloy. J. Appl. Phys., 35 (2): 444-451

Nakaya, U. and Matsumoto, A., 1954. Simple experiment showing the existence of "liquid water" film on the ice surface. J. of Colloid Sci. 9: 41-49

Nye, J.F., 1976. Water flow in glaciers: Jökulhlamps, tunnels and veins. J. Glaciology, 17 (76): 181-207

Nye, J.F. and Frank, F.C., 1973. Hydrology of the intergranular veins in a temperate glacier. Symposium on the Hydrology of Cambridge 1969. Publ. No. 95, International Association of Scientific Hydrology

Osterkamp, T.E. and Hareison, W.D., 1982. Temperature measurements in subsea permafrost of the coast of Alaska. H. French, ed. Proceedings of the Fourth Canadian Permafrost Conference: Ottawa, National Research Council of Canada: 238-248

Paterson, W.S.B., 1980a. The Physics of Glaciers. Pergamon, Oxford pp 380

Paterson, W.S.B., 1980b. Ice sheets and ice shelves. In Dynamics of Snow and Ice Masses (ed. S.C. Colbeck). Academic, New York: 1-78

Perla, R.I., 1980. Avalanche release, motion, and impact. In Dynamics of Snow and Ice Masses, ed. S.C. Colbeck. Academic, New York: 397-456, New York: Academic

Perla, R.I. and La Chapelle, E.R., 1970. A theory of snow slab failure. J. Geophys. Res. 75: 7619-27

Perla, R., Cheng, T.T. and McClung, M., 1980. A two-parameter model of snow avalanche motion. J. of Glaciology, 26, (94): 197-207

Robin, G. de Q., 1979. Formation, flow and disintegration of ice shelves. J. Glaciology, 24: 259-271

Robin, G. de Q. and Weertman, J., 1973. Cyclic surging glaciers. J. Glaciology 12: 3-18

Röthlisberger, H., 1972. Water pressure in intraglacial and subglacial channels. J. Glaciology, 11: (62): 177-203

Röthlisberger, H., 1978. Eislawinen und Ausbrüche von Gletscherseen. Jahrbuch der Schweizerischen Naturforschenden Gesellschaft, wissenschaftlicher Teil: 170-212

Savage, S.B., 1982. Granular flows at high shear rates. In: R.E. Meyer (Ed.) Theory of Dispersed Multiphase Flow, Academic Press, New York

Savage S.B., 1983. Granular flows down rough inclines-review and extension. In: Mechanics of Granular Materials: New Models and Constitutive Relations, edited by J.T. Jenkins and M. Satake, Elsevier, Amsterdam: 261-282

Savage S.B. and Jeffrey, D.J., 1981. The stress tensor in an granular flow at high shear rates. J. Fluid Mech. 110: 255-272

Scriven, L.E., 1960. Dynamics of a fluid interface. Chem. Engng. Sci. 12: 98-108

Sellmann, P.V., 1980. Regional distribution and characteristics of bottom sediments of arctic coastal waters of Alaska. Review of current literature: U.S. Army Cold Regions Research and Engineering Laboratory. Special Report 80-19

Scheiwiller, T., 1984. Binary mixture models of powder snow avalanches, theory and experiments (working title of a forthcoming Ph.D-dissertation)

Scheiwiller, T. and Hutter, K., 1982. Lawinendynamik, Uebersicht über Experimente und theoretische Modelle von Fliess- und Staublawinen. Mitteilung Nr. 58 der Versuchsanstalt für Wasserbau, Hydrologie und Glaziologie an der ETH Zürich

Shreve, R.L., 1972. Movements of water in glaciers. J. Glaciology, 11, (26): 205-214

Simpson, J.E., 1982. Gravity currents in the laboratory, atmosphere and ocean. Ann. Rev. Fluid Mech. 14:213-34

Slattery, J.C., 1967. General Balance Equation for a Phase Interface. I&EC Fundamentals 6: 108-115

Smith, G.D. and Morland, L.W., 1981. Viscous relations for the steady creep of polycrystalline ice. Cold Reg. Sci. Tech., 5: 141-150

Smith, G.F., 1965. On Isotropic Integrity Basis. Arch. Rat. Mech. Anal. 18: 282-292

Spring, U. and Hutter, K., 1981. Numerical studies of Jökulhlanps. Cold Regions Sci. Tech. 4: 227-244

Spring, U. and Hutter, K., 1982. Conduit flow of a fluid through its solid phase and its application to intraglacial channel flow. Int. J. Eng. Sci, 20: 327-363

Swift, D.W., Harrison, W.D. and Osterkamp, T.E., 1984. Heat and salt transport processes in thawing subsea permafrost at Prudhoe Bay, Alaska. Proceedings of the Fourth International Conference on Permafrost, Fairbanks, Alaska (in press)

Swift, D.W. and Harrison, W.D., 1984. Convective transport of brine and thaw of subsea permafrost: Results of numerical simulations. J. Geophys. Res. (in press)

Steinemann, S., 1958. Experimentelle Untersuchungen zur Plastizität Eis. Beiträge zur Geologie der Schweiz, Hydrologie No. 10.

Tochon-Danguy, J.C., 1977. Etude des courants de gravité sur forte pente avec application aux avalanches poudreuses. Thèse, Grenoble

Thomas, R.H., 1979. Ice shelves: a review. J. Glaciology, 24: 273-286

Verbitsky M.Ya., Govorukha, L.S. Thermohydrodynamic calculations of the modern regime of the Vavilov glacial dome in Severnaya Zemlya. In: Materials of glaciological researches. Chronicles, discussions, issue 36, M., 1979, pp. 81-86 (in Russian)

Verbitsky M.Ya. and Chalikov D.V., 1980. Thermohydrodynamic model of a large ice sheet. Diklady AN SSSR. 250, (1): 52-62 (in Russian)

Verbitsky M.Ya., 1981. Numerical modelling of sheet glaciation evolution. Doklady AN SSSR. 256, (6): 1333-1337 (in Russian)

Verbitsky M.Ya., 1981. Numerical experiments in hydrodynamics of the Eastern Antarctic and Greenland glacial shields. In: Materials of glaciological researches. Chronicles, discussions, issue 40, M: 51-58 (in Russian)

Voellmy, A., 1955. Ueber die Zerstörungskraft von Lawinen. Schweiz. Bauz. 73, (12): 159-62; (15): 212-17; (17): 246-49; (19): 280-85. Transl., 1964, in USDA Forest Serv. Alta Avalanche Study Cent. Transl. No. 2

Voight, B. (Ed.), 1978. Rockslides and Avalanches. I Natural Phenomena. Elsevier Amsterdam

Weertmann, J., 1979. The unsolved general glacier sliding problem. J. Glaciology 23: 97-115

Yakowitz, S., Hutter K. and Szidarovszky, F., 1984. Toward computation of steady-state profiles of ice sheets. Z. für Gletscherkunde (in press)

Theoretical and Applied Mechanics
F.I. Niordson and N. Olhoff
Elsevier Science Publishers B.V. (North-Holland)

NONLINEAR WAVE MOTION AND HOMOCLINIC BIFURCATION

Klaus Kirchgässner

Universität Stuttgart

Dedicated to Professor Dr.Dr.h.c. Henry Görtler on the occasion of his 75th. anniversary.

Research supported by the Deutsche Forschungsgemeinschaft under Nr. Ki 131/3-1

Internal- and surface solitary waves can be understood as homoclinic orbits in infinite-dimensional space. The influence of localized and periodic external forces is studied for different physical models. The mathematical method used leads to a low order ordinary differential equation describing all steady flows of small but finite amplitude.

1. Introduction.

In recent years it has been discovered that steady nonlinear water waves as described by the full Eulerian equations are amenable to a rigorous geometric interpratation [11]. Not only does this point of view shed new light on old and well understood problems, but it leads also the way to solving problems of current interest such as the influence of external forces on solitary waves. The main mathematical insight can be condensed into the following points: Solitary waves are homoclinic orbits in function spaces, for small amplitudes, all solutions lie on a low-dimensional manifold and thus can be described by an ordinary differential equation of correspondingly low order. External forces may break these homoclinic orbits and transverse homoclinic points can appear. Thus chaotic solution behavior exists for the steady Eulerian equations.

In order to outline the method we will treat a sequence of problems of increasing complexity but also probably of increasing interest. We restrict our consideration totally to solitary waves and their behavior under external forcing and to the analogous problem of the flow through a channel of varying density. We impose external forces of local- and of global nature (e.g. pressure distributions having compact support or being period in space). A lot could be said about the behavior in the large of the solutions obtained; but a comprehensive analysis would break the frame of this survey. Therefore we advise the interested reader to [2], [4].

The theory of nonlinear dispersive water waves has a long and glamorous history and its main results carry names of great renown. We mention some of these results when they appear in the analysis outlined here. In recent years the interest in the long standing open questions have revived under the influence of new powerful mathematical tools and thus finally, the Stokes conjecture for surface waves of extreme form has been finally resolved in a series of beautiful papers by Amick, Fraenkel, Mc Leod and Toland.

In this contribution we cannot give the sometimes quite involved details of the mathematical reasoning. The interested reader is therefore referred to the original literature or to a forthcoming extended survey [12]. Here we strive for a comprehensive introduction into the methods and an outline of its various consequences. We first treat the case of internal waves in a density stratified inviscid fluid under the influence of gravity. External forces are added when the fluid is conducting and a magnetic field is imposed. The situation then is quite similar to the steady flow through a two-dimensional channel, again with nondiffusive density variations throughout. External forcing is introduced by variation of the channels' depth. We treat in particular a case first studied by A. Mielke [14] of a small bump in the bottom and display the full small amplitude picture. Finally we describe the flow of nonlinear waves under local- and global pressurewaves on the surface of an inviscid fluid. This extension of the method has been started by Amick and the author [1] for treating solitary water waves under surface tension.

2. The Reduction Principle.

Before treating the various physical applications we describe a method to reduce the local study of nonlinear elliptic boundary value problems in strip-like domains to ordinary differential equations. Although the method works in the general elliptic case [6], we explain it here in the simple form used in most of the examples later. Consider

$$\text{(2.1)} \qquad \begin{aligned} \Delta\Phi + f(\lambda,y,\Phi) &= \varepsilon F(x,\Phi) \\ \Phi|_{\partial\Omega} &= o, \quad \Phi, \nabla\Phi \text{ bounded in } \Omega \end{aligned}$$

where $\Omega = \{ (x,y) \mid x \in \mathbb{R}, y \in (o,1) \}$ $\varepsilon, \lambda \in \mathbb{R}$ are real parameters and f, F are smooth functions of their arguments; Δ denotes the 2-dimensional Laplacean. Assume further

$$f(\lambda,y,\Phi) = a(\lambda,y)\,\Phi + g(\lambda,y,\Phi), \quad g = O(\Phi^2)$$

where $a(\lambda,y)$ is positive and increases strictly with λ. Observe that f does not explicitly depend on x. Thus, for $\varepsilon = o$, if $\Phi(x,y)$ is a solution then $\Phi(-x,y)$ is a solution as well. We call this property *reversibility*. In general, for non-even F, reversibility is broken when ε differs from o.

There is a simple observation yielding the key to the complete local analysis of (2.1). It consists in writing (2.1) as evolution equation in x. For the linearized part and $\varepsilon = o$ this reads

$$\text{(2.2)} \qquad \frac{d^2\Phi}{dx^2} - \left(\frac{d^2\Phi}{dy^2} - a(\lambda,\cdot)\,\Phi \right) = \frac{d^2\Phi}{dx^2} - T(\lambda)\,\Phi$$

where

$$D(T) = H^2(o,1) \cap \overset{\circ}{H}{}^1(o,1)$$

and T is a self-adjoint operator; H^j denoting the usual Sobolev-spaces $W^{j,2}$, and suffix o indicates the vanishing on y = o,1. The reduction principal claims that all small bounded solutions of (2.1) lie on a manifold which is modelled on the eigenspaces of T corresponding to eigenvalues on the negative real axis and close to o. (Essential is the fact that the square-root of these eigenvalues lie close to the imaginary axis). The eigenvalues of T are real and simple, and we denote them and the corresponding eigenfunctions by

$$\sigma_o(\lambda) < \sigma_1(\lambda) < \ldots ; \quad \varphi_o, \varphi_1, \ldots .$$

Choose λ_o so that

$$\sigma_o(\lambda_o) = o \quad , \quad \sigma_o'(\lambda_o) < o$$

then, the "critical" eigenspace is 1-dimensional. Decompose $\Phi = \alpha\varphi_o + \Phi_1$, where

$$(\varphi_o, \Phi_1) = \int_o^1 \varphi_o(y)\Phi_1(x,y)dy = o$$

For small bounded solutions of (2.1) Φ_1 is a point-wise function of α, α' and the parameters ε, λ. Let us consider Φ as a function of x with values in D(T). Then, if ε_o, $\lambda_o - \lambda$, δ are sufficiently small positive numbers, we have

$$(2.3) \qquad \Phi_1(x) = h(\lambda,\alpha(x),\alpha'(x)) + k(\varepsilon,\lambda,x,\alpha(x),\alpha'(x))$$

if only

$$\| \Phi(x) \|_{D(T)} < \delta \qquad \text{for all } x \in \mathbb{R}.$$

Here h and k are smooth bounded functions of their arguments which can be constructed to any order of its Taylor-expansion about $\alpha = \alpha' = o$ (c.f. [14]). Moreover h represents the reversible part and k the x-dependent part of (2.1). We will use the following simple properties

$$(2.4) \qquad \begin{aligned} h(\lambda,\alpha,\alpha') &= h(\lambda,\alpha,-\alpha') = O(\alpha^2 + \alpha'^2) \\ k(\varepsilon,\lambda,x,\alpha,\alpha') &= O(\varepsilon) \end{aligned}$$

where the O-terms are uniform in $x\varepsilon \mathbb{R}$ and in bounded sets of $\varepsilon,\lambda,\alpha,\alpha'$.

Thus we have reduced completely the local study of (2.1) to the ordinary differential equation

$$(2.5) \qquad \alpha'' - \sigma_o(\lambda)\,\alpha + g_o(\lambda,\alpha,\alpha') = F_o(\varepsilon,\lambda,\cdot,\alpha,\alpha')$$

$$g_o = (g(\lambda,\cdot,\alpha\varphi_o + h), \varphi_o)$$

$$F_o = (\varepsilon\, F(\cdot,\alpha\varphi_o + h + k) + g(\lambda,\cdot,\alpha\varphi_o + h) - g(\lambda,\cdot,\alpha\varphi_o + h + k), \varphi_o)$$

In the last example, where we treat the case of a free surface, the reduction yields a 3rd-order system; but its derivation is in the same spirit as that of (2.5). Observe that g_o is even in α' and $F_o = O(\varepsilon)$. The reduced equations for the following examples are derived by keeping the lowest order terms in $\alpha,\alpha',\lambda_o - \lambda$ and ε. The justification of this truncation requires more or less sophisticated arguments which will be suppressed here.

3. Internal Waves.

We investigate the influence of an external magnetic wave $\underline{H}$ moving with speed c in horizontal direction over a channel of width h. Its influence on the flow of an inviscid, electrically charged fluid with density ρ (for mass) and e (for charge) is studied for a local and a global case, i.e. when the amplitude B of $\underline{H}$ has compact support or is periodic. The density ρ of the fluid varies through the channel Ω, the flow is supposed to be nondiffusive, i.e. $\nabla\rho \cdot \underline{u} = o$ holds, where $\underline{u} = (u,v)$ denotes the velocity field. The density distribution is described far upstream (at $x = -\infty$) as q(y). In a coordinate system moving with speed c in x-direction the governing equations are in dimensionless form

$$\rho(\underline{u} \cdot \nabla)\, \underline{u} \;+\; \nabla p \;+\; \lambda\rho \underline{e}_2 \;=\; \varepsilon\, B(x)(v,-u)$$

$$(3.1) \qquad \nabla \cdot \underline{u} = o \quad , \quad \underline{u} \cdot \nabla\rho = o \qquad \text{in } \Omega$$

$$v|_{\partial\Omega} = o \quad , \quad \lim_{x\to-\infty} \underline{u}(x,y) = \underline{e}_1 + O(\varepsilon) \;,\quad \underline{u} \text{ bounded}$$

where we used the following notations

$$\Omega = \{ (x,y) \in \mathbb{R}\times(o,1) \} \qquad \partial\Omega = \text{bdry } \Omega$$

$$\lambda = \frac{gh}{c^2} \quad , \quad \varepsilon = \frac{\mu heB(o)}{cq(o)} \quad , \quad \mu = \text{magnetic permeability}$$

$$\underline{e}_1 = (1,o) \quad , \quad \underline{e}_2 = (o,1)$$

and where we neglected the magnetic field induced by the flow. The condition at $x = -\infty$ is vague for $\varepsilon \neq o$. It will be clarified when we see that there is a unique asymptotic value for $\varepsilon \neq o$ to be naturally imposed on $\underline{u}$.

The derivation of the modified Long - Yih's equations which govern the flow is well known (c.f. [11]). Therefore we simply indicate the main steps. Defining Ψ as

$$(\Psi_y,-\Psi_x) = \sqrt{\rho}\,(u,v) \quad , \quad \Psi_o(y) = \int_0^y q(s)^{1/2}\, ds$$

$$p_\infty(y) = -\lambda \int_0^y q(s)\, ds \quad , \quad \Psi = \Psi_o + \Phi$$

we obtain $\rho(\Psi) = q(y(\Psi))$, where $y(\Psi)$ is the inverse function of $\Psi = \Psi_o(y)$. Using Bernoulli's equation and (3.1) one arrives at the following boundary value problem for Φ

$$\Delta\Phi + a(\lambda,y)\,\Phi + r(\lambda,y,\Phi) = \varepsilon\, B(x)\, \rho^{-1/2}(\Psi_o + \Phi)$$

$$(3.2) \qquad \Phi|_{\partial\Omega} = o \quad , \quad \Phi,\ \nabla\Phi \text{ bounded in } \Omega$$

$$\Phi = O(\varepsilon) \quad \text{for } x = -\infty$$

Here a, r are given by the expressions

$$a(\lambda,\cdot) = -\lambda s' - \frac{1}{4} s'^2 - \frac{1}{2} s'' \quad , \quad s = \log q$$

$$b(\lambda,\cdot) = -\frac{1}{4\sqrt{q}} (s''' + s's'' + 4\lambda s'' + \lambda s'^2)$$

Henceforth we assume

$$s'(y) < o \quad , \quad s'^2 + 2s'' < 4\pi^2$$

which guarantees that $\sigma_o(\lambda_o) = o$ for some $\lambda_o > o$, where σ_o denotes again the smallest eigenvalue of $-\varphi'' + a(\lambda,\cdot)\varphi = \sigma\varphi$. Now, we apply the reduction method as described in the last section and obtain

$$\alpha'' - \sigma_o(\lambda)\alpha + \tilde{b}(\lambda)\alpha^2 + O(\alpha^3 + \alpha\alpha'^2) = \varepsilon F_{oo}(x) + O(\varepsilon\alpha+\varepsilon\alpha'^2+\varepsilon^2) \tag{3.3}$$

where

$$\tilde{b}(\lambda) = \int_o^1 b(\lambda,y)\varphi_o^3(y)\, dy \quad , \quad F_{oo}(x) = (B(x)\rho^{-1/2}(\Psi_o),\varphi_o)$$

The solution of (3.3) is obtained differently for the two cases to be considered. Therefore we treat first the <u>local case</u> : B has compact support in $\mathbb{R}$, i.e. it vanishes identically outside <u>a bounded</u> interval. In (3.2) the boundary condition at $x = -\infty$ reads now $\Phi = o$ and thus $\alpha(-\infty) = o$. Assume $\tilde{b} \neq o$ and scale the variables as follows

$$\xi = \mu x \quad , \quad \mu^2 = \sigma_o(\lambda) \quad , \quad \varepsilon = \mu^3\delta, \quad \mu > o$$
$$\alpha(x) = \mu^2 A(\xi) \quad , \quad F_{oo}(x) = G(\xi)$$

Here we follow the work of Mielke [14] and obtain

$$A'' - A + \tilde{b}\, A^2 - \frac{\eta}{\mu} G(\xi) + O(\mu) = o \tag{3.4}$$

For μ tending to o, $G(\xi)/\mu$ approaches o if $\xi \neq o$; in addition

$$\frac{1}{\mu}\int_{\mathbb{R}} G(\xi)\, d\xi = \int_{\mathbb{R}} F_{oo}(x)\, dx = \{F_{oo}\}$$

Thus, the solutions to perturb from, satisfy

$$A'' - A + \tilde{b}\, A = o \qquad \text{for } \xi \neq o$$

$$[A'] \quad = \quad A'(+o) \quad - \quad A'(-o) \quad = \quad \eta\{F_{oo}\} = : \delta / b$$

They are obtained by intersecting the unstable manifold of $A = o$ and a jump in A' with the bounded solutions for $\xi > o$. Figure 1 shows the procedere. Depending on the sign of the jump and its magnitude, the multiplicity of solutions varies (see Table 1)

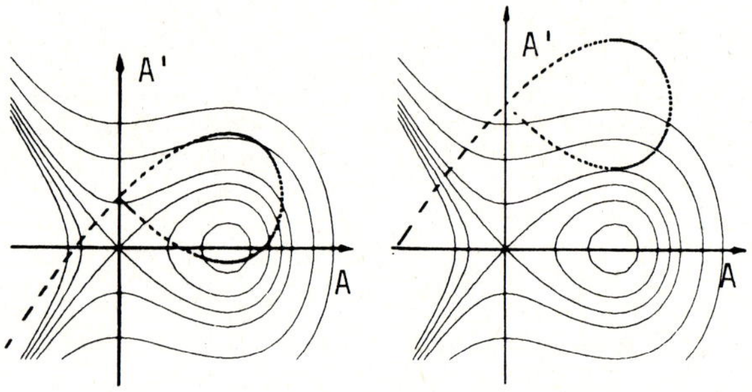

Figure 1: Solutions of (3.4) constructed in the phaseplane

δ	$(-\infty, -2/\sqrt{3})$	$(-2/\sqrt{3}, 0)$	$(0, 2/\sqrt{3})$	$> 2/3$
homoc.	1	3	2	0
period.	0	∞	∞	0

Table 1 : homoc.: number of homoclinic solutions

period: number of solutions with periodic wake

Every approximate solution thus obtained by a transversal intersection is extentable to a smooth solution of (3.4) for $\mu \neq o$ ([14]). Therefore, the real solutions of (3.4) are smooth perturbations of the curves given in Figure 2. Via the reduction principle they yield strict solutions of (3.2) and thus - since Φ, $\nabla\Phi$ are small - solutions of (3.1).
In particular we see that for some δ there are infinitely many solutions with periodic wake, and the period has a lower positive bound. The upstream influence is seen in those curves having extremas far apart and close to $\xi = o$.

A completely analogous problem arises when the steady flow through a density stratified channel is studied, given an arbitrarily chosen velocity profile at $x = -\infty$. For special datas, where the resulting Long-Yih-equation is linear has been studied by Grimshaw [8]. The general case was solved in [14], where it was also shown that no solutions have been lost through the special scaling employed above. Streamline pictures are shown in the following Figur 3 and 4

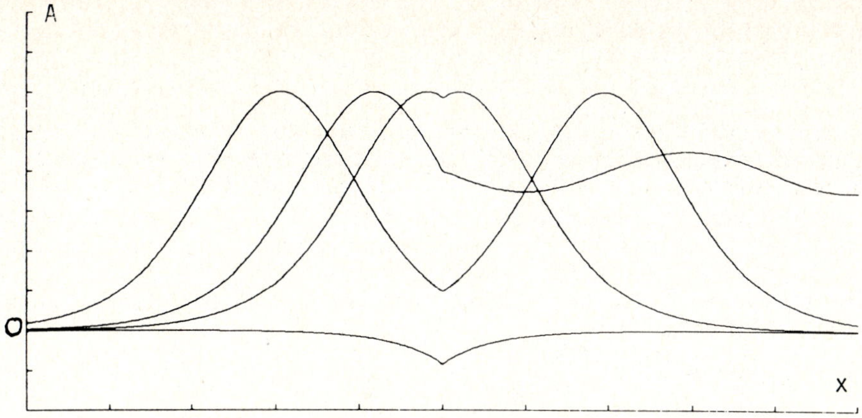

Figure 2 : Solutions of (3.4) for $\mu = 0$.

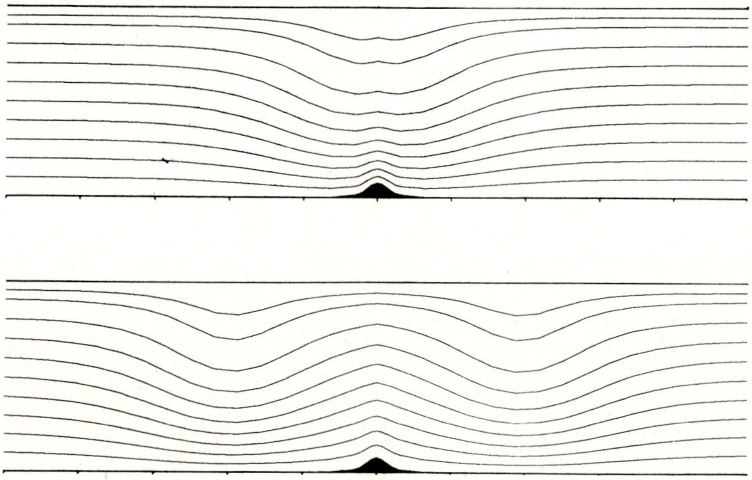

Figure 3 : Steady flow through a channel with a hump: Streamlines

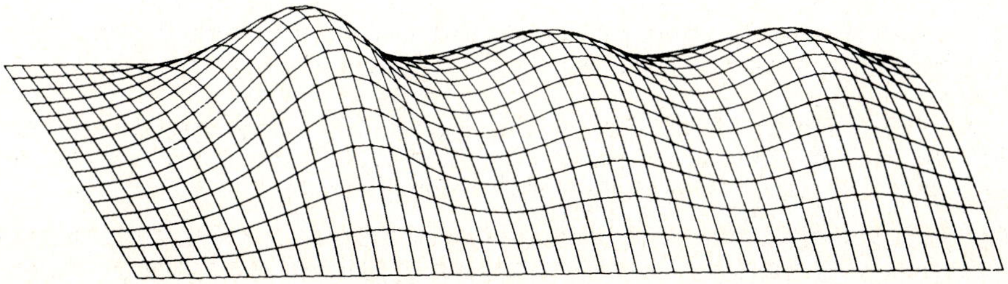

Figure 4 : Streamfunction

The global case : Now B is assumed to be 1-periodic in x. This implies that also the forcing term in (3.3) is 1-periodic. Neglecting the remainder-terms in that equation, one obtains an equation well studied in the literature [5], [9]. To estimate the effect of the residual terms we go back to (2.5). We choose

$$\lambda < \lambda_o \quad , \quad \mu^2 = \sigma_o(\lambda) \quad , \quad \mu > o$$

Observe that εF_o has a unique response of order ε, i.e. (2.5) has a unique bounded solution of order ε; this solution, which we denote by $\alpha^*(\varepsilon)$, is 1-periodic in x. This function provides the natural asymptotic condition for α at $x = -\infty$. Using the reduction-principle we obtain a corresponding solution $\Phi^*(\varepsilon)$, and (3.2) has to be supplemented by the requirement

$$\lim_{x \to -\infty} \left(\Phi(x,\cdot) - \Phi^*(x,\cdot,\varepsilon)\right) = o$$

Having found the perturbation of $\alpha = o$ under εF_o, we construct now perturbations of the "homoclinic" solution, denoted by p, of (3.3). This solution is the image of the solitary wave solution for $\varepsilon = o$ in the reduced phase space, as it was originally discovered by Ter Krikorov [16] and Benjamin [3]. Without loss of generality we can assume p to be even. Since we have freedom in phase, we can seek solutions near p for $\varepsilon \neq o$ via the ansatz

$$\alpha(x) = p(x + \beta) + z(x + \beta)$$

where β measures the phase. In order to control the dependence on μ we scale as follows

$$z(x) = \mu^2 Z(\mu x) \quad , \quad p(x) = \mu^2 P(\mu x) \quad , \quad \varepsilon = \mu^4 \delta$$

Insertion into (2.5) yields

$$L(P)\, Z = Z'' - Z + Dg(P)(Z,Z') = M(\delta,\beta,\lambda,Z,Z')$$

where

$$Dg(P)(Z,Z') = \mu^{-2} \{\partial_\alpha g_o(\lambda,\mu^2 P,\mu^3 P')\, Z + \partial_{\alpha'} g_o(\lambda,\mu^2 P,\mu^3 P')\mu Z'\}$$

$$M = \mu^{-4} \{F_o(\varepsilon,\lambda,x-\beta,\mu^2(P+Z),\mu^3(P'+Z')-g_o(\lambda,\mu^2(P+Z),\mu^3(P'+Z')) + g_o(\lambda,\mu^2 P,\mu^3 P') + Dg(P)(Z,Z')\}$$

We need the following oder - estimates

$$M = \delta F_{oo} + O(\mu^2\delta) + O(Z^2+\mu^2 Z'^2) \tag{3.7}$$

Nullspace and corange of L(P) are one-dimensional; they are spanned by P' resp.

$\tilde{P} = P' + O(\mu^2)$, the latter annihilating the adjoint of L(P) relative to the scalar-product

$$[Z,W] = \int_{-\infty}^{\infty} Z(\xi)\, W(\xi)\, d\xi$$

Define the projection into the nullspace of L(P) by

$$\Pi Z = [Z,\tilde{P}]\, P' \quad , \quad [\tilde{P},P'] = 1$$

and decompose $Z = \gamma P' + Z_1$, then (3.6) can be solved in the complement of P' uniquely for bounded Z_1, when M is replaced by (id - Π) M. Thus, for small $|\delta|$, $|\gamma|$, μ one obtains from (3.6) and (3.7)

$$Z_1 = Z_1^*(\delta,\beta,\mu,\gamma) = O((\delta+\gamma^2)(1+\mu^2))$$

Therefore, the solvability of (3.6) is guaranteed if the "Melnikov-condition" is satisfied

$$(3.8) \qquad \delta\, [\tilde{P},F_{oo}(\cdot-\beta)] + O(\delta\mu^2+\gamma^2(1+\mu^2)) = o$$

Divide by δ , use $\tilde{P} = P' + O(\mu^2)$ and set $\gamma = c\delta$, c fixed, then (3.8) reads

$$(3.9) \qquad m_o(\beta) + O(\mu^2+\delta) = o \quad , \quad m_o(\beta) = [P',F_{oo}(\cdot-\beta)],$$

If $m_o(\beta)$ has a simple zero at some β_o, one obtains $\beta(\delta,\mu)$ with $\beta(o,o) = \beta_o$ in some neighborhood of $\delta = \mu = o$ so that (3.8) holds identically in δ and μ. Moreover it is well known [5], that $m_o'(\beta) \neq o$ implies that $\alpha(o)$, given by (3.5) defines a transverse homoclinic point with respect to $\alpha^*(\varepsilon)$, $\varepsilon = \delta\mu^4$ for small $|\delta|$ and $|\mu|$. Therefore, from the solution constructed so far, one obtains, using the shadowing lemma [9], a multitude of solutions with sensitive dependence on initial conditions, typical for a chaotic solutionset.

Since P' in (3.9) can be replaced, up to the order μ^2, by the derivative of

$$P_o(\xi) = \frac{3}{2\tilde{b}\cosh^2(\xi/2)}$$

and since F_{oo} is known through the given datas B and q by (3.3), the solvability condition can be verified explicitly. We leave it as an easy task to the reader to construct meaningful concrete examples.

Proposition: Assume that $m_o(\beta)$ in (3.9) has a simple zero for some β_o. Then there exist, for each sufficiently small μ and δ, $\varepsilon = \delta\mu^4$, infinitely many solutions of (3.2), satisfying

$$\lim_{x \to -\infty} (\Phi(x,\cdot) - \Phi^*(x,\cdot;\varepsilon)) = o$$

In particular, for every $n \in \mathbb{N}$, there is a solution Φ whose component $\Phi_0 = (\Phi,\varphi_0)$ has at least n extremas.

The proof follows from our previous discussion, the reduction principle and from well known results in the theory of dynamical systems [9]. The following figure shows a solution for n = 3.

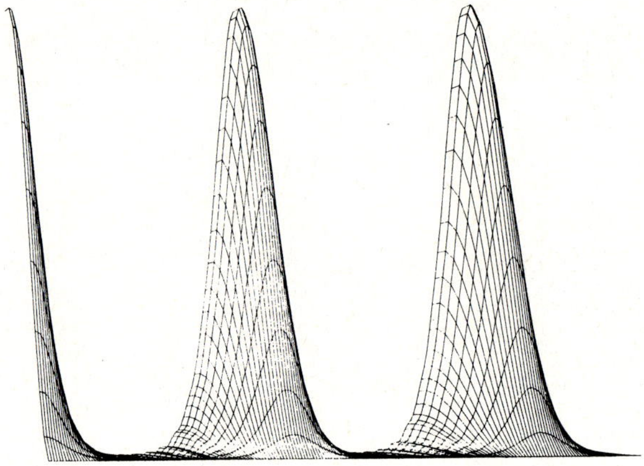

Figure 5

4. Surface Waves and External Waves.

In this last section we indicate a procedure for studying the influence of external pressure waves on gravity-driven solitary surface wave. The first formulation in this spirit appears in some work of Amick and the author [1] on solitary surface waves with surface tension. The present formulation is due to Mielke and the author and is particularly suited to the problem of external forcing. The reduction principle requires some extension from semilinear elliptic equations, as proved in [10],[6],[14], to quasilinear equations. We restrict ourselves to the presentation of the technical procedere and to the derivation of the final reduced equation. Since the conclusions to the local - and global case are similar to those considered in section 3, we leave the interpretations to the interested reader.

Imagine a pressure wave $\varepsilon p(x + ct)$ moving from right to left on the surface of inviscid, irrotational fluid under the influence of gravity (see Figure 6).

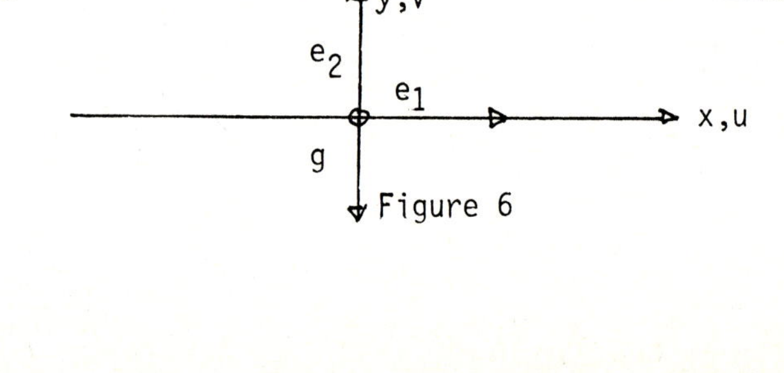

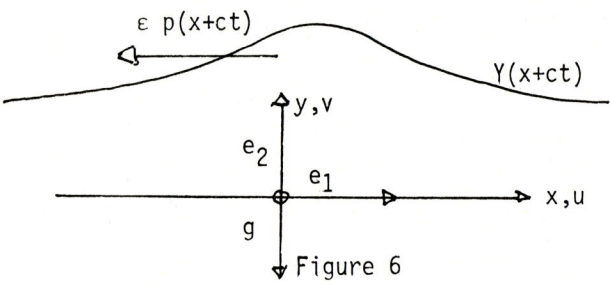

Figure 6

Write the basic equations in dimensionless form, $\lambda = gh/c^2$, h mean depth, and set x for x + ct, $\underline{u}(x,y) - \underline{e}$ for $\underline{u}(x + ct,y)$ etc., $\underline{u} = (u,v)$. Introduce a stream's function Ψ for $\underline{u}$ and apply the transformation (c.f.[3])

$$\xi = x \quad , \quad \eta = \Psi(x,y)$$

Setting $u = 1 + w$, $w(x,y) = W(\xi,\eta)$, $v(x,y) = V(\xi,\eta)$, $p(x) = P(\xi)$, we make a further - almost identical - transformation

$$w = W + \frac{1}{2}(W^2 + V^2) \quad , \quad v = \frac{V}{1+W}$$

Then Y' = v. From differentiating Bernoulli's law for $\eta = 1$, and from the equations for irrotational, solenoidal flow, we obtain for $\underline{w} = (w_1, w, v)$

(4.1) $$\underline{w}_\xi = \mathcal{A}\,\underline{w} + \mathcal{F}(\underline{w}) + \varepsilon\mathcal{J}(x,\underline{w})$$

where

$$\mathcal{A}\underline{w} = \begin{pmatrix} -\lambda v(1) \\ -v_\eta \\ w_\eta \end{pmatrix} \quad , \quad \mathcal{J}(x,\underline{w}) = \begin{pmatrix} \varepsilon P' \\ o \\ o \end{pmatrix}$$

$$\mathcal{F}(\underline{w}) = \begin{pmatrix} o \\ v_\eta + vw_\eta(1+W) - (1+W)^3 v_\eta \\ -w_\eta + \dfrac{w_\eta}{1+W} + vv_\eta\,(1+W) \end{pmatrix}$$

$$(1+W)^2 = \frac{1+2w}{1+v^2}$$

We treat (4.1) for $\underline{w} \in \mathbb{R}\times L_2(o,1)\times L_2(o,1)$ and in the domain of :
$D(\mathcal{A}) = \{ \mathbb{R}\times H^1(o,1)\times H^1(o,1) \mid v(o) = o\ ,\ w(1) = w_1\}$. Set $\lambda = 1-\mu$, $\mu > o$.

The spectrum of for $\mu = o$ consists of the eigenvalue o with multiplicity 3 and the simple eigenvalues $\sigma \neq o$ satisfying $\sigma \cos\sigma = \sin\sigma$. The critical part sits in $\sigma = o$ with the generalized eigenfunctions

$$\underline{\varphi}_{o1} = \begin{pmatrix} 1 \\ 1 \\ o \end{pmatrix} , \quad \underline{\varphi}_{o2} = \begin{pmatrix} o \\ o \\ \eta \end{pmatrix} , \quad \underline{\varphi}_{o3} = \begin{pmatrix} -\frac{1}{2} \\ -\frac{\eta^2}{2} \\ o \end{pmatrix}$$

A lengthy but elementary calculation yields the reduced equations on this three-dimensinal space

$$\underline{w} = \alpha_1 \underline{\varphi}_{o1} + \alpha_2 \underline{\varphi}_{o2} + \alpha_3 \underline{\varphi}_{o3} + O(|\underline{\alpha}|^2)$$

We obtain

$$\alpha_1' = (1 - \frac{3\mu}{5})\,\alpha_2 + \frac{6}{5}\alpha_1\alpha_2 + \frac{9}{25}\alpha_2\alpha_3 - \frac{3\varepsilon}{5} P'$$

$$\alpha_2' = \alpha_3 - \frac{2}{3}(\alpha_1\alpha_3 + \alpha_2^2)$$

$$\alpha_3' = 3\mu\alpha_2 + 9\alpha_1\alpha_2 + \frac{11}{5}\alpha_2\alpha_3 + 3\varepsilon P'$$

up to the order $O(|\underline{\alpha}|^3)$. For $\mu > o$, the eigenvalue o splits into 3 simple eigenvalues, namely o , $\pm(3\mu)^{1/2}$ with the eigenvectors

$$\underline{\varphi}_o = \begin{pmatrix} 1 \\ o \\ o \end{pmatrix}, \quad \underline{\varphi}_1 = \begin{pmatrix} 1 - \frac{3\mu}{5} \\ (3\mu)^{1/2} \\ 3\mu \end{pmatrix}, \quad \underline{\varphi}_2 = \begin{pmatrix} 1 - \frac{3\mu}{5} \\ -(3\mu)^{1/2} \\ 3\mu \end{pmatrix}$$

Setting

$$\underline{\alpha} = \beta_o \underline{\varphi}_o + \beta_1 \underline{\varphi}_1 + \beta_2 \underline{\varphi}_2$$

we obtain for

$$A = \beta_1 + \beta_{-1}, \quad B = \beta_1 - \beta_{-1}, \quad C = \beta_o + \beta_1 + \beta_2$$

$$C' = (3\mu)^{1/2} B, \quad A = C + \frac{3}{2\mu} C^2 + \frac{\varepsilon}{\mu} P$$

$$C'' = 3\mu C + \frac{9}{2} C^2 + 3\varepsilon P + O(\mu^2)$$

The last equation yields, for $\varepsilon = o$ the homoclinic solution which corresponds to the solitary wave solution obtained by Lavrentiev [13], Friedrichs and Hyers [7], Its perturbation for $\varepsilon \neq o$ follows the same laws as those discussed in section 3, and similar conclusions can be drawn.

REFERENCES:

1 Amick, C.J. and Kirchgässner, K., Solitary waves with surface tension, manuscript.

2 Amick, C.J. and Toland, J.F., On solitary water-waves of finite amplitude, **Arch. Rat.** Mech. Anal. 76 (1981) 9.

3 Benjamin, T.B., Internal waves of finite amplitude and permanent form, J. Fluid Mech. 25 (1966) 241.

4 Bona, J.L., Bose, D.K., Turner, R.E.L., Finite-amplitude steady waves in stratified fluids, J. Math. pures et appl. 62 (1983) 389-439.

5 Chow, S., Hale, J.K. and Mallet-Paret, J., An example of bifurcation to homoclinic orbits, J. Diff. Equ. 37 (1980) 351-373.

6 Fischer, G., Zentrumsmannigfaltigkeiten bei elliptischen Differentialgleichungen, Math. Nachrichten 115 (184) 137-157.

7 Friedrichs, K.O. and Hyers, D.H., The existence of solitary waves, Communs. Pure Appl. Math. 7 (1954) 517.

8 Grimshaw, R., A note on steady two dimensional flow of a stratified fluid over an obstacle, J. of Fluid Mech. 33 (1968) 293.

9 Guckenheimer, J., Bifurcations of dynamical systems, in: C.I.M.E. Lectures, Birkhäuser 1980.

10 Kirchgässner, K., Wave solutions of reversible systems and applications, J. Diff. Equ. 45 (1982) 113-127.

11 Kirchgässner, K., Nonlinear waves and homoclinic bifurcation, manuscript, to appear in Transact. Mech.

12 Kirchgässner, Nonlinear waves under external forcing, Advanced Mechanics, to appear.

13 Lavrentief, M.A. On the theory of long waves; A contribution to the theory of long waves 1947, in: Amer. Math. Soc. Translation 102 (1954), Providence, RI.

14 Mielke, A., Stationäre Lösungen der Eulergleichung in Kanälen variabler Tiefe, Dissertation Universität Stuttgart (1984).

15 Turner, R.E.L., Internal waves in fluids with rapidly varying density, Annali della Scuola Normale-Pisa, Ser. IV, Vol. 8 (1981) 513.

16 Ter-Krikorov, A.M., Théorie exacte des ondes longues stationnaires dans un liquide hétérogêne, J.d. Mécanique 2 (1963), 351-376.

Theoretical and Applied Mechanics
F.I. Niordson and N. Olhoff
Elsevier Science Publishers B.V. (North-Holland)

FROM PERIODICITY TO CHAOS IN HYDRODYNAMIC SYSTEMS

A. Libchaber

James Franck and Enrico Fermi Institutes and
Department of Physics
The University of Chicago
Chicago, IL 60637

In recent years a number of fluid experiments have been performed to study the relevance of dynamical system theory to some simple fluid flows. We will review here some of those aspects, for Rayleigh Benard experiments only.

INTRODUCTION

In small aspect ratio cells, Rayleigh Benard convection has led to interesting studies of the transition from periodic time dependent flow to chaotic behavior, as a function of the Rayleigh number. If the cell is small enough so that a few convective cells are present, an ordered pattern may exist on which time dependent states will be characterized as dissipative nonlinear systems with only a few degrees of freedom. In particular period doubling[1] and intermittency[2] may precede the onset of chaos. More complex quasiperiodic states may also be present and the transition to chaos may develop through one or two more oscillators being destabilized[3]. Mappings of the interval[4] appears as a useful model to describe some of those transitions,in particular as concerns the scaling behavior.

Another interesting model is related to nonlinear mapping of the circle into itself[5]. This pertains to quasiperiodic flows which trajectories define an invariant torus in phase space. The Poincare cross section of such a flow is an invariant circle, which is mapped into itself. In this case the evolution, as the flow is driven harder, leads to a breakdown of the circle, which occur in a universal way, in the sense that it does not depend on the exact structure of the mapping.

All those various routes to chaos have been observed in flow experiments and we will try to sum up here the observations. Various fluids with a large scale of Prandtl numbers have been used by different experimental groups and the results are fairly convincing for small aspect ratio cells.

PERIOD DOUBLING SEQUENCE

This route pertains mainly to low Prandtl number fluid. It has been observed in Mercury[6] Helium[7] and Water[8].

In the case of the very low Prandtl number fluid, Mercury, the complete scenario[6] has been observed: the direct cascade, the reverse one in the chaotic region and the universal sequence of laminar regions in the chaotic regime.

For low Prandtl numbers the onset of the various bifurcations occurs at very low Rayleigh numbers starting from a simple time dependent state, the socalled oscillatory instability. For higher Prandtl numbers where the onset is pushed to higher values of the control parameter, non Boussinesq effects may be relevant to see this route.

INTERMITTENCY

The experimental observation goes as follows: starting from a limit cycle and increasing the control parameter, one reaches a critical value R_0, where occasionally a burst of noise appears in time recording. For $R>R_0$ the bursts of noise become more frequent and finally the time recording becomes totally noisy. In the model proposed by Pomeau and Manneville[2], the interesting features of this intermittent transition to chaos are their scaling properties. The laminar periods (period between two noise bursts) diverge as one approaches R_0, following a law

$$T \simeq (R - R_0)^{-\alpha}$$

for $R>R_0$, the value of α depending on the model. For the simplest case $\alpha = 1/2$.

Intermittency has been observed in silicon oil[9] and Helium[7]. A detailed experimental analysis is given by Dubois[9].

FROM QUASIPERIODICITY TO CHAOS VIA A SCALING ROUTE

Within the framework of a quasiperiodic state with two frequencies present, a very general route to chaos has been described[5], via the mapping on the circle. As the control parameter is increased the torus can break down in various locking states where the winding number (ratio of the two frequencies) is a rational number. A period doubling sequence will then follow. If, in an experiment, one can keep the winding number constant and close to an irrational number, then the torus will break down by a fractalisation process, which will follow a precise scaling law. The first experimental observation of this route has been recently performed by Gollub and coworkers[10]. In the oscillatory state of the fluid, a second oscillator is introduced by A.C. heating of the lateral walls. The ratio of the oscillatory frequency to the imposed frequency is kept constant, as the Rayleigh number is increased. The observations seem to follow, at least qualitatively, the scenario.

OTHER CHAOTIC ROUTES, FROM QUASIPERIODICITY

One of the most frequent scenario observed from a quasiperiodic state seems to follow the Ruelle Takens[3] proposal. As the control parameter is increased one bifurcates to torus with higher dimensions, but the chaotic state is reached with a finite and small number of oscillators present. There is no definite prediction concerning this route, and the type and dimension of the strange attractor defining the chaotic state. In a recent experiment by Berge[11] the dimension of the attractor has been measured.

ACKNOWLEDGEMENT

This work was supported by the Materials Research Laboratory at The University of Chicago under Grant NSF DMR 82-16892 and also NSF Grant DMR 83-16204.

REFERENCES

[1] Eckmann, J. P., Rev. Mod. Phys. 53 (1981) 643.

[2] Pomeau, Y. and Manneville, P., Comm. Math. Phys. 74 (1980) 189.

[3] Ruelle, D., E. Fermi School of Physics, Varenna (1983).

[4] Collet, P. and Eckmann, J.P., Interated Maps on the interval as dynamical systems, Birkhauser (1980).

[5] Feigenbaum, M.J., Kadanoff, L. P. and Shenker, S. J., Physica 5D (1982) 370; Ostlund, S, Rand, D, J. Sethna and Siggia, E., Physica 8D (1983) 303.

[6] Libchaber, A., Fauve, S. and Laroche, C., Physica 7D (1983) 73.

[7] Libchaber, A. and Maurer, J., Nonlinear phenomena at phase transitions and instabilities (T. Riste ed. Plenum Press, 1982).

[8] Giglio, M., Muzatti, S. and Perini, V., Phys. Rev. Lett. 47 (1981) 243.

[9] Dubois, M., Rubio, M. A. and Bergé, P., Phys. Rev. Lett. 51 (1983) 1446.

[10] Fein, A. P., Heutmaker, M. S. and Gollub, J. P., to appear in Physica Scripta.

[11] Malraison, B., Atten, P., Bergé, P. and Dubois, M., J. Phys.(Paris) Lett. 44 (1983) L897.

Theoretical and Applied Mechanics
F.I. Niordson and N. Olhoff
Elsevier Science Publishers B.V. (North-Holland)

CONTINUUM FORMULATIONS FOR STOCHASTIC MULTICOMPONENT SYSTEMS

John J. McCoy

The Catholic University of America
Washington, D.C. 20064
USA

The macroscale response of a substance that is randomly heterogeneous on a microscale depends on the detailed statistics of the microstructure. This dependence results in differences both in the structure of an appropriate formulation for predicting the macroscale response, and in the values of effective material parameters required by these formulations. Calculations are presented that emphasize the dependence of the structure of continuum formulations on the microstructure statistics. Progress on the incorporation of higher-order statistical information of microstructure heterogeneity into effective material property bounds is reviewed.

INTRODUCTION

The statistical continuum calculations to be presented accept a specific model for the systems to be described, when observed on a microscale. This model is that of a linear, classical continuum with material heterogeneity that can be described by stochastic processes. The objectives of the calculations are (1) to obtain continuum formulations that will provide estimates of ensemble averaged field response measures for experiments in which these averaged measures vary on a macroscale, and (2) to relate both the structure of the formulations obtained and the effective material parameters contained therein to suitable descriptors of microstructural heterogeneity.

The calculations will be presented in two parts. One part will consider the structure of an efficient continuum formulation for determining the macroscale response and how this structure depends on microstructural heterogeneity. The second part will consider a specific continuum formulation for determining the macroscale response, that of a classical effective homogeneous continuum, and discuss the relationship between the material properties of this effective homogeneous continuum and microstructural heterogeneity.

The motivation for the studies presented is predicated on the position that the detailed variations of material property measures on the microscale can have significant effects on macroscale response measures and that this dependence can be exploited either in the design of materials with desirable macroscale response characteristics or for the inspection of a microscale based on macroscale response data. Logically, then, it is necessary that we refrain from implicitly restricting the type microstructure to which our calculations apply unless we can provide mathematical measures to delineate the restriction. Specifically, we must eschew assumptions based on self consistency which, at best, implicitly restrict the "higher-order" statistics of the material heterogeneity.

With reference to the first part of the presentation modern continuum formulations that can be termed extensions of the classical formulations accomplished in the nineteenth century, follow from one of two distinctly different abstractions [1]. One abstraction achieves the extension by allowing more degrees of freedom to a single homogeneous continuum than is allowed in the classical formulation. Thus, we are led to strain gradient continuum theories, nonlocal continuum theories, etc. Such theories are conveniently grouped as being applicable to continua with microstructure. In the second abstraction, the extension is achieved by increasing the number of interacting homogeneous continua that simultaneously occupy the entire region of the modelled substance, without allowing for more degrees of freedom in any one continuum. The theories that follow from this second abstraction are descriptively referred to as mixture theories. Clearly, a generalization that subsumes both of the above would be based on an abstraction of a mixture of homogeneous continua, all of which have many degrees of freedom. In application, however, generality usually must give way to efficiency and an applied question of some interest is to ask for which substances should a classical continuum formulation be extended by increasing the number of degrees of freedom for a single continuum and for which substances should the extension be in increasing the number of superimposed continua with no increase in the number of degrees of freedom.

The answer to the applied question rests in the underlying physical process which the theory is intended to describe. The answer cannot be obtained from an axiomatic development which does not consider the underlying process. The answer can only be obtained by modelling the process and by testing the efficacy of the continuum theory in the context of the process. In the first part of the presentation we consider this question for a microstructural model that is described by a linear, classical continuum with material heterogeneity. This question is considered in more detail in a forthcoming paper [2].

With reference to the second part of the presentation, we shall accept that the macroscale response field is to be determined using a classical effective homogeneous continuum model. The prediction problem is, then, to relate the effective material property measures to suitably described descriptors of the microstructural heterogeneity. This is a problem that has been studied extensively in the last three decades and we shall follow the approach of deriving bounds on the effective material property measures. The significance of the bound approach is that it is the only approach that allows for the incorporation of <u>partial</u> information of the microstructural heterogeneity and for obtaining an estimate of the significance of the information not incorporated. This estimate is provided by the width of the bounds. From the perspective of the design of material microstructure to obtain desirable effective property measures, widely separated bounds suggest that significant changes can be achieved by controlling the higher-order statistics not incorporated in the bounds.

Specifically we shall discuss the calculation of bounds for a particular type microstructural heterogeneity, that provided by a two-phase suspension of inclusions distributed throughout a homogeneous matrix. The objective is to incorporate in a hierarchy of bounds, measures of inclusion shape, of inclusion size and orientation statistics, and of the positional information of the inclusions. Bounds can be written that incorporate such information either directly or through multipoint correlation functions. A discussion of the first type was presented in an available paper [3] and will not be discussed here. Bounds that incorporate information of microstructural heterogeneity through functionals of multipoint point correlation functions have long appeared in the literature [4-10]. To specialize these bounds for application to two-phase suspensions requires the systematic resolution of the multipoint correlation functions into numerical measures of inclusion shape, of inclusion size and orientation statistics, and of the positional information of the inclusions. We shall present some results from an ongoing study of this aspect of the problem since these results hold promise of obtaining specific bounds that will be useful for the design of materials.

A FORMULATION ON THE AVERAGED RESPONSE FIELD

Restricting consideration to linear systems and introducing the notion of a reference material, we assume that the point by point response of the two-phase material, on the subscale, is governed by the symbolic equation,

$$u = u_o + G_o \delta C u. \tag{1}$$

Here, u denotes the response field; u_o denotes the response field in the reference material, assumed to be deterministic; δC denotes a local operator that represents the interaction of the response field and material "heterogeneity"; and, G_o denotes a nonlocal operator that describes the "propagation" of the effects of the local interaction due to material heterogeneity, to points removed from the center of interaction. The "forcing" of the problem appears via u_o, the response field in the reference material. The G_o operator is also defined for the reference material; algorithmically it is described by a Green's function and hence incorporates homogeneous boundary conditions applied to the material specimen. Randomness enters the formulation through the local interaction operator, δC.

The tensorial rank of the response field, the nature of the local interaction operator, and the dimensions of the space on which the formulation applies all depend on the application. For example, one algorithmic prescription of Eq. (1) in the context of the statical response of randomly heterogeneous, linearly elastic solids would have for the response field measure, $\epsilon_{ij}(x)$, the strain (a second rank tensor) field; for the interaction operator, $\delta C_{ijkl}(x)$, an algebraic operator described by a randomly varying fourth-rank tensor field, the material "stiffness" tensor; and the nonlocal G_o operator described by a Green's function with a tensorial rank of four. A derivation of the formulation expressed in is given in Reference [4].

A formulation on the average of the field response, across an ensemble of random materials can be obtained from Eq. (1) by a technique that can be termed a projection [11-15]. To proceed we first average, or project, Eq. (1) and write

$$\begin{aligned} \langle u \rangle &= u_o + G_o \langle \delta C u \rangle \\ &= u_o + G_o \langle \delta C \rangle \langle u \rangle + G_o \langle \delta C' u' \rangle, \end{aligned} \tag{2}$$

where the angular brackets denote the averaging and a prime denotes the fluctuating part of the indicated quantity. We can write, for example,

$$\langle u \rangle \equiv P u \,,$$

and (3)

$$u' \equiv (I - P) u \,,$$

where P denotes the projection operator and I the identity operator. Equation (2) is not a closed equation on ⟨u⟩, presenting the familiar closure problem. In the projection method this problem is solved by first operating on Eq. (1) using (I-P), to obtain

$$u' = G_o \delta C' \langle u \rangle + G_o \langle \delta C \rangle u' + G_o (I - P) \delta C' u', \tag{4}$$

which we interpret as an equation on the fluctuating, u', response field. The forcing in this equation is the $G_o \delta C' \langle u \rangle$ term. Solving Eq. (4) would give u' in terms of ⟨u⟩ which can then be used to form $\langle \delta C' u' \rangle$ in Eq. (2). The result of these operations is the desired closed equation on ⟨u⟩.

Accomplishing this program we write

$$u' = N G_0 \delta c' \langle u \rangle, \tag{5}$$

where the N operator is defined as

$$N = (I - G_0 \langle \delta c \rangle - G_0 (I - P) \delta c')^{-1}. \tag{6}$$

Next we form

$$\langle \delta c' u' \rangle = \langle \delta c' N G_0 \delta c' \rangle \langle u \rangle, \tag{7}$$

and finally write for the equation on $\langle u \rangle$,

$$\langle u \rangle = u_0 + G_0 M \langle u \rangle, \tag{8}$$

where the M operator is defined as

$$M = \langle \delta c \rangle + \langle \delta c' N G_0 \delta c' \rangle. \tag{9}$$

To be noted is the obvious conclusion that, in general, the M operator will be nonlocal.

Comparing Eqs. (1) and (8) allows the further conclusion that the equations that govern the averaged response field for a material with a random substructure, differ from those that govern the random response field only in that a random, local, interaction operator is to be replaced by a deterministic, nonlocal, interaction operator. It is natural then to term the M operator in Eq. (8), an effective interaction operator and to term the formulation, an effective interaction formulation.

It is obvious that the derivation of the effective interaction formulation is formal. Any attempt at a strict mathematical justification of the manipulations involved would necessitate statements as to the nature of the operators that describe the underlying physical process. Referring to Eq. (6), the presence of the projection operator would seem to make clear that N can only be made explicit in certain asymptotic regimes. Still, the result of the derivation is physically suggestive of a continuum formulation of the averaged field response, which is the use to which we wish to put it.

The above formulation makes no specific reference to the nature of the microstructural heterogeneity. Further, the formulation applies to any forcing field and is expressed in terms of the unconditionally averaged response field. Consider now a specific microstructure, that obtained on mixing two homogeneous phases, and a specific forcing, a point source forcing. Then, the projection method outlined can be generalized to accomodate conditional averages in the projection operator. In this way, closed systems of equations expressed in $\langle u_i \rangle$; i=1,2, the average of the response field taken over a subensemble for which the observation point is in the i phase, and in $\langle u_i^{(j)} \rangle$; i=1,2; j=1,2, the average of the response field taken over a subensemble for which the observation point is in the i phase and the source point is in the j phase, are obtained. The details are accomplished in Ref [2]. The structure of the closed systems of equations are written

$$\langle u_i \rangle = u_0 + G_0 \sum_{k=1}^{2} M_{ik} \langle u_k \rangle \quad ; \quad i = 1, 2 \,, \tag{10}$$

and

$$\langle u_i^{(j)} \rangle = u_0 + G_0 \sum_{k=1}^{2} M_{ik}^{(j)} \langle u_k^{(j)} \rangle \quad ; \quad i = 1, 2 \;; \; j = 1, 2 \,. \tag{11}$$

The M_{ik} and $M_{ik}^{(j)}$ are obvious extensions of the concept of effective interaction operators. The off-diagonal terms describe a coupling across the index locating the observation point phase; there is no corresponding coupling across the index locating the source point phase.

Equations (10) can be interpreted in terms of a mixture theory since the two conditionally averaged response fields are defined for each point covered by the mixture. Equations (11) are expressed in terms of four conditionally averaged response fields, the conditions refer to the locations of both the observation point and the source point.

The three formulations, Eqs. (8), (10) and (11), do not contradict each other; they must be consistent. In Ref [2], it was demonstrated that starting with Eqs. (11), one can derive a system of equations that can be identified with Eqs. (10), from which one can derive an equation that can be identified with Eq. (8). In this way equations can be written expressing the M_{ik} operators in terms of the $M_{ik}^{(j)}$ operator and the M operator in terms of the M_{ik} operators. We write

$$M_{ij} = \sum_{k=1}^{2} \sum_{m=1}^{2} M_{im}^{(k)} T_{mj}^{(k)} , \tag{12}$$

and

$$M = \sum_{i=1}^{2} \sum_{j=1}^{2} M_{ij} T_j , \tag{13}$$

where the $T_{mj}^{(k)}$ and T_j are identified as certain "resolution" operators, which in the first instance map the $\langle u_i \rangle$ fields into $\langle u_i^{(j)} \rangle$ fields and in the second instance map the $\langle u \rangle$ field into $\langle u_i \rangle$ fields.

The questions to be asked are not questions of mathematical consistency. The questions arise when we consider accepting the derived formulations, not as partial solutions of a mathematical problem, but as the basis of continuum "theories" for predicting the macroscale response. In this interpretation suitably parameterized effective interaction operators are to be accepted as phenomenological constants to be determined by matching physical data with predictions of the theory, in a limited number of canonical experiments. This interpretation is discussed in Ref. [2] and the argument is presented that it is desirable to have effective interaction operators that are local in the two-scale, macroscale/microscale, limit. Assuming that the M_{ij} operators of Eqs. (10) have this property, Eq. (13) can be used to argue that so will the M operator of Eq. (8) provided the T_j operators that resolve the $\langle u \rangle$ field into $\langle u_i \rangle$ fields are local in the limit. Two-phase substances for which this is true were termed strongly-coupled, with the implication being that an appropriately chosen field response due to a point disturbance introduced in one of the two phases will show little dependence on the specific phase of the disturbance except in the immediate vicinity of the disturbance. For a strongly-coupled, two-phase substance, then, any nonlocality in the formulation expressed in the unconditionally averaged field response would simply mirror a similar nonlocality in a formulation expressed in the conditionally averaged field response. There is no advantage, therefore, in extending the local effective homogeneous material concept by accepting a more complicated mixture theory formulation. For weakly-coupled, two-phase substances on the other hand, a nonlocality in the average field response formulation could arise as a result of nonlocal resolution operators. There would be an advantage in accepting a local mixture theory formulation for such a substance.

An example of a weakly-coupled, two-phase substance is a two-phase acoustic medium, with widely differing material stiffnesses for the component phases. An investigation of the limit in which the ratio of the two stiffnesses is large identifies two modes of momentum transport, which are weakly-coupled, and associates a particular mode with a particular phase in which the source point is located. It is the existence of these two modes of momentum transport, distinguished from each other in characteristic wave speeds, decay rates, etc. that requires a nonlocal M operator if all of physics that can be observed on the macroscale is to be incorporated.

BOUNDING EFFECTIVE MATERIAL PROPERTY MEASURES

All bounds on an effective property measure of a heterogeneous material are based on a definition of the measure in terms of an averaged energy stored in the material. Thus, for example, the effective thermal conductivity makes reference to a specimen comprised of the heterogeneous material, which is large enough to be representative of a collection of like manufactured specimens. (Mathematically, the thermal conductivity problem is identical to the dielectric and differs from the elasticity problem in the tensorial rank of the response measure.) The specimen is envisioned to be forced, at the boundary surfaces, in a manner that would result in homogeneous, i.e., constant heat flux $\underset{\sim}{q}(\underset{\sim}{x})$, and temperature gradient, $\underset{\sim}{T}(\underset{\sim}{x})$, fields if applied to a similarly shaped specimen but composed of a homogeneous material. The effective thermal conductivity is defined by equating the averaged thermal energy stored in a homogeneous, effective material specimen to that stored in the given, heterogeneous material specimen.

Reducing the definition to mathematical formulas, we have

$$\frac{1}{2}\,\underset{\sim}{q}_o\cdot\underset{\sim}{T}_o = \frac{1}{2V}\int \underset{\sim}{q}(\underset{\sim}{x})\cdot\underset{\sim}{T}(\underset{\sim}{x})\,d\underset{\sim}{x}, \tag{14}$$

where $\underset{\sim}{q}_o$ and $\underset{\sim}{T}_o$ denoted the constant heat flux and temperature gradient fields in the homogeneous, effective material specimen and $\underset{\sim}{q}(\underset{\sim}{x})$ and $\underset{\sim}{T}(\underset{\sim}{x})$ denote the spatially varying fields in the heterogeneous material specimen. The volume integral is over the extent of the specimen. Based on Eq. (14), we can write two expressions for computational definitions of k*, the effective thermal conductivity.

$$\frac{1}{2}\,k^*\underset{\sim}{T}_o\cdot\underset{\sim}{T}_o = \frac{1}{2V}\int k(\underset{\sim}{x})\,\underset{\sim}{T}(\underset{\sim}{x})\cdot\underset{\sim}{T}(\underset{\sim}{x})\,d\underset{\sim}{x}, \tag{15a}$$

and

$$\frac{\underset{\sim}{q}_o\cdot\underset{\sim}{q}_o}{2k^*} = \frac{1}{2V}\int \frac{\underset{\sim}{q}(\underset{\sim}{x})\cdot\underset{\sim}{q}(\underset{\sim}{x})}{k(\underset{\sim}{x})}\,d\underset{\sim}{x}. \tag{15b}$$

Here, $k(\underset{\sim}{x})$ is the spatially varying conductivity for the heterogeneous material specimen.

The bounds on k* are obtained by making use of two, complementary variational formulations of the heat conduction problem. Thus, the actual temperature gradient field in the specimen is the member of a class of precisely prescribed vector fields, which minimizes the right-hand side of Eq. (15a). The class of trial fields are described as vector fields that are derived from a scalar potential which is required to satisfy prescribed temperature conditions applied at the boundary of the specimen. In the context of the thought experiment in which k* is to be determined, the foregoing requirement on the boundary condition to be satisfied by an appropriate trial function can be translated into a condition that the spatial average of the trial function must equal the temperature gradient in the homogeneous, effective material specimen; i.e., $\underset{\sim}{T}_o$. On introducing $\underset{\sim}{T}_T(\underset{\sim}{x})$ to denote a generic member of the allowable class of vector fields, i.e., to denote a trial function, the variational formulation leads to an upper bound on k*, i.e.

$$k^*\underset{\sim}{T}_o\cdot\underset{\sim}{T}_o \le \frac{1}{V}\int k(\underset{\sim}{x})\,\underset{\sim}{T}_T(\underset{\sim}{x})\cdot\underset{\sim}{T}_T(\underset{\sim}{x})\,d\underset{\sim}{x}. \tag{16a}$$

The lower bound makes use of a complementary variational principle. The actual heat flux in the specimen is the member of a class of precisely prescribed vector fields that minimizes the right-hand side of Eq. (15b). The class of trial functions is now described as vector fields that are derived from vector potentials, and that satisfy prescribed heat flux conditions applied at the boundary of the specimen. The boundary condition to be satisfied, in the context of the present problem, is translated into a condition that the spatial average of the trial function must equal the heat flux in the homogeneous, effective material specimen, i.e., $\underset{\sim}{q}_o$. On introducing $\underset{\sim}{q}_T(\underset{\sim}{x})$ to denote a trial function, the variational principle leads to the lower bound.

$$\frac{\underset{\sim}{q}_o \cdot \underset{\sim}{q}_o}{k^*} \leq \frac{1}{V} \int \frac{\underset{\sim}{q}_T(\underset{\sim}{x}) \cdot \underset{\sim}{q}_T(\underset{\sim}{x})}{k(\underset{\sim}{x})} \, d\underset{\sim}{x} \, . \tag{16b}$$

The most elementary, or classical bounds follow immediately from Eqs. (16a) and (16b) on noting that the constant $\underset{\sim}{T}_o$ and $\underset{\sim}{q}_o$ fields are appropriate trial functions for Eqs. (16a) and (16b), respectively. Thus we write

$$\langle 1/k \rangle^{-1} \leq k^* \leq \langle k \rangle , \tag{17}$$

where the angular brackets are used to denote a spatial average taken over the specimen. It has been noted by a number of researchers, that the law of mixtures, equating the effective property measure to a spatial average of the heterogeneous property measure, actually provides a rigorous upper bound to the effective property measure. Further, it is well appreciated, for heterogeneous materials that are described as suspensions of an inclusion phase dispersed throughout a matrix phase, that the nature of the geometric information in $\langle k \rangle$ and in $\langle 1/k \rangle$ is the relative amount of the two phases, i.e., volume fraction information.

Improved bounds can be obtained by choosing trial functions that more nearly reproduce the actual temperature gradient and heat flux fields. This is accomplished by approximately solving for the temperature gradient and heat flux fields. For the bounds to remain rigorous, it is of course necessary that the trial functions satisfy exactly the conditions dictated by the variational principle on which the bound is founded. Two hierarchies of improved bounds can be obtained by using trial functions motivated by perturbation calculations and by dilute suspension calculations. These reported studies on improved bounds accept a statistical interpretation of the problem; an interpretation that replaces the spatial averages in Eqs. (16a) by statistical, or ensemble averages. In this statistical interpretation the information in $\langle k \rangle$ and $\langle 1/k \rangle$ is termed one-point information and can be obtained by dropping points, at random, on a substructure photograph.

Improved bounds collect additional information of the substructure geometry in a form that is determined by the trial functions upon which the bounds are derived. For trial functions that are motivated by perturbation calculations [4-10], the additional information is that contained in multipoint correlation functions, of the stochastic thermal conductivity field, $k(\underset{\sim}{x})$. We write, for example, the following bounds due to Beran [5].

$$k_L \leq k^* \leq k_U \tag{18}$$

where in three dimensions,

$$k_L = \left(\langle 1/k \rangle - \frac{4 \langle k'/k \rangle^2}{9 \langle k \rangle \left(\frac{1}{3} \langle k'/k \rangle^2 / \langle k \rangle + J \right)} \right)^{-1} , \tag{19}$$

and

$$k_U = \langle k \rangle - \frac{\frac{1}{3} \langle k'^2 \rangle / \langle k \rangle}{1 + 3 \langle k \rangle I / \langle k'^2 \rangle} \, . \tag{20}$$

In Eqs. (19) and (20), the factors denoted by I and J are the following functionals of three-point correlation functions,

$$I = \frac{1}{16\pi^2 \langle k \rangle^2} \iint \frac{\underset{\sim}{r}}{r^3} \cdot \frac{\underset{\sim}{s}}{s^3} \frac{\partial^2}{\partial r_3 \partial s_3} \langle k(0) k'(\underset{\sim}{r}) k'(\underset{\sim}{s}) \rangle \, d\underset{\sim}{r} \, d\underset{\sim}{s} \,, \tag{21}$$

and

$$J = \frac{1}{16\pi^2 \langle k \rangle^2} \iint \frac{\underset{\sim}{r}}{r^3} \cdot \frac{\underset{\sim}{s}}{s^3} \frac{\partial^2}{\partial r_3 \partial s_3} \left\langle \frac{k'(\underset{\sim}{r}) k'(\underset{\sim}{s})}{k(0)} \right\rangle d\underset{\sim}{r} \, d\underset{\sim}{s} \,. \tag{22}$$

In writing the bounds we have assumed material isotropy both on the microscale and on the macroscale, and have arbitrarily chosen the constant $\underset{\sim}{q}_0$ and $\underset{\sim}{T}_0$ fields to be directed along the x_3 coordinate axis.

To make the bounds specific, i.e. to reduce them to a pair of numbers, requires specification of the I and J measures of microstructural heterogeneity in addition to the ensemble averages of k and 1/k. For a completely general microstructure it is this additional information that results in a closer bound pair, than that given by Eqs. (17). That is, the wider separation of the simpler bounds is, for a completely general microstructure, due to incorporating less information in the bound pair and is not due to any weakness in the analysis that resulted in the simpler bounds. The bounds given above can be termed first order, since they are based on a first-order perturbation solution as a trial function. Second- and higher-order bounds can be written using second- and higher-order perturbation solutions as trial functions. This program was carved out to second-order by Elsayed [8], who obtained bounds that incorporate information of four- and five-point correlation functions. This additional information, when specified, will result in better, i.e. closer, bounds than those given by Eqs. (18-22).

The above described bounds apply to any type microstructure, i.e. arbitrary $k(\underset{\sim}{x})$. For the special case of a two-phase suspension, the higher-order (than one) multipoint correlation functions are determined by more intuitive descriptors of the microstructure geometry, such as the shapes and size distribution of the inclusions as well as statistical measures of their orientations and relative positions. An interesting problem, which we now consider, is to determine the relationship between the analytical measures and the intuitive geometrical descriptors. The basis for our discussion is the studies of Miller [7], Brown [16], Milton [9] and Torquato and Stell [17] and is due to Gillette [10].

Following Gilette, a three-point cross-correlation function of stochastic "property" fields $a(\underset{\sim}{x})$, $b(\underset{\sim}{x})$ and $c(\underset{\sim}{x})$ can, for a two-phase material, be written

$$\langle a(\underset{\sim}{x}_1) b(\underset{\sim}{x}_2) c(\underset{\sim}{x}_3) \rangle = \sum_{i=1}^{2} \sum_{j=1}^{2} \sum_{k=1}^{2} a_i b_j c_k f_{ijk}(\underset{\sim}{x}_1, \underset{\sim}{x}_2, \underset{\sim}{x}_3) \,, \tag{23}$$

where $(a_i, b_i, c_i,\ i=1,2)$ are the two values of the three property fields and $f_{ijk}(x_1, \underset{\sim}{x}_2, \underset{\sim}{x}_3)$ are the three point probabilities of finding the $\underset{\sim}{x}_1$ coordinate in phase i (=1 or 2), the $\underset{\sim}{x}_2$ coordinate in phase j and the $\underset{\sim}{x}_3$ coordinate in phase k. Equation (23) thus separates the material property measures from the geometric measures, in determining the correlation function.

Next, we make use of the constraints that the three point probabilities must add up to appropriate two-point and one-point probabilities and that the sum all three-point probabilities must be unity, to express the three-point probabilities in terms of one-point, two-point, and three-point probabilities for only one of the two phases. Thus, we write

$$\begin{aligned} \langle a(\underset{\sim}{x}_1) b(\underset{\sim}{x}_2) c(\underset{\sim}{x}_3) \rangle = {} & A_0 + A_1 f_1 + A_2 f_{11}(\underset{\sim}{x}_1, \underset{\sim}{x}_2) + A_3 f_{11}(\underset{\sim}{x}_1, \underset{\sim}{x}_3) \\ & + A_4 f_{11}(\underset{\sim}{x}_2, \underset{\sim}{x}_3) + A_5 f_{111}(\underset{\sim}{x}_1, \underset{\sim}{x}_2, \underset{\sim}{x}_3) \end{aligned} \tag{24}$$

where the A's are combinations of the a_i's, b_i's and c_i's and $f_{\prime}$ and $f_{\prime\prime}$ are one and two point probability functions. In writing Eq. (24) we have arbitrarily chosen to describe the geometry of the two-phase material in terms of the 1 phase.

The final resolution is to identify the 1 phase as the inclusion phase and to express the probabilities of locating two (or three) points in this inclusion phase in terms of probabilities of locating the two (or three) points in either the same inclusion or in two (or two or three) different inclusions. Thus, for example, we write

$$f_{111}(\underset{\sim}{x}_1,\underset{\sim}{x}_2,\underset{\sim}{x}_3) = f_1 g_1(\underset{\sim}{x}_1,\underset{\sim}{x}_2,\underset{\sim}{x}_3) + f_1\big(g_2(\underset{\sim}{x}_1;\underset{\sim}{x}_2,\underset{\sim}{x}_3) + g_2(\underset{\sim}{x}_2;\underset{\sim}{x}_1,\underset{\sim}{x}_3) + g_2(\underset{\sim}{x}_3;\underset{\sim}{x}_1,\underset{\sim}{x}_2)\big) + f_1 g_3(\underset{\sim}{x}_1,\underset{\sim}{x}_2,\underset{\sim}{x}_3). \tag{25}$$

The $g_i(x_1,x_2,x_3)$ are conditional probabilities with the subscript indicating the number of inclusions involved. In writing the g_2 probabilities the notation indicates the specific coordinate that is isolated in a single inclusion.

Combining the above we write the following expression for $\langle a(x_1)b(x_2)c(x_3)\rangle$,

$$\begin{aligned}\langle a(\underset{\sim}{x}_1)b(\underset{\sim}{x}_2)c(\underset{\sim}{x}_3)\rangle &= A_0 + f_1 A_1 \\ &+ f_1\big(A_2 g_1(\underset{\sim}{x}_1,\underset{\sim}{x}_2) + A_3 g_1(\underset{\sim}{x}_1,\underset{\sim}{x}_3) + A_4 g_1(\underset{\sim}{x}_2,\underset{\sim}{x}_3) + A_5 g_1(\underset{\sim}{x}_1,\underset{\sim}{x}_2,\underset{\sim}{x}_3)\big) \\ &+ f_1\big(A_2 g_2(\underset{\sim}{x}_1,\underset{\sim}{x}_2) + A_3 g_2(\underset{\sim}{x}_1,\underset{\sim}{x}_3) + A_4 g_2(\underset{\sim}{x}_2,\underset{\sim}{x}_3) + A_5\big(g_2(\underset{\sim}{x}_1;\underset{\sim}{x}_2,\underset{\sim}{x}_3) \\ &\qquad + g_2(\underset{\sim}{x}_2;\underset{\sim}{x}_1,\underset{\sim}{x}_3) + g_2(\underset{\sim}{x}_3;\underset{\sim}{x}_1,\underset{\sim}{x}_2)\big)\big) \\ &+ f_1 A_5 g_3(\underset{\sim}{x}_1,\underset{\sim}{x}_2,\underset{\sim}{x}_3).\end{aligned} \tag{26}$$

Equation (26) traces the contributions of substructure material, through the A's, and substructure geometry, through the g_i's, to the three point correlation function. Further, the measures of substructure geometry trace the contributions from the inclusions taken one at a time, through the g_1's, taken two at a time; through the g_2's and taken three at a time through the g_3's.

Expressing the correlation functions that appear in I and J in the form of Eq. (26) we obtain the following expressions for the first-order bounds applied to a two-phase suspension.

$$k_L = \left(\langle 1/k\rangle - \frac{4(k_1k_2)^{-1}(k_1-k_2)^2 f_1(1-f_1)^2}{3(\langle k\rangle + k_1)(1-f_1) + 9(k_1-k_2)(G_1+G_2+G_3)}\right)^{-1} \tag{27}$$

and

$$k_U = \left(\langle k\rangle - \frac{(k_1-k_2)^2 f_1(1-f_1)^2}{3k_2(1-f_1) + 9(k_1-k_2)(G_1+G_2+G_3)}\right), \tag{28}$$

where

$$G_1 = \frac{1}{(4\pi)^2}\iint \frac{\underset{\sim}{r}}{r^3}\cdot\frac{\underset{\sim}{s}}{s^3}\frac{\partial^2}{\partial r_3\partial s_3} g_1(0,\underset{\sim}{r},\underset{\sim}{s})\,d\underset{\sim}{r}\,d\underset{\sim}{s},$$

$$G_2 = \frac{1}{(4\pi)^2}\iint \frac{\underset{\sim}{r}}{r^3}\cdot\frac{\underset{\sim}{s}}{s^3}\frac{\partial^2}{\partial r_3\partial s_3}\big(g_2(0;\underset{\sim}{r},\underset{\sim}{s}) + g_2(\underset{\sim}{r};0,\underset{\sim}{s}) + g_2(\underset{\sim}{s};0,\underset{\sim}{r})\big)\,d\underset{\sim}{r}\,d\underset{\sim}{s} \tag{29}$$

and

$$G_3 = \frac{1}{(4\pi)^2}\iint \frac{\underset{\sim}{r}}{r^3}\cdot\frac{\underset{\sim}{s}}{s^3}\frac{\partial^2}{\partial r_3\partial s_3} g_3(\underset{\sim}{0},\underset{\sim}{r},\underset{\sim}{s})\,d\underset{\sim}{r}\,d\underset{\sim}{s}.$$

As is well appreciated, the assumption of isotropy results in first order bounds that are independent of any two point correlation function.

The geometry of the suspension is described in the bounds by the three functionals denoted by G_i. Again the subscript refers to the number of inclusions that are significant in determining a specific value. Thus, G_1, is determined by the inclusions taken singly. It can be shown that G_1 can be designated a shape factor. The G_2 functional is determined by the inclusions taken two at a time. It contains, therefore, some information of the positional statistics of the inclusions up to the order of the pairwise distribution function. It also can vary with inclusion shape and with inclusion size distribution. Finally G_3 can vary with changes in the triplet distribution function.

Equations (27) and (28) trace the several separate geometric factors that can effect the rigorously derived first order bounds and therefore accomplish one of our objectives. A second objective is to develop a procedure for calculating the G_i functionals for inclusions of prescribed shape and size distribution and, in the case of G_2 and G_3, of prescribed pairwise and triplet distribution functions. This, in turn, requires a procedure for determining the conditional probability density functions in their definition. One might envision a Monte Carlo approach to this problem, in which triangles are figuratively droped at random in a model of the material and the probability functions determined as a ratio of the number of occurrences of random events. Gillette [10] formulated the problem, however, by fixing the locations of the three vertices of the triangle and by figuratively dropping an inclusion; or an inclusion pair; or an inclusion triplet. In this way he was able to express the probability functions as a ratio of volumes of regions within which a reference point in the dropped inclusion, or inclusion pair or triplet must fall for a desribed statistical event to occur. This enabled his reducing the computations required to calculate a specific g_i to a point in which they only require a micro-computer. Taken together, then, we now have the capability of constructing rigorous numerical bounds to effective material property values for suspensions which incorporate partial information of inclusion shape, size distribution and orientation statistics and some information of inclusion positioning. This capability is presently being applied to computer generated composite materials. The results will be reported in future articles. The application of this capability to photomicrographs of physical composites would then be an obvious next step.

SUMMARY

In summary, then, the calculations presented were intended to emphasize that the macroscale response of a substance with a random microstructure is dependent on the detailed statistics of microstructure. Moreover, this dependence can occur not only in a vlaue of an effective material property measure but can also occur in the structure of a continuum formulation for estimating the macroscale response. Further, we reviewed some recent progress on constructing numerical bounds.

REFERENCES

[1] Bedford, A. and Drumheller, D. S., Theories of immiscible and structured mixtures, Int. J. Engr. Sci 21 (1983) 863-960.

[2] McCoy, J. J., Continuum formulations for weakly-coupled, two-phase materials with random substructures, to be submitted for publication.

[3] McCoy, J. J., Bounds on the transverse effective conductivity of computer-generated fiber composites, J. Appl. Mech. 49 (1982) 319-326.

[4] A discussion of an explicitly statistical formulation of the average field response of materials with a random substructure, including a detailed discussion of effective material property bounds is contained in J. J. McCoy, Macroscopic response of continua with random microstructures, in: Nemat-Nasser, S (ed.), Mechanics Today 6 (Pergamon Press, USA, 1981).

[5] Beran, M., Use of a variational approach to determine bounds for the effective permittivity of a random medium, Il Nuovo Cimento 38 (1965) 771-782.

[6] McCoy, J. J., A new set of bounds for the effective permittivity of a random medium, Il Nuovo Cimento 57 (1968) 139-153.

[7] Miller, M. N., Bounds for effective bulk modulus of heterogeneous materials, J. Math. Phys. 10 (1969) 2005-2013.

[8] Elsayed, M. A., Bounds for effective thermal, electrical, and magnetic properties of heterogeneous materials using high order statistical information, J. Math. Phys. 15 (1974) 2001-2015.

[9] Milton, G. W., Bounds on the elastic and transport properties of two-component composites, J. Mech. Phys. Solids 30 (1982) 177-191 and references therein.

[10] Gillette, G. J., Effective properties of two-phase random suspensions using statistics of the inclusion phase geometry, Ph.D. Dissertation to be submitted to the Dept. of Civil Engr., The Catholic University of America.

[11] Keller, J. B., Wave propagation in random media, Proc. Symp. Appl. Math. 13, Amer. Math. Soc., Providence (1964) 227-246.

[12] Frisch, U., Wave propagation in random media, in: Bharucha-Reid, A. T. (ed.), Probabilistic methods in applied mathematics 1 (Academic Press, N.Y.,1968).

[13] Beran, M. J., and McCoy, J. J., Mean field variations in a random medium, Quar. Appl. Math. 28 (1970) 245-258.

[14] Beran, M. J., and McCoy, J. J., Mean field variations in a statistical sample of heterogeneous linearly elastic solids, Int. J. Solids Struc. 6 (1970) 1035-1954.

[15] Fishman, L. and McCoy, J. J., Homogeneization and smoothing: A unified view of two derivations of effective property theories and extensions, J. Appl. Mech. 47 (1980) 679-682.

[16] Brown, W. F., Jr., Effective properties of two-phase "cell materials," J. Math. Phys. 15 (1974) 1516-1524.

[17] Torquato, S. and Stell, G., Microstructure of two-phase random media I, J. Chem. Phys. 77 (1982) 2071-2077.

Theoretical and Applied Mechanics
F.I. Niordson and N. Olhoff
Elsevier Science Publishers B.V. (North-Holland)

RECENT STUDIES ON OCEAN WAVE SPECTRA

Hisashi Mitsuyasu

Research Institute for Applied Mechanics
Kyushu University
Kasuga, 816
JAPAN

This paper presents some recent results concerning properties of ocean wave spectra. A number of topics are discussed: (a) the dispersion relation of spectral components of wind waves, (b) general properties of wind wave spectrum in a growth stage, (c) important relations among spectral parameters, (d) similarity form of the wind wave spectrum, which covers idealized form of the frequency spectrum, the equilibrium spectrum, high-frequency spectrum and angular distribution function.

INTRODUCTION

When wind blows over a still water surface, wind waves are generated and develop with time and space. After sufficiently long time, the wind waves attain to a statistically stationary state, while they still develop with distance (fetch) and finally become huge ocean waves at large fetches. Such wind-generated ocean waves show random properties and they are described by the two-dimensional wave spectrum.

Studies on ocean waves cover very wide area; fundamental problems such as, the generation of wind waves, coupling between wind and waves, nonlinear interaction among spectral components, dissipation of wave energy by wave breaking, structure of wave spectrum, and practical problems such as wave forecasting, wave statistics (short time and long time), marine-structure wave interaction, remote sensing of ocean surface phenomena. In the present paper, topics are limited to various problems of ocean wave spectrum which is considered to be important to the marine-structure wave interaction.

The dispersion relation for the spectral components is first discussed, which is one of the foundations of spectral model of wind waves. Then we discuss more practical problem such as similarity forms of wind wave spectrum in a dominant frequency region and in a high frequency region. The spectral form in the former region is important to the marine-structure wave interaction, and that in the latter region is important to the remote sensing of ocean surface phenomena. Finally we discuss the angular distribution function in the wave spectrum.

SPECTRAL MODEL OF OCEAN SURFACE WAVES

To a first approximation, ocean surface waves may be regarded as a linear superposition of statistically independent free waves with random phase and are consequently described by their two-dimensional wave spectra $\Psi(f,\theta)$. Such a spectral model of the ocean surface waves has been generally accepted but sometimes has undergone critical discussion.

One of the most critical results has been presented by Ramamonjiarisoa (1974) [1], which have shown a nondispersive nature of the spectral components at a high frequency region of the spectrum. Such a finding has stimulated, on one side, to construct a new model of ocean surface waves [2], [3].

On the other side, traditional weak-nonlinear theory for the ocean surface waves has been studied to solve the controversial problem [4], [5], [6], [7], [8]. The results of the latter studies are summerized as follows; For laboratory wind waves in decay area, spectral components near the spectral peak frequency f_m are largely due to free waves which satisfy a linear dispersion relation, but those at high frequency region ($f>1.8f_m$) are largely due to nonlinear bounded waves and contributions of free waves are relatively small (Figure 1). The peculiar characteristics of the phase velocity as observed by Ramamonjiarisoa (1974) are largely attributed to the effects of the nonlinear bounded waves [6]. Even for the wind waves in a generation area the linear dispersion relation is satisfied in a dominant frequency region ($0.7f_m \lesssim f \lesssim 1.6f_m$), if we consider the effects of drift current and angular distribution of the spectral components [7], [9]. More than 90% of the spectral energy is contained in the frequency region $0.7f_m \lesssim f \lesssim 1.8f_m$. Therefore, we can conclude that the ordinary spectral model can be applied for various problems as far as the energy containing dominant frequency region is concerned. We need to consider the nonlinear interaction among spectral components for evaluating a long-time evolution of the wave spectrum, but this can be treated within a frame of weak-nonlinear theory for the ocean surface waves. Furthermore, since the effects of nonlinearity is relatively small for ocean waves as shown later, the linear dispersion relation is approximately satisfied even for high frequency region up to $3f_m$ [8].

However, at very high frequency region, apparent phase speeds of the spectral components approach the orbital velocity of the dominant waves. Here again, the apparent phase speed becomes independent of the frequency, though it does not imply a lack of dispersion of the high frequency spectral components [10].

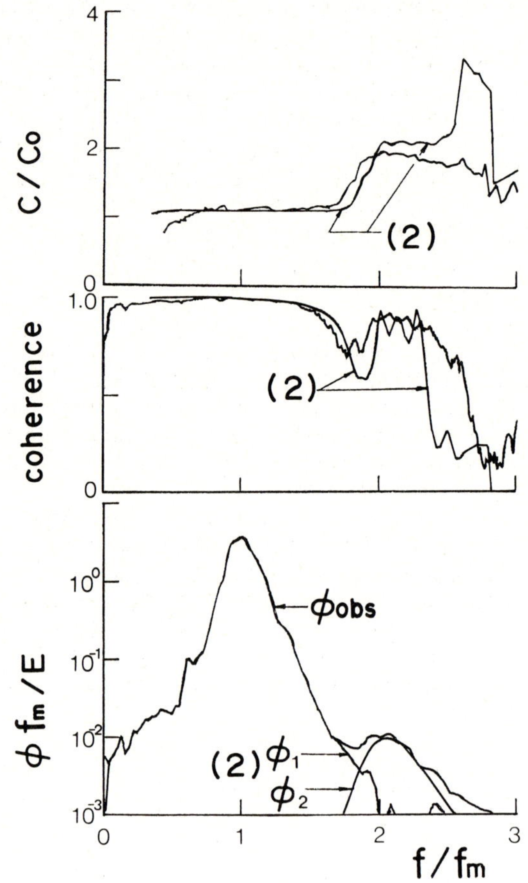

Figure 1
Comparison of the nonlinear theory with observational data; normalized phase velocity C/C_0 coherence and normalized frequency spectrum $\phi f_m/E$. Curve (2) is a second-order approximation to the nonlinear theory, ϕ_{obs} a observed spectrum, ϕ_1 the spectrum of free waves and ϕ_2 the spectrum of nonlinear bounded waves. C_0 is a phase velocity of long-crested linear waves [6].

GROWTH OF THE WAVE SPECTRUM

Generation of wind waves When wind begins to blow over still water, a surface drift current is first generated by wind shear stress. A short time later tiny ripples are generated by a shear flow instability mechanism of two-layer viscous fluid (air and water) [11], [12]. The initial wavelets have a long-crested nature and develop exponentially with time but without initially changing their spectral peak frequency. After a short duration, a fairly abrupt transition from regular and long-crested initial wavelets to irregular and short-crested wind waves takes place [12]. Following this transition, the wind waves show an ordinary growth pattern with the spectral peak frequency shifting towards successively lower frequencies with the increase in spectral energy. Quite similar evolution of wind waves is also observed near the upwind boundary in the process of the generation and growth with fetch under steady condition [13].

Growth of the wind wave spectrum After the transition from the initial wavelets to the wind waves, wind wave spectrum develops with time keeping approximately similar spectral forms [14], and finally attains to a saturated spectrum (steady state). Even after the steady state the wave spectrum develops with fetch as shown in Figure 2(a). The wave spectrum develops also with wind speed (Figure 2(b)), though the growth pattern at high frequencies is slightly different from that observed in the wave spectra developing with fetch under a fixed wind speed.

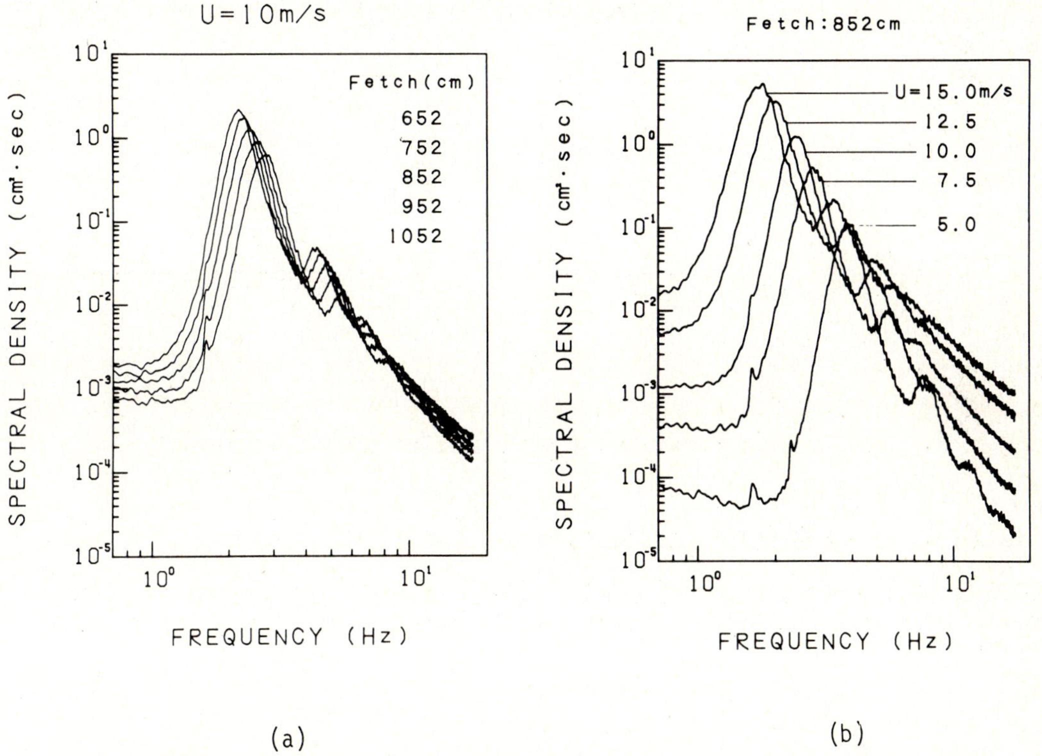

Figure 2
(a) growth of the wave spectrum with fetch for a fixed wind speed.
(b) growth of the wave spectrum with wind speed at a fixed fetch.

The wave spectra shown in Figure 2 are normalized in the form

$$\phi f_m / E = \Phi (f / f_m) ,$$

$E = \int_0^\infty \phi(f)df (\equiv \overline{\eta^2}, \overline{\eta^2}$: variance of surface elevation η). The following properties for the growth of the wind wave spectrum are apparent in Figures 2 and 3: (1)

Spectral energy increases mainly in a frequency region lower than the spectral peak frequency f_m and the spectral peak frequency shifts gradually to the low frequency side (Figure 2). (2) Spectral forms at each growth stage show approximately similar form (Figure 3), though some wind-speed dependence is seen at high frequency region (Figure 3(b)).

The similar properties are seen for the growth of wind wave spectra in the ocean [15], [16], though the normalized spectra show slightly different forms [16], [17].

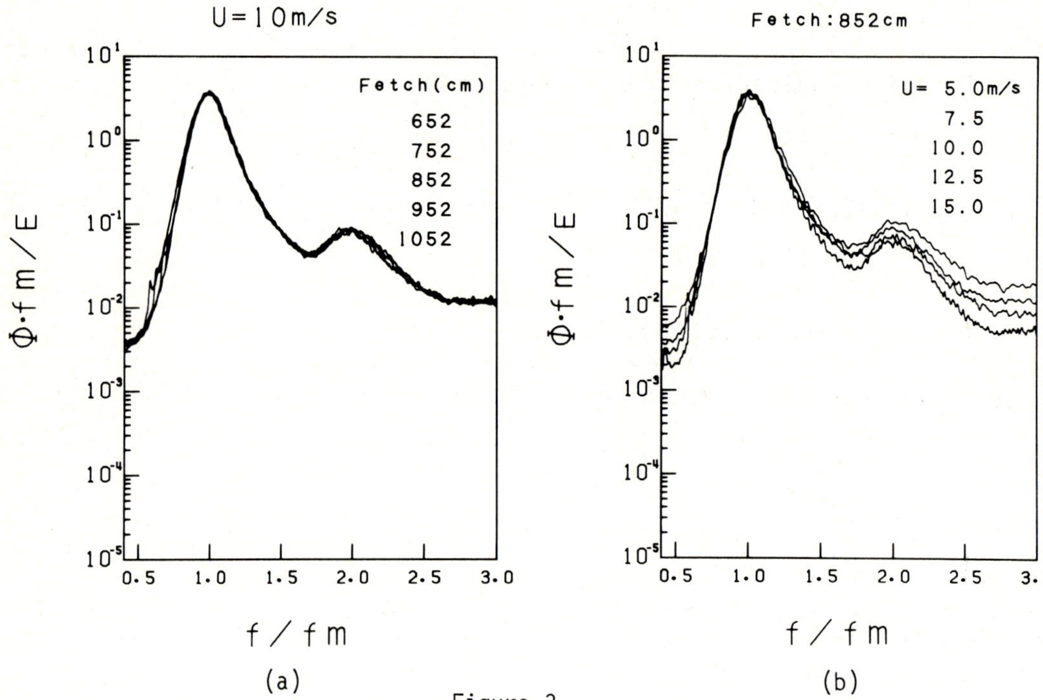

Figure 3
Normalized forms of the wave spectra shown in Figure 2.

FETCH RELATIONS FOR SPECTRAL PARAMETERS

For fetch-limited wind waves, the total energy of the wave spectrum E and the spectral peak frequency f_m are related to the friction velocity of the wind U_* and the fetch F

$$\frac{g\sqrt{E}}{U_*^2} = 1.31 \times 10^{-2} \left(\frac{gF}{U_*^2} \right)^{0.504}, \qquad (1)$$

$$\frac{U_* f_m}{g} = 1.00 \left(\frac{gF}{U_*^2} \right)^{-0.330}, \qquad (2)$$

where g is the acceleration of gravity [18].

The relations (1) and (2) are modified as follows by approximating the constants as 0.504 → 1/2, 0.330 → 1/3, converting the friction velocity U_* into the wind speed U at 10m height by

$$U_*^2 = C_D U^2 , \tag{3}$$

and assuming the drag coefficient as $C_D=1.6\times10^{-3}$;

$$\tilde{E}^{1/2} = 5.24 \times 10^{-4} \tilde{F}^{1/2} , \tag{4}$$

$$\tilde{f}_m = 2.92 \tilde{F}^{-1/3} . \tag{5}$$

Here $\tilde{E}$, $\tilde{F}$ and $\tilde{f}m$ represent the dimensionless variables

$$\tilde{E} = g^2E / U^4 , \tilde{F} = gF / U^2 \text{ and } \tilde{f}m = Ufm / g. \tag{6}$$

Similary, if we assume $C_D=1.0\times10^{-3}$ the relations (1) and (2) can be writen as

$$\tilde{E}^{1/2} = 4.14 \times 10^{-4} \tilde{F}^{1/2} , \tag{7}$$

$$\tilde{f}_m = 3.16 \tilde{F}^{-1/3} , \tag{8}$$

which are quite similar to the following relations obtained by Hasselmann et al. (1973) [15],

$$\tilde{E}^{1/2} = 4.0 \times 10^{-4} \tilde{F}^{1/2} , \tag{9}$$

$$\tilde{f}_m = 3.5 \tilde{F}^{-0.33} . \tag{10}$$

In fact, they assumed $C_D=1.0\times10^{-3}$. Therefore, the relations (9) and (10) become almost the same to the relations (1) and (2), if the wind speed U is converted to the friction velocity of the wind U_*.

From the relations (4) and (5), fetch relations for various important parameters characterizing the wind waves are obtained;

Nonlinearity parameter $E\,\omega_m{}^4/ g^2$,[19]

$$E\omega^4{}_m / g^2 = 3.11 \times 10^{-2} \tilde{F}^{-1/3} , \tag{11}$$

significant slope §,[20]

$$§ \equiv E^{1/2} / \lambda_0 = 2\pi f_m{}^2 E^{1/2} / g = 2.81 \times 10^{-2} \tilde{F}^{-1/6} \tag{12}$$

Three-second power law,[21]

$$Ef_m{}^3 / gU = \tilde{E}\tilde{f}_m{}^3 = 6.84 \times 10^{-6} . \tag{13}$$

The relations (11) and (12) mean that nonlinearity of the wind waves decreases with increasing the dimensionless fetch. That is, the nonlinearity of the ocean waves are relatively small as compared to that of laboratory wind waves. The relation (13) is a universal relation which is satisfied irrespective of the dimensionless fetch.

SIMILARITY FORM OF THE WAVE SPECTRUM

Idealized form of the frequency spectrum The two-dimensional wave spectrum $\Psi(f,\theta)$ is usually expressed in the form

$$\psi(f,\theta) = \phi(f)\, G(\theta, f), \tag{14}$$

where $\phi(f)$ is the (one-dimensional) frequency spectrum and $G(\theta,f)$ is the angular distribution function.

For the frequency spectrum of fully developed wind waves in the ocean, Pierson & Moskowitz (1964) [22] proposed a fetch independent idealized form, well known Pierson-Moskowitz spectrum. On the other hand, for the frequency spectrum of fetch-limited wind waves, Hasselmann et al. (1973) [15] proposed the JONSWAP spectrum,

$$\phi(f) = \alpha(2\pi)^{-4} g^2 f^{-5} \exp\left[-\frac{5}{4}\left(\frac{f}{f_m}\right)^{-4}\right] \cdot \gamma^{\exp\frac{-(f/fm-1)^2}{2\sigma^2}} \tag{15}$$,

where α and f_m are fetch-dependent scale parameters, and γ and σ are shape parameters. According to Hasselmann et al. (1973), the scale parameters α and f_m are given respectively by

$$\alpha = 7.6 \times 10^{-2} \tilde{F}^{-0.22}, \tag{16}$$

$$\tilde{f}_m = 3.5 \tilde{F}^{-0.33}, \tag{17}$$

and mean values of the shape parameters, γ and σ are given respectively by

$$\gamma = 3.3, \qquad \sigma = \begin{cases} 0.07 & \text{for } f \leqslant f_m \\ 0.09 & \text{for } f > f_m \end{cases}. \tag{18}$$

Recently, Mitsuyasu et al. (1980) [17] applied the spectral form (15) to ocean wave spectra observed in generation area in the open ocean and found quite good agreements ((a) and (b) in Figure 4). However, the scale parameter α and the shape parameter γ showed the following forms,

$$\alpha = 8.17 \times 10^{-2} \tilde{F}^{-2/7} \tag{19}$$

$$\gamma = 7.0 \tilde{F}^{-1/7}. \tag{20}$$

which are different from the results of Hasselmann et al. (1973)

In Figure 5 the relations (16) and (19) are compared with various measurements, where the following empirical relations are also included;

$$\alpha = 8.1 \times 10^{-2} \tilde{F}^{-0.308}, \quad [23] \tag{21}$$

$$\alpha = 8.0 \times 10^{-2} \tilde{F}^{-0.25}, \quad [24] \tag{22}$$

It can be seen from Figure 5 that the relation (19) shows the best overall fit to the data.

According to (20), the parameter γ decreases with the increase of $\tilde{F}$ and takes the value 3.3 near $\tilde{F}=2\times10^2$ and approaches to 1 near $\tilde{F}=10^6$. For laboratory wind waves ($\tilde{F}=1$) γ approaches to 7.

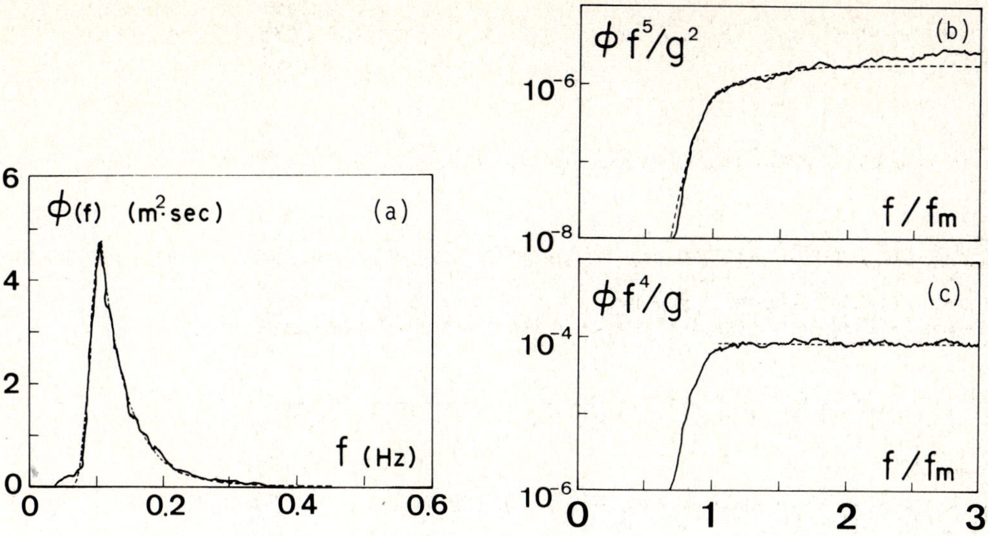

Figure 4
The figure (a) is the frequency spectrum $\phi(f)$ and (b) is its normalized form $f^5\phi/g$. Short-dash line is (15). The figure (c) is the same spectrum shown in a different form, where short-dash line is (24).

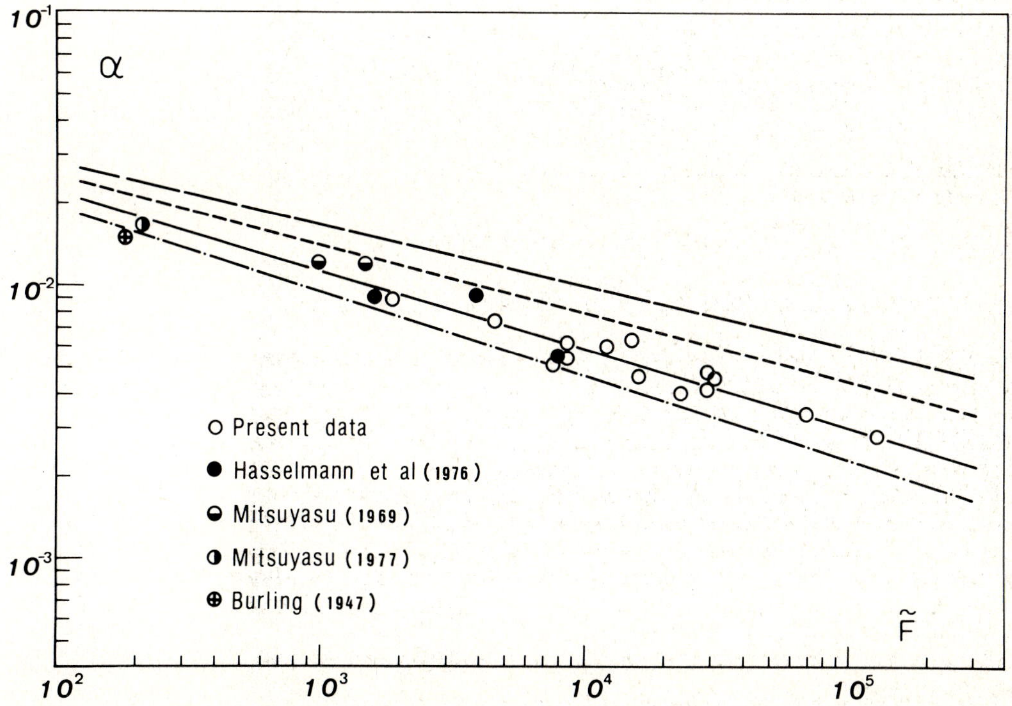

Figure 5
The scale parameter α versus the dimensionless fetch $\tilde{F}$; long-dash line (16), solid line (19), dot-dash line (21), short-dash line (22).

The equilibrium form of the wave spectrum Phillips (1958) [25] proposed the existence of the saturation range in a high frequency region of the wind wave spectrum, where the spectrum is limited only by breaking under the influence of gravity. On dimensional ground, he delived the following equilibrium form of the wave spectrum

$$\phi(f) = \alpha (2\pi)^{-4} g^2 f^{-5} \quad \text{for } f_m < f \ll f_\gamma , \tag{23}$$

where α is a dimensionless constant (Phillips constant), and f_γ is the frequency where the effects of gravity and surface tension are equall. At high frequency region of the spectrum, both the Pierson-Moskowitz spectrum and the JONSWAP spectrum (15) tend to the equilibrium form (23), though α is not an universal constant as previously shown in Figure 5.

Recently, however, a different equilibrium form of the wind wave spectrum has been presented by several authors [17], [26], [27], [28], [29], [30], which is given by

$$\phi(f) = \alpha'(2\pi)^{-3} g u_* f^{-4} \quad \text{for } f_m < f , \tag{24}$$

where α' is a dimensionless constant. The existence of such a spectral form was originally conjectured by Kitaigorodskii (1963) [31] from similarity considerations, but experimental verification was not done untill the study of Toba (1973) [26]. Recently Mitsuyasu et al. (1980) [17] has shown that α' depends very weakly on the dimensionless fetch $\tilde{F}$ as

$$\alpha' = 5.57 \times 10^{-2} \tilde{F}^{1/21} , \tag{25}$$

The spectral form (24) is compared in Figure 4 (c), with the measured spectrum which is the same to that used in the figures (a) and (b) [17].
As shown in Figure 4, both the JONSWAP spectrum (15) and the new equilibrium spectrum (24) fit quite well to the measured spectrum in arange, $1 \lesssim f/f_m \lesssim 2$, but (24) seems to fit up to higher frequencies. From this fact we can suggest the following new standard form of the wave spectrum,

$$\phi(f) = \alpha'(2\pi)^{-3} g u_* f^{-4} \exp\left[-\frac{5}{4}\left(\frac{f}{f_m}\right)^{-4}\right] \cdot \gamma^{\exp \frac{-(f/f_m-1)^2}{2\sigma^2}} \tag{26}$$

where α' is given by (25).

High-frequency wave spectrum Recently much attention has been focused on the high-frequency wave spectrum despite its neglisible energy. This is mainly due to the fact that the high-frequency waves that have the same wave length as the radar signals play an important roll in using the radar for the remote sensing of ocean parameters, such as ocean waves and ocean surface winds [32]. Particularly the wind-speed dependence of the high-frequency wave spectrum is very important for the remote measurement of the ocean surface wind by a satellite scatterometer.

After the pioneering study of Pierson & Stacy (1973) [33] many experimental studies have been carried out on the high-frequency wave spectrum in a gravity-capillary and capillary range [34], [35], [36], [37]. The results of these studies are summerized in a form

$$\phi f^5 / g^2 = D (u_* f / g)^a (\nu u_* / \gamma_s)^b , \tag{27}$$

where D, a and b are dimensionless constants, ν is the kinematic viscosity of the water and $\gamma_s (\equiv \sigma_s/\rho_w)$ is the ratio of surface tension σ_s to water density ρ_w. From (27) various empirical formulae can be obtained;

Mitsuyasu & Honda (1974) spectrum [34]

$$\phi(f) = \alpha g (2\pi)^{-3} g_* u_* f^{-4} \tag{28}$$

$$g_* = g + \gamma_s k^2 , \qquad (k\text{: wave number})$$

for $a = 1, \; D(\nu u_* / \gamma_s)^b = \alpha g (2\pi)^{-3} g_* / g ,$

Lleonart & Blackman (1980) spectrum [36]

$$\phi(f) = 4.12 \times 10^{-3} u_*^2 (\nu u_* / \gamma_s)^{1/2} f^{-3}$$

for $b = \frac{1}{2}, \; D = 4.12 \times 10^{-3}$ (29)

Liu & Lin (1982) spectrum [37]

$$\phi(f) = \beta \gamma_s^{2/3} f^{-7/3} \tag{30}$$

for $a = \frac{8}{3}, \; b = -\frac{2}{3}, \; D = (g\nu / u_*^3)^{2/3} \beta$

Although the spectral forms (28), (29) and (30) are apparently different in each other they show similar spectral form and its wind-speed dependence [36], [37]. However, in order to obtain more consistent results further studies are needed. Furthermore, studies on the vector wavenumber spectrum in high-wavenumber region are very important. Because the vector wavenumber spectrum is directly connected with the interaction between water waves and electromagnetic waves and conversion from the frequency spectrum to the wavenumber spectrum needs the dispersion relation and the angular distribution function of high-frequency waves, both of which are still not clear.

Angular distribution function For wind waves in generation area their angular distribution function can be approximated by the following form [16], [38];

$$G(\theta, f) = G'(S) \left| \cos \frac{\theta - \bar{\theta}}{2} \right|^{2s} \tag{31}$$

where G'(s) is a normalizing function

$$G'(S) = \frac{2^{2s}}{2\pi} \frac{\Gamma^2(s+1)}{\Gamma(2s+1)} \tag{32}$$

$\bar{\theta} = \bar{\theta}(f)$ is the mean direction of a spectral component and $s = s(f)$ is a parameter which controlles the concentration of the angular distribution of the spectral energy. In (32) Γ is a gamma function.

Generally the parameter s is considered to be a function of f/f_m and U/C_m or equivalently $\tilde{F}$:

$$S = S(f / f_m, U / C_m) \quad \text{or} \quad S = S(f / f_m, \tilde{F}) . \tag{33}$$

Here, $C_m = g/2\pi f_m$ is the phase velocity of the frequency component at the spectral peak frequency f_m. In fact, according to Mitsuyasu et al. (1975) [16] the parameter s shows maximum value s_m near $f/f_m = 1$ and it decreases rapidly towared higher ($f/f_m>1$) and lower ($f/f_m<1$) frequencies. An empirical relation for the parameter s is given approximately by

$$S / S_m = (f / f_m)^{\mu} , \tag{34}$$

where

$$S_m = 11.5 (U / C_m)^{-2.5} , \quad \text{for} \quad 0.7 \lesssim U / C_m \lesssim 1.2 \tag{35}$$

$$\mu = \begin{cases} \mu_a = 5 & \text{for} \quad f / f_m \lesssim 1 \\ \mu_b = -2.5 & \text{for} \quad f / f_m \gtrsim 1 \end{cases} \qquad (36)$$

The parameterization (34), (35) and (36) can be rewritten as

$$S = 11.5\,(U / C_m)^{-7.5} (U / C)^5 \quad \text{for} \quad f / f_m \lesssim 1 \qquad (37)$$

$$\text{and} \quad 0.7 \lesssim U / C_m \lesssim 1.2$$

$$S = 11.5\,(U / C)^{-2.5} \quad \text{for} \quad f / f_m \gtrsim 1 , \qquad (38)$$

Recently Hasselmann et al. (1980)[39] proposed a slightly different parameterization for s. They used the same form (34), but sm and μ are given as follows;

$$\left.\begin{aligned} S_m &= 6.97 \pm 0.83 \\ \mu &= \mu_a = 4.06 \pm 0.22 \end{aligned}\right\}, \quad \text{for} \quad f / f_m \lesssim 1.06 , \qquad \begin{matrix}(39a)\\(39b)\end{matrix}$$

and

$$S_m = 9.77 \pm 0.43 , \qquad (40a)$$

$$\mu = \mu_b = -(2.33 \pm 0.06) - (1.45 \pm 0.45)\,(U / C_m - 1.17) , \qquad (40b)$$

$$\text{for} \quad f / f_m \gtrsim 1.05 .$$

The parameterization (39) and (40) is different in forms from (35) and (36) but their original data is not so far from (37) and (38) (Figure 2 in [39]). For practical applications of these empirical formulas it will be safe to use them within each limited range of U/Cm , because both (35) and (40) give unbelievable values of s for very large U/Cm.

CONCLUDING REMARKS

The ocean surface waves can be described effectively by an ordinary spectral model due to their weak nonlinearity; the linear model and weakly nonlinear model have been applied widely and successfully to various problems. However, in-order to clarify the evolution of the wave field, strongly nonlinear phenomena such as wave instability and succeeding wave breaking must be investigated. For they are responsible for the dissipation of the wave energy.

The ocean wave spectra generated by strong and steady wind of long duration have similarity forms, whose parameters are interrelated with one another by simple formulae. It should be kept in mined, however, that more complicated wind conditions yield various kinds of different spectral forms.

Finally it should be noted that much progress has been made recently on (1) the fundamental dynamics of the evolution of the ocean wave spectra including the strongly nonlinear phenomena, and also on (2) the practical description of shallow-water wave spectra, although they are not discussed in this paper.

REFERENCES

[1] Ramamonjiarisoa, A., Contribution a l'étude de la structure statistique et des mécanismes de génération des vagues de vent. Thesis, Universite de Provence (Inst. Mech. Stat. de la Turbulence, no. A.O. 10. 023. (1974)

[2] Lake, B.M., and Yuen, H.C., A new model for nonlinear wind waves, Part 1 Physical model and experimental evidence, J. Fluid Mech., 88 (1978) 33-62.

[3] Mollo-Christensen, E., and Ramamonjiarisoa, A., Modeling the presence of wave groups in a random wave field, J. Geophys. Res., 83 (1978) 4117-4122.

[4] Barrick, D.E. and Weber, B.L., On the nonlinear theory of gravity waves on the ocean's surface. Part 2, Interpretation and Applications, J. Phys. Oceanogr., 7 (1977) 11-22.

[5] Masuda, A., Kuo, Yi-Yu and Mitsuyasu, H., On the dispersion relation of random gravity waves, Part 1, Theoretical framework, J. Fluid Mech., 92 (1979) 717-730.

[6] Mitsuyasu, H., Kuo Yi-Yu and Masuda, A., On the dispersion relation of random gravity waves, Part 2, An experiment, J. Fluid Mech., 92 (1979) 731-749.

[7] Kuo, Yi-Yu, Mitsuyasu, H. and Masuda, A., Experimental study on the phase velocity of wind waves, Part 1, Laboratory wind waves, Rep. Res. Inst. Appl. Mech., Kyushu Univ., 83 (1979a) 1-19.

[8] ——————————————————, Experimental study on the phase velocity of wind waves, Part 2, Ocean waves, Rep. Res. Inst. Appl. Mech., Kyushu Univ., 84 (1979b) 47-66.

[9] Plant, W.J. and Wright, J.W., Spectral decomposition of short gravity wave systems, J. Phys. Oceanogr., 9 (1979) 621-624.

[10] Phillips, O.M., The dispersion of short wavelets in the presence of dominant long wave, J. Fluid Mech., 107 (1981) 465-485.

[11] Valenzuela, G.R., The growth of gravity-capillary waves in a coupled shear flow, J. Fluid Mech., 76 (1976) 229-250.

[12] Kawai, S., Generation of initial wavelets by instability of a coupled shear flow and their evolution to wind waves, J. Fluid Mech., 93 (1979) 661-703.

[13] Ramamonjiarisoa, A., Baldy, S. and Choi, I., Laboratory Studies on wind-wave generation, amplification and evolution, in: Favre, A. and Hasselmann, K. (eds), Turbulent fluxes through the sea surface, wave dynamics, and prediction (Plenum, N.Y. 1978).

[14] Mitsuyasu, H. and Rikiishi, K., The growth of duration-limited wind waves, J. Fluid Mech., 85 (1978) 705-730.

[15] Hasselmann, K., Barnett, T.P., Bouws, E., Carlson, H., Cartwright, D.E., Enke, K., Ewing J.A., Gienapp, H., Hasselmann, D.E., Kruseman, P., Meerburg, A., Muller, P., Olber, D.J., Richter, K., Sell, W., and Walden, H., Measurement of wind wave growth and swell decay during the Joint North Sea Wave Project (JONSWAP), Dtsch. Hydrogr. Z., Suppl. A, No.12 (1973) 95pp.

[16] Mitsuyasu, H., Tasai, F., Suhara, T., Mizuno, S., Ohkusu, M., Honda, T. and Rikiishi, K., Observation of the directional spectrum of ocean waves using a cloverleaf buoys, J. Phys. Oceanogr., 5 (1975) 750-760.

[17] ——————, Observation of the power spectrum of ocean waves using a cloverleaf buoy, J. Phys. Oceanogr., 10 (1980) 286-296.

[18] Mitsuyasu, H., On the growth of the spectrum of wind-generated waves (1), Rep. Res. Inst. Appl. Mech., Kyushu Univ., 16 (1968) 459-482.

[19] Masuda, A., Nonlinear energy transfer between wind waves, J. Phys. Oceanogr., 10 (1980) 2082-2093.

[20] Huang, N.E., Long, S.R., Tung, C.C., Yuen, Y. and Bliven, L.F., A unified two-parameter wave spectral model for a general sea state, J. Fluid Mech., 112 (1981) 203-224.

[21] Toba, Y., Local balance in the air-sea boundary process 1, J. Oceanogr. Soc. Japan, 28 (1972) 109-120.

[22] Pierson, W.J. and Moskowitz, L., A proposed spectral form for fully developed wind seas based on the similarity theory of S.A. Kitaigorodskii, J. Geophys. Res., 69 (1964) 5181-5190.

[23] Mitsuyasu, H., One dimensional wave spectra at limited fetch, Rep. Res. Inst. Appl. Mech., Kyushu Univ., 20 (1973) 37-53.

[24] Liu, P.C., Normalized and equilibrium spectra of wind waves on Lake Michigan, J. Phys. Oceanogr., 1 (1971) 249-257.

[25] Phillips, O.M., The equilibrium range in the spectrum of wind-generated waves, J. Fluid Mech., 4 (1958) 426-434.

[26] Toba, Y., Local balance in the air-sea boundary process III, On the spectrum of wind waves, J. Oceanogr. Soc. Japan, 29 (1973) 209-220.

[27] ———, Stochastic form of the growth of wind waves in a single parameter representation with physical implications, J. Phys. Oceanogr., 8 (1978) 494-507.

[28] Forristal, G.Z., Measurements of a saturated range in ocean wave spectra, J. Geophs. Res., 86 (1981) 8075-8084.

[29] Kahma, K.K., A study of the growth of the wave spectrum with fetch, J. Phys. Oceanogr., 11 (1981) 1503-1515.

[30] Kitaigorodskii, S.A., On the theory of the equilibrium range in the spectrum of wind-generated gravity waves, J. Phys. Oceanogr., 13 (1983) 816-827.

[31] ——————, Applications of the theory of similarity to the analysis of wind-generated wave motion as a stochastic process, Izv. Geophys. Ser. Acad. Sci., USSR., 1 (1962) 105-117.

[32] Pierson, W.J., The theory and application of ocean wave measuring systems at and below sea surface on the land, from the aircraft and from spacecraft, NASA Contractor Rep. no. CR-2646 (1976)

[33] Pierson, W.J.,and Stacy R.A., The elevation, slope, and curvature spectra of wind roughened sea surface, NASA Contractor Rep. no. CR-2247 (1973)

[34] Mitsuyasu, H. and Honda, T., The high frequency spectrum of wind-generated waves, J. Oceanogr. Soc. Japan, 30 (1974) 185-198.

[35] Mitsuyasu, H., Measurement of the high-frequency spectrum of ocean surface waves, J. Phys. Oceanogr., 7 (1977) 882-891.

[36] Lleonart, G.T. and Blackman, D.R., The spectral characteristics of wind-generated capillary waves, J. Fluid Mech., 97 (1980) 455-479.

[37] Liu, Hsein-Ta and Lin, Jung-Tai, On the spectra of high-frequency wind waves, J. Fluid Mech., 123 (1982) 165-185.

[38] Longuet-Higgins, M.S., Cartwright, D.E. and Smith N.D., Observations of the

directional spectrum of sea waves using the motion of a floating buoy, in: Ocean Wave Spectra (Prentice Hall, 1963)

[39] Hasselmann, D.E., Dunkel, M. and Ewing, J.A., Directional wave spectra observed during JONSWAP 1973, J. Phys. Oceanogr., 10 (1980) 1264-1280.

Theoretical and Applied Mechanics
F.I. Niordson and N. Olhoff
Elsevier Science Publishers B.V. (North-Holland)

NUMERICAL AND ANALYTICAL ASPECTS OF HELMHOLTZ INSTABILITY

D. W. Moore

Imperial College
London

Recent analytical and numerical work on the evolution of plane vortex sheets is examined and the role of discrete Helmholtz instability in numerical computations is examined. The evidence that non-linear Helmholtz instability leads to the formation of a singularity in the shape of the sheet is reviewed.

§1 INTRODUCTION

I am concerned with the evolution of a discontinuity in an unbounded plane irrotational flow of inviscid, incompressible, homogeneous fluid. This discontinuity, or vortex sheet, is a popular approximation to a thin shear layer in a viscous fluid at high Reynolds number and recently computation of the vortex sheet at the interface between immiscible homogeneous fluids has proved an effective way of studying large amplitude waves or instabilities. However, I do not consider two-fluid problems here.

The tangential velocity is discontinuous across the vortex sheet and the jump γ measures the strength of the sheet. An element pp' of the sheet of length ds contains circulation $d\Gamma = \gamma ds$ and it can be shown that, if the end points p and p' move with the mean of the fluid velocity on the two sides, $d\Gamma$ is invariant. Provided $\gamma > 0$, the net circulation is a monotone increasing function of s and can thus replace s as an intrinsic coordinate. The circulation Γ has the great merit of being Lagrangian and if the shape of the sheet is described by a complex coordinate $z(\Gamma,t)$, the comparatively simple Rott-Birkhoff equation (2.1) results.

The straight uniform infinite vortex sheet was shown by Helmholtz to be unstable to all disturbances and a sinusoidal wave of length λ has growth rate $\pi\gamma/\lambda$.

The purpose of this article is to trace out some of the consequences of this pathological instability. I have analysed the numerics and argued (Moore 1981) that this pathology is the cause of many of the problems associated with Rosenheads (1931) point vortex method, difficulties first noted in important early papers by Birkhoff and Fisher (1959) and Birkhoff (1962). In particular I show in §2 that the discretisation of the principal value integral in (2.1) is not the cause of the difficulty.

Since a shear layer of finite thickness and finite vorticity will not exhibit pathological instability it is natural to consider ways of removing it from the vortex sheet and I discuss this in §3. Finally in §4 I examine the analytical consequences of the pathology and review evidence that a singularity forms.

§2 NUMERICAL ASPECTS OF THE INITIAL-VALUE PROBLEM

I consider a vortex sheet in the form of a simple initially analytic closed curve. The total circulation is Γ_o, so that $z(o,t) = z(\Gamma_o,t)$, and the initial-value problem is that of solving the Rott-Birkhoff equation

$$\frac{\partial \bar{z}(\Gamma,t)}{\partial t} = -\frac{i}{2\pi}\int_0^{\Gamma_o}\frac{d\Gamma'}{z(\Gamma,t) - z(\Gamma',t)} \qquad (0 \leqslant \Gamma \leqslant \Gamma_o), \qquad (2.1)$$

subject to given initial shape $f(\Gamma)$, so that

$$z(\Gamma,o) = f(\Gamma) \qquad (0 \leqslant \Gamma \leqslant \Gamma_o). \qquad (2.2)$$

I examine here only the point-vortex method, with various refinements, because this method is general and thus I do not examine methods, such as that of Meiron, Baker and Orszag (1982) or Schwarz (1981), which depend on properties special to the particular initial-value problems studied. Nor do I discuss other types of discrete representation.

To evaluate the principal-value integral, one can proceed in two distinct ways.

(a) A cancellation function, which has exactly the same residue at $\Gamma = \Gamma'$ as the integrand in (2.1), but which can be integrated exactly, can be introduced. This was done by Van der Vooren[1] (1980) for periodic disturbances to an infinite, straight, uniform vortex sheet. It appears that the cancelled integrand is analytic and so any of the standard integration formulae can be employed. The problem can be discretised by dividing Γ_o into N equal portions $\Delta\Gamma$ and defining

$$\Gamma_s = s\,\Delta\Gamma \qquad (s = 1,2,\ldots,N) \qquad (2.3)$$

and

$$z_s(t) = z(\Gamma_s,t) \qquad (2.4)$$

If the trapezium rule is used, equation (2.1) is replaced by

$$\frac{d\bar{z}_s}{dt} = -\frac{i}{2\pi}\sum_r{}' \frac{\Delta\Gamma}{z_s - z_r} - \frac{i\Delta\Gamma}{4\pi}\frac{D^2 z_s}{(Dz_s)^2} \qquad (s = 1,2,\ldots N) \qquad (2.5)$$

where D denotes a centred-difference operator.

If the second term on the right is dropped, (2.5) reduces to Rosenheads (1931) point-vortex method, which thus emerges as a leading order approximation. The second term is Van der Voorens correction. In general, (2.5) is second-order accurate, although for the special case of a smooth closed vortex sheet, for which trapezoidal integration of the cancelled integrand is exponentially accurate, the accuracy is controlled by the order of the differences used in the correction term. Fornberg (private communication) has pointed out that, with neglect of $O(\Delta\Gamma^3)$, the correction term can be written in the form

$$-\frac{i\Delta\Gamma}{4\pi}\left(\frac{1}{z_s - z_{s-1}} + \frac{1}{z_s - z_{s+1}}\right)$$

which is equivalent to an increase of the contribution of the two neighbouring point vortices. Clearly the point vortex method can be made second order accurate at negligible computational cost.

(b) A second strategy for the accurate evaluation of the integral, which has been employed by Kuwahara and Takami (1973), Chorin and Bernard (1973) and elaborated in the context of smooth patches of vorticity by Beale and Majda (1982) is to replace the integrand in (2.1) by a bounded function

$$(\bar{z} - \bar{z}')\, f(|\bar{z} - \bar{z}'|;\, \delta)$$

where $f(o;\delta) \neq \infty$ and $f(u;\delta) \sim u^{-2}$ when $u \gg \delta$. The integral can now be evaluated by any standard formulae and the trapezium rule gives

$$\frac{d\bar{z}_s}{dt} = -\frac{i}{2\pi}\sum_r \Delta\Gamma(\bar{z}_s - \bar{z}_r)\, f(|z_s - z_r|;\delta) \qquad (2.6)$$

which is Rosenhead's method for modified point vortices.

The method can be applied to the uniform circular vortex sheet of unit radius and unit strength. The velocity of any point is of magnitude 0.5 while (2.6) gives a speed Q_N where (for odd N)

$$Q_N = \frac{4}{N}\sum_{s=1}^{\frac{N-1}{2}} \mathrm{Sin}^2\,\frac{s\varphi}{2}\; f(2\,\mathrm{Sin}\,\frac{s\varphi}{2};\delta) \qquad (2.7)$$

where $\varphi = 2\pi/N$. I illustrate by choosing (for simplicity)

$$f(u;\delta) = \delta^{-2} \quad (u \leqslant \delta);$$
$$f(u;\delta) = u^{-2} \quad (u \geqslant \delta). \qquad (2.8)$$

If $\delta \ll 1$ (as it must be if sizeable errors are not to be introduced) it is easy to show that

$$Q_N = \frac{1}{2N}\left(N - 1 - 2\,s_o + \frac{4}{3}\,\frac{\pi^2}{N^2\delta^2}\,(2s_o+1)(s_o+1)s_o\right)$$

where

$$s_o = [N\delta/2\pi]$$

If $s_o = 0$, corresponding to a core size less than the inter vortex spacing

$$Q_N = \tfrac{1}{2}(1-\frac{1}{N})$$

which is the usual point vortex approximation. If s_o is not large, corresponding to $N\delta/2\pi$ of order unity, the method gives comparable accuracy to the Rosenhead point vortex method, but if $N\delta >> 1$, corresponding to a core size of many intervortex spacings,

$$Q_N \approx \tfrac{1}{2}(1 - \frac{2\delta}{3\pi})$$

Clearly, then, δ must be carefully chosen. Baker and Beale (private communication) have shown that the need to choose δ carefully is more acute when more sophisticated forms of f are chosen and that an initial choice can become inappropriate as the vortex sheet evolves, leading to sizeable errors.

Both methods (a) and (b) lead to autonomous systems of ordinary differential equations so I must next ask if there are any special problems attached to the time integration of these systems. If the correction term in (2.5) is dropped one is dealing with a set of point vortices of equal strength. The invariants of this system are well known (Lamb 1932 p229) and in particular

$$\Delta\Gamma \sum_s |z_s|^2 = \text{const} \qquad (2.9)$$

and

$$\frac{\Delta\Gamma^2}{2\pi} \sum_{\text{all pairs}}\sum \ln|z_s - z_r| = \text{const} \qquad (2.10)$$

Clearly if $z_s \to z_r$ for some pair of values, the second invariant requires that at least one other point vortex recedes to ∞. But this is ruled out by the first invariant (2.9). Thus, in the case considered collisions are impossible and time integration presents no special difficulty. However, even though the autonomous system remains regular and standard methods can be applied, the system possess, if $N >> 1$, a wide range of inherent time scales. I take up this question in the following section.

§3 ILL-POSEDNESS AND SMOOTHING

I have tried to show that there is no great difficulty in evaluating accurately, the Cauchy principal-value integral in equation (2.1). The difficulty arises because the resulting discrete inital value problem can be ill-posed, which means that small changes in the initial conditions will produce finite changes in the solution at a finite time. No general discussion will be attempted, but special cases are highly suggestive. I examine first (2.5), dropping the correction term for simplicity.

Consider a straight infinite array of equally spaced point vortices of equal strength $\Delta\Gamma$, the intervortex distance being h. This configuration is an exact solution of (2.5) and its stability was discussed by Van Karman (described in Lamb 1932 p225). A spatially periodic disturbance of wavelength Rh grows with growth rate σ_R given by

$$\sigma_R = 4(R-1)/TR^2 \qquad (R = 2,3,4,\ldots) \tag{3.1}$$

where T is a characteristic time given by

$$T = 4h^2/\pi\Delta\Gamma \tag{3.2}$$

Clearly σ_R is a decreasing function of R for $R \geqslant 2$ so shorter waves grow faster, just as in Helmholtz instability. The shortest wave possible in the array has wavelength 2h and growth rate T^{-1} (which is just half the growth rate of a wave of length 2h on the continuous sheet the array is representing, a point to which I will return).

The ill-posedness of the initially value problem for the uniform array is now clear. Noise of amplitude 10^{-n} h will excite the shortest wave at comparable amplitude producing a growing disturbance 10^{-n} h exp(t/T). This will become apparent when $t/T \approx 2.3n$. I have confirmed this picture for uniform circular vortex sheet (Moore 1981) by time integration of (2.5) and spectral analysis of the results. The presence of discrete Helmholtz instability was confirmed by verifying that the numerically determined growth rates agreed with those found analytically for a regular polygon of equal point vortices by Havelock (1931).

The ill-posedness is introduced by the step of replacing a thin shear layer by a vortex sheet. A shear layer of finite thickness will not be unstable to waves much shorter than its thickness and this suggests trying to suppress the shorter waves on the discrete array.

The simplest method is to apply a smoothing formulae in which, after each time step, the coordinate of each point vortex is replaced by a linear combination of its near neighbours. For example Longuet-Higgins and Cokelet (1976) recommend

$$z_s \rightarrow (-z_{s-2} + 4z_{s-1} + 10z_s + 4z_s - z_{s+2})/16 \ ; \tag{3.3}$$

this was used by them in their calculations of breaking waves[2]. This can be applied to the initial value problem for the uniform array. If the time step is δt and $\delta t << T$ short waves R = 2,3, are heavily damped and waves with R >> 1 have a growth rate $\bar{\sigma}_R$ given by (Moore l.c.)

$$\bar{\sigma}_R = 4/RT - \frac{\pi^4}{R^4\delta t} \tag{3.4}$$

so that waves with $R < 2.89\,(T/\delta t)^{1/3}$ will not grow.

The re-positioning technique of Fink and Soh (1978) in which equal spacing of the point vortices is demanded can be shown to reduce discrete Helmholtz instability both by its direct effect on the growth rate (Moore 1981) and by ensuring that h does not become small in regions where the sheet is being compressed.

In a periodic problem the Fast Fourier transform can be used to analyse the shape of the vortex sheet, the amplitude of unwanted modes set equal to zero and the shape re-constructed by an inverse transform Fornberg (private communication), Krasny (1983).

Endowing the point vortex with a finite core, as in strategy (b) of §2, can also provide damping of short waves. It is a simple matter to repeat von Karman's calculation with (2.6) as the governing equation, to show that the growth rate μ_R is given by

$$\mu^2_R = \frac{\Delta\Gamma^2}{\pi^2} \quad s_1 s_2 \qquad (3.5)$$

where

$$s_1 = \sum_1^\infty (1 - \cos \frac{2\pi n}{R}) \; f(nh;\delta) \qquad (3.6)$$

and

$$s_2 = \sum_1^\infty (1 - \cos \frac{2\pi n}{R}) \; (-\frac{d}{ds}(s \; f(s;\delta))|_{s=nh}) \qquad (3.7)$$

Now s $f(s;\delta)$ is the swirl velocity in the modified vortex which is increasing outwards in the core region $s \sim \delta$ Thus if δ is large enough, s_2 can be negative, leading to stability. However, detailed calculation is needed to decide if a given mode is stabilized and comprehensive results will be reported elsewhere.

One can of course compute with thin shear layers directly and promising results have been obtained by Pozrikidis and Higdon(1984), for the case of uniform vorticity.

§4 THE FORMATION OF A SINGULARITY

I start with a reminder that small sinusoidal disturbances to a straight infinite uniform vortex sheet of strength γ have a growth rate $\pi\gamma/\lambda$. As was first pointed out Birkhoff and Fisher (l.c.) this dependence of growth rate on wavelength can lead, on linear theory, to the formation of a singularity at a finite time. In the periodic case, a growing disturbance takes the form

$$\sum_1^\infty A_n \quad \exp(n\pi\gamma t/\lambda) \; \sin(n\pi x/\lambda)$$

where A_n is the nth Fourier coefficient of the initial disturbance. If $A_n = e^{-\sqrt{n}}$ the initial disturbance is analytic, but convergence is lost at t = 0+, while if $A_n = e^{-n} \, n^{-p}$ (p>0) a singularity, whose detailed form depends on p, will form when $\pi\gamma t/\lambda = 1$.

The governing equations are non-linear so that, even if the initial disturbance contains only a single Fourier component, non-linear interactions will excite the entire spectrum. The crucial point is the strength of the excitation of higher modes; if (for example) A_n decayed as fast as e^{-n^2}, no singularity would form.

I have (Moore 1979 - M) tackled the initial value problem for the case of a sinusoidal initial disturbance of unit wave number imposed on a straight infinite vortex sheet of unit strength; thus I consider

$$\frac{\partial \bar{z}}{\partial t}(\Gamma,t) = \frac{1}{2\pi i}\int_{-\infty}^{\infty} \frac{d\Gamma'}{z(\Gamma,t) - z(\Gamma',t)} \tag{4.1}$$

with initial condition

$$z(\Gamma,0) = \Gamma + i\varepsilon \sin \Gamma \tag{4.2}$$

Here ε is the semi-amplitude of the sinusoidal disturbance. Some analytical progress is possible if $\varepsilon << 1$.

The method of solution is to write

$$z(\Gamma,t) = \Gamma + 2i \sum_{1}^{\infty} A_n(t) \sin n\Gamma \tag{4.3}$$

so that, in view of (4.2)

$$A_n(0) = \tfrac{1}{2}\varepsilon \quad \delta_{n1} \tag{4.4}$$

The key step is to realise that the expansion is of the row-echelon structure

$$A_n = \varepsilon^n A_{n0} + \varepsilon^{n+2} A_{n2} + \ldots\ldots \qquad (n = 1,2,3,\ldots) \tag{4.5}$$

This echelon structure enables A_{n0} to be determined recursively and for $n >> 1$

$$\varepsilon^n A_{n0} \approx t^{-1}(2\pi)^{-\frac{1}{2}} (1+i) n^{-5/2} \exp\left(n(1+\tfrac{1}{2}t + \ln(\tfrac{1}{4}\varepsilon t))\right), \tag{4.6}$$

so that exponential decay of the coefficients is lost when $t = t_c$, where

$$1 + \tfrac{1}{2}t_c + \ln t_c = \ln(4/\varepsilon) \tag{4.7}$$

A weak singularity forms in the shape of the sheet at $\Gamma = 0, \pm 2\pi, \ldots$ and explicitly

$$z(\Gamma,t) = \Gamma + \frac{2\sqrt{3}}{3t}(1+i)\left\{ (1-e^{i\Gamma}\varepsilon\Theta)^{3/2} - (1-e^{-i\Gamma}\varepsilon\Theta)^{3/2}\right\} + \text{less singular terms}, \tag{4.8}$$

where

$$\Theta = \tfrac{1}{4}t \exp(\tfrac{1}{2}t + 1) \tag{4.9}$$

For $t < t_c$, $\varepsilon\Theta < 1$, so that the solution has branch point singularities at $\Gamma = \pm i \ln(1/\varepsilon\Theta)$ which approach the real axis as $t \to t_c - 0$.

As pointed out by Merion, Baker and Orszag (1982) (subsequently referred to as MBO) the form of singularity is revealed most clearly by the distribution of strength $\gamma(\Gamma)$. If $t \to t_c - 0$ and $|\Gamma| << 1$, then (4.6) yields

$$\gamma(\Gamma) = 1 - \frac{\sqrt{3}}{t_c}(t_c - t + ((t_c - t)^2 + 4\Gamma^2)^{\frac{1}{2}})^{\frac{1}{2}} + 0(\frac{1}{t_c{}^2}) \tag{4.10}$$

so that while γ remains bounded at the critical time t_c, the distribution has a cuspidal form.

The problem can be approached in a different way, which makes the appearance of the singularity more natural. If one retains only A_{n0}, then

$$z = \Gamma + \sum_1^\infty A_{n0}\, \varepsilon^n e^{in\Gamma} - \sum_1^\infty A_{n0}\, \varepsilon^n e^{-in\Gamma} \tag{4.11}$$

Now the singularity forms at the points

$$\Gamma = \pm i\alpha(t) + 2N\pi \qquad (\alpha \text{ real}) \tag{4.12}$$

which suggest one puts $x = \varepsilon e^{i\Gamma}$ examine the generating function

$$G(x) = \sum_1^\infty A_{n0}\, x^n \qquad (x \text{ real}) \tag{4.13}$$

Provided x is real, the function $G(x,t)$ can be shown to satisfy the partial differential equation

$$\frac{\partial \bar{G}}{\partial t} = -\tfrac{1}{2} i\, x \frac{\partial G}{\partial x} \Big/ (1 + i\, x \frac{\partial G}{\partial x}) \tag{4.14}$$

with initial condition

$$G(x,0) = \tfrac{1}{2}x \tag{4.15}$$

It is convenient to put $x = \varepsilon\, e^y$ and define $f(y) = G(\varepsilon\, e^y)$, so that the problem is to solve

$$\frac{\partial \bar{f}}{\partial t} = -\tfrac{1}{2}\, i \frac{\partial f}{\partial y} \Big/ (1 + i\, \frac{\partial f}{\partial y}) \tag{4.16}$$

subject to initial condition

$$f(y,0) = \tfrac{1}{2}\varepsilon\, e^{y}. \tag{4.17}$$

My colleague Professor J.T. Stuart has transformed this problem, in a way which makes its structure clear, by writing

$$f = -\tfrac{1}{2}t + iy - \frac{i}{\sqrt{2}} \int \sqrt{h}\, e^{-\frac{1}{2}ig}\, dy. \tag{4.18}$$

This yields the non-linear hyperbolic system

$$\begin{aligned} \frac{\partial h}{\partial t} &= \frac{\partial g}{\partial y} \\ \frac{\partial g}{\partial t} &= \frac{1}{h^2}\frac{\partial h}{\partial y} \end{aligned} \tag{4.19}$$

with initial values

$$\begin{aligned} h(y,0) &= 2 + \tfrac{1}{2}\varepsilon^{2} e^{2y} \\ g(y,0) &= -2 \tan^{-1} (\tfrac{1}{2}\varepsilon\, e^{y}) \end{aligned} \tag{4.20}$$

The equivalent Riemann-invariant form is easily seen to be

$$\begin{aligned} g - \ln h &= \text{const} \quad \text{on} \quad \frac{dy}{dt} = \frac{1}{h} \\ g + \ln h &= \text{const} \quad \text{on} \quad \frac{dy}{dt} = -\frac{1}{h} \end{aligned} \tag{4.21}$$

and the initial condition shows that neither invariant can be constant in space and time. Thus one expects a shock to form, although detailed calculation is necessary to verify this.

If the fact that $\varepsilon \ll 1$ is exploited, an approximation to the characteristics can easily be calculated and those springing from the point $(y_o,0)$ in the (y,t) plane are

$$y = y_o + \tfrac{1}{2}t + \tfrac{1}{4}\varepsilon\, e^{y_o} (e^{t} - t - 1) + 0(\varepsilon^2)$$

and

$$y = y_o - \tfrac{1}{2}t - \tfrac{1}{4}\varepsilon\, e^{y_o} (t + e^{-t} - 1) + 0(\varepsilon^2)$$

The second family has an envelope in the domain $t > 0$, on which the solution has a singularity. This envelope has equation

$$y = \ln \left[\frac{4}{\varepsilon(t + e^{-t} - 1)}\right] - \tfrac{1}{2}t - 1, \tag{4.22}$$

so that, for small t, y >> 1 and the singularity is far out along the imaginary axis in the Γ plane. As t increases it moves in and reaches $\Gamma = 0$ when $x = \varepsilon$ or $y = 0$. According to (4.22) the singularity reaches $y = 0$ when

$$\tfrac{1}{2}t + 1 + \ln(t + e^{-t} - 1) = \ln\left(\frac{4}{\varepsilon}\right)$$

which agress with (4.7), neglecting $(\ln(\frac{1}{\varepsilon}))^{-1}$.

This analysis is not rigorous and, in any cases, is tied to a particular initial value problem. Numerical work on (4.1) with various initial conditions has been carried out by MBO, by Krasny (1983) (K) and Higdon and Pozrikidis (1984) (HP). MBO noted that the initial condition defined parametrically by

$$\begin{aligned} \Gamma &= e - \varepsilon \sin e \\ z &= e \end{aligned} \tag{4.23}$$

enabled the algebraic structure of the problem to be simplified and they were able to compute the terms in the power series representation

$$z(e,t) = e + \sum_{1}^{\infty} t^n z_n(e)$$

up to t^{38} (in some cases). Knowledge of the coefficients in the power series enables the formation of a singularity to be studied by standard methods; a singularity formed in all the cases studied (ε = 1/16, 1/8, 1/4, 1/2, 1.) and the formation times were in qualitative agreement with (4.7). The Fourier coefficients behaved like n^{-p} with $p = 2.7 \pm 0.2$, whereas M gave $n = 2.5$.

The same initial condition (4.23) as used by MBO was studied by HP, with $\varepsilon = 1$ throughout. A discrete method in which the sheet is represented by circular arcs was used. The results for the singularity are similar to MBO, but the singularity formation time is larger and $p = 2.3 \pm 0.2$. The larger formation time may be due to the underestimation of the growth of short waves characteristic of discrete methods.

A significant difference is that HP find that the sheet strength $\gamma(\Gamma)$ becomes infinite, while MBO find, in agreement with M, that the strength remains bounded at the critical time, although $\gamma(\Gamma)$ developes a cusp. HP also report the developement of a small and shrinking double-branched spiral in the later stages of the evolution.

K used the point vortex method, with a spectral smoothing technique. The bulk of the work was done with the initial condition

$$z(\Gamma,0) = \Gamma + \varepsilon(1 - i) \sin \Gamma \tag{4.24}$$

with $\varepsilon = .063$ so that comparison with other authors is not possible, although the results were generally similar. However p was estimated at -2.9. The calculation was continued past the singularity formation time, and a small growing double-branched spiral observed.

K carried out some runs with the initial condition (4.2). Good agreement with the singularity formation time predicted in M was obtained for $\varepsilon < .3$. When $\varepsilon = .5$ a new feature emerged in that the sheet rolled up at two points in each wavelength.

Singularity formation has been reported for initially circular non-uniform vortex sheets by HP; the results are rather like the periodic case. Schwarz (l.c.) has considered a finite vortex sheet in which initially

$$\gamma(x) = 3x(1 - x^2)^{\frac{1}{2}} \qquad (-1 < x < 1).$$

The self induced velocity at $x = \pm 1$ is finite because $\gamma \to 0$ sufficiently rapidly as $x \to \pm 1$. However a singularity forms at the ends of the sheet at a critical time and Schwarz suggests that an exponential spiral forms at this critical time.

If a spiral of vanishing size forms at $t = t_c$ in the case of closed or periodic roll up it would presumably be double branched and be tightly wound with an infinite number of turns (Pullin; quoted by K). Clearly, further work on this aspect of the problem is called for.

I would like to thank Professor G.R. Baker for some valuable discussions and Professor J.J.L. Higdon for preprints for his work.

FOOTNOTES

1. Van der Vooren's work appeared in report form in 1965 but it did not become widely known.

2. The instability here has a different cause.

REFERENCES

Beale, J.T. and Majda, A. 1982. Vortex methods II: Higher order accuracy in two and three dimensions. Maths. Comp. 39, 29.

Birkhoff, G. and Fisher, J. 1959. Do vortex sheets roll up? Rend Circ. Mat. Palermo, Sec 2, 8, 77.

Birkhoff, G. 1962 Helmholtz and Taylor instability Proc. Symp. Appl. Math. XIII AMS 55.

Chorin, A.J. and Bernard, P.S. 1973. Discretization of a vortex sheet with an example of roll up J. Comp. Phys. 13, 423.

Fink, P.T. and Soh, W.K. 1978. A new approach to roll-up calculations of vortex sheets. PRS A 362, 195.

Havelock, T.H. 1931. The stability of motion of rectilinear vortices in ring formation. Phil.Mag, II, 617.

Higdon, J.J.L. and Pozrikidis, C. 1984. The self induced motion of vortex sheets. Preprint.

Krasny, R. 1983. A numerical study of Kelvin-Hlemholtz instability by the point vortex method. U C Berkeley, PAM report 193.

Kuwahara, K and Takami, H 1973. Numerical studies of two-dimensional vortex motion by a system of point vortices. J. Phys. Soc. Japan 34, 247.

Lamb, H. 1932 Hydrodynamics C.U.P.

Longuet-Higgins, M.S. and Cokelet, E.L. 1976. The deformation of steep surface waves on water I. A numerical method of computation. PRS A 350, 1.

Meiron, D.J., Baker G.R., and Orszag, S.A. 1982 Analytical structure of vortex-sheet dynamics. Part I Kelvin-Helmholtz instability. JFM 114, 283.

Moore, D.W. 1979 The spontaneous appearance of a singularity in the shape of an evolving vortex sheet. PRS A 365, 105.

Moore, D.W. 1981 On the point vortex method. SIAM J.Sci. Stat. Comp. 2, 65.

Pozrikidis, C and Higdon J.J.L. 1984 Non-linear Kelvin-Helmholtz instability of a finite vortex layer. Preprint.

Rosenhead, L. 1931 The formation of vortices at a surface of discontinuity PRS A 134, 170.

Saffman, P.G. and Baker G.R. 1979 Vortex interactions. Ann. Rev. Fluid Mech. II, 95.

Schwarz, L. 1981 A semi-analytic approach to the self-induced , motion of vortex sheets JFM 111, 475.

Van der Vooren, A.I. 1980 A numerical investigation of the rolling up of vortex sheets. PRS A 373, 67.

Theoretical and Applied Mechanics
F.I. Niordson and N. Olhoff
Elsevier Science Publishers B.V. (North-Holland)

HYDRODYNAMIC SYNTHESIS OF MARINE STRUCTURES

Professor J. R. Paulling

Department of Naval Architecture and Offshore Engineering
University of California
Berkeley, California 94720 USA

The diverse and sometimes irregular geometry of offshore structures, together with their exposure to a severe dynamic marine environment, results in a system of hydrodynamic loads whose origins lie in several different fluid phenomena. The fundamental reasons for computing the hydrodynamic forces are twofold: first, they are needed for computing the motions of the floating structure in waves and second, they form an important part of the structural load system. The characteristics of structure geometry and motion environment which influence the above phenomena will be outlined. The various components of loads and their interaction will be discussed and we examine the techniques currently in use for design analysis.

INTRODUCTION

The term synthesis is defined as "the putting together of parts to form the whole" and is quite appropriately applied to the process of determining the hydrodynamic forces on marine structures of the type used in offshore oil drilling and other similar operations. In contrast to ships, which usually have a hydrodynamically smooth and regular surface, offshore platforms are constructed in the form of an irregular three-dimensional assembly of pontoons, columns and braces, often of cylindrical or near cylindrical configuration.

These offshore platforms may be subdivided into two very general categories of fixed and floating types, respectively. Fixed platforms are attached to the seafloor by a combination of their weight and by piling driven into the sea bed. Floating platforms may be held in place by means of an array of chains and cables with anchors at the sea bed, or they may be dynamically positioned, i.e., held in place by means of computer controlled thrustors which are activated in such a way as to maintain the platform in position over a fixed location on the bottom. Some platforms are supported by raising themselves above the sea surface on legs which extend to the sea floor. For transportation, such jackup platforms are lowered until they float on the main hull, which resembles a barge, and the legs are raised into an elevated position above the platform. A recent innovation in platform design is the tension leg platform or TLP. This is very similar to the moored platform with the exception that the buoyancy exceeds the weight and vertical force equilibrium is maintained by means of taut vertical mooring members extending from the platform to the sea floor.

In the present paper, we shall concern ourselves principally with the floating platforms. Fixed platforms of the pile-supported type are subject to many of the same wave-structure interactions with the exception of those due to the platform's own motion. We shall discuss some of the important considerations involved in making estimates of the hydrodynamic forces on such structures and deducing their response. Particular attention will be paid to those areas in which the designer and engineer is forced to make simplifications and approximations as a result of deficiencies in the relevant hydrodynamic theories.

The most common geometric configuration for both the semisubmersible and the TLP consists of a space-frame arrangement of tubular members of circular or oval cross-section. Platforms of somewhat simplified geometry but having proportions typical of modern offshore oil field drilling and production applications are shown in Figures 1 and 2. The twin hull arrangement has become, by far, the most widely used arrangement for the semisubmersible, dictated largely by considerations of low resistance to forward motion when in transit. The first full scale working TLP has recently been installed in the North Sea and operational experience with the concept is yet to be obtained. The TLP shown in Figure 2 is similar to several conceptual designs which have been developed by various oil companies. Since most of these have been intended for oil production rather than exploratory drilling, mobility has not influenced the arrangement to the same extent that it has for the semisubmersible. Considerations of optimum mooring tension response have resulted in somewhat larger vertical members and smaller horizontal members as compared to the semisubmersible.

Some important characteristic of both of these platforms are the following:

Much of the immersed volume is deeply submerged and the waterplane area is relatively small in relation to the total volume.

Most of the tubular members are relatively small in cross section compared to their own length and to the length of important wave components.

The spacing between members is large compared to the member cross section.

The first of the above characteristics gives the platform its "wave transparency" properties. The second and third may be utilized in deriving an analysis of the wave-induced forces and resultant motions which, although greatly simplified, nevertheless is quite accurate and informative in developing understanding of the principles upon which the two types of platforms may be designed.

GENERAL CONSIDERATIONS

In solving for the hydrodynamic loads and resulting motion of the platform, it is usually acceptable to treat the platform as a rigid body since the elastic deformations are small compared to the wave and body motions which give rise to the forces. The system of external loads acting on the body fall into several different categories, including:

(1) Environmental forces caused by waves, current and wind.

(2) Forces due to the platform motion.

(3) Restraint forces exerted by the mooring or positioning system.

(4) Forces associated with the platform operations such as heavy lifts, riser tension, drill pipe forces, and contact with attendant service ships.

The initial objective in the problem solution is to find the time history of the relationship between two coordinate systems, one fixed in the platform and the other fixed in space, and these are illustrated in Figure 3. After solving for the motions of the platform, we may add to the above system of external forces the internal inertia loads of the distributed mass of the platform and its contents to obtain a self-equilibrating system of structural loads. These loads may be used as input, for example, to a finite-element structural analysis.

The position of the platform is defined at any time by means of a set of three displacements of its origin along the space axes and a vector of three Euler angles representing rotations about three successive orientations of the body axes. At any instant of time, Newton's second law may be stated:

$$\underline{F} = \frac{d}{dt}(\underline{m}\ \underline{v})$$

$$\underline{M} = \frac{d}{dt}(\underline{I}\ \underline{\omega}) = \underline{I}\ \dot{\underline{\omega}} + \underline{\omega} \times (\underline{I}\ \underline{\omega}) \tag{1}$$

Here $\underline{m}$, $\underline{I}$ = the mass and moment of inertia matrices representing the physical mass of the platform itself.

$\underline{F}$, $\underline{M}$ = vectors of external forces and moments.

$\underline{v}$ = translatory velocity of the C.G. in global coordinate system.

$\underline{\omega}$ = angular velocity vector in the body coordinate system.

The equations of motion are nonlinear both as a result of the product of angular velocities in the second of (1) and through nonlinear relationships between the external forces and the platform motions, waves, currents and other influences. The force and moment terms include all of the external forces listed above, and much of the effort in obtaining the solution to (1) is devoted to the determination of these forces. In the design of platforms, several different approximations are used in order to obtain estimates of these forces. The method used in a given situation depends upon the geometry of the structure, the severity of the environment being simulated and the computational means at the disposal of the user.

In the case of small waves and small platform motions, two important simplifications are possible. First, the equations may be linearized, and second, the external hydrodynamic forces may be replaced by a simple sum of four terms proportional respectively to the absolute acceleration, the absolute velocity, the displacement from the mean position and a term independent of the body motion. The result is expressed as follows:

$$\underline{m}\ \ddot{\underline{x}} = -\underline{m}'\ \ddot{\underline{x}} - \underline{b}\ \dot{\underline{x}} - \underline{c}\ \underline{x} + \underline{F}_w(t) \tag{2}$$

Here $\underline{m}$ = composite matrix of mass and moment of inertia.

$\underline{x}$ = composite vector of displacements of c.g. and Euler angles.

$\underline{m}'$ = added mass matrix for structure.

$\underline{b}$ = damping matrix

$\underline{c}$ = restoring force matrix including hydrostatic and mooring effects.

$\underline{F}_w(t)$ = excitation force caused by external effects such as waves.

In general, the coefficients $\underline{m}'$, $\underline{b}$ and $\underline{F}_w$ depend upon the frequency of the incoming waves, ω. Consequently, equation (2) cannot be integrated directly for any arbitrary $\underline{F}_w(t)$. In the case of random excitation such as that due to realistic ocean waves, we may solve a separate equation (2) for each of the composite waves into which the random seaway may be decomposed and then superimpose these individual regular wave solutions. Thus, (2) reduces to an algebraic equation for the complex motion amplitude in the frequency domain. If we wish to deal with the complete nonlinear equation system (1) by a direct numerical integration scheme, it is necessary first to transform these frequency-dependent terms into an impulse response function in the time domain by evaluating their Fourier transform. A convolution integral is then employed to express the force in the time domain.

Much insight and understanding of the respone characteristics of ocean platforms may be gained by examining the solutions to (2). There are, however, several important phenomena and effects which are lost in the process of linearization and which may only be obtained by solving the exact equations, (1).

The response of the platform to waves, as in the case of any oscillating system, depends upon two quantities, the exciting force and the resonant response of the system to unit excitation. Both of these quantities are found to vary with the frequency of the incident waves and, consequently, the total response characteristics of the structure to waves depends upon the relationship between the two functions.

Three general procedures are in common use for the computation of the exciting forces and the coefficients in Equation (2). The first is three-dimensional potential flow theory applied to the entire platform. This is the most general method and yields the most complete results for situations in which viscous forces do not play an important role. Second, we may utilize a composite theory or synthesis involving the summation of forces computed for several simple geometric shapes into which the platform may be subdivided. These elementary solutions may be obtained by either potential theory solutions for elementary members such as horizontal or vertical circular cylinders, or by Morison's formula (Morison, et al, (1950)).

In the potential theory methods, the fluid forces are obtained by integrating a resultant pressure over the wetted surface of the platform and viscous shear forces are neglected. The process requires the solution for a velocity potential, $\phi(x,y,z,t)$, which satisfies the Laplace equation

$$\nabla^2\phi = 0 \tag{3}$$

in the fluid domain and which satisfies appropriate boundary conditions on the surface of the platform, on the free surface and at infinity. In the linear solution procedure, the potential function is expressed as the sum of three terms

$$\phi = \phi_I + \phi_D + \phi_M \tag{4}$$

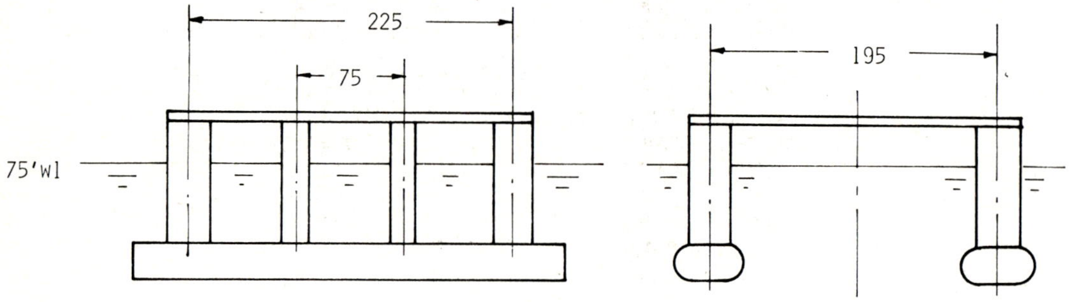

Main hulls 21 x 50 x 300 ft.
Main columns 30 ft. dia.
Intermediate columns 15 ft. dia.
Weight 48 x 10^6 lbs.

Figure 1
Twin Hull Semisubmersible

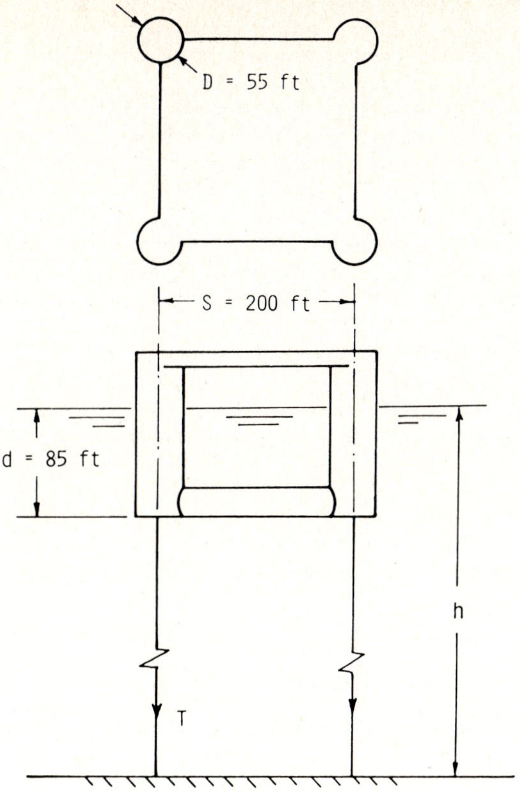

Figure 2
Four Column Tension Leg Platform

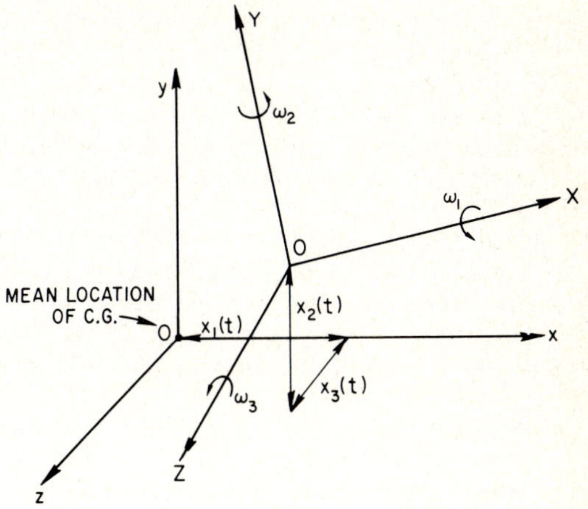

Figure 3
Coordinate Systems

Here, ϕ_I is the potential associated with the incident wave system, ϕ_D is the potential of a system of diffracted waves caused by the interaction of the platform with the incident waves, and ϕ_M is the potential which expresses the effect of the motion of the body. The pressure is given by the linearized Bernoulli equation

$$p = -\rho g y - \rho \frac{\partial \phi}{\partial t} \tag{5}$$

When the total force is evaluated, it is found that the first two terms of the potential function result in a time-varying exciting force which is independent of the platform motion. If sinusoidal motion is assumed, the third term, ϕ_M, results in two force components, one in phase with the body acceleration called the added mass and the other in phase with the velocity called the damping. As a result of this linear decomposition of the velocity potential, we see that the motion dependent forces are computed in the absence of incoming waves and the wave forces are computed as though the body were stationary.

The principal difficulty in applying the three-dimensional method to a practical offshore platform lies in finding the potential function which satisfies the necessary boundary conditions on the geometrically complex surface of the platform. The great advantage of the procedure is that it inherently contains the hydrodynamic interaction between the different members of the platform which may be lost in the other two methods. A number of different analytical or numerical procedures are used in obtaining the three-dimensional solutions. One in very common use is the distributed source procedure as published, for example by Garrison (1977), Faltinsen and Michelsen (1974) and Hogben and Standing (1974).

In some cases, regularity of the platform geometry may permit its being approximated by several individual members of simple geometry for which somewhat simpler potential solutions are available. Examples are the pontoon members of the platform which may be approximated by either a single cylinder or a pair of cylinders, and vertical cylindrical members. Solutions for the single horizontal cylinder have been given by Bolton and Ursell (1973). Solutions for slender horizontal members of arbitrary shape may be obtained by the Frank close fit method, Frank (1967) which has been extended by Lee (1971) to treat two parallel horizontal cylinders similar to the pontoons of a catamaran. This procedure has been embedded in a general platform motions program by Hong and Paulling (1977). Such solutions for horizontal cylinders may be combined to represent a member whose cross section varies along the length, resulting in the "strip theory" of ship motions in waves.

Solutions for a vertical circular cylinder are useful in treating the forces on the columns of the platform. Solutions for a vertical cylinder extending to the sea floor is given by McCamy and Fuchs (1954). Solutions for a cylinder of finite depth are given by Garrett (1971) and Eatock-Taylor (1979). A solution for a group of vertical cylinders is given by Ohkusu (1974).

The solutions for single horizontal or vertical cylinders would not contain the member-member interactions referred to above. The multi-cylinder solutions would contain such interactions, but in the strip theory solutions, this would be present only in the two-dimensional sense. In some cases, the interaction results in appreciable modification to the resultant combined fluid forces.

The Morison formula represents an intuitive attempt to include the viscous drag forces on the cylindrical member by means of a force proportional to the square of the relative velocity between fluid and cylinder. Such a viscous force is expected to be of importance in platforms having members of relatively small cross section. In this case, the spacing between members is often large compared to the member cross-sectional dimensions and the hydrodynamic disturbance introduced into the flow by the presence of other members may be neglected. The force on a single cylindrical member is then obtained by computing the fluid pressures, velocities and accelerations at the member location neglecting the presence of the other members. The flow properties are computed by an appropriate wave theory and the kinematics of the platform motion are added to express the motion relative to the member. A form of the Morison formula in terms of the relative motion is

$$\underline{F} = -\iint_s p\,\underline{n}\,ds + \int_l C_D \underline{u}_n |\underline{u}_n|\,dl + \int_l C_M \underline{a}_n\,dl \tag{6}$$

Here p = pressure which would exist in the undisturbed flow field at a point on the member surface.

$\underline{n}$ = unit outward normal vector.

C_D = quadratic drag coefficient for flow normal to the member centerline.

$\underline{u}_n$ = component of resultant velocity normal to the centerline of the member.

C_M = acceleration force coefficient.

$\underline{a}_n$ = component of resultant acceleration normal to the centerline of the member.

Note that both $\underline{u}_n$ and $\underline{a}_n$ are the resultant of the wave-induced fluid motion and the motion of the member which results from the rigid-body motion of the platform. The relative motion form of (6) is an intuitive extension of the formula originally proposed for computing the forces exerted by waves on a vertical stationary piling.

If the Morison formula is to be used in the linearized Equation (2), the quadratic drag term presents an obstacle. In this case, the quadratic term may be replaced by an equivalent linear drag force provided that the quadratic force is not inherently responsible for any effects such as sub- or superharmonic excitation which are not also contained in the linear form. In many but not all cases, this is a good assumption and we shall examine some of the exceptions later. The equivalent linear drag coefficient, C_{DL} , is defined as follows

$$C_{DL}\,\underline{u}_n = C_D\,\underline{u}_n\,|\,\underline{u}_n\,| \tag{7}$$

An optimal estimate of C_{DL} may be made by means of a procedure which minimizes the mean square error between the linear force and the "exact" quadratic force. For regular waves, the result is

$$C_{DL} = \frac{8u_o}{3\pi}\,C_D \tag{8}$$

where u_o = the amplitude of the sinusoidal relative velocity.

For random waves, the result is

$$C_{DL} = \sqrt{\frac{8}{\pi}}\,\bar{u}\,C_D \tag{9}$$

where $\bar{u}$ = RMS value of the randomly varying resultant normal velocity.

A similar procedure may be used as shown by Paulling (1980) in the case of a steady current superimposed on the oscillatory relative velocity. In this case, there is a steady mean force whose magnitude exceeds the drag force due to the current alone, and the equivalent oscillatory force exceeds the force which would result from the oscillatory component of velocity alone.

In all of these cases, the equivalent linear coefficient is found to depend on the magnitude of the relative velocity which, in turn, depends on both the wave-induced velocity and on the body motion. Since the latter is initially unknown and is to be found by solving Equation (2), an iterative solution procedure must be employed. Computational procedures based on the Morison formula have received wide acceptance in the offshore industry and are found to yield results in good agreement with experiments provided care is exercised in the choice of the coefficients of drag and added mass.

FIRST ORDER HEAVING RESPONSE OF SEMISUBMERSIBLES

We shall first examine in some detail the heave response of the semisubmersible platform consisting of vertical and horizontal members as depicted in Figure 1. A good approximation to the heave force exerted by waves of small amplitude on cylindrical members of small cross section is obtained from only the first and last terms of the Morison formula (6). Using this expression in conjunction with the pressure and acceleration obtained from linear infinitesimal wave theory, the vertical force exerted by beam seas on one of a pair of parallel horizontal circular cylinders will be given by

$$Y_H = -2\rho\,A_H\,L\,a\,\omega^2\,e^{-kd}\cos\,(kb-\omega t) \tag{10}$$

Here A_H = cross sectional area of one horizontal cylinder.
       L = length of cylinder.
       $2b$ = horizontal separation of the two cylinders.
       d = depth from mean water surface to centerline of cylinder.

The force on all of the vertical surface-piercing members on one side of the platform is given by

$$Y_v = \rho g\, A_v\, a\, e^{-kd} \cos(kb - \omega t) \tag{11}$$

Here A_v = total cross sectional area of all vertical members on one side.

It is assumed that the depth of the lower end of the vertical member is approximately equal to the depth of the centerline of the horizontal member and this is consistent with the assumption of small member diameter.

The total vertical force on all members comprising both sides of the platform is given by

$$\begin{aligned}Y &= 2\rho g\, a\, e^{-kd}(A_v - 2k\,L\,A_H)\cos kb \cos \omega t \\ &= F_o\, e^{-kd}(1 - kb\,A_o)\cos kb \cos \omega t\end{aligned} \tag{12}$$

Here $F_o = 2\rho g\, A_v\, a$

       $A_o = 2\dfrac{L}{b}\dfrac{A_H}{A_v}$

       $k = \dfrac{\omega^2}{g}$

The amplitude of this force is the product of three terms which vary with the wave frequency in such a manner that there is force cancellation at certain frequencies and reinforcement at others. The individual terms are illustrated in the upper part of Figure 4 and their product in the lower part of this figure. For practical platforms, the most important portion of this composite function lies in the region $0 \leq kb \leq \dfrac{3\pi}{2}$. The remaining portion of the range corresponds to short waves in which the assumption of small member diameter no longer is valid and in which the forces are found to be negligibly small.

A comparison of this simplified formulation with the more exact strip theory computations is given in Figure 5. The curves labelled "slender member" have been obtained by the complete Morison formula including linearized viscous drag and a modified added mass coefficient appropriate to the noncircular horizontal members. The curves labelled strip theory have been obtained by an ideal fluid theory which takes into consideration the hydrodynamic interaction between the horizontal members. Results are given here for the two pontoons alone (lower hulls) and for the composite structure containing lower hulls and columns. From a comparison of this figure and Figure 4, we may conclude that the heave force cancellation occurring at the value of $T = 25$ results from the phase difference in the forces on vertical and horizontal members while the cancellation at $T = 9$ results from the phase difference due to the transverse separation of the members. For the example platform, it is seen that the effect of finite member cross section and interference between members is quite small, although this is not the case for all twin hull semisubmersibles.

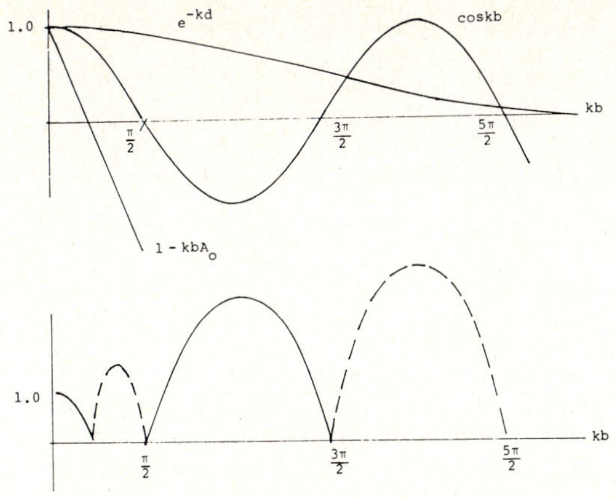

Figure 4
Heave Force on Twin Hull Semisubmersible in Beam Seas

The vertical or heaving motion of the semisubmersible is usually decoupled from the other motions as a result of symmetry or near symmetry of the platform in the transverse and longitudinal directions. We may, therefore, consider heaving as a single degree of freedom motion with typical response characteristics of such a system. The restoration of this motion is primarily hydrostatic with a small contribution from the anchoring system in some cases, and the effective mass includes the added mass term in Equation (2). Figure 6 presents the added mass and damping for this platform as estimated by the Morison formula (slender member) and the strip theory. The equivalent linear damping coefficients are shown for the platform motion generated by three different sea states.

The platform natural frequency in heave is given by

$$\omega_n = \sqrt{\frac{c}{m + m'}} \tag{13}$$

In a typical platform the added mass is approximately equal to displaced mass of the pontoons, $2\ \rho A_H L$. The spring constant in heave, neglecting mooring effects is $2\ \rho g A_v$. Substituting in (13) gives the following expression for the natural

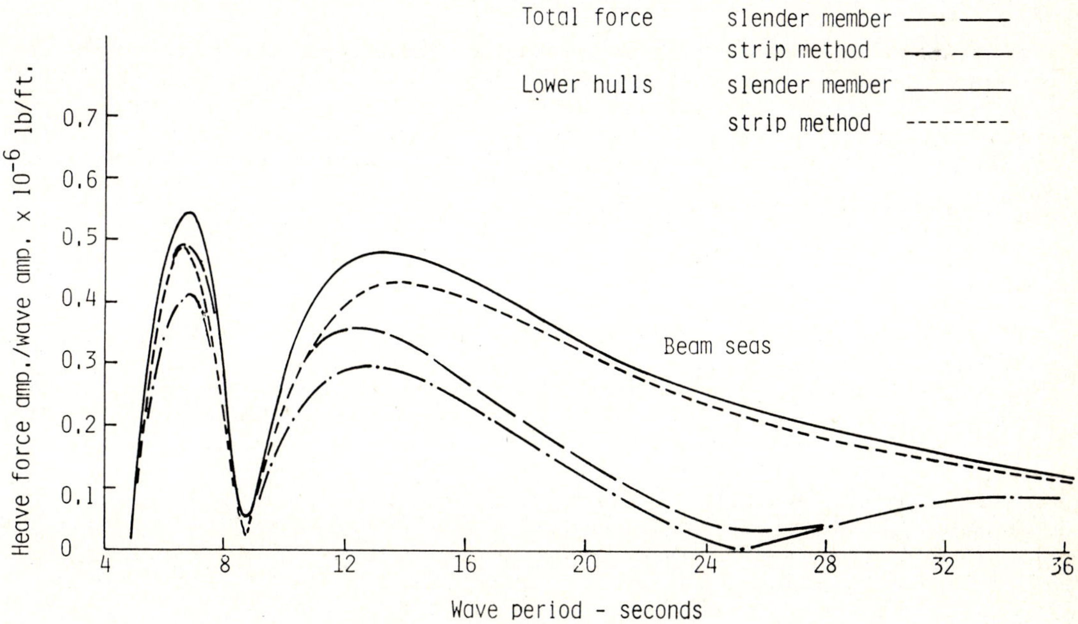

Figure 5
Twin Hull Semisubmersible Heave Exciting Force

frequency in heave.

$$\omega_n = \sqrt{\frac{2\rho g\, A_v}{2\rho(A_v\, d + 2\, A_H\, L)}} = \sqrt{\frac{g/d}{1 + 2\,\frac{V_H}{V_v}}} \tag{14}$$

Here V_v = the volume of vertical members.

V_H = the volume of horizontal members.

The resultant heave motion response to regular waves made nondimensional in terms of the wave amplitude is shown in Figure 7. The first minimum of the response to the left of the resonant peak corresponds to $kb = \pi/2$ and the second minimum to $kb = 1/A_o$. The remaining zeroes seen in Figure 4 are compressed into the left hand portion of the axis and, as noted, are of no practical importance.

The principles of force reinforcement and cancellation on members having different orientation in the platform structure plays an important role in the design of such platforms for optimum motion response in waves. By adjusting the ratios of horizontal to vertical surface-piercing member volume, the designer can adjust the frequencies of the response minima and control to a lesser extent the resonant peak. The amplitude of the latter peak and the minimum adjacent to it are strongly influenced by damping which may include important contributions from both viscous and radiation effects.

The typical ocean wave spectrum peak lies to the left of the resonance peak, and as a result, the mean response to waves is determined primarily by the peak just

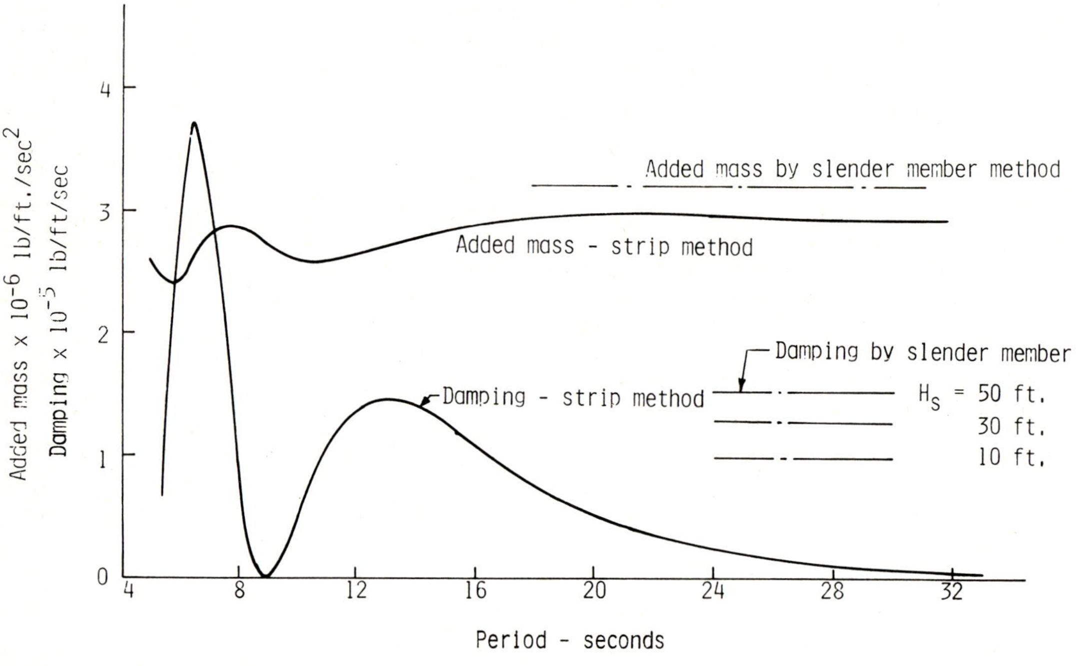

Figure 6
Twin Hull Semisubmersible - Heave Damping and Added Mass

below $kb = \pi/2$, together with the cancellation minimum at $kb = 1/A_o$. Common design practice is to choose the proportions of the platform such that the resonant frequency lies well outside the range of any appreciable wave energy.

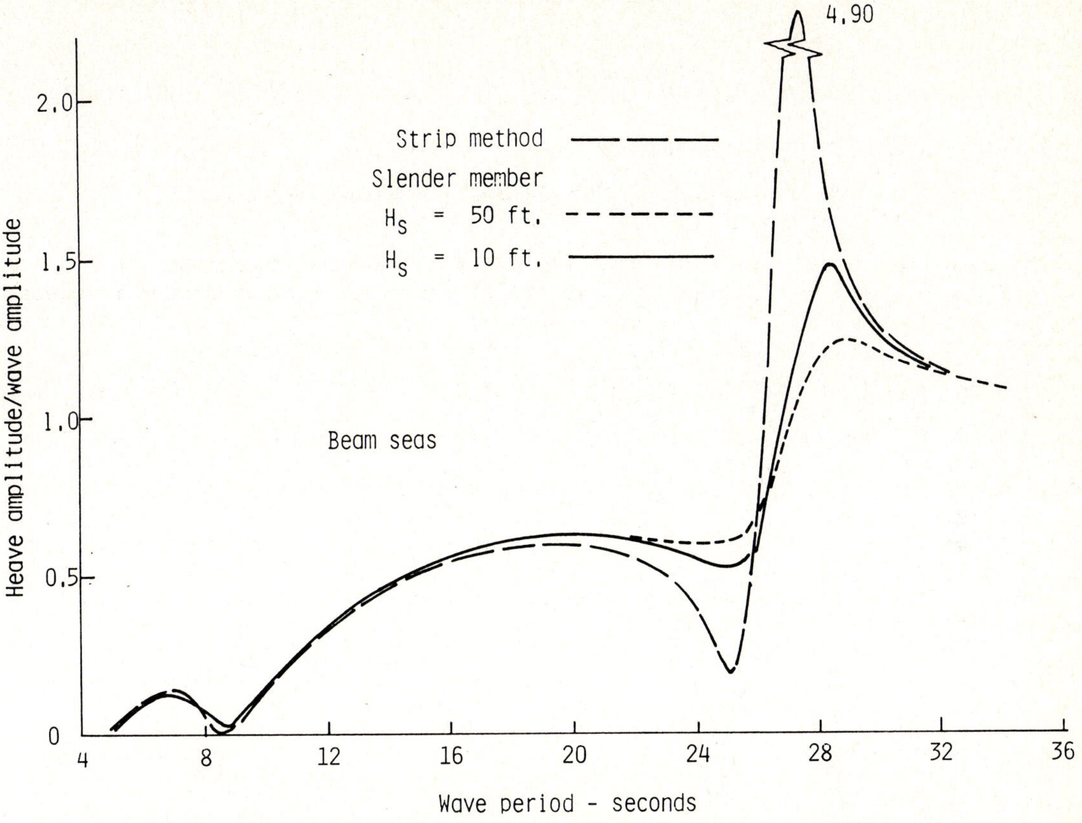

Figure 7
Twin Hull Semisubmersible Heave Response Function

RESPONSE OF TENSION LEG PLATFORMS

In contrast to the semisubmersible, the heave motion of the TLP is almost fully suppressed by the vertical mooring members. The lateral motions of surge, sway and yaw are only lightly restored by the vertical moorings. We may derive some important aspects of the TLP response by considering a simplified model somewhat similar to the previous semisubmersible space frame model. Figure 2 depicts a composite TLP similar to those which have been proposed for oil production in relatively deep water. An important design consideration in the TLP is the attainment of minimum wave-induced tension variation in the mooring members and it is found that this is obtained with relatively large vertical columns. The hydrodynamic interactions within groups of vertical cylinders, therefore, is important in TLP force computations. The horizontal (surge) exciting force is shown in Figure 8 as computed by a 3-D source distribution procedure and by the Morison formula. Two variations of the latter are shown. In the first, the fluid velocity and acceleration are computed on the member centerline as in the "pure" Morison formula. In the second, the average values of these parameters are computed over the member cross-section in order to provide an intuitive correction for the finite member dimensions.

Both the lateral motion (sway, surge) and the vertical (heave, roll) natural frequencies are of interest to the designer of the TLP. An estimate of the lateral natural frequency may be made as follows. From Figure 2 it is seen that the lateral restoring force coefficient is given by the net mooring tension divided by the length of the mooring lines. The tension must equal the difference between the platform buoyance and weight and, in deep water, the length of line is approximately equal to the water depth, h . The restoring coefficient is therefore given by

$$c = \frac{T}{h - d} = \frac{\Delta - w}{h} \tag{15}$$

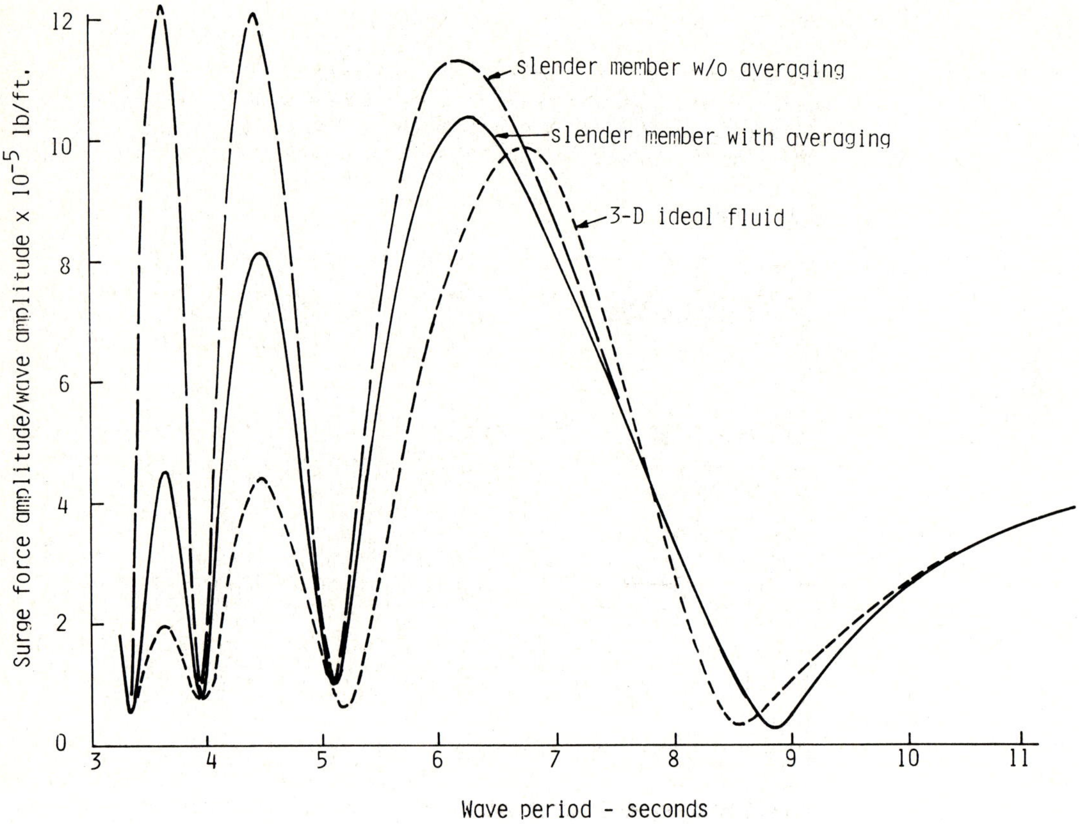

Figure 8
Four Column TLP Surge Force

For a typical TLP, the added mass in surge is approximately equal to the displaced water mass and the effective mass of the platform is therefore approximately equal to the sum of the physical mass and the displaced mass. Typical plots of the damping and added mass for the four column configuration are shown in Figures 9 and 10. Results are shown for the full four column configuration, a single column multiplied by four and for the frequency independent value given by the Morison formula. The effect of intermember interference is seen by a comparison of the former two values. The viscous damping by Morison's formula is seen to be small in comparison to the peak radiation damping in the range of short waves but is the only appreciable damping in the long wave range.

The natural frequency in sway or surge is given, approximately, by Equation (16)

$$\omega_n = \sqrt{\frac{cg}{\Delta + w}} = \sqrt{\frac{g}{h}\left(\frac{1 - w/\Delta}{1 + w/\Delta}\right)} \tag{16}$$

In heave, the restoration coefficient is associated with the elastic stiffness of the mooring members and hydrostatic effects are neglible. Let A_m represents the total cross-sectional area of all mooring members, and, as before let us assume deep water so that h is the member length. The stiffness in heave is then given by

$$c = \frac{EA_m}{h} \tag{17}$$

where E is the material modulus of elasticity.

The added mass in heave is usually somewhat less than the displaced water mass since the principal part of the volume is in the vertical column members. As a reasonable approximation, the effective mass may be taken to be twice the physical mass for typical platform proportions, resulting in a natural frequency of heave given by Equation (18).

$$\omega_n = \sqrt{\frac{cg}{2w}} = \sqrt{\frac{E A_m g}{2 h w}} \tag{18}$$

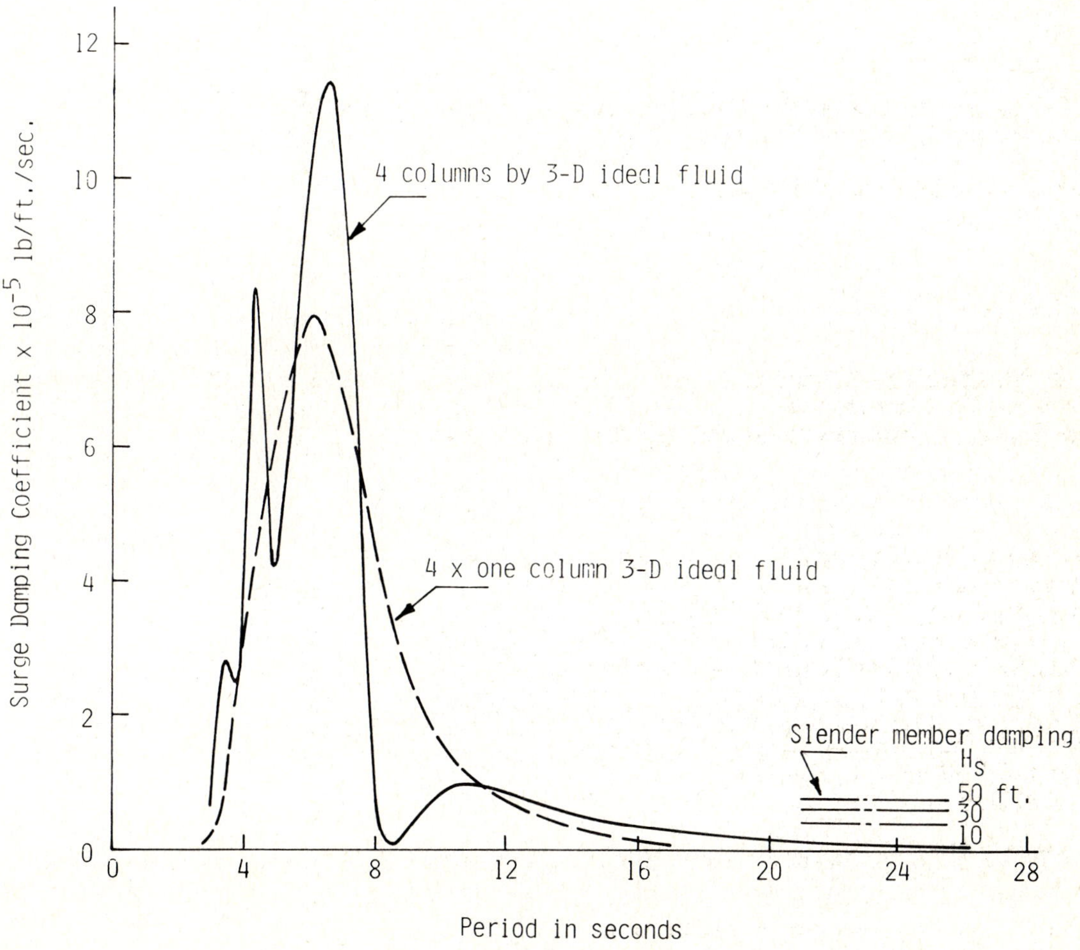

Figure 9
Four Column TLP Surge Damping

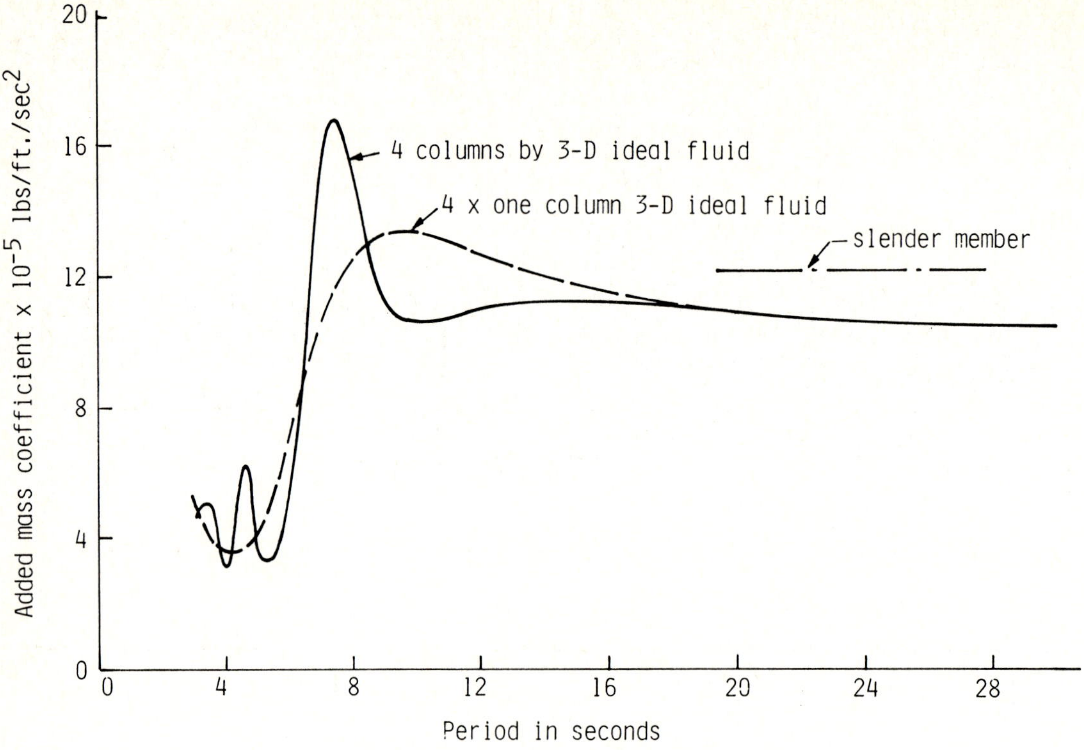

Figure 10
Four Column TLP Added Mass in Surge

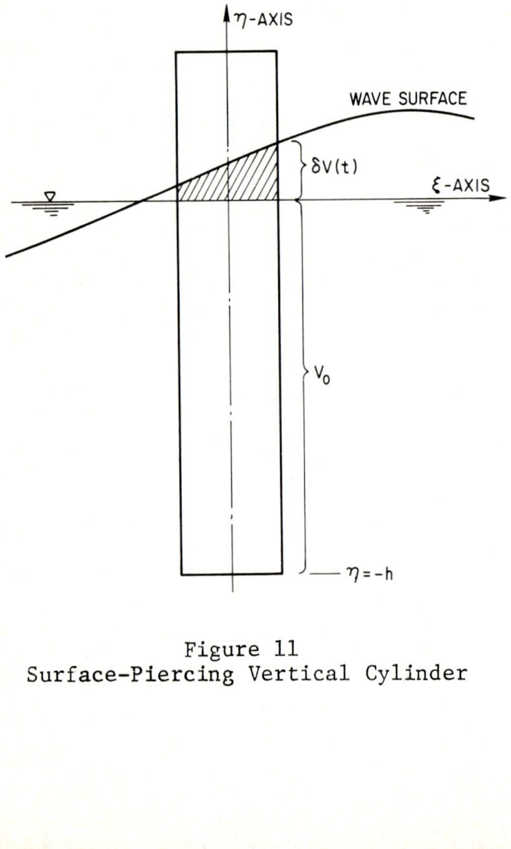

Figure 11
Surface-Piercing Vertical Cylinder

It is found that the frequency given by (16) is typical of all three lateral motions in surge, sway and yaw, while (18) is typical of the vertical plane motions of heave, roll and pitch. If we were to introduce the proportions of a typical (proposed) TLP for a water depth of 500 to 1000 meters, the natural periods of these categories of motions would be found to lie in the vicinity of 100 seconds and two to three seconds, respectively. These frequencies are outside the range of important wave frequencies and therefore, resonant amplification of the fundamental wave-induced response is usually of small concern. Since both the vertical and the lateral modes are lightly damped in the vicinity of their resonant frequencies, they can be troublesome due to excitation by nonlinear effects which result in forces which occur at either multiples or differences of important wave frequencies. Some of these effects will be discussed later and the accurate determination of the relevant forces forms one of the most important areas of needed research in the field.

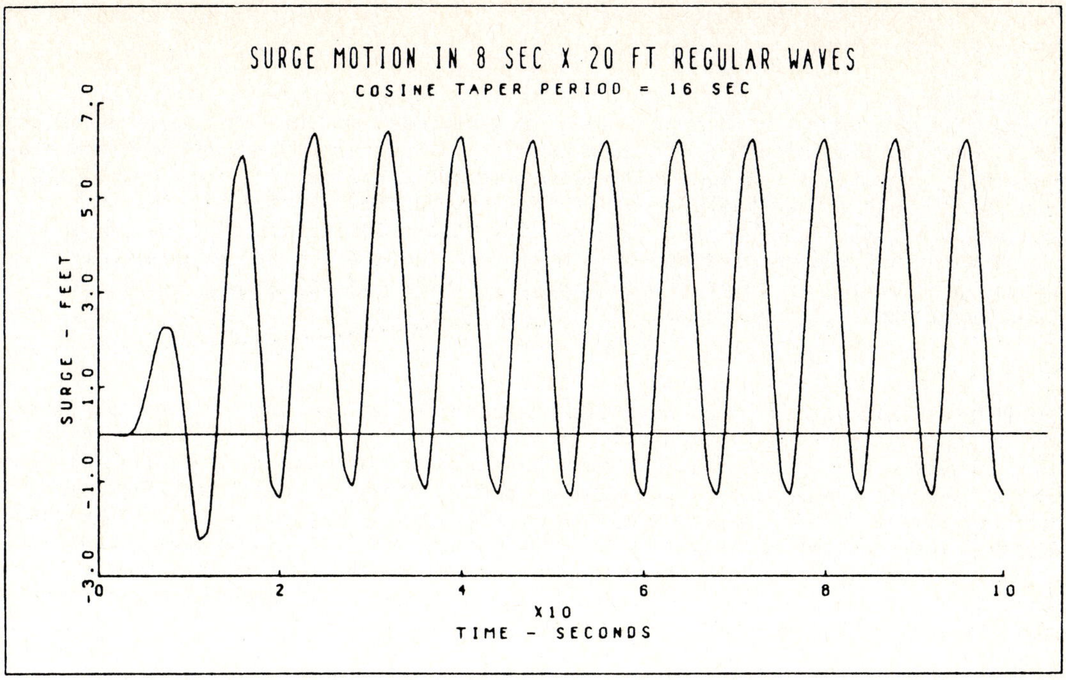

Figure 12
Surge Motion Time History for Experimental TLP

LOW AND HIGH FREQUENCY WAVE FORCES

Several unrelated fluid flow phenomena may contribute to wave-induced forces which have frequencies which differ by being integral multiples or fractions of the basic wave frequency. These second order-forces have been the subject of considerable recent investigation and details of their analyses may be found in papers by Pinkster (1979), Faltinsen (1978) and Newman (1967). Pinkster has computed the potential low frequency forces by integration of the second order pressure terms over the hull surface. It is shown that the most important contribution comes from the hydrostatic pressure integrated the region between the mean waterline and the instantaneous waterline. This effect is neglected in the usual linear analysis in which the integration is carried out only to the mean waterline.

A second source of such forces originates in the viscous drag on the alternately wet and exposed portion of the member projecting through the water surface. In Figure 11 we see a vertical surface-piercing cylindrical member exposed to wave action. When the member is in a wave crest, the drag force in the down wave direction acts on a wetted length of member extending up to the full height of the crest. In a wave trough, the force acts in the opposite direction but the wetted length on which the force acts is reduced by the wave height. In regular waves, the result consists of a steady force in the down wave direction plus an oscillatory force at twice the wave frequency. In random waves, the mean force becomes a slowly varying force having a frequency content similar to the wave envelope.

The characteristic high or low natural frequencies of the TLP as given by Equations (16) and (18) isolate the response from appreciable wave-frequency resonant amplification. The platform is, however, vulnerable to the high and low frequency effects described above. The high frequency effects are caused primarily by forces acting at the waterline and, as a result, these forces may exert an appreciable pitching moment on the platform. The resonant amplification of the lightly damped pitch motion, which has a natural frequency similar to that in heave, may result in severe oscillatory mooring member tensions with consequent accelerated fatigue damage in those members.

The low frequency and steady components of such forces may result in large offset of the platform from its mean position. Superimposed on this is the wave frequency oscillatory motion and effects caused by wind and current. In order for the platform to be capable of supporting drilling and other operations, this total offset must usually not exceed ten percent of the water depth. A time history of the motions of a small experimental TLP including the drift effect is shown in Figure 12. These results were obtained by a numerical integration of the complete equations of motion as described by Paulling (1977) and therefore include effects caused by large amplitude motions of the platform and waves.

CONCLUSIONS

The hydrodynamic forces on an offshore platform depend on both wave and platform motion and include both viscous and potential effects. The wave forces acting on different parts of the platform vary in phase and amplitude with wave frequency and, at certain frequencies, there may be either cancellation or reinforcement in their resultant. The platform response is strongly influenced by the relationship between these cancellation or reinforcement frequencies and the natural frequencies of the platform motions. The designer can sometimes choose the proportions of the platform in such a way as to take advantage of these force cancellation effects in order to achieve minimum motion response in a seaway.

In some cases, notably the tension leg platform, the natural frequencies are widely separated from the wave frequencies. Motion excitation may nevertheless occur as a result of second order low or high frequency effects. The accurate determination of these forces as well as the platform damping and other hydrodynamic characteristics in the range of very low and very high frequencies are important problem areas in which research is urgently needed.

REFERENCES

[1] Bolton, W. E. and Ursell, F., The Wave Force on an Infinitely Long Circular Cylinder in an Oblique Sea, J. Fluid Mech. (1973) v 57 pt 2, pp 241-256.

[2] Faltinsen, O. M. and Løken, A. E., Drift Forces and Slowly Varying Forces on Ships and Offshore Structures, Nor. Marit. Res., No. 1, (1978) pp 2-15.

[3] Faltinsen, O. M. and Michelsen, F., Motions of Large Structures in Waves at Zero Froude Number, Proc. Intl. Symp. on Dynamics of Marine Vehicles and Structures in Waves, Inst. Mech. Engrs., London, April 1974.

[4] Frank, W. , Oscillations of Cylinders in or Below the Free Surface of Deep Fluids, NSRDC Report 2375, Oct. 1967.

[5] Garrett, C. J. R., Wave Forces on a Circular Dock, J. Fluid Mech. (1971), v 46, pt 1, pp 129-139.

[6] Garrison, C. J., Hydrodynamic Interaction of Waves with a Large-Displacement Floating Body, U.S. Navy Postgraduate School, Report NPS-69Gm77091, Sept. 1977.

[7] Hammett, D. S., The First Dynamically Stationed Semisubmersible Sedco 709, Offshore Tech. Conf., Houston Paper No. 2972, 1977.

[8] Hogben, N. and Standing, R. G., Wave Loads on Large Bodies, Proc. Intl. Symp. on Dynamics of Marine Vehicles and Structures in Waves, Inst. Mech. Engrs., London, April 1974.

[9] Horton, E. E., McCammon, L. B., Murtha, J. P. and Paulling, J. R., Optimization of Stable Platform Characteristics, Offshore Tech. Conf., Houston, Paper No. 1553,1972.

[10] Lee, C. M., Jones, H. and Bedel, J. W., Added Mass and Damping Coefficients of Heaving Twin Cylinders in a Free Surface, NSRDC Report 3695, Aug. 1971.

[11] Lighthill, J., Waves and Hydrodynamic Loading, Proceedings BOSS 1979, pp 1-40.

[12] McCamy, R. C. and Fuchs, R. A., Wave Forces on Piles: A Diffraction Theory Beach Erosion Board, U.S. Army Corps of Engineers, Tech. Memo 69, 1954.

[13] Morison, J. R., O'Brien, M. P., Johnson, J. W. and Schaaf, S. A., The Force Exerted by Surface Waves on Piles, Pet. Trans., AIME, v 189, 1950.

[14] Newman, J. N., The Drift Force and Moment on Ships in Waves, J. Ship Research, 1967, pp 51-60.

[15] Ohkusu, M, On the Hydrodynamic Forces on Multiple Cylinders in Waves, Proc. Intl. Symp. on Dynamics of Marine Vehicles and Structures in Waves, University College London, 1974, pp 115-120.

[16] Paulling, J. R. and Horton, E. E., Analysis of the Tension Leg Platform, Offshore Tech. Conf., 1263, 1970.

[17] Paulling, J. R. and Hong, Y. S., Analysis of Semisubmersible Catamaran-type Platforms, Offshore Tech. Conf., Houston, Paper No. 2975, 1977.

[18] Paulling, J. R., Time-Domain Simulation of Semisubmersible Platform Motion with Application to the Tension-Leg Platform, SNAME Spring Meeting/STAR Symposium, San Francisco, 1977.

[19] Paulling, J. R., An Equivalent Linear Representation of the Forces Exerted on the OTEC CW Pipe by Combined Effects of Waves and Current, Ocean Energy for OTEC, OED, v 9, ASME 1980, pp 21-28.

[20] Paulling, J. R., The Sensitivity of Predicted Loads and Responses of Floating Platforms to Computational Methods, PROC. Conf. on Integrity of Offshore Structures, Glasgow, 1981, (Pub. by Applied Science Publishers, Essex, U. K.,) 1981.

[21] Pinkster, J. A., Mean and Low Frequency Wave Drifting Forces on Floating Structures, Ocean Engng., v 6, 1979, pp 593-616.

[22] Salvesen, N., et. al, Computations of Nonlinear Surge Motions of Tension Leg Platforms, Offshore Tech. Conf., 4394, 1982.

[23] Taylor, R. Eatock and Dolla, J. P., Hydrodynamic Loads on Vertical Bodies of Revolution, Trans. RINA, v 122, 1980, pp 285-298.

Theoretical and Applied Mechanics
F.I. Niordson and N. Olhoff
Elsevier Science Publishers B.V. (North-Holland)

STRUCTURAL DESIGN OF MARINE STRUCTURES

P. Terndrup Pedersen

Professor at the Institute of Ocean Engineering
The Technical University of Denmark
DK-2800 Lyngby
Denmark

The paper deals with structural design procedures for ships and offshore structures and emphasizes the contribution from applied mechanics research.

In the introduction a general description is given of the field of marine structures and an existing international research cooperation.

In the main part of the paper the structural analysis procedure is described in three sections. In the first section wave-induced loading on marine structures is discussed. It is shown that total loads may depend on the response of the system, therefore, besides waves the loads will depend on geometry, mass and stiffness properties. The second section deals more specifically with response analysis, and simplified mathematical models are discussed with indication of needed future applied mechanics research. In the third section a discussion on design criteria for thin-walled stiffened plate and shell structures is presented.

INTRODUCTION

The object of this paper is to provide a short summary of structural analysis procedures needed for design of marine structures. The presentation will outline the breakdown of the subject in different disciplines and emphasize the role of applied mechanics in the development of rational analysis procedures for ships as well as offshore structures. It is the author's hope that the paper will contribute to attract the interest of researchers from the fields of solid and structural mechanics so that in the future a larger number of scientists from this field will turn their interest to problems related to marine structures. As will be shown: interaction between waves and structures is an area where an abundance of unsolved interesting problems exists and where research is needed in order to explore and utilize those 70 per cent of our earth which is covered with water.

Structures For Sea Transportation:

The history of marine structures is mainly the history of ships.

Although sea-going ships have been with us for 8000 years ship designs are at present changing at a rapid rate, but ship design techniques are changing even more rapidly.

Some decades ago the basis for design of naval and merchant ships were rules giving acceptable profiles and thicknesses when the principal dimensions of the ships were known. The background for these rules were mainly empirical: experience had shown that an acceptable damage record was obtained if these scantlings were used. Those who used the rules had, however, little feeling about the loading and the structural behaviour of the ship.

As the general knowledge about environmental conditions, hydrodynamic loading and the strength of ships increased during the nineteensixties it became possible to base the structural design on analysis procedures which to an increasing extent became based on applied mechanics. It is this continuing development which makes marine structures an interesting subject for participants at an IUTAM Congress.

Due to these relatively new design procedures based on analysis it became possible around 1970 to increase dramatically the size of ships without experiencing an excessive number of structural failures. See Fig. 1.

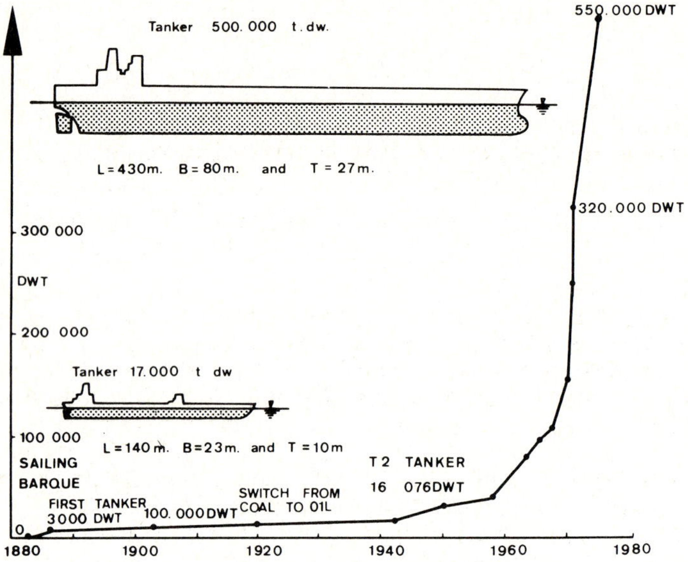

Fig. 1. Development in ship size.

Another major breakthrough for the applied mechanics approach to ship design was demonstrated during the seventies with construction of different types of ships for transportation of liquified natural gas (LNG). In these large ships the gas is transported at atmospheric pressure boiling at a temperature equal to -162° C. Since only the tank material is chosen such that it has sufficient toughness at these low temperatures then leakage from a tank can have catastrophic consequences due to the high energy content of the gas. To establish a "leak before failure" design for ships with independent tanks as the one shown in Fig. 2 relatively advanced fracture mechanics methods were applied.

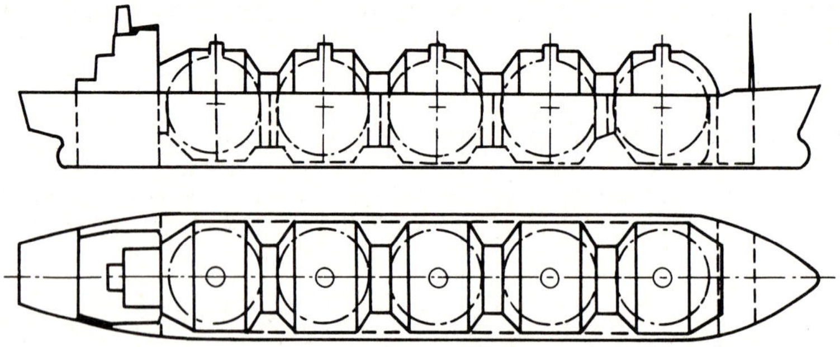

Fig. 2. Ship for transportation of 125.000 m^3 liquified natural gas (LNG) at -162° C.

Hydrodynamics had to be used to determine motions and accelerations of the ship in waves and to determine sloshing loads in the tanks. Initial postbuckling theories were used to determine whether the tanks had adequate safety against buckling.

Also in the future there will be a need for ships for transportation of bulk and general cargo. This can be predicted with some certainty when it is noticed that moving 1 ton of cargo 1 km costs about 0.3 US cent as an average on all ships in the world. This can be compared to 1.8 US cent for rail transport in North America or North Europe and 6.7 US cent per ton - km for a European truck transport.

There will of course be a continuing development towards new ship types. As an example it can be mentioned that at present projects are being evaluated for transportation of gas from the arctic regions in North America to the East Coast of Canada and US and to Europe. One project studied involved icebreaking LNG carriers with spherical cargo tanks containing 140.000 m^3 liquified gas. The propulsion power necessary to break the arctic ice between Canada and Greenland is estimated to be 150.000 hp. Such projects will increase the needs for better knowledge on ice mechanics, on fracture mechanics of welded structures at low temperatures, on response of large complicated structures to impulsive loading, etc. Another project involves transport below the ice in huge submarines. These vehicles will probably be about 450 m long, have a breadth about 75 m and a depth about 29 m.

Structures For Exploration and Exploitation of The Sea.

During the last two decades there has also been an enormous interest in marine structures intended for general exploration and exploitation of the resources in and below the sea.

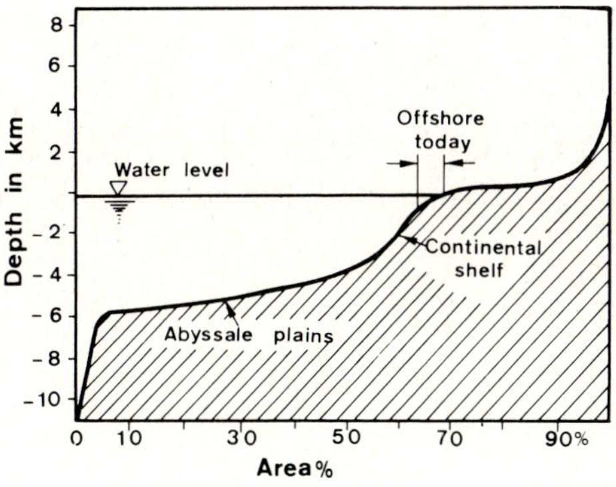

Fig. 3. Water depth distribution.

About 70% the surface of the earth is covered by the sea. See Fig. 3. Out of these 70% it is only about 5% where the water depth is less than 2-300 m. It is on this continental shelf that it is possible at present to exploit the resources of the sea within reasonable economic and technological frames.

Besides fish it is mainly oil, gas and sand which are extracted. Today's challenge to engineers is to move the economic exploitation limit from water depths of about 2-300 m to larger water depths.

Productions platforms for water depths less than 300 m are normally relatively stiff structures. An example from the Danish sector of the North Sea with water depth around 40 m is shown on Fig. 4. When the water depth exceeds about 300 m, fixed platforms become very expensive. Among the reasons for this are that in order to avoid dynamic magnification of first order wave induced forces it is necessary to avoid structures where the lowest natural period for global

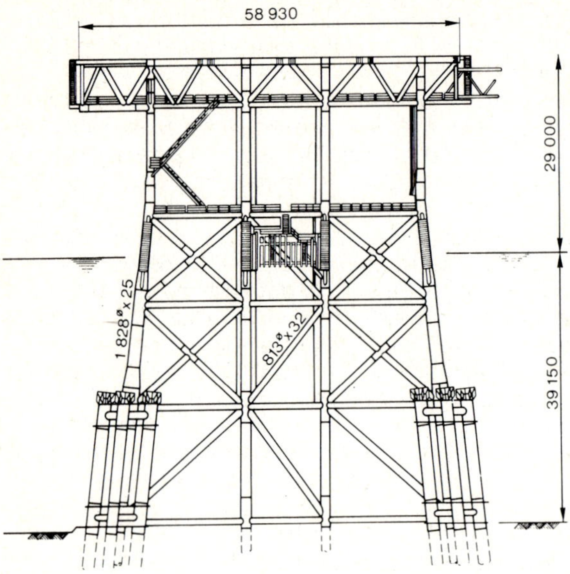

Fig. 4. Jacket and deck of platform in the North Sea (dimensions given in mm).

structural vibration is between 5 sec. and 25 sec. See Fig. 5. For fixed production platforms that means the lowest natural period for transverse vibrations must be less than around 5 sec. in order to avoid problems concerning fatigue cracks. Recently some new types of flexible structures have appeared where the natural period for transverse vibrations is larger than 25 sec. and the natural frequency for vertical vibrations is less than 5 sec. Such a structure is the tension leg platform (TLP) shown in Fig. 6. A TLP has a floating hull with so much excess buoyancy that it can maintain tension in a 6-700 m long system of cables between the platform and the ocean floor. Since there has been no previous experience with such structures it has been necessary to base their design on the existing theoretical knowledge, to some extent backed up by experiments.

It is characteristic for the offshore field that development proceeds so rapidly that often there is no time to await practical experience with one generation of structures before the next generation is being built. Therefore, reliable theoretical models for environmental conditions, wave load determination, structural response analysis, evaluation of design criteria and computerized design methods are paramount for design of marine structures.

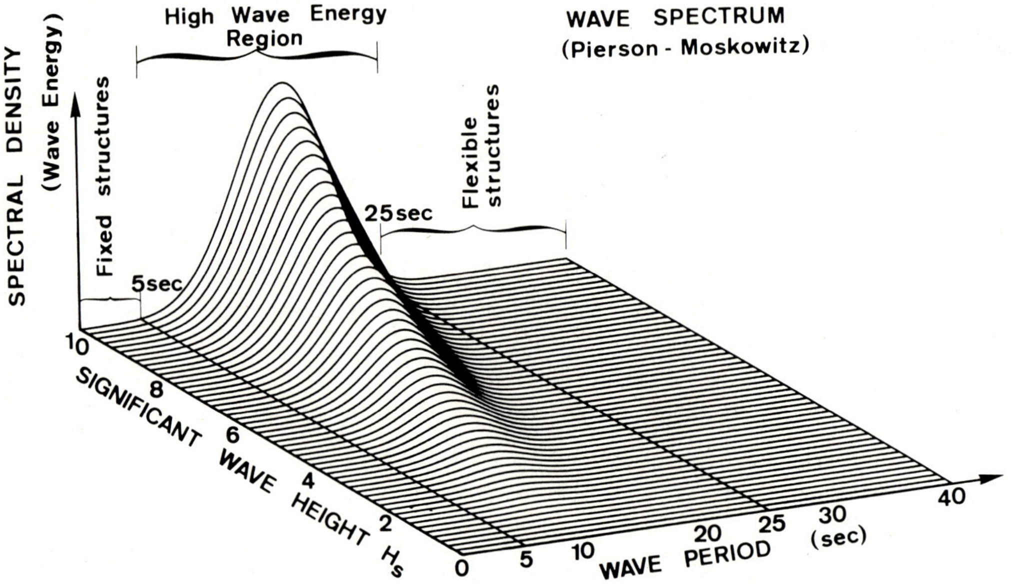

Fig. 5. Pierson-Moskowitz wave-spectra for different significant wave heights and wave periods typical for the North Atlantic Ocean.

International Ship Structures Committee (ISSC):

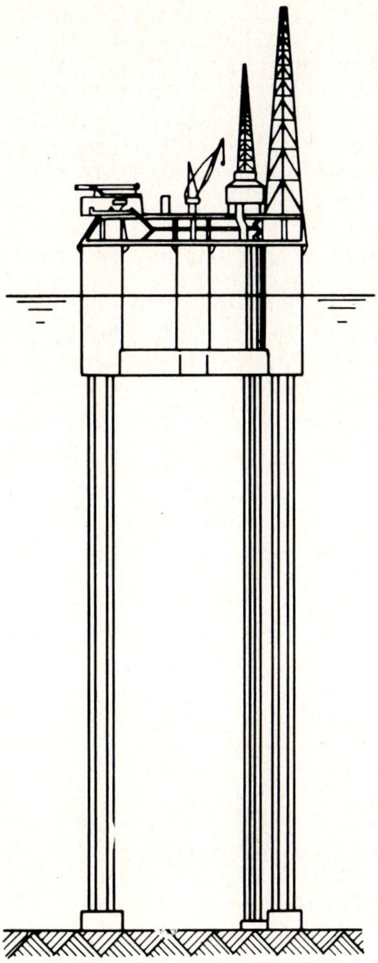

Fig. 6. A tension leg platform.

The ISSC has played a very important role as a link between the applied mechanics community and the engineering profession working with marine structures. ISSC was founded in 1959 and the aim is "to give experts in different countries engaged in research work on strength and structure problems related to ships and other structures used for sea transport and sea exploration an opportunity to meet at suitable intervals to discuss questions of common interest, to exchange information about preliminary results obtained and research in progress or contemplated in their respective countries and to recommend further research, especially where co-operation between different countries is of special interest or when a standard procedure is to be recommended".

Since the establishment ISSC has had 3-yearly meetings in different parts of the world with participation by invitation only. These meetings have fulfilled an important role in the development of analysis procedures for marine structures. Not in the least, the resulting published proceedings have provided an important source of references for researchers, designers and others [1]. The proceedings from the Congresses consist of reports from 12 technical committees each with 10 members who are active in research within the mandate of the committee. At present the subjects of the 12 committees are the following:

I: 1. Environmental Conditions: 2. Derived Loads.

II: 1. Elastic Response; 2. Non-Linear Structural Response; 3. Transient Dynamic Loading and Response; 4. Vibration and Noise.

III: 1. Ferrous materials; 2. Non-Ferrous and Composite Structures; 3. Fabrication and Service Factors.

IV: 1. Computation Means.

V: 1. Design Philosophy; 2. Applied Design.

In the present paper we shall discuss the elements which constitute a structural analysis procedure for man-made structures in the sea. We shall consider the following three main steps in the analysis of any marine structure:

a) Wave-induced loading

b) Response

c) Design criteria

Recently O. Hughes has published a textbook [2] which deals with these analysis techniques for ship structures.

Wave-Induced Loading

In the present context load will be defined as the action of sea waves on a marine structure (ship, offshore drilling platform, etc.). As will be shown total loads may depend on the response of the system, therefore, besides waves the loads will depend on geometry, mass and stiffness properties.

Bishop, Price and co-workers have in a series of papers and in a textbook [3] given a significant contribution to a consistent formulation of wave-induced loading on flexible marine structures. Within their notation some basic considerations regarding "loading", "properties" and "response" of marine structures are reproduced below. Let us consider a linear or linearized structural system. The dynamic properties of this mechanical system moving in vacuo is expressed in the form

$$[M]\{\ddot{u}\} + [C]\{\dot{u}\} + [K]\{u\} = \{F\} \tag{1}$$

For freely floating structures the left hand side represents rigid body motions as well as normal structural distortional vibration modes. The right hand side vector $\{F(t)\}$ denotes the loading of the system. This loading does not stem from waves only but also from the motions of the structure in water. In the case of a steady state linear system the equations of motion in water have the form

$$[M]\{\ddot{u}\} + [C]\{\dot{u}\} + [K]\{u\} = \{Q\} - [\mathcal{M}]\{\ddot{u}\} - [\mathcal{C}]\{\dot{u}\} - [\mathcal{K}]\{u\} \tag{2}$$

where $\{Q(t)\}$ is a vector giving the wave-induced forces, and the hydrodynamic forces due to the motion in the water is represented by an added mass matrix $[\mathcal{M}]$, a damping matrix $[\mathcal{C}]$, and a restoring matrix $[\mathcal{K}]$. The expression (2) indicates the reason for the traditional definition according to which the hydrodynamic loads derived from the flow potentials due to body motion in water are an inherent part of the problem of evaluating wave loads. Of course, for solution purposes a more convenient formulation of (1) and (2) is:

$$([M] + [\mathcal{M}])\{\ddot{u}\} + ([C] + [\mathcal{C}])\{\dot{u}\} + ([K] + [\mathcal{K}])\{u\} = \{Q\} \tag{3}$$

Properties Loading

As everybody knows who has watched waves on the sea, the seaway is far from being deterministic. Therefore, a probabilistic approach is needed to predict motions, accelerations, pressures, etc. The first step in such a probabilistic approach for a linear system is to assume that all wave disturbances and all structural responses may be regarded as ergodic random processes. That is, the random processes are assumed stationary so that statistical properties such as mean values and mean square values remain unchanges with time. In addition it is assumed that the statistical properties evaluated across the ensemble at some instant are the same as those derived by taking temporal averages along a single typical realisation. In nature these conditions are only approximately met during relatively short time intervals (of the order of magnitude hours). Therefore, such data determined from a single record of short duration are usually referred to as "short term statistics". For short term predictions, valid for a given sea state, see Fig. 5, one can then use the well-known result in linear random process theory that the mean square spectral density $S_R(\omega)$ of an output variable is related to that of an input, $S(\omega)$, by the equation

$$S_R(\omega) = |\Phi_R(\omega)|^2 \, S(\omega) \tag{4}$$

where $\Phi_R(\omega)$ is an appropriate transfer function determined from (3). In stochastic wave load analysis the effect of the directional spread of the waves may easily be included by introducing a directional distribution function in the spectral density function.

Based on calculated short term predictions and on oceanographic observation of the long term behaviour of specific ocean areas one can in principle make long term predictions by integration procedures. The long term distributions of stresses are then used for fatigue calculations and prediction of extreme values.

In order to illustrate the procedure we can consider a mathematical model for calculation of wave-induced stresses in the vertical plane of a symmetric ship hull. The response in the vertical plane of a symmetric ship hull is decoupled from the horizontal bending-torsion response. So, if we model the ship hull as a simple Timoshenko beam then the equations corresponding to eqs. (1) can be taken in the form

$$\frac{\partial}{\partial x}[EI(1+\eta\frac{\partial}{\partial t})\frac{\partial\varphi}{\partial x}] + \mu GA(1+\eta\frac{\partial}{\partial t})(\frac{\partial w}{\partial x} - \varphi) = m_s r^2 \frac{\partial^2\varphi}{\partial t^2}$$

$$\frac{\partial}{\partial x}[\mu GA(1+\frac{\partial}{\partial t})(\frac{\partial w}{\partial x} - \varphi)] = m_s \frac{\partial^2 w}{\partial t^2} - F(x,t) \qquad (5)$$

Here $EI(x)$ and $\mu GA(x)$ are the vertical bending and shear rigidities, respectively, $\varphi(t,x)$ is the slope due to bending, and $w(t,x)$ is the total deflection, η is an internal damping coefficient, and x is a longitudinal coordinate in an x,y,z coordinate system fixed with regard to the undisturbed ship so that the z-axis is in the vertical direction. See Fig. 7. Finally, $m_s(x)$ is the hull mass per unit length, $m_s r^2(x)$ is the mass moment of inertia about the horizontal y-axis, and $F(x,t)$ is the external force per unit length which is non-linear in w.

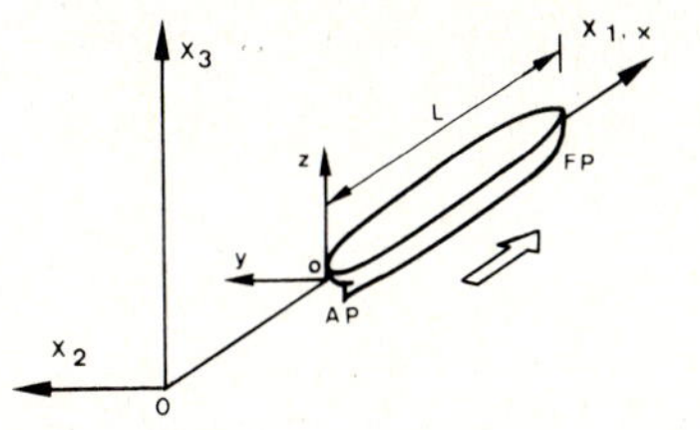

Fig. 7. Local and global coordinate systems.

The boundary conditions to the equations (5) express that the bending moments and shear forces are zero at the ends of the ship.

For ships the calculation of the hydrodynamic forces $F(x,t)$ can be based on a strip theory approximation. That is to say, any disturbance of the fluid caused by the motions of the ship section concerned propagates only in directions perpendicular to the axis Ox in the plane of symmetry. The upward force per unit length exerted by the fluid on the hull is according to a widely used, simple theory by Gerritsma and Beukelman [4] caused by a change in momentum of the added mass of water, damping due to energy dissipation in surface waves and a restoring term.

$$F(x,t) = -[\frac{D}{Dt}\{m(\tilde{z},x)\frac{D\tilde{z}}{Dt}\} + N(\tilde{z},x)\frac{D\tilde{z}}{Dt} + \int_{-T}^{-\tilde{z}} B(z,x)\frac{\partial p}{\partial z}\Big|_{z+w} dz] \qquad (6)$$

Here, the difference between the absolute displacement of the ship in the vertical direction, $w(x,t)$, and the surface of the ocean, $h(x,t)$ is denoted by $\tilde{z}(x,t)$, the added mass per unit length by m, and the damping by N. The operator

$\frac{D}{Dt}$ is the total derivative with respect to time t, that is

$$\frac{D}{Dt} \equiv \frac{\partial}{\partial t} - V\frac{\partial}{\partial x}$$

where V is the forward speed of the ship. The breadth of the ship is denoted by $B(\tilde{z},x)$; $T(x)$ denotes the draught, and finally, p is the Froude-Krylov fluid pressure that is fluid pressure in the wave in the absence of the ship. If we neglect the $\tilde{z}$ dependence in m, N and B, the force expression (6) corresponds to the linear strip theory proposed in [4].

It should be mentioned that a more consistent theory for calculation of first order (linear) hydrodynamic actions on ship hulls has been developed by Salvesen, Tuck and Faltinsen [5]. They determine the hydrodynamic forces from a potential flow theory in which the fluid is assumed irrotational, incompressible and inviscid. Only in the last stages of the analysis is a local two-dimensional approximation introduced. In the following paper by professor O. Faltinsen is presented a detailed discussion of different linear strip theories.

Therefore, here we shall mainly discuss the effect of inclusion of certain non-linear terms. In order to illustrate the nature of non-linearities in hydrodynamic loading we can evaluate (6) by a perturbational method, taking into account linear and quadratic terms in the relative displacement $\tilde{z}$ and thereby in the displacement of the hull w and the wave surface elevation h. In order to do this we can evaluate the water line breadth B, the added mass $\tilde{m}$, and the damping coefficient N around $\tilde{z}=0$, and include terms which are linear in $\tilde{z}$. This results in components of $F(x,t)$ which are linear and quadratic in $\tilde{z}$.

For deep water waves the wave surface elevation h and the pressure p, which approximately fulfills the free surface conditions, are also expressed as sums of a linear term and a quadratic term:

$$h(x,t) = h^{(1)} + h^{(2)}$$

or

$$h(x,t) = \sum_{i=1}^{n} a_i \cos\psi_i + \frac{1}{4} \sum_{i=1}^{n} \sum_{j=1}^{n} a_i a_j [(k_i+k_j)\cos(\psi_i+\psi_j) - |k_i-k_j|\cos(\psi_i-\psi_j)] \tag{7}$$

where a_i are wave amplitudes, and, considering head waves only

$$\psi_i = -k_i(x+Vt) - \sigma_i t + \theta_i$$

where k_i denotes the wave number in the x direction, and the frequency σ_i is given by $\sigma_i=\sqrt{gk_i}$ where g is the acceleration of gravity.

Altogether these approximations lead to

$$F(x,\tilde{z},t) = F^{(1)} + F^{(2)} \tag{8}$$

where $F^{(1)}$ are linear terms in the displacement $w^{(1)}$ and the wave surface amplitude a_i and $F^{(2)}$ are terms which are quadratic in these quantities.

For a given system of incoming waves the hull deflection $w(x,t)$; $\varphi(x,t)$ can be found from the equations (5) and (8) by a modal expansion procedure. If we introduce the amplitudes a_i and the phase lag θ_i of the regular sea wave through the variables ξ_i and ξ_{i+n} given by

$$\xi_i = a_i\cos\theta_i \quad \text{and} \quad \xi_{i+n} = a_i\sin\theta_i$$

and denote the frequencies of the corresponding wave encounters $\omega_i(=\sigma_i+\sigma_i^2V/g)$ then the wave induced bending moment in the hull beam can be expressed in the quadratic form

$$M(x,t) = \sum_{j=1}^{2n} \xi_j M_j(x,\omega_j)\cos\{\omega_j t+\varepsilon_j^1\} + \sum_{i,j=1}^{2n} [\xi_i\xi_j M_{ij}^+(x,\omega_j,\omega_i)\cos\{(\{\omega_j+\omega_i)t+\varepsilon_{ij}^+\} + \\ + \xi_i\xi_j M_{ij}^-(x,\omega_j,\omega_i)\cos\{(\omega_j-\omega_i)t+\varepsilon_{ij}^-\}].$$

where M_j and M_{ij} are known coefficients.

To model a stationary stochastic seaway the wave amplitude a_j and the phase lag θ_j are chosen so that $\xi_j=a_j\cos\theta_j$ and $\xi_{j+n}=a_j\sin\theta_j$ are jointly normally distributed with θ_j uniformly distributed, and so that the half mean sqared amplitudes $\frac{1}{2}a_j^2$ are equal to the wave energy in the associated range of frequencies.

Thus, for a given wave spectrum $S(\sigma)$, (see Fig. 5), σ being the wave frequency, the independent variables ξ_j and ξ_{j+n} $(j=1,2,\dots,n)$ are chosen so that they have the mean value zero and a variance given by

$$V_j = V_{j+n} = S(\sigma_j)\cdot\Delta\sigma_j = S^e(\omega_j)\cdot\Delta\omega_j$$

where $S^e(\omega)$ is the spectrum for the frequency of wave encounter ω.

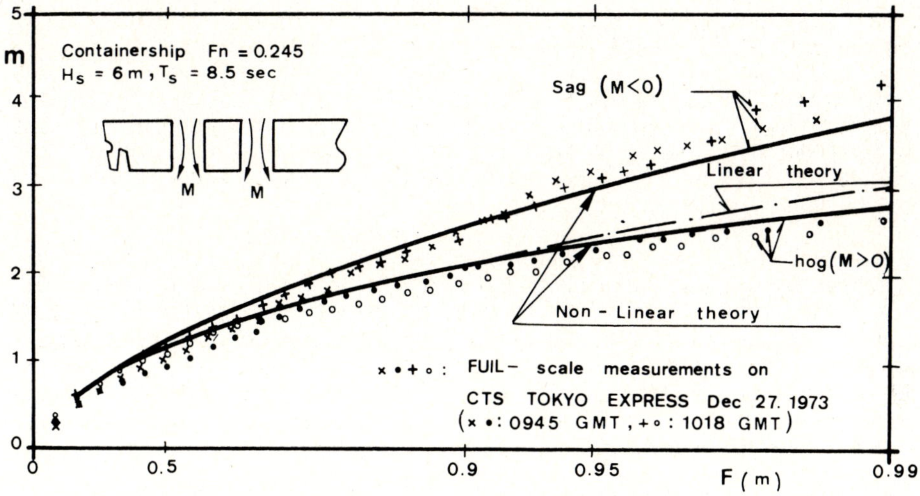

Fig. 8. Probability distribution function F for peak values of the non - dimensional midship bending moment $m = M/\lambda$ where λ is the standard deviation of the first-order (linear) contribution (λ_{calc} = 451 NMm). [6].

Fig. 8 shows a comparison between measured and predicted wave-induced bending moments in a containership of length $L = 270$ m, $B = 32.2$ m sailing the North Atlantic with a speed given by $F_n = V/\sqrt{gL} = 0.245$ in a sea state with a significant waveheight $H_s = 6$ m.

The bending moment is normalized by the standard deviation of the first order contribution in order to clearly identify the difference between the hogging and sagging bending moments. It must be noted that the mean level is taken as the average of the hogging and sagging amplitude which is exceeded by 80% of all amplitudes. It is seen that the agreement between the theoretical predictions and the full-scale measurements is reasonable for the non-linear theory.

Fig. 8 also shows a theoretical prediction based on a linear theory, that is neglecting the quadratic terms in (8). The linear theory has the shortcoming that it gives predictions which are independent of the sign of the peak values of the bending moment and leads to significant underestimation of the wave-induced sagging bending moments (compression stresses in the deck).

In conclusion it can be said that linear theories for wave-induced loading on ships and offshore structures are relatively well developed. Unfortunately, as for instance illustrated in Fig. 8, linear theories are only good mathematical models for small waves.

For moderate waves there is at present research going on along the lines indicated above, that is to get procedures in the frequency domain. But also more consistent and much more involved time-domain methods are being developed which may be applied to several actualizations of the random process. However, the analysis costs involved are very high.

For the extreme waves, that are large breaking waves caused by the "centenary storm", there are nearly no analytical tools available. For these waves there is no practical statistical description of the ocean wave elevation together with the pressure distribution below the ocean surface. When, sometime in the future, such a theory exists, there remains to be developed a procedure for calculation of the loads derived from such representation.

Research in this area is very important since it is these extreme sea states the designer wants his structure to withstand.

RESPONSE

Knowing the environmental loads the next analytical task for the designer is to perform a response analysis, which as indicated above cannot be completely separated from the load analysis.

Marine structures such as ships and offshore platforms are of great complexity and size. Perhaps of greater complexity than any other type of man-made structures. They are built of elements which possess unknown residual stresses before they are ever joined and in every step of the building process they are subjected to processes that introduce deformations, stresses and strains which excape computation by present analysis procedures. On top of that the system is launched and put into service in the hostile environment of the sea which, as shown above, can only be described by relatively crude statistical methods.

Considering these facts one is led to recognize that a fairly objective assessment of the strength of a given marine structure is a very difficult task.

Normally, a wave-response analysis is performed in three steps. In the first step an overall response analysis is performed. Most current methods used for practical calculation of the global structural response of ships and offshore structures in waves are based on linear, deterministic beam models. This overall analysis serves to give boundary conditions to a more detailed analysis of submodules.

A submodule is a region of structure in which a sufficient number of the scantlings are linked such that the region forms a logical entity. Examples on submodules may be one hold in a cargo ship or it may be a complex stiffened corner leg in a semi-submersible drilling platform. The submodule analysis often has to be non-linear in order to give a reasonably correct picture of the redistribution of stresses and strains which is caused by buckling and plasticity.

Again the submodule analysis serves to provide boundary conditions to the next step in the response analysis which consists of a load effect analysis of the principal strength members of the structure. Examples on principal members are stiffened panels, corrugated bulkheads, frames and stiffened cylinders in offshore structures.

It was mentioned above that the overall response analysis for most marine structures consists of a linear beam analysis. In spite of this apparently well known analysis procedure there is still room for significant contributions from research.

Let us, as one example among many, again consider the most common marine structure, that is a ship hull subjected to wave induced loading.

A modern container ship, see Fig. 9, is characterized by a fine body form and wide hatch openings. Both these characteristics result in poor torsional rigidity of the hull girder - a fact that has caused serious problems on many ships. Here, a reasonably accurate analysis based on a beam model can play a central role in the early design stages by providing the designer with information regarding the deformation and stresses of the hull in the seaway and the natural vibrations characteristics.

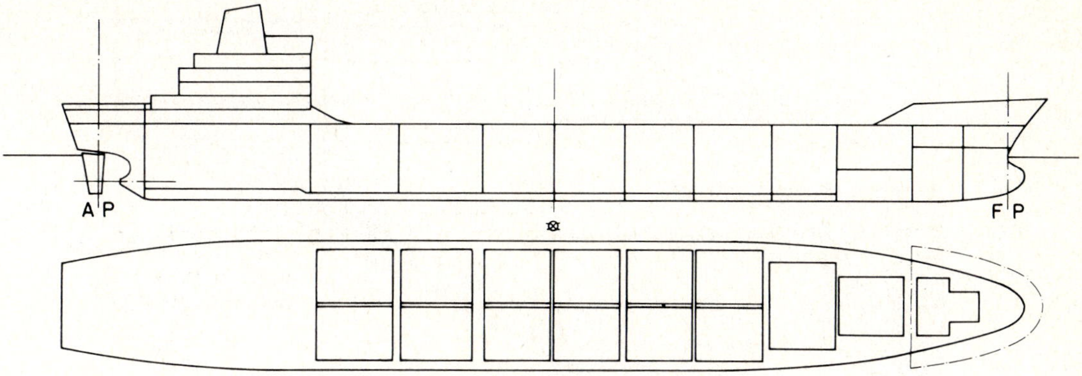

Fig. 9. A containership with large hatches.

The alternatives to beam analysis of the hull are model tests and finite element analysis using membrane and plate elements. There may, of course, be cases where these much more expensive methods provide the only true guarantee that a given design will be able to sustain the horizontal loads and torsional moments. However, the expense involved in the use of these methods considerably restricts their use. Also, most designers realize that large complex models are not always synonymous with correct results. Typically, the results obtained from these models are difficult to interpret, since the output may contain clusters of local and erroneous information which may tend to obscure the more important information of the overall behaviour.

Therefore, beam models are widely used for static and dynamic global analysis of stresses and deformations of ship hulls subjected to torsion and horizontal bending.

In order to show that derivation of such consistent beam theories for complex hulls is a worthy occupation for researchers in applied mechanics we shall briefly describe the state of the art and outline needed improvements.

As indicated in Fig. 10, a ship hull can be assumed to be composed of a series of connected hull segments with smoothly varying properties. To model the torsional-horizontal response of such a hull segment a beam theory is used which takes into account bending and warping deflections due to shear and rotatory inertia. As dependent variables is used the displacement $v(x,t)$ in the y-direction of points on the x-axis, the slope $\psi(x,t)$ due to horizontal bending, the angle of rotation $\Phi(x,t)$ about the x-axis, and finally the warping function $\chi(x,t)$ defined such that the deflection due to warping of the cross-section in the direction of the x-axis can be described by the product

$$u_x(x,s,t) = -\chi(x,t)\cdot\omega(s)$$

where $\omega(s)$ is the modified sectorial coordinate. The four dependent variables are related through four coupled differential equations.

Now, the abrupt changes in cross-sectional properties which occur at the transitions between open and closed parts of the ship hull and where deck beams are situated must be introduced as discontinuity conditions in the differential equations.

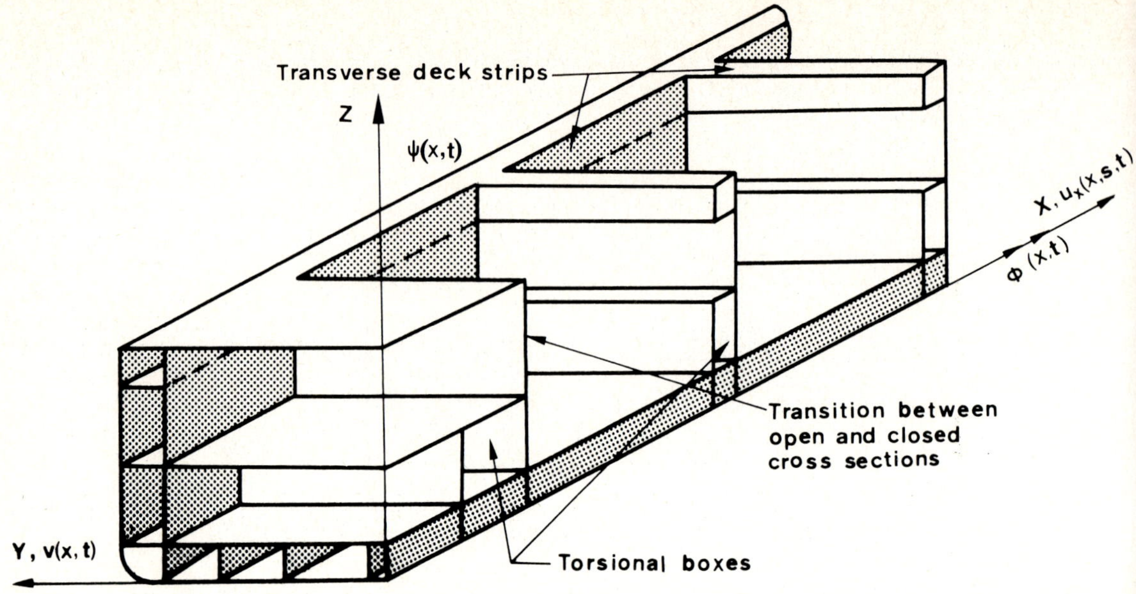

Fig. 10. Part of hull structure of a container ship.

Let us first consider discontinuity conditions at the boundary between two different cross-sections. An example is the transition from the closed cross-section in the aft part of the ship to the open cross-section in the cargo area at $x = x_i$. At such an intersection it is not possible to obtain complete compatibility since the two cross-sections in the general case will have different sectorial coordinate functions. In order to obtain compatibility in the mean, it is assumed that the warping function on the right hand side of the intersection $\chi(x^+_i)$ is related to the warping function on the left-hand side $\chi(x^-_i)$ by the form

$$\chi(x_i^+) = s_1^i \, \chi(x_i^-)$$

where s_1 is a warping compatibility factor. We also take into account the fact that an improved compatibility between the two cross-sections can be obtained by introduction of coupling between the bending angle ψ about the vertical z-axis and the warping function χ

$$\psi(x_i^+) = \psi(x_i^-) + s_2^i \, \chi(x_i^-)$$

The coupling coefficients s_1^i and s_2^i can be determined by imposing the condition that the displacement field given by the gap between the two cross-sections is orthogonal to the displacement fields produced by the generalized coordinates.

The effect of vertical or horizontal torsional boxes and transverse deck strips can be taken into account by introducing discrete "bimoment springs", that is, by relations between the added bimoment M_ω^* and the warping function $\chi(x)$ in the form

$$M_\omega(x_j^-) - M_\omega(x_j^+) = K_\omega^j \, \chi(x_j)$$

where K_ω^j is determined by the location and the stiffness properties of the restraining structure.

The beam model mentioned above for the container ship shown in Fig. 9 has three major cross-sectional discontinuities and seven deck beams. The calculation of the discontinuity condition and the sectional properties of the hull beam are based on the cross-sections indicated in Fig. 11 and a few more around the discontinuities. The stiffness contribution of the superstructure is neglected.

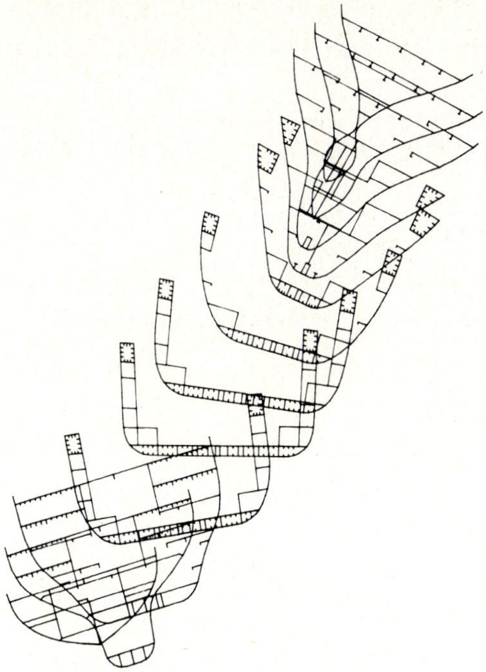

Fig. 11. First vibration mode (f_i=0.82 Hz) for containership vibrating in water [7].

The purpose of an overall torsional-bending analysis and a subsequent hull modul anasysis is to ensure that warping stresses in the hull are acceptable, that stresses at the transitions between deck beams and side shell are acceptable, that hatch opening diagonal elongations due to torsion and horizontal bending are within acceptable limits, and that the natural frequencies for the coupled torsional and horizontal bending modes are acceptable.

Beam models as the one described above are widely used and extremely helpful to give overall design guidance. However, there is a need for improved beam theories. For example one of the assumptions made in the theories used at present is that the cross-sectional variation in the longitudinal direction of the continuous parts of the hull beam can be neglected in comparison with the variation of the displacements. If this assumption is not made then a beam theory results with many more coupling terms in the governing differential equations. The drawback of such a theory is that the cross-sectional constants cannot be determined section by section. Instead, the hull has to be treated as a complicated three-dimensional structure in order to evaluate those cross-sectional constants which depend on derivatives with respect to the longitudinal coordinate. Such calculations would be extremely difficult to perform, but before abandoning the idea due to the difficulties involved, an analysis of the importance of these terms ought to be performed.

Furthermore, in the procedures used at present warping is represented by one generalized coordinate (χ). For long prismatic beams subjected to slowly varying torsional loads this has been shown to be adequate. However, for ships composed of an assemblage of relatively short prismatic thin-walled beams, it may be advantageous to develop a beam theory where the warping is described by a number of generalized coordinates. Such a method will also lead to better compatibility between different cross-sections.

This is just one area among many where applied mechanics research can give significant contribution to response analysis of large, complicated structures such as ships and offshore structures.

Design Criteria

The main goal of response analysis of marine structures is to determine wave-induced membrane stresses and bending moments acting together with local pressures on the strength elements of the structure. These resulting loading systems must then be compared to those levels and combinations of loads which cause structural failure in the form of large local yielding, instability or fracture.

It is in the development of rational failure criteria one could anticipate that research in structural and solid mechanics had contributed most to the field of ocean engineering since failure criteria have been the main research area for these branches of mechanics for centuries. But this is not so. It is within the field of failure criteria that designers of marine structures have been most reluctant to replace their empirically based rules by theoretically founded criteria.

Let us consider a simple example. The most common structural strength element in a marine structure is a stiffened panel. Based on linear plate theory and a first yield criterion the thickness requirement to a plate element in such a panel subjected to a pure lateral pressure, p, is

$$h = \frac{\sqrt{2}}{2}\, b\, \sqrt{p/\sigma_y}$$

where b is the stiffener spacing and σ_y is the yield strength of the plate material. However, even the most conservative empirical ship design criteria have requirements of the form

$$h = 0.5\, b \sqrt{p/\sigma_y}$$

and presently there is a tendency to allow for even smaller scantlings by relating the requirements to necessary plate thicknesses to acceptable permanent sets after unloading.

From this simple example it is obvious that a rationally based design criterion for a simple rectangular plate field subjected to pure lateral loading must take into account membrane and plasticity effects. On the other hand, for practical design relatively fast evaluation procedures are needed so the applied mechanics answer should not be a time consuming finite element analysis procedure.

Another example, the elastic-plastic buckling behaviour of stiffened panels subjected to in-plane loading has received much attention from researchers. Here reference can be made to one group of authors who have studied bifurcation, initial postbuckling behaviour and limit loads using some of the ideas behind Koiter's initial postbuckling theory for elastic structures [8]. Another group has developed comprehensive numerical procedures for buckling calculations of elastic-plastic, stiffened panels. Again, due to the complexity of analysis procedures proposed by both groups the design criteria used to determine limit loads in even advanced marine structures are still primarily based on the concept of columns with an effective width of plating using semi-empirical corrections for effects such as geometric imperfections and residual stresses due to welding [9].

But of course, there are also a number of examples on contributions from applied mechanics. During recent years elastic buckling of thin-walled shell structures has become a subject of interest in connection with design of marine structures. Here reference can be made to a variety of different designs for self-supporting tanks for ship transport of LNG, see Fig. 2, where elastic buckling is one of the main modes of failure. Initial post-buckling analysis procedures have played a significant role in determination of optimum scantlings for such tanks.

Another example on recent applied mechanics contributions to development of strength criteria is for offshore pipelines. Pipelines will usually be subjected to the most severe loads during the laying operation where large bending loads will be superimposed on the external hydrostatic pressure. The bending loads consist of a static part caused by the weight of the pipeline, which has to be large enough to ensure that the installed pipe does not move due to wave action, and a dynamic part due to wave induced motions of the system composed of the laying barge and the pipe. The highest bending stresses will occur in the pipeline sections in contact with the stinger (the overbend region) and in the section situated a small distance above the ocean bottom (the sagbend region). See Fig. 12.

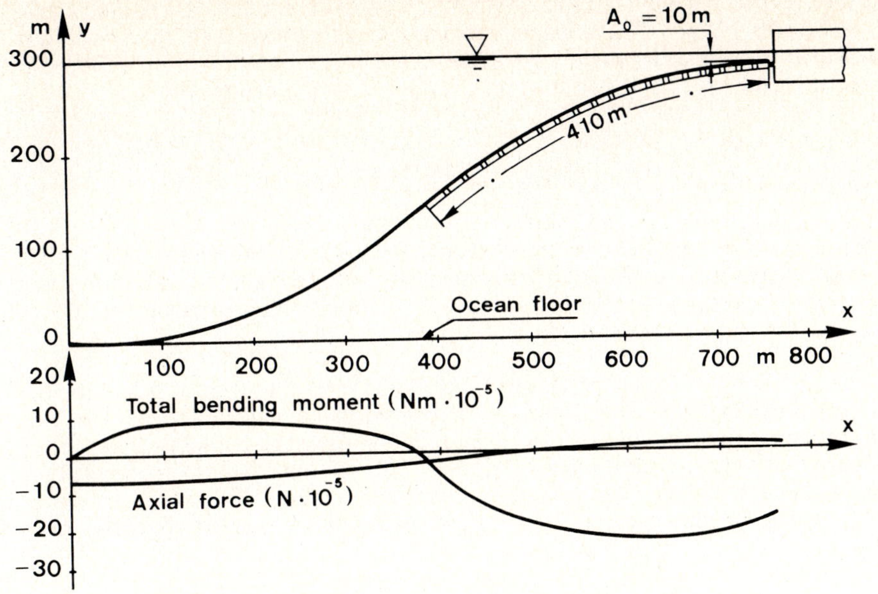

Fig. 12. Static equilibrium form and sectional forces in a pipeline during laying in the ocean.

In the analysis of the collapse strength of pipelines one can consider the length of the critical pipesection to be "infinite" with constant axial force, bending moment and external pressure [10].

Accurate theoretical predictions of collapse of long elastic-plastic pipes under combined bending and pressure have been presented recently in [11] and [12]. Both references make use of the principle of virtual work to derive a set of governing equations for the problem in question, which is non-linear with respect to geometry and material. The main difference in the two approaches is found in the choice of plasticity theory. In [11] a deformation plasticity theory is used in connection with a Ramberg-Osgood approximation of the experimental stress-strain curves. The analysis presented in [12] makes use of a small-strain J_2-flow theory with isotropic hardening together with a modified power law description of experimental stress-strain curves. The analysis presented in this reference shows that an elastic-plastic pipe subjected to combined bending and external pressure can be sensitive to geometric imperfections.

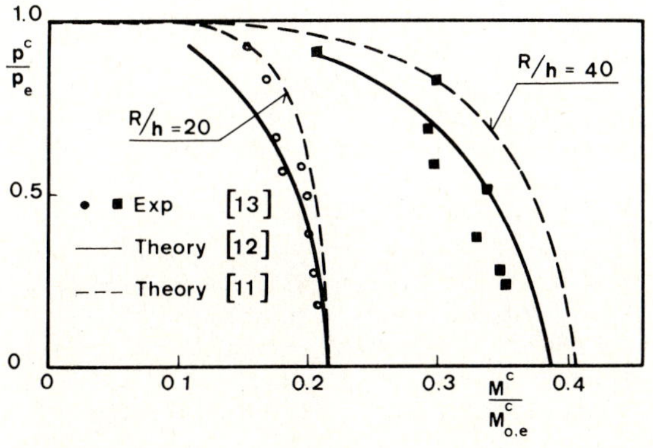

Fig. 13.

Steel tubes in combined bending and external pressure. Comparison between experiments [13] and theoretical findings [11], [12].

In Fig. 13 experimental results from [13] are compared with the theoretical findings in [11] and [12]. The external pressure is normalized by the elastic buckling pressure p_e and in order to separate the curves the limit moment is in Fig. 13 normalized by the elastic limit $M^c_{o,e}$ in pure bending:

$$p_e = \frac{E}{4(1-\nu^2)}\left(\frac{h}{R}\right)^3 \quad \text{and} \quad M^c_{o,e} = 1.02\ ERh^2$$

Since, as seen from Fig. 13, these consistent analysis procedures can predict the experimental collapse loads of pipes reasonable accurate it has been possible to use them with confidence to derive simple semi-empirical interaction formulas which are suited for practical design [14].

One of the reasons for striving at a small probability for local buckling of pipelines is the possibility of collapse of the pipe already laid on the ocean floor due to the propagation of a buckle along its length. Experiments in [15] show that pressures below a critical pressure p^* do not cause propagating buckles while at any external hydrostatic pressure above p^* a buckle once initiated, for example due to bending, will run dynamically collapsing the entire length of the pipe.

Many attempts have been made to determine this critical pressure theoretically. However, recently a very elegant solution was presented to this important problem in [16]. The solution is based on a calculation of the buckling and post-buckling behaviour of a ring under plane strain deformations. The outcome of such a relatively simple calculation is indicated in Fig. 14 in the form of the external pressure p per unit length as function of the area reduction. In [16] it is shown that the energy balance requirement is that the work done by the critical pressure p^* must equal the change of strain energy ΔW absorbed in the pipe in a unit advance of the transition front between buckled and unbuckled pipe:

$$p^*(\Delta A_D - \Delta A_U) = W$$

If $p(\Delta A)$ denotes the relation of pressure to area reduction for the ring under plane strain deformations the critical buckling pressure can be determined simply by

$$p^* = \int_{\Delta A_U}^{\Delta A_D} p(\Delta A)\, d\Delta A / (\Delta A_D - \Delta A_U)$$

A graphical interpretation is indicated in Fig 14.

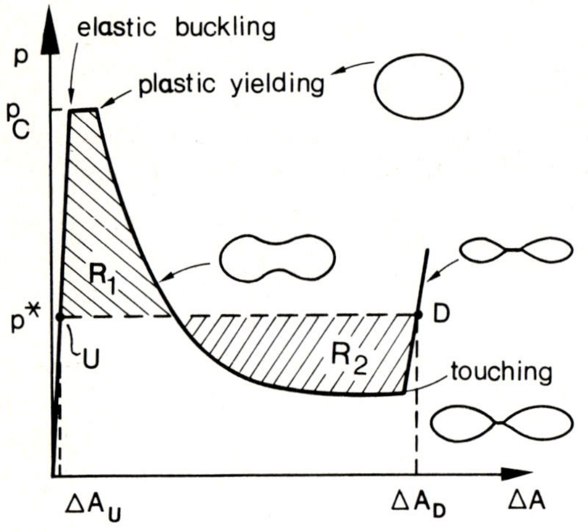

Fig. 14. Buckling and post-buckling behaviour of a ring undergoing plane strain deformation. Quasi-static propagation pressure p^* is given by the condition $R_1=R_2$ for a pipe of material characterized by deformation theory [16].

In conclusion it can be said that in recent years consistent applied mechanics procedures have been presented which has improved immensely the basis for design of pipelines against buckling failure.

However, for more complicated stiffened plate and shell structures subjected to complex loading conditions there is a need for similar development of rational analysis procedures. Such procedures must take into account residual stresses from welding and geometrical imperfections considerably larger than those encountered in aerospace applications of shell structures. Due to lack of contributions from the applied mechanics community in the form of consistent and reliable analysis procedures design of the huge thin-walled stiffened buoyancy cylinders in tension leg platforms, see Fig. 6, have at present to be based on semi-empirical methods like for example the effective width concept [17]. Important questions related to the residual strength of columns having relatively large dents caused by ship collissions cannot at present be given a satisfactory answer.

CONCLUSIONS

Applied mechanics research has had a dramatic influence on design of marine structures by providing basic understanding and consistent theoretical analysis models for environmental conditions, wave load determination, structural response analysis, evaluation of design criteria, and computerized design methods.

However, considering the needs for research related to marine structures and the enormous mass of published research results from the field of applied mechanics applicable to marine structural design then there is scope for more interaction. In the present paper it is shown that applied mechanics research in fields relevant to marine structures often only to a very modest extent benefits those who have the primary task of developing design basis for marine structures. More effort should be done from both sides to harness these research results for application in the important and exciting field of marine structures.

REFERENCES

[1] Proceedings of the International Ship Structures Congress: (A) First Congress. Glasgow 1961; (b) Second Congress, Delft 1964; (c) Third Congress, Oslo, 1967; (d) Fourth Congress, Tokyo, 1970; (e) Fifth Congress, Hamburg, 1973; (f) Sixth Congress, Boston, 1976; (g) Seventh Congress, Paris, 1979; (h) Eighth Congress, Gdansk, 1982.

[2] Hughes, O.: Ship Structural Design (J. Wiley and Sons, 1983).

[3] Bishop. R.E.D. and Price, W.G.: Hydroelasticity of Ships (Cambridge University Press, 1979).

[4] Gerritsma, J. and Beukelman, W.: The Distribution of the Hydrodynamic Forces on a Heaving and Pitching Ship Model in Still Water. Figth Symposium on Naval Hydrodynamics (1964).

[5] Salvesen, N., Tuck, E.O. and Faltinsen, O.: Ship Motions and Sea Loads, Trans. SNAME, 78, (1970), 250-287.

[6] Jensen, J. Juncher and Pedersen, P. Terndrup: Bending Moments and Shear Forces in Ships Sailing in Irregular Waves. Jrnl. of Ship Research, Vol. 25, 4, (Dec. 1981), 243-251.

[7] Pedersen, P. Terndrup: A Beam Model for the Torsional-Bending Response of Ship Hulls. Transaction of the Royal Institution of Naval Architects (1982).

[8] Tvergaard, V.: Buckling Behaviour of Plate and Shell Structures. Proceedings IUTAM Congress (North Holland, Delft, 1976).

[9] Faulkner, D.: Compression Test on Welded Eccentrically Stiffened Plate Panels. Proceedings Symp. Steel Plated Structures (Crosby Lockwood Staples, London, 1977).

[10] Axelrad, E.L. and Emmerling, F.A.: Grosse Verformungen und Traglasten Elastischer Rohre unter Biegung und Aussendruck, Ingenieurarchiv, 53, (1983, 41-52.

[11] Kyriakides, S. and Shaw, P.K.: Response and Stability of Elastic-Plastic Circular Pipes under Combined Bending and External Pressure. Int. J. Solids Structures 18, No. 11, pp. 957-973, (1982).

[12] Fabian, O.: Elastic-Plastic Collapse of Long Tubes under Combined Bending and Pressure Load. Ocean Engineering 8, (1981), 295-330.

[13] Johns, T.G. et al: Inelastic Buckling of Pipelines under Combined Loads, Offshore Technology Conference, Houston, USA, No. OTC 2209, (1975).

[14] Jensen, J. Juncher and Pedersen, P. Terndrup: Interaction Formulas for Buckling of Shell Structures Subjected to Combined Loads. Tokyo and Seoul, PRADS 83, (1983), 483-492.

[15] Kyriakides, S. and Babcock, D.C.: Experimental Determination of the Propagation Pressure of Circular Pipes. Jrnl. of Pressure Vessel Technology, Trans. of the ASME, Vol. 103, (1981), 328-336.

[16] Chater, E. and Hutchinson, J.W.: On the Propagation of Bulges and Buckles. Harvard Un., Report Mech. 44, Div. of Applied Sci. (June 1983).

[17] Faulkner, D., Chen, Y.N. and de Oliveira, J.G.: Limit State Design Criteria for Stiffened Cylinders of Offshore Structures. 4th National Congress on Pressure Vessel and Piping Technology, Portland, Oregon, (1983).

Theoretical and Applied Mechanics
F.I. Niordson and N. Olhoff
Elsevier Science Publishers B.V. (North-Holland)

HYDROELASTICITY OF MARINE STRUCTURES

W.G. Price
Wu Yousheng
Department of Mechanical Engineering,
Brunel University,
Uxbridge, Middlesex,
U.K.

Hydroelasticity is that branch of science which is concerned with the motion of deformable bodies through liquids. A floating or fixed marine structure is a flexible body which deforms due to the actions of fluid loadings or other externally applied forces. This paper discusses a general three dimensional hydroelastic theory applicable to marine structures which may be moving or fixed in regular sinusoidal waves. Applications of the theory are illustrated with reference to ships, multi-hulls (SWATHS, catamaran, semi-submersible), jack-up rig, fixed plate, etc.

INTRODUCTION

Consider a naval architect who is about to design a marine structure e.g. a conventional or unconventional ship, hovercraft, hydrofoil, offshore floating or fixed platform etc. that is not some close relative of an existing structure. The creation must meet specific requirements i.e. an initial cost, safety, reliability maintainability, performance etc. There is no possibility of prototype testing and misjudgement could produce horrendous consequences. It is against this background that one of the naval architect's problems is to be discussed. That is *the response of the marine structure to waves.*

Historically, the naval architect has been involved mainly in the design of ships, though in recent years interest has shifted to the design of offshore structures. He has the responsibility of ensuring that the designed ship has good seakeeping qualities. In brief, this seakeeping or seaworthiness may be defined as the ability of the ship to travel safely in the roughest seas and proceed on course with the minimum delay. The problem has changed with time - from sailing vessels which followed the prevailing winds to the advent of moderately powered steam ships capable of travelling directly to windward. Though structural damage occurred to the early steam ships, full engine power could be used in almost any weather condition even though ship speed was reduced by the actions of wind and sea.

For modern fast ships travelling in rough seas, the engine power available to any ship's captain is excessive and must be reduced voluntarily to avoid severe structural damage to the hull. Thus the decision of a ship's captain to maintain a particular schedule now depends on the behaviour of the ship's response in the seaway as well as the amount of available power at his command. Thus if bodily motions become too great because of the wave excitations, ship operations are greatly reduced because of excessive motions, deck wetting, maintaining course, speed reductions, slamming with consequential high stresses, etc. Alternatively, since the ship's structure is flexible, distortions may become excessive with resultant cracking or buckling of plates, pipes, etc. If all these possibilities in modern ship design and operations are untoward, they are not immediately catastrophic. But catastrophes do happen as a result of excessive wave excited hull responses. For example, tankers have split in waves, hatch covers have been dislodged and washed overboard, some types of hulls twist and crack, and smaller vessels, such as trawlers, capsize for reasons that are not fully understood.

In recent years, with the ever increasing demand for petroleum products, offshore exploration has become common place in deeper and deeper waters. This has demanded the building and development of large specialised offshore structures which may be fixed or floating. Again it is of the utmost importance that the naval architect ensures that the design can withstand the roughest of seas without impairing the safety of personnel and the structural integrity of the offshore structure.

In contrast to ships, these offshore structures can be either floating (slow moving or stationary) or fixed to the sea bed with very large geometrical dimensions and shapes which are far from slender. However the naval architect is faced with a plethora of similar design problems whether he is dealing with ship like bodies or offshore structures but the emphasis and importance of a particular aspect i.e. forward speed, arbitrary geometry, operations etc. may be different.

In order to assess the wave excited responses of a flexible marine structure, the naval architect has to be able to describe and identify the form of the structure and the fluid actions applied to it. He needs techniques which are the products of structural analysis and naval hydrodynamics (see, for example, figure 1).

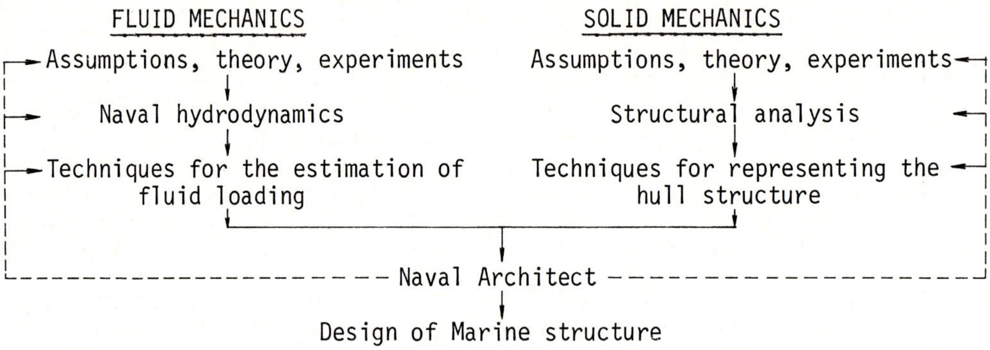

Figure. 1. Flow chart illustrating information transfer in the design process.

These have their origins in the theories of solid mechanics and fluid mechanics respectively. This does not mean that it is customary for designers to employ methods they receive ready made from classical physicists and applied mathematicians who may specialise in the relevant branches of applied mechanics and hydromechanics. Far from it, the techniques adopted are fashioned by naval architects since there is a strong feed back from actual designs. In addition naval architects tend to specialise either in structures or in naval hydrodynamics and seldom in both.

When the structural and hydrodynamic theories are available, the naval architect has a very difficult task in reconciling the two theories to describe the overall behaviour of the marine structure in rough seas since the theories are of necessity idealised. However, if one accepts the predictions of fluid actions offered by naval hydrodynamics and the representation of the marine structure as some form of elastic structure proposed by the structural analyst, the estimation of the responses associated with the marine structure becomes in effect a refined vibration problem. However, because of the physical nature of the sea environment, in general the vibrations excited in the structure are random and the oscillating marine structure - wave system constitutes a non-conservative system.

There are no new ideas introduced in discussing the marine structure as a more or less elastic body excited by waves but it is not common to estimate the responses of the structure in any unified way using the estaiblished techniques of dynamics. Therefore the aim of this paper and lecture is to show that by generalising the approach of Bishop and Price (1979) a suitable unifying hydroelasticity theory involving both the techniques of structural analysis and naval hydrodynamics may be developed and this describes in a realistic manner *the responses of an arbitrary shaped, flexible marine structure travelling with forward speed in the free surface of a seaway.*

In fact, the theoretical model describing the responses associated with a fixed marine structure i.e. offshore platform or a moving, totally submerged structure are special cases of the general problem just posed. For example, a deflection associated with a floating, flexible marine structure will involve a combination of bodily motions and distortions of the structure whilst for a fixed body only distortions of the structure are admitted into the theory. However, because of the imposed geometric conditions relating to the arbitrary shape of the marine structure, it becomes necessary to adopt and develop suitable three dimensional structural and hydrodynamic theories. Thus from the same basic theory and set of assumptions the responses of a rigid body, flexible ship-like body, semi-submersible/SWATHS/catamaran hull, fixed or moving jack-up rig, fixed offshore platform etc. may be determined and discussed.

GENERALISED EQUATIONS OF MOTION

In the hydroelasticity theory of ships developed by Bishop and Price (1979) the hull is assumed beamlike and its vibration characteristics in vacuo are determined in the absence of damping and external forces. By adopting a suitable beam theory a set of principal modes and natural frequencies of the dry ship may be determined. When allowance is made for the water in which the dry hull is placed all forces are treated as external actions applied to the hull. From the relevant mathematical model, resonance frequencies and responses (i.e. displacement, bending moment, shearing force, twisting moment, etc.) at any arbitrary position in the hull may be determined.

Whether symmetric and/or antisymmetric responses of the hull are under investigation, it has been shown that the generalised linear equations of motion describing the responses of the flexible dry hull may be expressed in the form

$$\underset{\sim}{a}\ddot{\underset{\sim}{p}}(t) + \underset{\sim}{b}\dot{\underset{\sim}{p}}(t) + \underset{\sim}{c}\underset{\sim}{p}(t) = \underset{\sim}{Z}(t)$$

where

$\underset{\sim}{a}$ is the real inertia matrix of the dry hull, which is diagonal apart from off diagonal antisymmetric rigid body contributions,

$\underset{\sim}{b}$ is the structural damping matrix of the dry hull,

$\underset{\sim}{c}$ is the diagonal stiffness matrix of the dry hull with elements $c_{ss}=\omega_s^2 a_{ss}$.

The column matrices $\underset{\sim}{p}(t)$ and $\underset{\sim}{Z}(t)$ represent the responses of the hull and input loading respectively. In fact, it can be shown that this input loading may be represented in the form

$$\underset{\sim}{Z}(t) = -\underset{\sim}{A}\ddot{\underset{\sim}{p}}(t) - \underset{\sim}{B}\dot{\underset{\sim}{p}}(t) - \underset{\sim}{C}\underset{\sim}{p}(t) + \underset{\sim}{\Xi}(t)$$

where $\underset{\sim}{A}$, $\underset{\sim}{B}$ and $\underset{\sim}{C}$ are square matrices associated with the motions of the flexible body in the fluid and $\underset{\sim}{\Xi}(t)$ is a column matrix representing the wave excitation.

To describe non-beamlike structures which may be floating or fixed a more general mathematical model must be developed. One alternative approach to the two dimensional beamlike model is to assume a finite element discretisation of the three dimensional structure.

By adopting the techniques developed in finite element theory (see, for example, Zienkiewitz (1977)), it has been shown by Bishop, Price and Wu (1984) that if the continuous dry structure is discretised with m degrees of freedom then the (mx1) column nodal displacement vector $\underset{\sim}{U}=\{U_1,U_2,\ldots,U_m\}$ is a solution of the set of equations describing the responses of the dry structure. Namely,

$$M\ddot{U} + C\dot{U} + KU = P \tag{1}$$

where the (mxm) square matrices M, C, K describe the system mass, damping, stiffness properties respectively and the (mxl) column matrix P represents a general external loading applied to the structure. The matrices M and K are semi-definite or definite depending on the boundary restraints imposed on the structure.

For free motion P=0=C, a solution of the reduced matrix equation is sought in the general form

$$U = \Delta e^{i\omega t}$$

where Δ denotes a principal eigenvector matrix and the eigenvalue ω represents a natural frequency of the dry structure. At each natural frequency $\omega=\omega_r(r=1,2,.,m)$ there exists solutions for U and Δ. These may be represented as

$$U_r = D_r e^{i\omega_r t}$$

for each r=1,2,...,m with the column matrices $U_r=\{U_1,U_2,..,U_m\}_r$ and the characteristic vector of the rth principal mode $D_r=\{D_1,D_2,...,D_m\}_r$ where the element

$D_{rj} = \{u_r,v_r,w_r,\theta_{xr},\theta_{yr},\theta_{zr}\}_j$

represents the displacement vector of the rth modal shape at the jth node (i.e. u,v,w translations and $\theta_x,\theta_y,\theta_z$ rotations about the equilibrium coordinate axes Ox, Oy, Oz respectively, see fig. 2.)

If a submatrix of D_r corresponding to the rth modal shape of all the nodes within one particular finite element is denoted by d_r then, as shown by Bishop, Price and Wu (1984), the rth principal modal shape at any arbitrary point within that finite element may be expressed as

$$u_r = \ell^T N L d_r = \{u_r,v_r,w_r\} \quad . \tag{2}$$

Here $\ell^T NL$ denotes a transformation matrix between the local coordinate system defined within the finite element and the equilibrium axes system used to define the overall mode shape of the structure and as will be discussed later, also the external fluid loading.

For forced motion P≠0≠C, according to a theorem by Rayleigh (1894) any distortion of the structure may be expressed as an aggregate of distortions in the principal modes. That is to say, the nodal displacement in the forced solution may be expressed in the form

$$U = \sum_{r=1}^{m} D_r p_r(t) = Dp(t) \tag{3}$$

where the (mxl) column vector p represents the m principal coordinates associated with the principal modes of the dry structure and the (mxm) square matrix $D = \{D_1,D_2,...,D_m\}$ is the principal mode matrix.

Thus the displacement at any arbitrary point within the structure may be written in the form

$$u = \{u,v,w\} = \sum_{r=1}^{m} u_r p_r(t) = \sum_{r=1}^{m}\{u_r,v_r,w_r\}p_r(t)$$

where u_r denotes the rth principal mode shape vector associated with the dry structure.

Using derived orthogonality relationships, the symmetric principal mass matrix, a and stiffness matrix, c, may be expressed as

$$a = D^T M D \ , \qquad c = D^T K D$$

where the superscript T denotes a transpose matrix and the element $c_{ss}=\omega_s^2 a_{ss}$. After substituting for U from equation (3), premultiplying by D^T and using these orthogonality conditions, the equation of motion (1) may be caste into the familiar form,

$$\underset{\sim}{a}\ddot{\underset{\sim}{p}}(t) + \underset{\sim}{b}\dot{\underset{\sim}{p}}(t) + \underset{\sim}{c}\underset{\sim}{p}(t) = \underset{\sim}{Z}(t) \tag{4}$$

where the (mx1) column matrix

$$\underset{\sim}{Z}(t) = \underset{\sim}{D}^T\underset{\sim}{P}(t) = \{Z_1(t),Z_2(t),\ldots,Z_m(t)\} \tag{5}$$

represents the generalised external force arising from the waves, mooring lines, propeller, gravity, etc., and

$$\underset{\sim}{b} = \underset{\sim}{D}^T\underset{\sim}{C}\underset{\sim}{D}$$

is the generalised structural damping matrix which is usually assumed diagonal. This simplification is introduced because of the great scarcity of information on the distribution of damping throughout the structure.

Under these assumptions, the generalised linear equations of motion may be expressed as

$$a_{rr}\ddot{p}_r(t) + b_{rr}\dot{p}_r(t) + c_{rr}p_r(t) = Z_r(t) \tag{6}$$

for r=1,2,..,m.

THE FLUID-STRUCTURE PROBLEM

A flexible, floating structure travelling with constant forward speed $\bar{U}$ at an arbitrary heading angle χ (=180°, head waves) in deep water, regular sinusoidal waves of amplitude a, frequency ω, wave number k or wavelength $\lambda(=2\pi/k)$ experiences waves of encounter frequency

$$\omega_e = \omega - \bar{U}k\cos\chi = \omega - (\bar{U}\omega^2\cos\chi)/g.$$

The freely floating structure, excited by the waves, responds in both rigid body and distortion modes. The rigid body responses are associated with the first six principal modes of the dry structure which form a subset of the infinite number of modes describing the dynamic characteristics of the dry flexible structure. These rigid body responses are discussed extensively in seakeeping analyses and only a resumé of the underlying theory is presented here before embarking on a general hydrodynamic - structures theory associated with the flexible structure.

For a fixed flexible structure, rigid body modes do not exist and only the distortion responses of the structure need be considered.

AXES AND GENERAL FORMULATION

Figure 2 illustrates the three right hand axes systems used to define the fluid actions. $Ax_0y_0z_0$ is a fixed spatial set of axes, Oxyz is an equilibrium set of axes moving with forward speed $\bar{U}$ and remaining parallel to $Ax_0y_0z_0$ and O'x'y'z' is an axis system fixed in the arbitrary shaped structure and, prior to any disturbance, coincides with Oxyz.

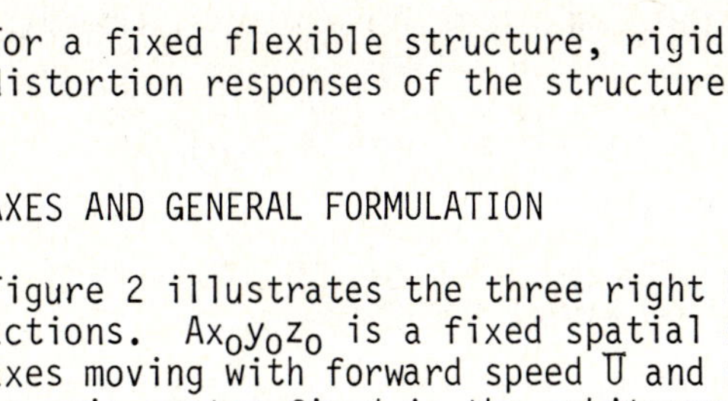

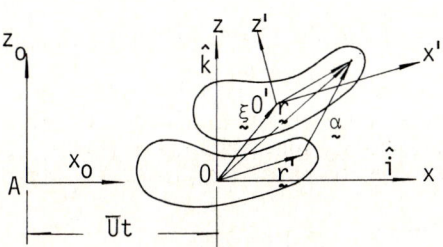

Figure 2. Axes systems.

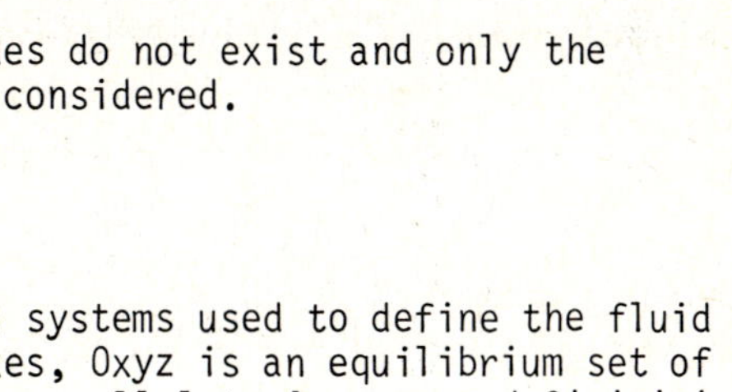

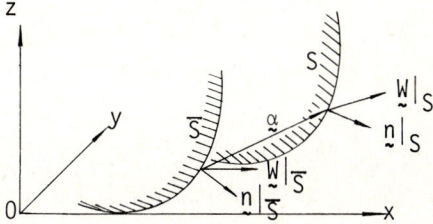

Figure 3. Instantaneous positions of unit normal vectors and steady flow velocity vectors on the wetted surfaces.

Assuming the fluid to be incompressible, inviscid and irrotational, there exists a potential function $\Phi(x_0,y_0,z_0,t)$ such that the fluid velocity $\underset{\sim}{V}(x_0,y_0,z_0)$ is

$$\underset{\sim}{V} = \text{grad}\ \Phi = \nabla\Phi$$

and throughout the fluid, the Laplace condition $\nabla^2\Phi = 0$ is valid.

Newman (1978) in his many extensive studies of rigid ship motions has shown that this potential satisfies the following boundary conditions:
(i) On the free surface, $z_0=\zeta$,

$$\Phi_{tt} - 2\nabla\Phi.\nabla\Phi_t + \tfrac{1}{2}\nabla\Phi.\nabla(\nabla\Phi.\nabla\Phi) + g\Phi_{z_0} = 0 \quad , \tag{7}$$

where $\Phi_{tt} = \partial^2\Phi/\partial t^2$ etc. and ζ is the elevation of the wave disturbance.
(ii) On the sea bed, $z_0=-d$,

$$\Phi_{z_0} = 0 \quad . \tag{8}$$

(iii) A suitable far field boundary condition.
(iv) On the instantaneous total wetted surface area S of the structure

$$\partial\Phi/\partial n = \Phi_n = \underset{\sim}{V}^S.\underset{\sim}{n} \tag{9}$$

where $\underset{\sim}{V}^S$ denotes the local velocity on the wetted surface S and $\underset{\sim}{n}$ is the outward drawn unit normal vector into the fluid.

From figure 2 it is seen that if $\underset{\sim}{\xi}$ represents a translation of O' from O and $\underset{\sim}{\Omega}$ denotes a rotation of the rigid structure, the displacement of the point $\underset{\sim}{r}'(=x',y',z')$ relative to O is

$$\underset{\sim}{\alpha} = \underset{\sim}{\xi} + \underset{\sim}{\Omega}x\underset{\sim}{r}' \tag{10}$$

and the velocity of this point is

$$\underset{\sim}{V}^S = \bar{U}\hat{i} + \dot{\underset{\sim}{\alpha}} = \bar{U}\hat{i} + \dot{\underset{\sim}{\xi}} + \dot{\underset{\sim}{\Omega}}x\underset{\sim}{r}' \quad . \tag{11}$$

Figure 3 illustrates the effect of the body disturbance on the change in the direction of the unit vector $\underset{\sim}{n}$ relative to the equilibrium axes. This direction is unaffected by a pure translation but rotations of the structure must be accounted for. Thus if $\underset{\sim}{n}|_S$, $\underset{\sim}{n}|_{\bar{S}}$ denote the unit vectors relating to the disturbed and steady state condition respectively it follows that to a first approximation,

$$\underset{\sim}{n}|_S = \underset{\sim}{n}|_{\bar{S}} + \underset{\sim}{\Omega}x\underset{\sim}{n}|_{\bar{S}} \ . \tag{12}$$

The total potential Φ may be represented in the equilibrium frame of axes as

$$\Phi(x_0,y_0,z_0,t) \equiv \bar{U}\bar{\phi}(x,y,z) + \phi(x,y,z,t) \tag{13}$$

where $\bar{\phi}$, ϕ denote velocity potential components due to the steady motion of the structure in calm water and the unsteady motion in waves respectively.

When only the steady motion is considered, the velocity of the steady flow relative to the moving equilibrium frame of axes is

$$\underset{\sim}{W} = \bar{U}\text{grad}(\bar{\phi} - x) \tag{14}$$

and the boundary condition (iv) takes the form $\underset{\sim}{W}.\underset{\sim}{n}=0$ on the mean wetted surface $\bar{S}$.

However, when the structure is disturbed, the flow velocity associated with the instantaneous wetted surface S i.e. $\underset{\sim}{W}|_S$ may be expressed in terms of the steady flow velocity i.e. $\underset{\sim}{W}|_{\bar{S}}$ and a contribution due to the disturbed displacement of the structure. That is, to a first approximation the variation of $\underset{\sim}{W}|_S$ about $\underset{\sim}{W}|_{\bar{S}}$ due to a parasitic disturbance $\underset{\sim}{\alpha}$ may be written as

$$\underset{\sim}{W}|_S = [1 + (\underset{\sim}{\alpha}.\nabla)]\underset{\sim}{W}|_{\bar{S}} = [1 + (\underset{\sim}{\alpha}.\nabla)]\{\bar{U}\nabla(\bar{\phi}-x)\}|_{\bar{S}} = \bar{U}\nabla(\bar{\phi}-x)|_S \tag{15}$$

since for any vector e.g. $\nabla\bar{\phi}$, it follows from the above that

$$\nabla\bar{\phi}|_S = [1 + (\underset{\sim}{\alpha}.\nabla)]\nabla\bar{\phi}|_{\bar{S}} \qquad \text{etc.} \tag{16}$$

Using these previous relationships, i.e. equations (9, 11, 13, 14) the boundary condition (iv) on S may be expressed as

$$\partial\Phi/\partial n = \bar{U}\partial\bar{\phi}/\partial n + \partial\phi/\partial n = (\bar{U}\nabla\bar{\phi} + \nabla\phi).\underset{\sim}{n} = (\bar{U}\hat{i} + \dot{\underset{\sim}{\alpha}}).\underset{\sim}{n}$$

or

$$\underset{\sim}{W}.\underset{\sim}{n} + \nabla\phi.\underset{\sim}{n} = \dot{\underset{\sim}{\alpha}}.\underset{\sim}{n} \quad \text{on S.}$$

After the necessary substitutions for $\underset{\sim}{W}|_S$, $\underset{\sim}{n}|_S$ etc. it follows that to a first order approximation

$$\partial\phi/\partial n = [\dot{\underset{\sim}{\alpha}} + \underset{\sim}{\Omega}x\underset{\sim}{W} - (\underset{\sim}{\alpha}.\nabla)\underset{\sim}{W}].\underset{\sim}{n} \quad \text{on } \bar{S} . \tag{17}$$

This is the familiar relationship derived by Timman and Newman (1962) in the development of the mathematical model to describe the rigid body motions of a ship moving in waves.

RIGID BODY MOTIONS - FLUID ACTIONS

When investigating the behaviour of the six bodily motions of a floating rigid ship-like structure, Salvesen, Tuck and Faltinsen (1970) expressed the unsteady oscillatory potential ϕ as

$$\phi = (\phi_O + \phi_D + \sum_{r=1}^{6}\eta_r\phi_r)e^{i\omega_e t} \tag{18}$$

where η_r represents the amplitude of motion in surge, r=1; sway, r=2; heave, r=3; roll, r=4; pitch, r=5 and yaw, r=6. The incident wave potential amplitude

$$\phi_O(x,y,z) = \frac{iga}{\omega}\exp\{kz - ik(x\cos\chi - y\sin\chi)\} , \tag{19}$$

ϕ_D is the diffracted wave potential amplitude and ϕ_r is the radiation potential amplitude due to a motion of unit amplitude in each of the six body modes of motion in calm water. According to Newman (1978), these linear velocity potentials satisfy the following boundary conditions:

(i) On the free surface all the potentials ϕ_O, ϕ_D or ϕ_r (r=1,2,...,6) satisfy the linearised boundary condition

$$\bar{U}\partial^2\phi/\partial x^2 - 2i\omega_e\bar{U}\partial\phi/\partial x - \omega_e^2\phi + g\partial\phi/\partial z = 0 \quad \text{on } z=0 \tag{20}$$

where ϕ represents either ϕ_O, ϕ_D or ϕ_r.

(ii) Suitable bottom and radiation conditions at infinite distance from the oscillating, translating structure.

(iii) The incident and diffraction potentials satisfy the relationship

$$\partial\phi_D/\partial n = -\partial\phi_O/\partial n \quad \text{on } \bar{S} . \tag{21}$$

(iv) The radiation potentials are governed by the body boundary condition derived from equation (17), i.e.

$$\partial\phi_r/\partial n = i\omega_e n_r + \bar{U}m_r \qquad \text{on } \bar{S} \tag{22}$$

for r=1,2,...,6 and the components

$$\underset{\sim}{n} = (n_1,n_2,n_3),\ \underset{\sim}{r}x\underset{\sim}{n} = (n_4,n_5,n_6),\ \underset{\sim}{r} = (x,y,z),\ (n.\nabla)\underset{\sim}{W} = -\bar{U}(m_1,m_2,m_3)$$

$$(\underset{\sim}{n}.\nabla)(\underset{\sim}{r}x\underset{\sim}{W}) = -\bar{U}(m_4,m_5,m_6).$$

This formulation implies that the steady motion problem in calm water must be initially solved before these boundary conditions can be properly defined. However, as shown by Inglis and Price (1980), this complication greatly increases the complexity of the solution for the linear velocity potentials. As a simplification, if it is assumed that the perturbation of the steady flow due to the presence of the body is negligible then

$$\underset{\sim}{W} = -(\bar{U},0,0) = -\underset{\sim}{U} = -\bar{U}\hat{i} \tag{23}$$

and this approximation allows the unsteady motion problem to be derived independent of the description of the steady motion problem in calm water. In this situation, the boundary condition in equation (22) greatly simplifies since now

$$m_1=0=m_2=m_3=m_4 \quad , \quad m_5=n_3, \quad m_6=-n_2 \quad . \tag{24}$$

It has been shown by Inglis and Price (1981, 1982a,b), Price and Wu (1983a,b) that the velocity potential of a pulsating or, and translating source satisfies the free surface boundary condition and together with the other boundary conditions, the fluid actions associated with the structure oscillating and moving in the free surface may be determined. In fact, the total oscillatory hydrodynamic force amplitude, Z_r on the structure may be obtained by integrating over the mean wetted surface $\bar{S}$ the unsteady pressure component defined by Bernoulli's equation. That is

$$Z_r(t)=e^{i\omega_e t}Z_r=e^{i\omega_e t}\rho\iint_{\bar{S}} n_r\{i\omega_e-\underset{\sim}{U}.\nabla\}[(\phi_0+\phi_D)+\sum_{k=1}^{6}\eta_k\phi_k]dS= e^{i\omega_e t}(\Xi_r + H_r) \tag{25}$$

where Z_r (r=1,2,3) represent the total hydrodynamic components of force in the Ox, Oy, Oz directions respectively and the remaining terms r=4,5,6 represent the moments about these axes. The exciting force and moment due to the waves is

$$\Xi_r = \rho\iint_{\bar{S}} n_r\{i\omega_e - \underset{\sim}{U}.\nabla\}(\phi_0 + \phi_D)dS = \Xi_{0r} + \Xi_{Dr} \tag{26}$$

where Ξ_{0r} represents the Froude-Krylov component of the excitation and reflects that the presence of the structure does not influence the pressure distribution in the incident wave. The diffraction force, Ξ_{Dr}, accounts for the scattering of the incident waves by the structure. For $\underset{\sim}{U} = (\bar{U},0,0)$, the Froude-Krylov contribution reduces to

$$\Xi_{0r} = \rho\iint_{\bar{S}} n_r\{i\omega_e - \underset{\sim}{U}.\nabla\}\phi_0 dS = -i\rho\iint_{\bar{S}} n_r\omega\phi_0 dS \tag{27}$$

and is independent of forward speed. Thus in the wave exciting force all speed dependence arises from the diffraction component.

The force and moment due solely to the rigid body motions of the structure in calm water are given by

$$H_r = \rho\iint_{\bar{S}} n_r\{i\omega_e - \underset{\sim}{U}.\nabla\}\sum_{k=1}^{6}\eta_k\phi_k dS = \sum_{k=1}^{6} T_{rk}\eta_k \tag{28}$$

where

$$T_{rk} = \omega_e^2 A_{rk} - i\omega_e B_{rk} \tag{29}$$

denotes the hydrodynamic force and moment in the rth direction per unit oscillatory displacement in the kth mode. The terms A_{rk} and B_{rk} (r,k=1,2,...,6) represent added mass or inertia and damping coefficients respectively associated with these displacements.

FLEXIBLE BODY THEORY

For a free floating flexible body, the rigid body modes discussed previously are only the first six principal modes of the infinite number of modes of the dry structure. The fluid actions associated with the structure depend on the distortions of the structure as well as the rigid body motions. Therefore a general and rational velocity potential theory must be developed to include the effects of distortion and rigid body motions, forward speed and account for arbitrary shaped three dimensional structures. That is the conventional potential theory adopted in seakeeping theory is only a small portion of the general linear theory for flexible structures.

PRINCIPAL COORDINATES AND DISPLACEMENTS

According to a theorem due initially to Rayleigh (1894), see also Timoshenko (1955), any distortion of the structure may be expressed as an aggregate of distortions in its principal modes. That is the components of the deflection defined in the equilibrium coordinate axis system Oxyz may be expressed as

$$u(x,y,z,t)=\sum_{r=1}^{\infty}p_r(t)u_r(x,y,z)\ ,\quad v(x,y,z,t)=\sum_{r=1}^{\infty}p_r(t)v_r(x,y,z),$$
$$w(x,y,z,t)=\sum_{r=1}^{\infty}p_r(t)w_r(x,y,z) \tag{30}$$

where $p_r(t)$ is the rth principal coordinate and u_r, v_r, w_r are the rth principal modes of the dry hull. These latter functions are defined with respect to the mean equilibrium position of the floating structure in which initially the axes systems Oxyz and O'x'y'z' are coincident. By adopting a suitable transformation, these mode shapes may be expressed as functions of (x',y',z'), and by representing the principal mode in the vector form

$$\underset{\sim}{u}_r(x',y',z')=u_r\hat{i}+v_r\hat{j}+w_r\hat{k}=\{u_r,v_r,w_r\}$$

such that the displacement

$$\underset{\sim}{u}(x',y',z',t)=u\hat{i}+v\hat{j}+w\hat{k}=\sum_{r=1}^{\infty}p_r(t)\underset{\sim}{u}_r\ ,$$

the velocity of any point (x',y',z') on the surface of the structure travelling with forward velocity $\bar{U}$ can be expressed as

$$\underset{\sim}{V}^s(x',y',z',t)=\bar{U}\hat{i}+\dot{\underset{\sim}{u}}=\bar{U}\hat{i}+\sum_{r=1}^{\infty}\dot{p}_r(t)\underset{\sim}{u}_r\ . \tag{31}$$

In a similar manner the rotation vector at any point (x',y',z') is given by

$$\underset{\sim}{\theta}(x',y',z',t)=\sum_{r=1}^{\infty}p_r(t)\underset{\sim}{\theta}_r \tag{32}$$

where

$$\underset{\sim}{\theta}_r(x',y',z',t)=\{\theta_{xr},\theta_{yr},\theta_{zr}\}=\tfrac{1}{2}\mathrm{curl}\,\underset{\sim}{u}_r=\tfrac{1}{2}[(\frac{\partial w_r}{\partial y'}-\frac{\partial v_r}{\partial z'})\hat{i}+(\frac{\partial u_r}{\partial z'}-\frac{\partial w_r}{\partial x'})\hat{j}+(\frac{\partial v_r}{\partial x'}-\frac{\partial u_r}{\partial y'})\hat{k}].$$

By repeating and extending the arguments used in the rigid body theory, the unit normal vector on the instantaneous wetted surface area S may be written as

$$\underset{\sim}{n}|_S=\underset{\sim}{n}|_{\bar{S}}+\underset{\sim}{\theta}\times\underset{\sim}{n}|_{\bar{S}} \tag{33}$$

whilst the velocity of the steady flow $\underset{\sim}{W}$ due to the deflection of the structure is

$$\underset{\sim}{W}|_S=[1+(\underset{\sim}{u}.\nabla)]\underset{\sim}{W}|_{\bar{S}}\ . \tag{34}$$

Now in the body fixed axis system O'x'y'z' the rigid body modes of the dry structure may be expressed as

$$\underset{\sim}{u}_1=(1,0,0)\ ,\ \underset{\sim}{u}_2=(0,1,0)\ ,\ \underset{\sim}{u}_3=(0,0,1)\ ,\ \underset{\sim}{u}_4=(0,-z',y'),\ \underset{\sim}{u}_5=(z',0,-x'),\ \underset{\sim}{u}_6=(-y',x',0) \tag{35}$$

and these correspond to the principal coordinates $p_1(t)$, $p_2(t)$, ..., $p_6(t)$. If we denote

$$\underset{\sim}{\eta}=\{p_1(t),p_2(t),p_3(t)\},\quad \underset{\sim}{\Omega}=\{p_4(t),p_5(t),p_6(t)\}\quad\text{and}\quad \underset{\sim}{\alpha}=\underset{\sim}{\eta}+\underset{\sim}{\Omega}\times\underset{\sim}{r}'$$

then it follows that the deflection

$$\underset{\sim}{u}=\underset{\sim}{\alpha}+\sum_{r=7}^{\infty}p_r(t)\underset{\sim}{u}_r\ ,$$

rotation

$$\underset{\sim}{\theta}=\underset{\sim}{\Omega}+\sum_{r=7}^{\infty}p_r(t)\underset{\sim}{\theta}_r\ ,$$

velocity

$$\underset{\sim}{V}^S=\bar{U}\hat{i}+\dot{\underset{\sim}{\alpha}}+\sum_{r=7}^{\infty}\dot{p}_r(t)\underset{\sim}{u}_r\ ,$$

unit normal

$$\underset{\sim}{n}|_S=\underset{\sim}{n}|_{\bar{S}}+\underset{\sim}{\Omega}\times\underset{\sim}{n}|_{\bar{S}}+\sum_{r=7}^{\infty}p_r(t)\underset{\sim}{\theta}_r\times\underset{\sim}{n}|_{\bar{S}}\ ,$$

and the velocity of the steady flow

$$\underset{\sim}{W}|_S=[1+(\underset{\sim}{\alpha}.\nabla)]\underset{\sim}{W}|_{\bar{S}}+\sum_{r=7}^{\infty}p_r(t)(\underset{\sim}{u}_r.\nabla)\underset{\sim}{W}|_{\bar{S}}\ .$$

It is immediately clear that if the body has no distortion i.e. $p_r(t)=0$ for $r\geqslant7$ then the rigid body theory is obtained.

VELOCITY POTENTIAL

The unsteady component of the velocity potential function must include contributions accounting for the distortions of the structure in the fluid. Thus as a generalisation of the formulation adopted previously in the rigid body theory, the total velocity potential is again of the form

$$\Phi(x_0,y_0,z_0,t) \equiv \overline{U}\,\overline{\phi}(x,y,z) + \phi(x,y,z,t) \tag{36}$$

where now, for the flexible body, the unsteady component

$$\phi(x,y,z,t) = \phi_0(x,y,z,t) + \phi_D(x,y,z,t) + \sum_{r=1}^{\infty}\phi_r(x,y,z,t) \quad . \tag{37}$$

In a linear structural dynamics theory the deflection of the structure may be expressed by a series of distortions in the principal modes and a similar interpretation may be adopted for the unsteady velocity potential. That is, the unsteady potential may be described by a series of potentials $\phi_1, \phi_2, \ldots, \phi_6, \phi_7, \phi_8, \ldots$ each component corresponding to each of the principal modes of the dry structure. The subscript numbers 1 to 6 relate to rigid body modes and 7 onwards to the distortions of the dry structure. Therefore no distinction need be made between rigid and flexible body radiation potentials $\phi_r(r=1,2,\ldots,\infty)$, each being treated equally in the mathematical model and satisfying the same set of boundary conditions (i.e. free surface, bottom and radiation conditions, body etc.) .

By analogy with the rigid body theory, each of these radiation potentials may be written in the form

$$\phi_r(x,y,z,t) = \phi_r(x,y,z)p_r(t) \quad \text{for } r=1,2,\ldots,\infty \ . \tag{38}$$

Thus the principal coordinate $p_1(t)$ relates to surge motion; $p_2(t)$, sway; $p_3(t)$, heave; $p_4(t)$, roll; $p_5(t)$, pitch; $p_6(t)$, yaw and $p_7(t)$, $p_8(t)$,... to the distortions of the structure. Further in a linear theory for a sinusoidal excitation of frequency ω_e the principal coordinates are given by

$$p_r(t) = p_r e^{i\omega_e t} \tag{39}$$

where the amplitude p_r of the principal coordinate may be complex in form.

GENERALISED TIMMAN-NEWMAN RELATIONSHIPS

The boundary condition applicable on the instantaneous wetted surface area S of the flexible body is again of the form

$$\partial\Phi/\partial n = \underset{\sim}{V}^S.\underset{\sim}{n} \quad .$$

After substituting from equations (14), (31) and (36) this reduces to

$$(\overline{U}\nabla\overline{\phi} + \nabla\phi).\underset{\sim}{n} = (\overline{U}\hat{i} + \dot{\underset{\sim}{u}}).\underset{\sim}{n} \quad \text{or} \quad \partial\phi/\partial n = (\dot{\underset{\sim}{u}} - \underset{\sim}{W}).\underset{\sim}{n} \ \text{ on S} \quad .$$

However, since

$$\underset{\sim}{W}|_S = [1 + (\underset{\sim}{u}.\nabla)]\underset{\sim}{W}|_{\overline{S}}$$

etc. it follows, after neglecting the second order terms in ϕ, $\underset{\sim}{u}$ and $\underset{\sim}{\theta}$, that the linearised boundary condition may be reduced to the form

$$\partial\phi/\partial n = [\dot{\underset{\sim}{u}} + \dot{\underset{\sim}{\theta}}\times\underset{\sim}{W} - (\underset{\sim}{u}.\nabla)\underset{\sim}{W}].\underset{\sim}{n} \ \text{ on } \overline{S} \quad . \tag{40}$$

Further substitution of these quantities in series form gives

$$\sum_{r=1}^{\infty}[\partial\phi_r/\partial n - i\omega_e\underset{\sim}{u}_r.\underset{\sim}{n} - \underset{\sim}{\theta}_r\times\underset{\sim}{W}.\underset{\sim}{n} + (\underset{\sim}{u}_r.\nabla)\underset{\sim}{W}.\underset{\sim}{n}]p_r e^{i\omega_e t} = 0 \ \text{ on } \overline{S}$$

and this boundary condition must be satisfied for any arbitrary combination of p_r. This is always true if the condition is satisfied for each p_r. That is

$$\partial\phi_r/\partial n = [i\omega_e\underset{\sim}{u}_r + \underset{\sim}{\theta}_r\times\underset{\sim}{W} - (\underset{\sim}{u}_r.\nabla)\underset{\sim}{W}].\underset{\sim}{n} \quad \text{on } \overline{S} \tag{41}$$

for each $r=1,2,\ldots,\infty$. This expression is a generalisation of the Timman-Newman relationships derived previously for the rigid body modes. This is immediately seen since for modes $r=1,2,\ldots,6$ the variables $\underset{\sim}{\alpha}$ and $\underset{\sim}{\Omega}$ used in the rigid body theory have their equivalents in $\underset{\sim}{u}$ and $\underset{\sim}{\theta}$. Thus this body surface boundary condition is valid for all modes - rigid body modes or flexible modes.

In the special (though usually taken) case, when the steady flow is approximated to $\underset{\sim}{W} = -\bar{U}\hat{i}$, the body boundary condition reduces to

$$\partial\phi_r/\partial n = i\omega_e(u_r n_1 + v_r n_2 + w_r n_3) + \bar{U}/2[n_3(\partial u_r/\partial z' - \partial w_r/\partial x') - n_2(\partial v_r/\partial x' - \partial u_r/\partial y')] \text{ on } \bar{S} \quad (42)$$

for each $r=1,2,\ldots,\infty$. After substituting for the dry hull body mode shapes $r=1,2,\ldots,6$ in equation (35) the body surface boundary condition derived in the rigid body theory section is again obtained.

LINEARISED FLEXIBLE BOUNDARY CONDITIONS

The linear velocity potentials associated with the flow around the flexible body satisfy the following boundary conditions:

(i) On the free surface the incident, diffracted and radiation potentials i.e. ϕ_0, ϕ_D and ϕ_r $(r=1,2,\ldots,6,7,\ldots,\infty)$ respectively satisfy the linearised boundary condition

$$\bar{U}^2\partial^2\phi/\partial x^2 - 2i\omega_e\bar{U}\partial\phi/\partial x - \omega_e^2\phi + g\partial\phi/\partial z = 0 \quad \text{on } z = 0 \quad (43)$$

where ϕ represents either ϕ_0, ϕ_D or ϕ_r.

(ii) Suitable bottom and radiation conditions at infinite distance from the oscillating, translating structure.

(iii) The incident and diffracted potentials satisfy the relationship

$$\partial\phi_D/\partial n = -\partial\phi_0/\partial n \quad \text{on } \bar{S}\ . \quad (44)$$

(iv) The radiation potentials are governed by the body boundary condition

$$\partial\phi_r/\partial n = [i\omega_e\underset{\sim}{u}_r + \underset{\sim}{\theta}_r \times \underset{\sim}{W} - (\underset{\sim}{u}_r.\nabla)\underset{\sim}{W}].\underset{\sim}{n} \quad \text{on } \bar{S}\ ,$$

or when $\underset{\sim}{W} = -\bar{U}\hat{i}$,

$$\partial\phi_r/\partial n = i\omega_e(u_r n_1 + v_r n_2 + w_r n_3) + \bar{U}/2[n_3(\partial u_r/\partial z' - \partial w_r/\partial x') - n_2(\partial v_r/\partial x' - \partial u_r/\partial y')] \text{ on } \bar{S} \quad (45)$$

for each $r=1,2,\ldots,6,7\ldots,\infty$.

PRESSURE DISTRIBUTION

The fluid pressure acting on the instantaneous wetted body surface S during the oscillatory motion of the flexible structure is according to Bernoulli's formulation given by

$$p = -\rho[\partial\phi/\partial t + \underset{\sim}{W}.\nabla\phi + \tfrac{1}{2}(W^2 - \bar{U}^2) + \tfrac{1}{2}\nabla\phi.\nabla\phi + gz]. \quad (46)$$

Unfortunately to determine this expression a knowledge of the position of S is necessary and as discussed by Newman (1978) this difficulty may be satisfactorily overcome by relating the pressure on the surface S to the pressure on the body mean surface $\bar{S}$ by means of a Taylor series expansion. Thus, for the flexible structure, it follows that

$$p|_S = [1 + (\underset{\sim}{u}.\nabla) + \tfrac{1}{2}(\underset{\sim}{u}.\nabla)^2 + \ldots\ ]p|_{\bar{S}}\ . \quad (47)$$

If it is assumed that the oscillatory motion of the body and parasitic flow are small i.e. neglecting the second order terms of the unsteady component, then the pressure on the wetted surface S becomes

$$p|_S = -\rho[\partial\phi/\partial t + \underset{\sim}{W}.\nabla\phi + \{\tfrac{1}{2}(W^2 - \bar{U}^2) + gz'\} + \{gw + \tfrac{1}{2}(\underset{\sim}{u}.\nabla)W^2\}]_{\bar{S}}\ . \quad (48)$$

This implies that the oscillatory flow and the motion of the structure are linearised but the steady flow due to the steady forward motion remains non-linear.

Table 1 illustrates a comparison of the orders of magnitude of the terms involved in this formulation i.e. equation (48) based on differing body geometry assumptions. For example, in the three dimensional case in column 1, the geometry of the structure is such that the three main dimensions are of the same order. Thus it follows that

$$(x,y,z)=\{o(1),o(1),o(1)\}, \quad \underset{\sim}{n}=(n_1,n_2,n_3)=\{o(1),o(1),o(1)\}$$

and

$$\bar{\phi} = o(L) = o(1) \ , \quad \underset{\sim}{W} = \bar{U}\{o(1),o(1),o(1)\} \text{ on } S \text{ and } \bar{S} \ .$$

For the unsteady potential ϕ, the incident wave potential ϕ_0 is independent of body geometry and its order of magnitude depends on the wave amplitude $\bar{a}=a/L$ and wave frequency ω. Because of the fluid boundary conditions on the body, the diffraction potential is of the same order as ϕ_0 and the radiation potential

$$\phi_R = \sum_{r=1}^{\infty}\phi_r = \phi - (\phi_0 + \phi_D) = o(|\underset{\sim}{u}|L\omega_e) \ .$$

The order of magnitude of each term is shown in Table 1, column 1 and the terms are non-dimensionalised with respect to the four parameters $\bar{a}=a/L$, $\bar{u}=|\underset{\sim}{u}|/L$, $\delta=\omega_e\sqrt{(L/g)}$ and Froude number $Fn=\bar{U}/\sqrt{(gL)}$.

If it is assumed that $Fn=o(1)$ and $\bar{a}=o(1)$ then the leading order of terms in the pressure equation (48) is of $o(\bar{u})$ and no term can be ignored. If $\delta=o(\varepsilon^{-\frac{1}{2}})$ the last terms in the pressure equation are of $o(\bar{u})$ and the remaining terms are of $o(\bar{u}\varepsilon^{-\frac{1}{2}})$ or $o(1)$. According to Newman (1978), this is the resonant frequency region for the rigid body motions (i.e. heave, pitch and roll) so that $\bar{u}=o(1)$ and all the terms in the pressure equation must be retained. However in the high frequency region i.e. $\delta \gtrsim o(\varepsilon^{-1})$ the resonant frequencies of the structure are to be found - though this naturally depends on the type and flexibility of the structure - and the magnitude of the distortions are unlikely to be of $o(1)$. In this case the leading order of terms in the pressure equation is of $o(1)$ and reduces to

$$p = -\rho[\partial\phi/\partial t + \underset{\sim}{W}.\nabla\phi + \tfrac{1}{2}(W^2-\bar{U}^2) + gz]_{\bar{S}} \ . \tag{49}$$

Table 1. Comparison of the orders of magnitude of the terms involved in the expression for the pressure on the body surface in equation (48).

Item	General 3D case	Slender body	Thin body	Flat body
Length of body L	α	α	α	α
Beam of body B	α	β	β	α
Draught T	α	β	α	β
$\underset{\sim}{r}=(x,y,z)$	(α,α,α)	(α,β,β)	(α,β,α)	(α,α,β)
$\underset{\sim}{n}=(n_1,n_2,n_3)$	(α,α,α)	(β,α,α)	(β,α,β)	(β,β,α)
$\underset{\sim}{r}\times\underset{\sim}{n}=(n_4,n_5,n_6)$	(α,α,α)	(β,α,α)	(α,β,α)	(α,α,β)
$\bar{\phi}$	$o(L)$	$o(\varepsilon^2L)$	$o(\varepsilon^2L)$	$o(\varepsilon^2L)$
$\underset{\sim}{W}=\bar{U}(\partial\bar{\phi}/\partial x-1,\partial\bar{\phi}/\partial y,\partial\bar{\phi}/\partial z)$	$\bar{U}(\alpha,\alpha,\alpha)$	$\bar{U}(\alpha,\beta,\beta)$	$\bar{U}(\alpha,\beta,\gamma)$	$\bar{U}(\alpha,\gamma,\beta)$
$\phi_R=\sum_{r=1}^{\infty}\phi_r$	$o(\|\underset{\sim}{u}\|L\omega_e)$	$o(\|\underset{\sim}{u}\|L\varepsilon\omega_e)$	$o(\|\underset{\sim}{u}\|L\varepsilon\omega_e)$	$o(\|\underset{\sim}{u}\|L\varepsilon\omega_e)$
$\partial/\partial t(\phi_0+\phi_D)/gL$	$o(\bar{a})$	$o(\bar{a})$	$o(\bar{a})$	$o(\bar{a})$
$\underset{\sim}{W}.\nabla(\phi_0+\phi_D)/gL$	$o(\bar{a}Fn\delta)$	$o(\bar{a}Fn\delta)$	$o(\bar{a}Fn\delta)$	$o(\bar{a}Fn\delta)$
$\partial/\partial t\phi_R/gL$	$o(\bar{u}\delta^2)$	$o(\bar{u}\varepsilon\delta^2)$	$o(\bar{u}\varepsilon\delta^2)$	$o(\bar{u}\varepsilon\delta^2)$
$\underset{\sim}{W}.\nabla\phi_R/gL$	$o(\bar{u}Fn\delta)$	$o(\bar{u}\varepsilon Fn\delta)$	$o(\bar{u}\varepsilon Fn\delta)$	$o(\bar{u}\varepsilon Fn\delta)$
$(W^2-\bar{U}^2)/gL$	$o(Fn^2)$	$o(Fn^2\varepsilon^2)$	$o(Fn^2\varepsilon^2)$	$o(Fn^2\varepsilon^2)$
z'/L	α	β	α	β
w/L	$o(\bar{u})$	$o(\bar{u})$	$o(\bar{u})$	$o(\bar{u})$
$(\underset{\sim}{u}.\nabla)W^2/gL$	$o(\bar{u}Fn^2)$	$o(\bar{u}\varepsilon Fn^2)$	$o(\bar{u}\varepsilon Fn^2)$	$o(\bar{u}\varepsilon Fn^2)$

$[\alpha \equiv o(1), \quad \beta \equiv o(\varepsilon), \quad \gamma \equiv o(\varepsilon^2)]$

When the body is slender, thin or flat, the leading order of terms in the pressure equation (48) is of $o(\bar{u}\varepsilon)$ provided that $\bar{a}=o(1)$ and the Froude number $Fn=o(1)$. In this case, Table 1 indicates that the pressure equation reduces to

$$p|_S = -\rho[\partial\phi/\partial t + \underset{\sim}{W}.\nabla\phi + \tfrac{1}{2}(W^2-\overline{U}^2) + gz' + gw]_{\overline{S}} \; . \tag{50}$$

The third and fourth terms describe contributions from steady state and hydrostatic components respectively. They modify the generalised still water forces and in turn the equilibrium position of the body.

Based on the approximation that the steady flow $\underset{\sim}{W}=-\overline{U}\hat{i}$, the previous pressure equation on S reduces to

$$p = -\rho[\partial\phi/\partial t - \overline{U}\partial\phi/\partial x]_{\overline{S}} - \rho g[z'+ w]_{\overline{S}} \; . \tag{51}$$

GENERALISED FLUID FORCES

The rth component of the generalised external force $\underset{\sim}{Z}$ acting on the flexible structure due to the fluid only may be expressed in the form

$$Z_r(t) = -\iint_S \underset{\sim}{n}^T.\underset{\sim}{u}_r p dS \tag{52}$$

where $\underset{\sim}{n}^T$ is the transpose of the unit normal vector pointing out of the body surface into the fluid and $\underset{\sim}{u}_r$ is the rth principal mode vector associated with the dry structure. This integration extends over the instantaneous body wetted surface $\tilde{S}$.

Expression (52) may be shown to be equivalent to the rth generalised fluid force defined previously in equation (5). That is

$$Z_r = \underset{\sim}{D}_r^T\underset{\sim}{P} = \sum_e \underset{\sim}{d}_r^T\underset{\sim}{P}_e$$

where the summation $\sum_e$ includes contributions from all the elements within the instantaneous wetted surface S and the vector P_e denotes the distribution of the externally applied fluid force over the element and is defined with respect to the equilibrium axes.

By use of the principle of virtual work it has been shown by Bishop, Price and Wu (1984) that the generalised force related to the hydrodynamic pressure p can be expressed as

$$\overline{P}_e = -\iint_{S_e} p\underset{\sim}{N}^T.\overline{\underset{\sim}{n}}dS$$

where $\overline{n}$ is the outward drawn unit normal vector on the wetted finite element surface S_e measured in the local element coordinate system. $\overline{P}_e$ is associated with the nodal displacements and is equivalent to a set of concentrated forces or moments acting at the nodes of the elements doing an equivalent amount of work to the pressure p during the motion of the flexible structure. In addition transformations $\underset{\sim}{L}^T$ and $\underset{\sim}{\ell}$ may be defined such that

$$\underset{\sim}{P}_e = \underset{\sim}{L}^T\overline{\underset{\sim}{P}}_e \quad , \qquad \overline{\underset{\sim}{n}} = \underset{\sim}{\ell}\underset{\sim}{n} \quad ,$$

and these relate quatities between the local element coordinate system and the global or equilibrium axes systems.

Thus it follows from these previous results that the rth generalised fluid force can be written as

$$Z_r = -\sum_e \underset{\sim}{d}_r^T\underset{\sim}{L}^T\iint_{S_e}\underset{\sim}{N}^T\underset{\sim}{\ell}\underset{\sim}{n}pdS \; .$$

Taking the transpose of this expression, using the definition of the rth principal mode shape vector u_r for the overall structure and the fact that Z_r is scalar, then it follows that

$$Z_r = -\sum_e\iint_{S_e}\underset{\sim}{n}^T(\underset{\sim}{\ell}^T\underset{\sim}{N}\underset{\sim}{L}\underset{\sim}{d}_r)pdS = -\iint_S\underset{\sim}{n}^T.\underset{\sim}{u}_r pdS$$

which agrees with the expression in equation (52).

Substituting the general expression i.e. equation (48) for the pressure p into equation (52), and including also the contribution from the generalised gravitational force, the rth generalised external force may be expressed in the component form

$$Z_r(t) = \Xi_r(t) + H_r(t) + R_r(t) + \bar{R}_r \qquad \text{for } r=1,2,..,m. \tag{53}$$

and the rth generalised wave exciting force is defined as

$$\Xi_r(t) = \rho\iint_S \underset{\sim}{n}^T.\underset{\sim}{u}_r[\partial/\partial t + \underset{\sim}{W}.\nabla](\phi_o + \phi_D)dS \quad , \tag{54}$$

the rth generalised radiation force

$$H_r(t) = \rho\iint_S \underset{\sim}{n}^T.\underset{\sim}{u}_r[\partial/\partial t + \underset{\sim}{W}.\nabla]\sum_{k=1}^{m} p_k(t)\phi_k \; dS \quad , \tag{55}$$

the rth generalised restoring force

$$R_r(t) = \rho\iint_S \underset{\sim}{n}^T.\underset{\sim}{u}_r[gw + \tfrac{1}{2}(\underset{\sim}{u}.\nabla)W^2]dS \quad , \tag{56}$$

and the rth generalised hydrostatic force

$$\bar{R}_r = \rho\iint_S \underset{\sim}{n}^T.\underset{\sim}{u}_r[gz' + \tfrac{1}{2}(W^2 - \bar{U}^2)]dS - \iiint_{V} \rho_b g w_r dV \quad . \tag{57}$$

The first three components are associated with unsteady deflections of the structure whilst the hydrostatic contribution is independent of all unsteady motions. The latter contains components arising from the generalised hydrostatic fluid action, generalised forces due to the structure travelling with constant forward speed in calm water and the generalised gravitational force described by the volume integral.

GENERALISED WAVE FORCE

In regular sinusoidal waves the incident and diffraction potentials are both sinusoidal functions of encounter frequency ω_e. Hence the rth generalised wave exciting force may be written as

$$\Xi_r(t) = \Xi_r e^{i\omega_e t} = (\Xi_{or} + \Xi_{Dr})e^{i\omega_e t} \quad ,$$

where from equation (54) the rth generalised Froude-Krylov contribution is

$$\Xi_{or} = \rho\iint_S \underset{\sim}{n}^T.\underset{\sim}{u}_r[i\omega_e + \underset{\sim}{W}.\nabla]\phi_o dS \tag{58}$$

and the rth generalised diffraction force accounting for the scattering of the incident wave due to presence of the flexible structure is

$$\Xi_{Dr} = \rho\iint_S \underset{\sim}{n}^T.\underset{\sim}{u}_r[i\omega_e + \underset{\sim}{W}.\nabla]\phi_D dS \quad . \tag{59}$$

When $\underset{\sim}{W} = -\bar{U}\hat{i}$, the rth generalised Froude-Krylov contribution reduces to

$$\Xi_{or} = \rho\iint_S \underset{\sim}{n}^T.\underset{\sim}{u}_r \omega\phi_o dS \tag{60}$$

and is independent of forward speed for each r=1,2,...,m.

GENERALISED RADIATION FORCE

Assuming a sinusoidal solution of the rth principal coordinate in the form

$$p_r(t) = p_r e^{i\omega_e t} \quad ;$$

it follows from equation (55) that the rth generalised radiation force may be written as

$$H_r(t) = \sum_{k=1}^{m} p_k T_{rk} e^{i\omega_e t} = \sum_{k=1}^{m} p_k(\omega_e^2 A_{rk} - i\omega_e B_{rk}) e^{i\omega_e t} \qquad (61)$$

for r=1,2,...,m. The coefficients (r=1,2,...,m; k=1,2,...,m)

$$A_{rk} = (\rho/\omega_e^2)\ \mathrm{Re}\{\iint_S \underset{\sim}{n}^T . \underset{\sim}{u}_r [i\omega_e + \underset{\sim}{W}.\nabla]\phi_k dS\} \qquad (62)$$

is in phase with the acceleration whilst

$$B_{rk} = (-\rho/\omega_e)\ \mathrm{Im}\{\iint_S \underset{\sim}{n}^T . \underset{\sim}{u}_r [i\omega_e + \underset{\sim}{W}.\nabla]\phi_k dS\} \qquad (63)$$

is in phase with the velocity.

The terms A_{rk} and B_{rk} represent added mass or inertias and damping coefficients respectively associated with the rth mode per unit oscillatory distortion in the kth mode. The theory indicates that these coefficients may be determined experimentally by forced oscillation of the flexible structure in a prescribed dry hull mode shape at the arbitrary frequency ω_e as the structure travels with constant speed in calm water.

GENERALISED RESTORING FORCE

Since the displacement at any arbitrary point in the structure is given by

$$\underset{\sim}{u} = \{u,v,w\} = \sum_{k=1}^{m} \underset{\sim}{u}_k p_k e^{i\omega_e t} = \sum_{k=1}^{m} \{u_k, v_k, w_k\} p_k e^{i\omega_e t} ,$$

the rth generalised restoring force in equation (56) may be written as

$$R_r(t) = -\sum_{k=1}^{m} p_k C_{rk} e^{i\omega_e t} \qquad (64)$$

where the generalised restoring force coefficient

$$C_{rk} = -\rho \iint_S \underset{\sim}{n}^T . \underset{\sim}{u}_r [g w_k + \tfrac{1}{2}(\underset{\sim}{u}_k . \nabla) W^2] dS$$

for r=1,2,...,m and k=1,2,...,m. The second term in this integral produces contributions to the rth generalised restoring force due to unit unsteady motions of the structure within the steady pressure field. For a general three dimensional body of arbitrary shape, this contribution is of the same order as obtained from the first term. However, for slender, thin or flat structures the second term is an order smaller than the first term and may be neglected by comparison. Thus for these types of structures the generalised restoring force coefficient reduces to the simpler form

$$C_{rk} = -\rho g \iint_S \underset{\sim}{n}^T . \underset{\sim}{u}_r w_k dS \qquad \text{for } r=1,2,...,m;\ k=1,2,...,m. \qquad (65)$$

It can be easily shown that this formulation includes a description of the restoring coefficients usually associated with the rigid body modes only.

GENERALISED EQUATIONS OF MOTION

From the theory presented, the generalised linear equations describing the responses of the floating flexible marine structure may be expressed in the general form

$$\omega_r^2 a_{rr} p_r(t) + \sum_{k=1}^{m} [a_{rk}\ddot{p}_k(t) + b_{rk}\dot{p}_k(t)] = Z_r(t) = \Xi_r(t) + H_r(t) + R_r(t) + \overline{R}_r \qquad (66)$$

for r=1,2,...,m. In this formulation the generalised mass a_{rk} is symmetric since, it must be remembered that there exists the possibility of off-diagonal contributions occurring in the rigid body modes.

STEADY EQUATIONS

For the structure in calm water, there exists a steady state solution

$$p_r(t) = \bar{p}_r$$

satisfying the equation

$$a_{rr}\omega_r^2\bar{p}_r = \sum_{k=1}^{m}\bar{p}_r C_{rk} + \bar{R}_r \qquad \text{for } r=1,2,\ldots, m. \tag{67}$$

FLOATING (FREE-FREE STRUCTURE)

For the *floating* structure moving or stationary in waves and in the absence of any steady state conditions, the generalised linear equation of motion may be written in the form

$$\omega_r^2 a_{rr}p_r(t) + \sum_{k=1}^{m}[(a_{rk}+A_{rk})\ddot{p}_k(t) + (b_{rk}+B_{rk})\dot{p}_k(t)+C_{rk}p_k(t)] = \Xi_r e^{i\omega_e t} \tag{68}$$

for r=1,2,...m. Alternatively in matrix form this set of equations becomes

$$(\underset{\sim}{a} + \underset{\sim}{A})\ddot{\underset{\sim}{p}}(t) + (\underset{\sim}{b} + \underset{\sim}{B})\dot{\underset{\sim}{p}}(t) + (\underset{\sim}{c} + \underset{\sim}{C})\underset{\sim}{p}(t) = \underset{\sim}{\Xi}(t) = \underset{\sim}{\Xi}e^{i\omega_e t} \tag{69}$$

where

$\underset{\sim}{a}$ is the inertia matrix of the dry structure and is diagonal apart from possible off-diagonal elements occurring in the rigid body modes (r=1,2,...,6).
$\underset{\sim}{A}$ is the hydrodynamic inertia matrix .
$\underset{\sim}{b}$ is the structural damping matrix of the dry structure and is usually assumed diagonal.
$\underset{\sim}{B}$ is the hydrodynamic damping matrix.
$\underset{\sim}{c}$ is the diagonal stiffness matrix with elements $c_{ss}=\omega_s^2 a_{ss}$. The natural frequency $\omega_s=0$ for the rigid body modes s=1,2,...,6.
$\underset{\sim}{C}$ is the hydrodynamic restoring or stiffness matrix.

Thus assuming a solution of the principal coordinate matrix in the form

$$\underset{\sim}{p}(t) = \underset{\sim}{p}e^{i\omega_e t} \quad ,$$

equation (69) reduces to

$$\underset{\sim}{I}\underset{\sim}{p} = \frac{\text{adj } \underset{\sim}{D}}{\det \underset{\sim}{D}}\,\underset{\sim}{\Xi} \tag{70}$$

where $\underset{\sim}{I}$ is a unit matrix,

$$\underset{\sim}{D} = -\omega_e^2(\underset{\sim}{a} + \underset{\sim}{A}) + i\omega_e(\underset{\sim}{b} + \underset{\sim}{B}) + (\underset{\sim}{c} + \underset{\sim}{C}) \ , \tag{71}$$

and the matrices $\underset{\sim}{A}$, $\underset{\sim}{B}$ and $\underset{\sim}{D}$ are functions of the frequency of encounter ω_e.

Hence from a knowledge of the principal mode shapes of the dry structure and the determined principal coordinates the displacement at any position in the structure is given by

$$\underset{\sim}{u}(x,y,z,t) = \sum_{r=1}^{m}\underset{\sim}{u}_r(x,y,z)p_r e^{i\omega_e t} \quad . \tag{72}$$

The bending moments, shearing forces, twisting moments and any other relevant response may be similarly determined using the appropriate principal mode shape of the dry structure.

FIXED (CLAMPED, PINNED ETC. STRUCTURES)

For a *fixed* flexible structure, rigid body modes no longer exist and only the distortion modes r=7,8,... are applicable and need be included in the theory. Naturally the shapes of these modes depend on the imposed end conditions i.e. clamped, pinned etc. In fact for a fixed cantilevered vertical structure (say). idealised end conditions may be imposed i.e. clamped-free and these mode shapes adopted in the theory. However, in reality such idealised end conditions may not be a ~~true~~ relfection of the actual end conditions due say to structure-soil interaction. The theory presented remains valid and by a suitable description of the externally applied generalised force a theoretical model based on the idealised clamped mode shapes may be developed which accounts for the more realistic physical end condition. From such a model the responses of the fixed structure may be determined once again. Thus for the *fixed* structure in waves of frequency ω the generalised linear equations describing the principal coordinate response now take the form

$$a_{rr}[\ddot{p}_r(t) + \omega_r^2 p_r(t)] + \sum_{k=7}^{m}[A_{rk}\ddot{p}_k(t) + (b_{rk}+B_{rk})\dot{p}_k(t)+C_{rk}p_k(t)] = \Xi_r e^{i\omega t} \quad (73)$$

for r=7,8,...,m. By a simple process of redefining and renumbering (i.e. let r=1 denote the first distortion mode, etc.) this set of equations may be caste into the matrix form discussed previously.

MODIFICATIONS (MOORING LINES, NON-LINEARITIES ETC.)

Further modifications may be introduced into the existing theory. For example, if mooring lines are attached to the structure additional external concentrated applied loadings need be included in the description of the rth generalised external applied force. In a simple model, the mooring forces will provide an additional component to the restoring force term in the steady state condition and the general equation describing the unsteady motion may be written in the form

$$(\underset{\sim}{a} + \underset{\sim}{A})\ddot{\underset{\sim}{p}}(t) + (\underset{\sim}{b} + \underset{\sim}{B})\dot{\underset{\sim}{p}}(t) + (\underset{\sim}{c} + \underset{\sim}{C} + \Delta\underset{\sim}{C})\underset{\sim}{p}(t) = \underset{\sim}{\Xi}(t) \quad (74)$$

where the generalised restoring matrix $\Delta\underset{\sim}{C}$ describes the linear contribution arising from the mooring lines.

An underlying assumption in all the theory presented relates to a linear description of the dry structural model so that mode shapes may be determined. However no such assumption need be imposed on the rth generalised external force although in this paper a linear description is only discussed. However with suitable modifications, a non-linear description of the fluid actions may be included in the theory and the equations describing the responses of the structure may be expressed in the form

$$(\underset{\sim}{a} + \underset{\sim}{A})\ddot{\underset{\sim}{p}}(t) + (\underset{\sim}{b} + \underset{\sim}{B})\dot{\underset{\sim}{p}}(t) + (\underset{\sim}{c} + \underset{\sim}{C})\underset{\sim}{p}(t) = \underset{\sim}{\Xi}(t) + \underset{\sim}{\mathcal{F}}(\underset{\sim}{p},\dot{\underset{\sim}{p}},\ldots) \quad (75)$$

where the matrix $\mathcal{F}$ denotes a generalised contribution resulting from the non-linear external fluid actions and contains elements involving products of the coordinate $\underset{\sim}{p}(t)$. Such a mathematical model includes the Morison type formulation used to describe the non-linear fluid actions applied to fixed structures. Unfortunately when introducing such modifications, a more semi-empirical approach becomes necessary before the problem can be readily solved.

COMPUTATIONS AND EXAMPLES

It has been shown by Brard (1972) that when applying a singularity distribution method to a surface piercing structure with forward speed a line integral contribution must be included in the expression for the velocity potential at any point (x_1,y_1,z_1) in the fluid. That is

$$\phi(x_1,y_1,z_1)=\iint_S Q(x,y,z)G(x_1,y_1,z_1;x,y,z)dS+\frac{\overline{U}^2}{g}\oint_{\overline{C}} Q(x,y,0)G(x_1,y_1,z_1;x,y,0)n_1(x,y,0)dC$$

where (x,y,z) denotes a point on the structure, the contour $\overline{C}$ is the intersection of the structure's surface and the mean calm water surface, Q is the source density on the structure's surface and G is the appropriate Green's function.

For structures with port and starboard symmetry, Price and Wu (1983a,b) have developed a composite source distribution method to evaluate the unknown velocity potentials associated with the rigid body motions of a multi-hull body. Bishop, Price and Wu (1984) have extended this method to determine the required velocity potentials when the structure is flexible. This method is adopted to determine the hydrodynamic coefficients, responses etc. of the structures discussed in the following examples. The interested reader may refer to the cited references for more information about the numerical techniques used.

(i) UNIFORM SHIP

Figure 4 illustrates a typical set of calculated results for a flexible ship travelling with Froude number Fn=0.2 in unit amplitude sinusoidal head waves. In this case the monohull structure is idealised by a uniform beam of ship like proportions. The mean wetted surface area is discretised by 180 surface panel elements and a limited comparison is included between predictions based on the present three dimensional theory and the two dimensional strip-beam approach of Bishop and Price (1979). Figures 4(a,b) illustrate the variation with frequency of the generalised hydrodynamic coefficients and wave exciting modes r=3(heave), r=5(pitch) and the first two distortion modes r=7 and 9. The pitch principal coordinate $|p_5|$ in figure 4(c) shows the dominant response occurring in the low frequency range and the coupling existing between the principal coordinates due to the fluid-structure interaction. Figure 4(d) shows the affect of including different numbers of modal contributions in the summation of the vertical distortion calculated at position x=0.25L measured from the stern and figures 4(e,f) show the variation in the amplitudes of vertical deflection and bending moment respectively along the uniform beam at different wave encounter frequencies.

(ii) MULTI-HULL

Figure 5(a) illustrates a hypothetical multi-hull structure which in this case represents a small water-plane area twin hull ship (SWATHS), though with minor modifications it could easily be a semisubmersible rig, catamaran, etc. Figure 5(b) illustrates some calculated dry hull modal characteristics (i.e. mode shapes, bending moments, principal stresses) based on a beam-plate discretisation of the structure. Generalised hydrodynamic coefficients and wave forces are illustrated in figures 5(c,d) respectively for the SWATHS travelling with Froude number Fn=0.223 in regular sinusoidal head waves. When compared with figures 4(a,b) for the monohull, the fluid-structure interaction between the port and starboard pontoons is clearly visible in the results. For an imposed vertical oscillatory deflection, a more detailed analysis reveals the generation of a standing wave symmetric about the centreline of the rig with a maximum amplitude at the centreline and a frequency in the region where the curves show fluctuations. Figure 5(e) illustrates a typical principal coordinate response, and figures 5(f,g) show the variation in the amplitude of principal stress with frequency and position along the hull.

(iii) JACK-UP RIG

For a hypothetical jack-up rig travelling at Froude number Fn=0.1 in regular sinusoidal head waves, figure 6 illustrates a selection of results relating to leg A. Although only one dry principal mode shape is shown an extensive set of results was obtained from a discretisation of the structure involving beam and plate elements. The structure is now far from beamlike or slender, however by adopting the three dimensional approach generalised hydrodynamic coefficients, wave exciting forces etc. were determined for this floating structure before a dynamic analysis of the leg was performed. It is easy to visualise that a similar analysis may be undertaken to discuss the dynamics of a mast or superstructure fixed to a ship etc.

(iv) FIXED STRUCTURE

Figure 7 illustrates the first three symmetric dry principal modes (r=1,3,5) of a cantilevered square plate (10x10x0.1m). Also included are calculated values for the first five natural and resonance frequencies. For the plate fixed horizontally in water at a depth of 5m below the free surface, the generalised hydrodynamic properties may be evaluated as discussed previously and the principal coordinates determined for an imposed oscillatory generalised force. A typical principal coordinate i.e. $|p_3|$ is shown. Naturally the same approach may be utilised to discuss the influences of water depth, free surface etc. on the responses as well as considering different plate configurations (i.e. vertical partly immersed, geometry, etc.).

Although only a simple fixed structure is considered the analysis may be readily extended to more complex fixed structures (i.e. jack-up rig, platforms, single-point mooring) with little difficulty.

CONCLUSIONS

A general unified hydroelasticity theory has been developed for fixed or floating marine structures which may be either stationary or moving with constant forward speed in regular sinusoidal waves. Based on the techniques of structural dynamics a consistent appoach has been outlined for the evaluation of the responses (i.e. displacements, distortions, bending moments, shearing forces, twisting moments, stress, etc.) of a marine structure to waves. The coupling existing between the flexible structure and fluid has been highlighted in the development of the mathematical model to determine the fluid loadings arising from incident waves, diffracted waves, bodily motions and distortions of the structure etc.

In this paper it is shown that the theory has a wide range of application being able to analyse the responses of several marine structures each having a different geometric configuration. However, the techniques and analysis developed are not restricted to marine structure problems only but may be easily modified or extended to other fields of engineering in which the description of the interaction between fluid and structure is of great importance.

ACKNOWLEDGEMENTS

One of us (WY) greatly acknowledges the Chinese Ship Scientific Research Centre, Wuxi, China for the opportunity to study in Brunel University during the past three years. We are both extremely grateful to our colleagues Professor R.E.D. Bishop, Dr. P. Temarel, Fu Yuning, J.J.M. Baar and Mrs. J.M. Price for their continual encouragements and efforts in the preparation of this text and the examples given.

REFERENCES

1. Bishop, R.E.D. and Price, W.G., Hydroelasticity of ships. (Cambridge, 1979).
2. Bishop, R.E.D., Price, W.G. and Wu, Y., A general linear hydroelasticity theory of floating bodies moving in a seaway. (To be published, 1984).
3. Brard, R., The representation of a given ship form by singularity distribution when the boundary condition on the free surface is linearised. J. Ship Res. 16 (1972), 79-92.
4. Inglis, R.B. and Price, W.G., The hydrodynamic coefficients of an oscillating ellipsoid moving in a free surface. J. Hydronautics 14 (1980), 105-110.
5. Inglis, R.B. and Price, W.G., Calculation of the velocity potential of a translating, pulsating source. Trans. RINA 123 (1981), 163-175.
6. Inglis. R.B. and Price, W.G., A three dimensional ship motion theory:Comparison between theoretical predictions and experimental data of the hydrodynamic coefficients with forward speed. Trans. RINA 124 (1982), 141-157.
7. Inglis, R.B. and Price, W.G., A three dimensional ship motion theory;Calculation of wave loading and responses with forward speed. Trans. RINA 124(1982),183-192.
8. Newman, J.N., Marine hydrodynamics (Cambridge (Mass): MIT Press, 1977).
9. Price, W.G. and Wu, Y., Hydrodynamic coefficients and responses of semi-submersibles in waves. Second International Symposium on Ocean Engineering and Ship Handling, SSPA, Gothenburg (1983), 393-415.
10. Price, W.G. and Wu, Y., Fluid interaction in multi-hull structures travelling in waves. Second International Symposium on Practical Design in Shipbuilding, Tokyo and Seoul (1983), 251-263.
11. Rayleigh, Lord, The theory of sound. (2nd edition, Macmillan, 1894).
12. Salvesen, N., Tuck, E.O. and Faltinsen, O., Ship motions and sea loads. Trans. SNAME 78 (1970), 250-287.
13. Newman, J.N., The theory of ship motions. Advances in Applied Mechanics 18 (1978), 221-283.

14. Timoshenko, S., Vibration problems in engineering. (Van Nostrand, 1955).
15. Timman, R. and Newman, J.N., The coupled damping coefficients of a symmetric ship. J. Ship Res. 5 (1962), 1-7.
16. Zienkiewitz, O.C., The finite element method. (McGraw Hill, 1977).

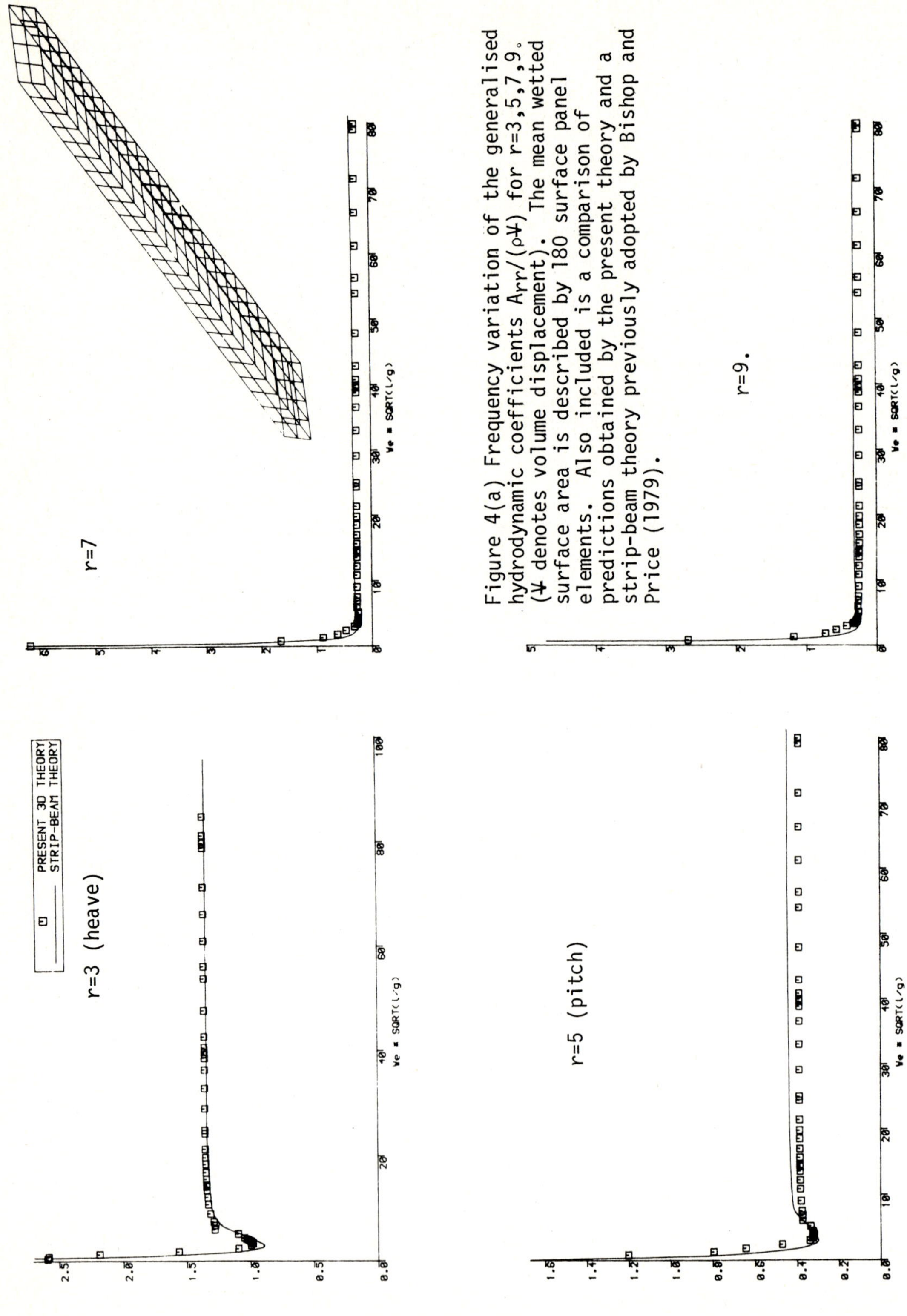

Figure 4(a) Frequency variation of the generalised hydrodynamic coefficients $A_{rr}/(\rho\forall)$ for r=3,5,7,9. ($\forall$ denotes volume displacement). The mean wetted surface area is described by 180 surface panel elements. Also included is a comparison of predictions obtained by the present theory and a strip-beam theory previously adopted by Bishop and Price (1979).

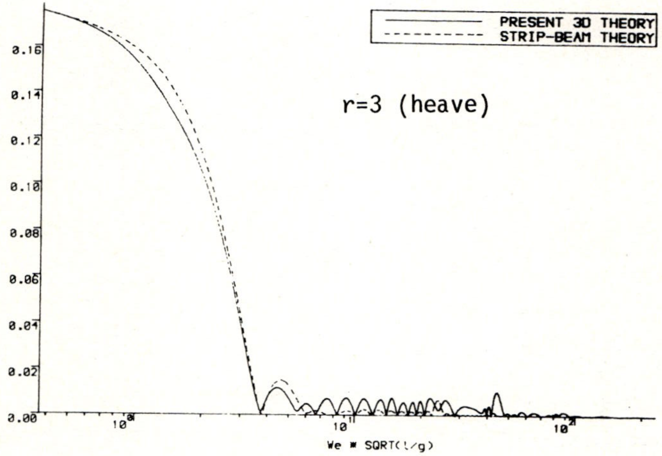

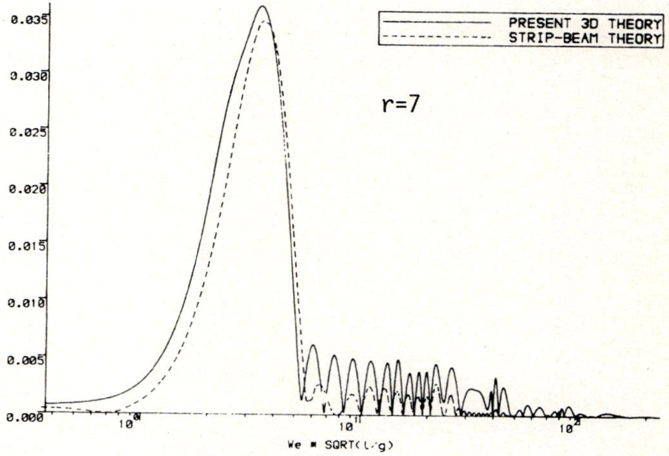

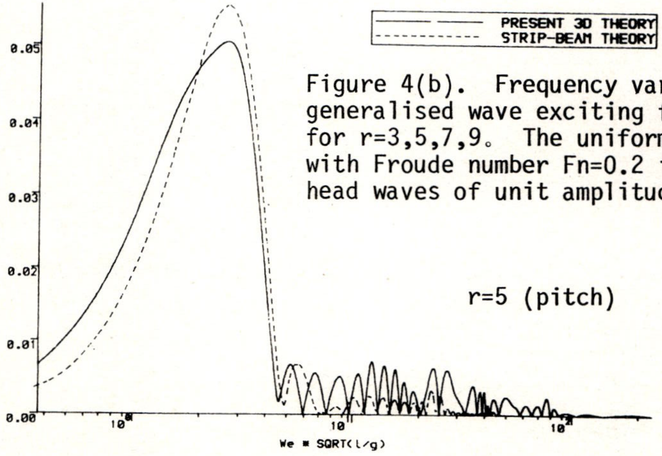

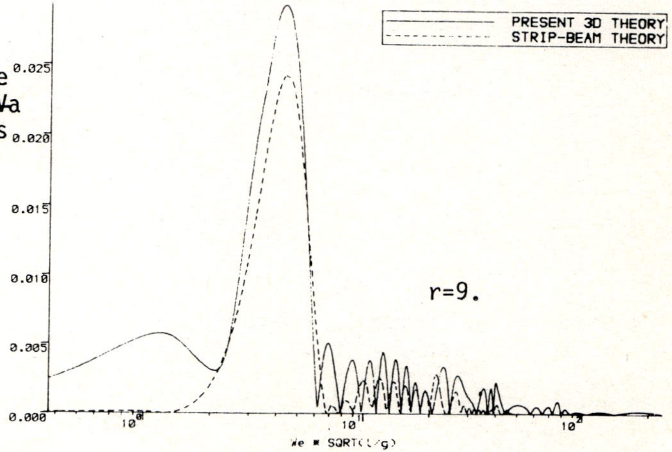

Figure 4(b). Frequency variation of the generalised wave exciting force $|\Xi_r|/\rho g \nabla a$ for r=3,5,7,9. The uniform beam travels with Froude number Fn=0.2 in sinusoidal head waves of unit amplitude.

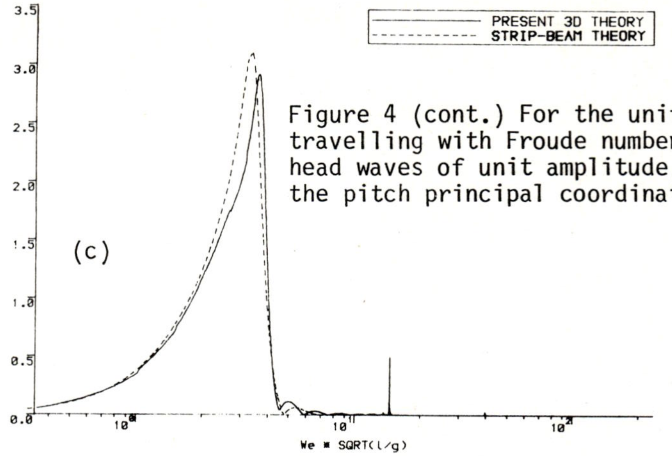

Figure 4 (cont.) For the uniform beam travelling with Froude number Fn=0.2 in head waves of unit amplitude, (c) describes the pitch principal coordinate $|p_5|/a$,

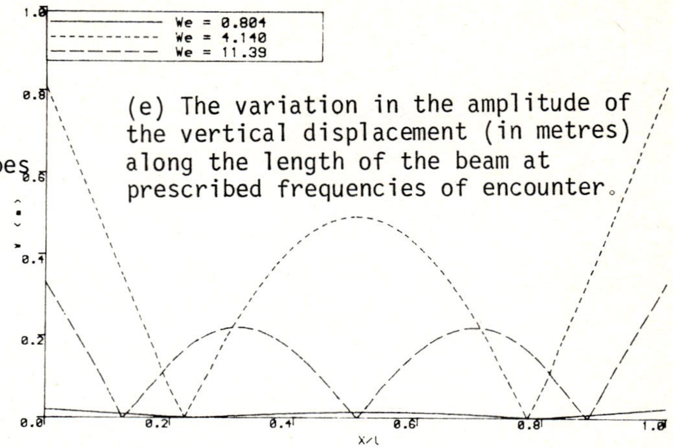

(e) The variation in the amplitude of the vertical displacement (in metres) along the length of the beam at prescribed frequencies of encounter.

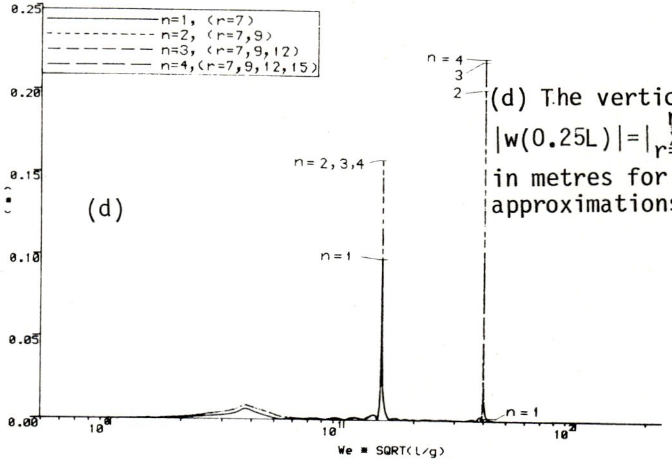

(d) The vertical distortion $|w(0.25L)| = |\sum_{r=7}^{n} w_r(0.25L)p_r|$ in metres for various approximations to n,

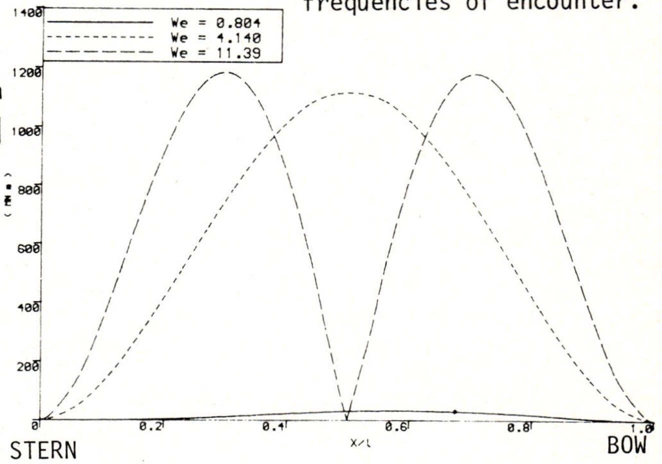

(f) The vertical bending moment (MNm) along the length of the beam at prescribed frequencies of encounter.

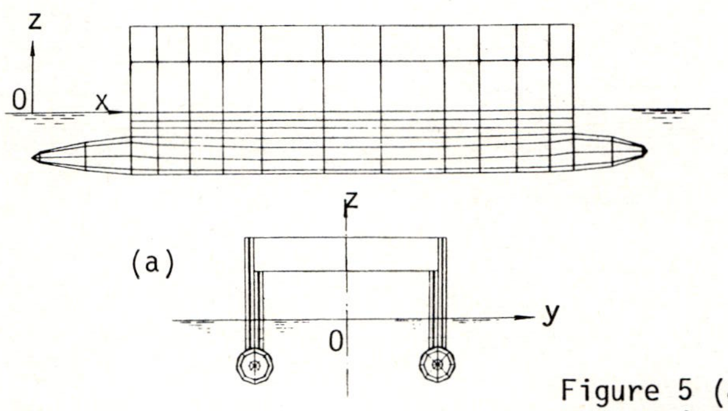

(a)

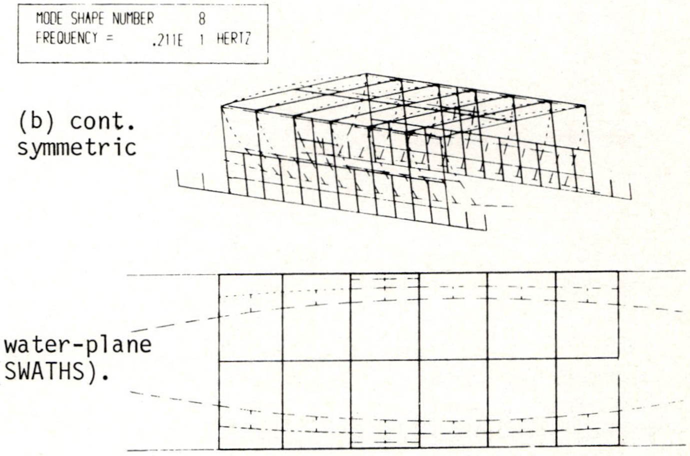

(b) cont.
symmetric

Figure 5 (a) A small water-plane area twin hull ship (SWATHS).
(b) Dry hull dynamic characteristics.

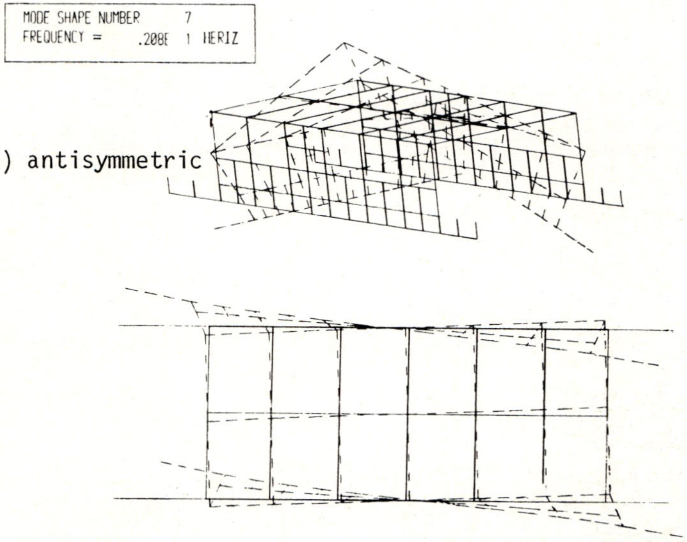

(b) antisymmetric

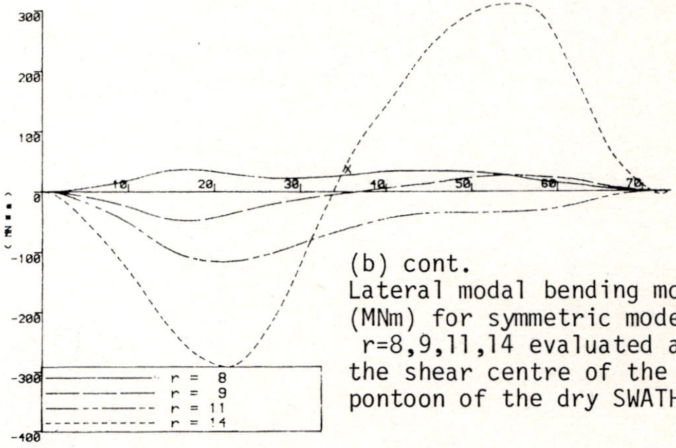

(b) cont.
Lateral modal bending moment (MNm) for symmetric modes r=8,9,11,14 evaluated along the shear centre of the port pontoon of the dry SWATHS.

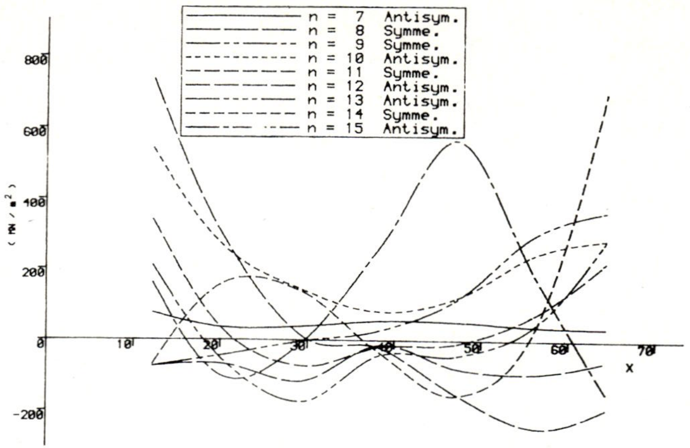

(b) cont. Amplitudes of the dry principal stress mode shapes (MN/m^2) on the outer surface of the port strut along the join with the bridge structure.

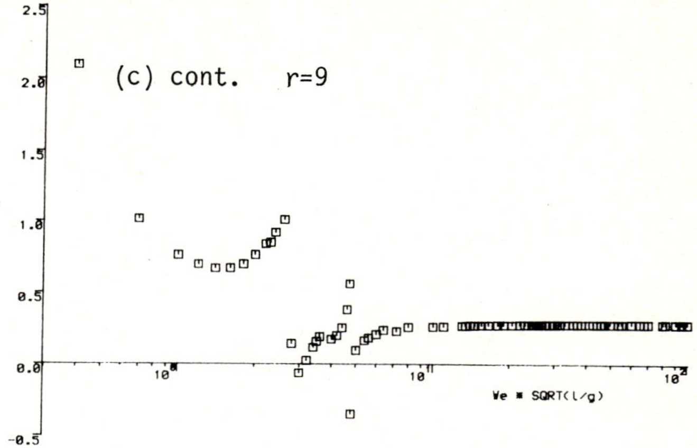

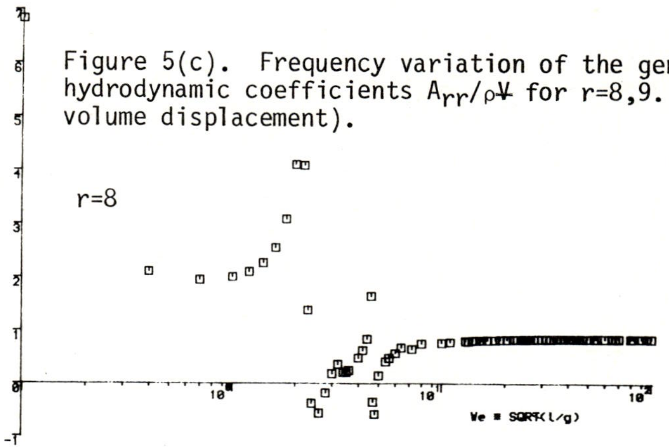

Figure 5(c). Frequency variation of the generalised hydrodynamic coefficients $A_{rr}/\rho\forall$ for r=8,9. ($\forall$ denotes volume displacement).

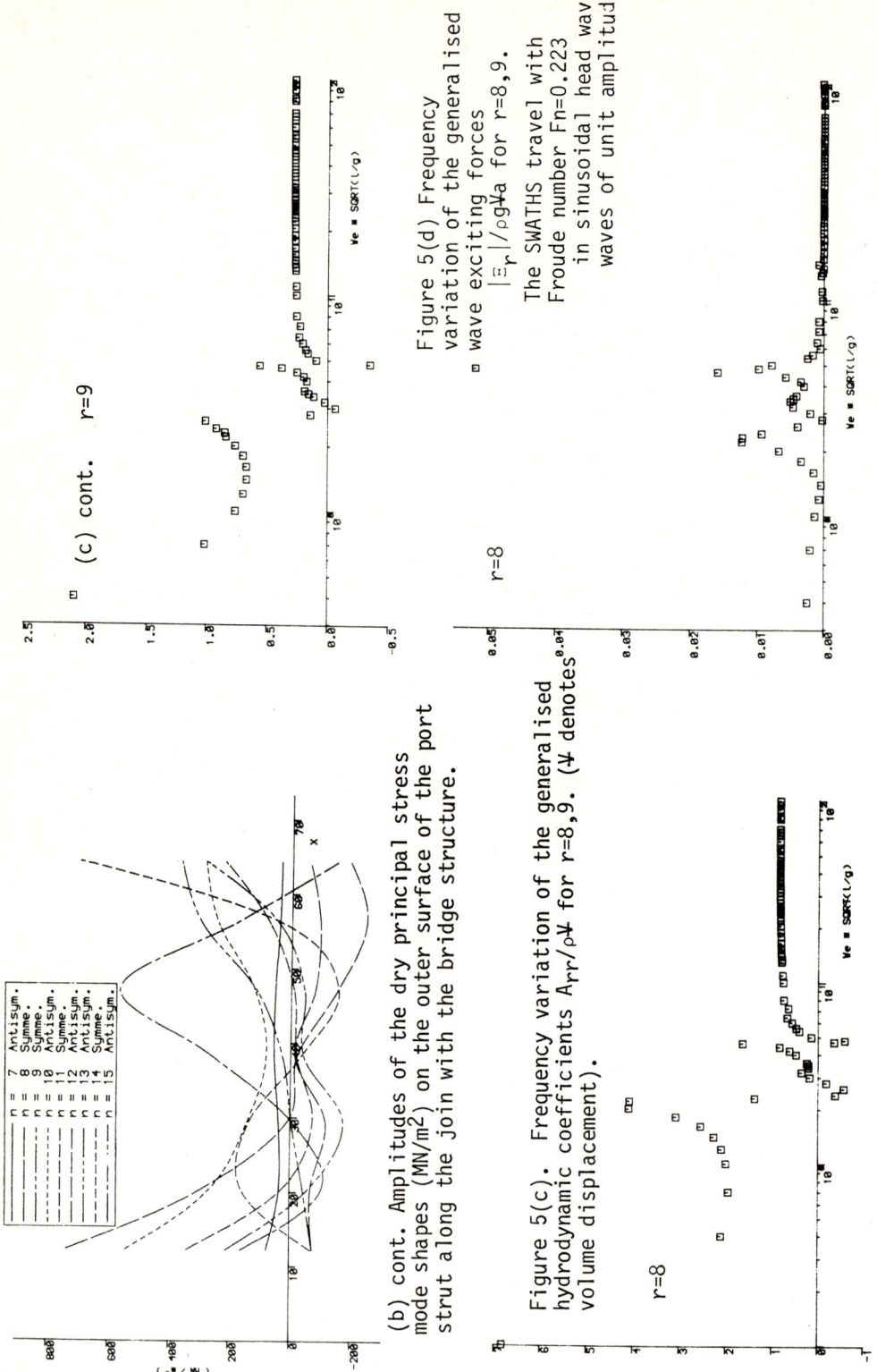

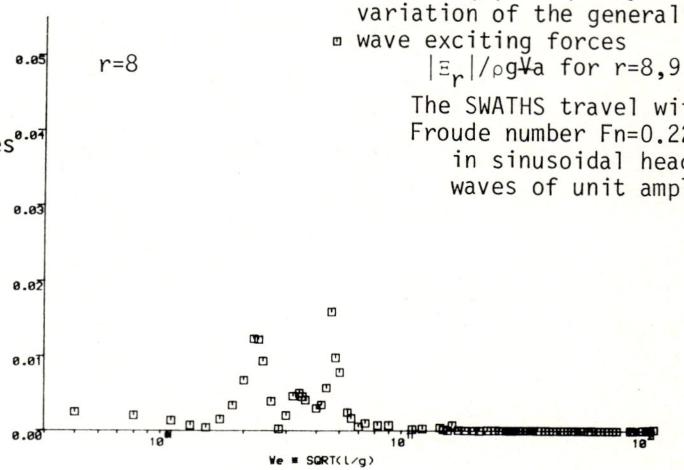

Figure 5(d) Frequency variation of the generalised wave exciting forces $|\Xi_r|/\rho g \forall a$ for r=8,9.

The SWATHS travel with Froude number Fn=0.223 in sinusoidal head wave waves of unit amplitude

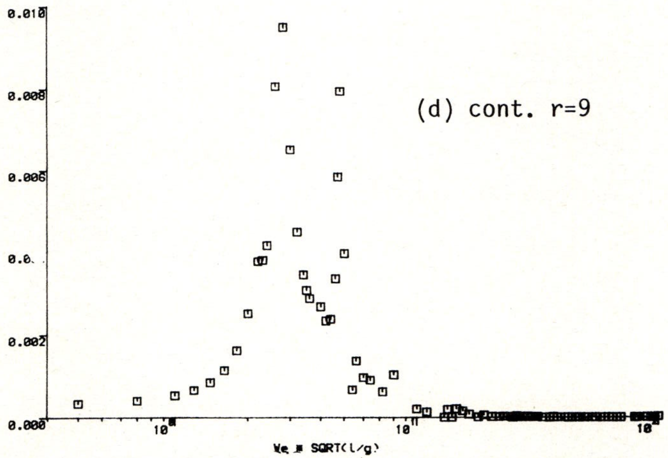

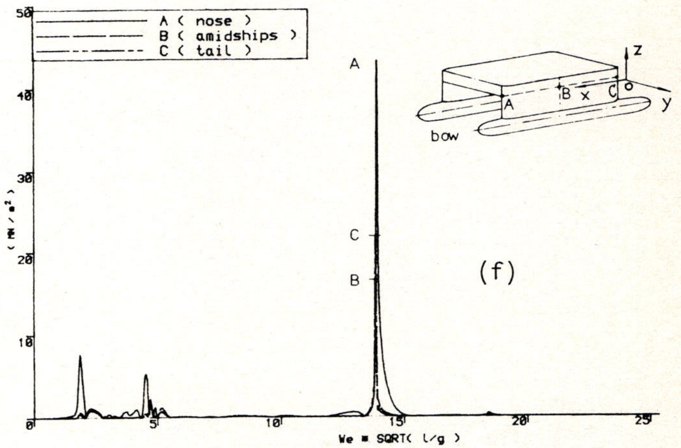

Figure 5(cont.) For the SWATHS travelling with Froude number Fn=0.223 in sinusoidal head waves of unit amplitude (e) describes the principal coordinate $|p_8|/a$; the variation in the principal stress amplitude (MN/m^2) (f) at points A, B, C on the outer surface as indicated and (g) along the line ABC at prescribed frequencies of encounter.

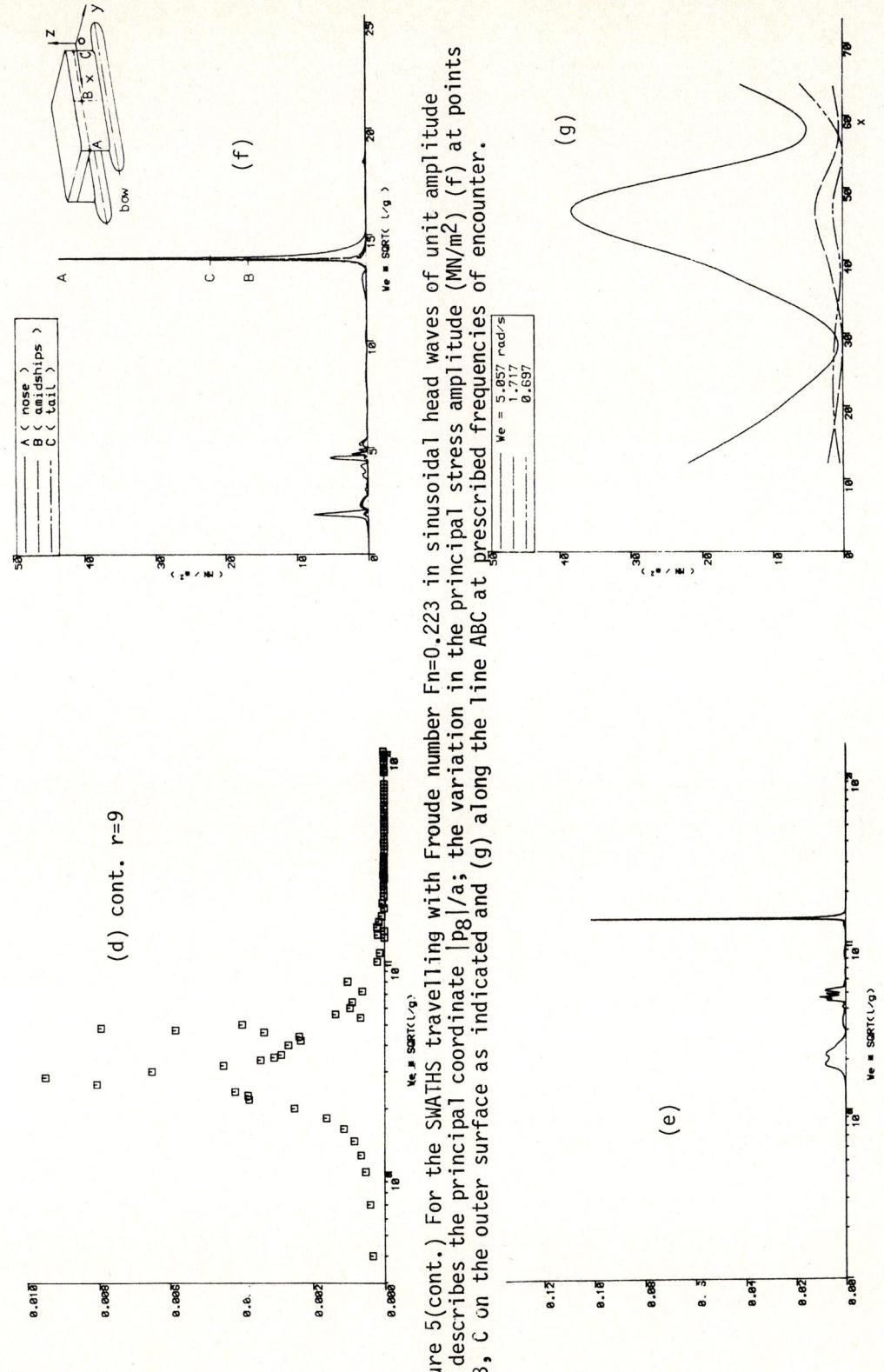

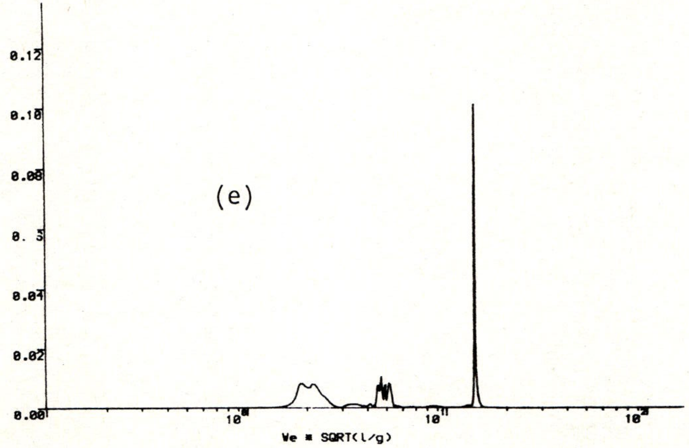

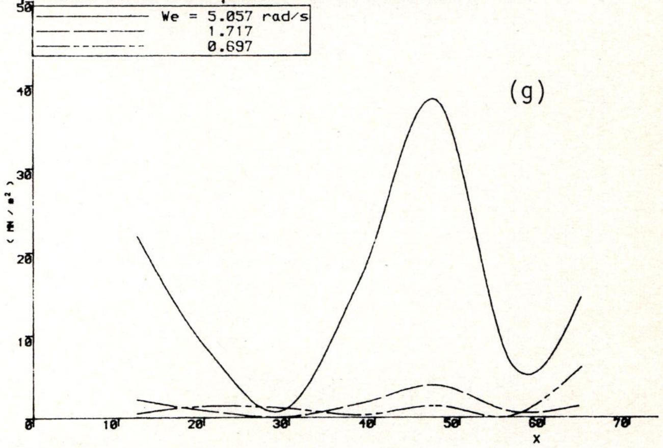

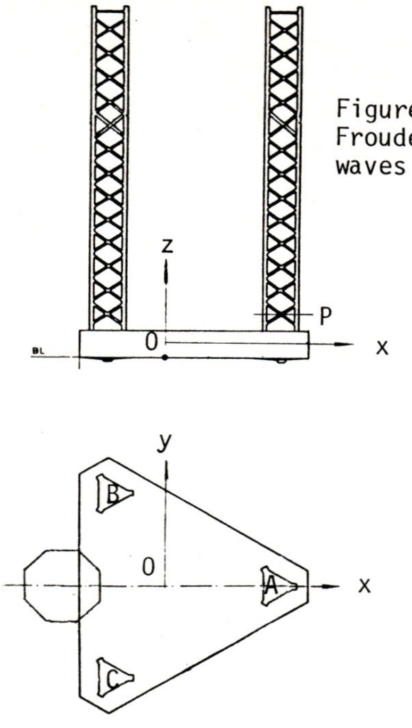

Figure 6. A jack-up rig in transit with Froude number Fn=0.1 in sinusoidal head waves of unit amplitude.

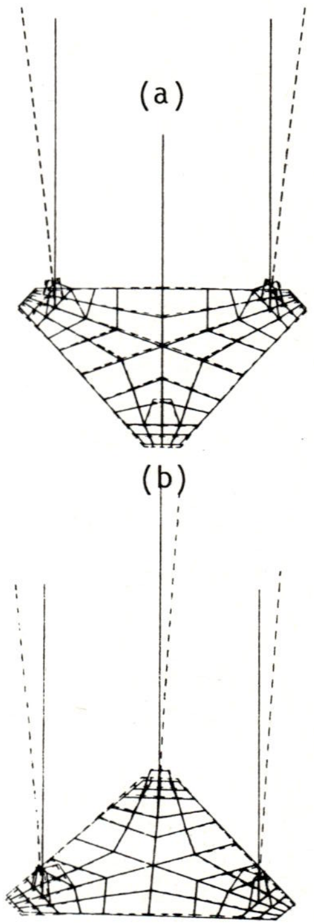

The principal dry modes of the rig in vacuo
(a) r=9 (symmetric), ω_r=2.47 rad/s.
(b) r=8 (antisymmetric), ω_r=2.29 rad/s.

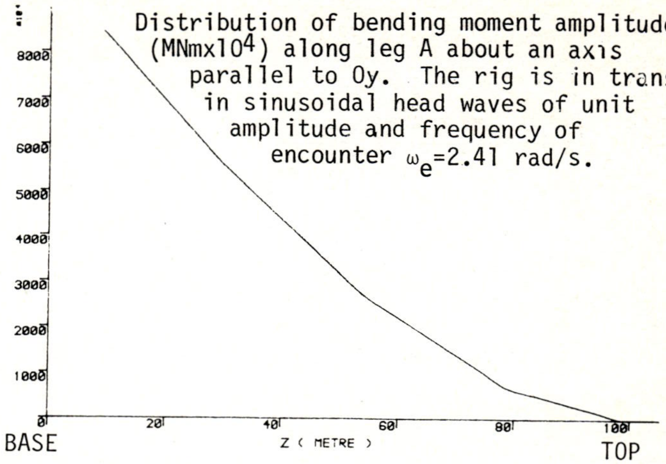

Distribution of bending moment amplitude (MNmx10^4) along leg A about an axis parallel to Oy. The rig is in transit in sinusoidal head waves of unit amplitude and frequency of encounter ω_e=2.41 rad/s.

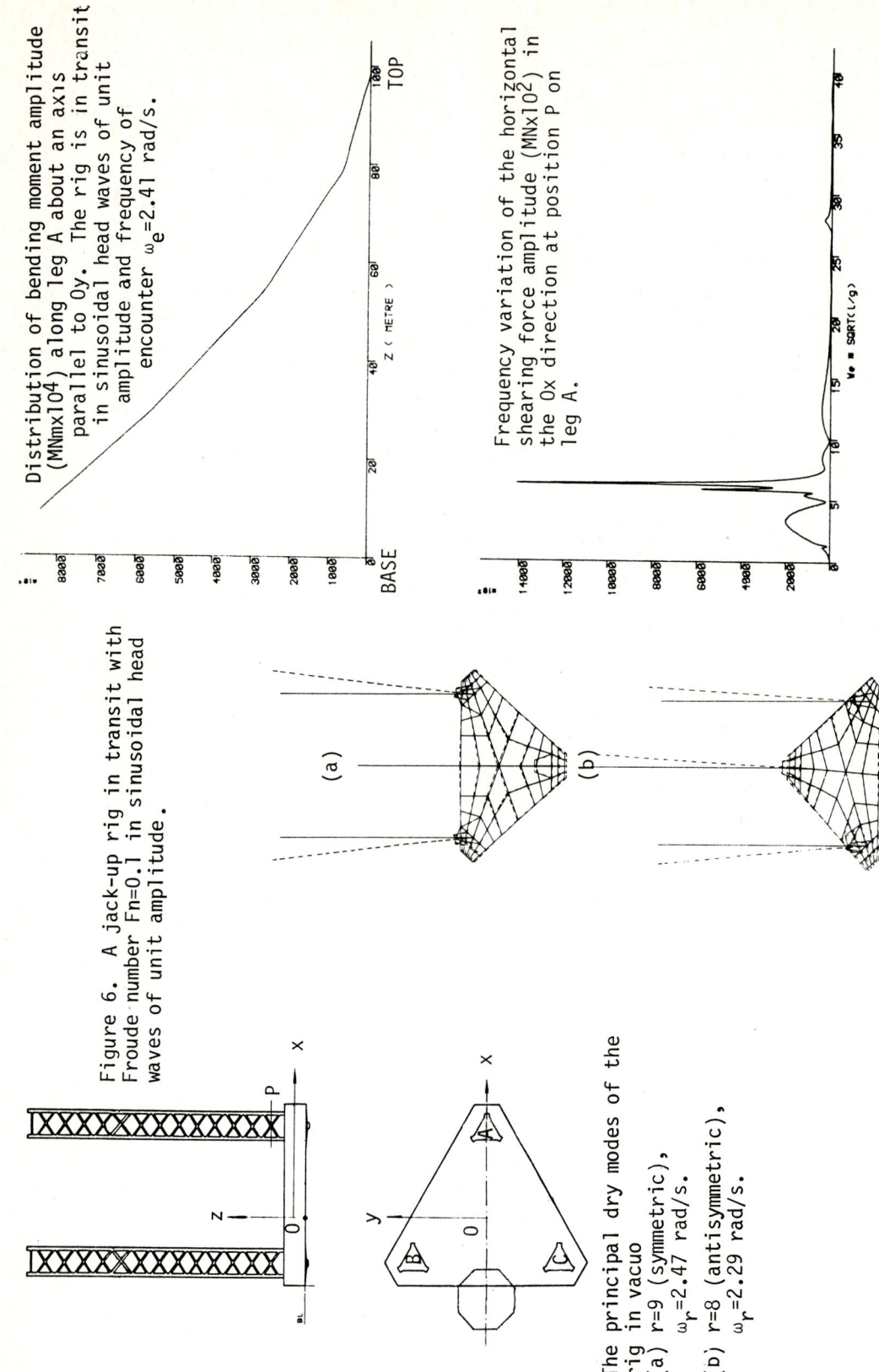

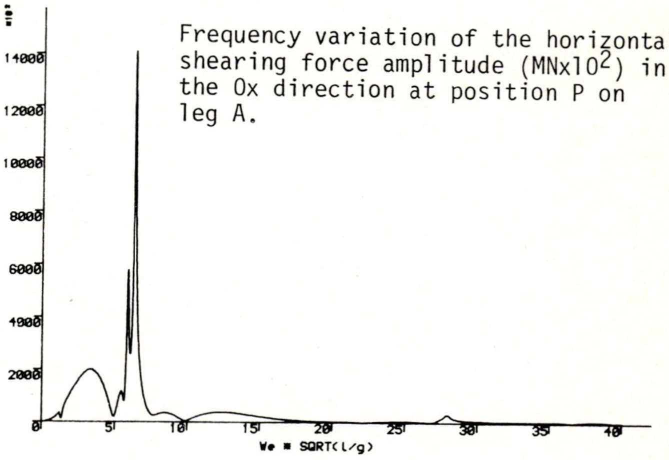

Frequency variation of the horizontal shearing force amplitude (MNx10^2) in the Ox direction at position P on leg A.

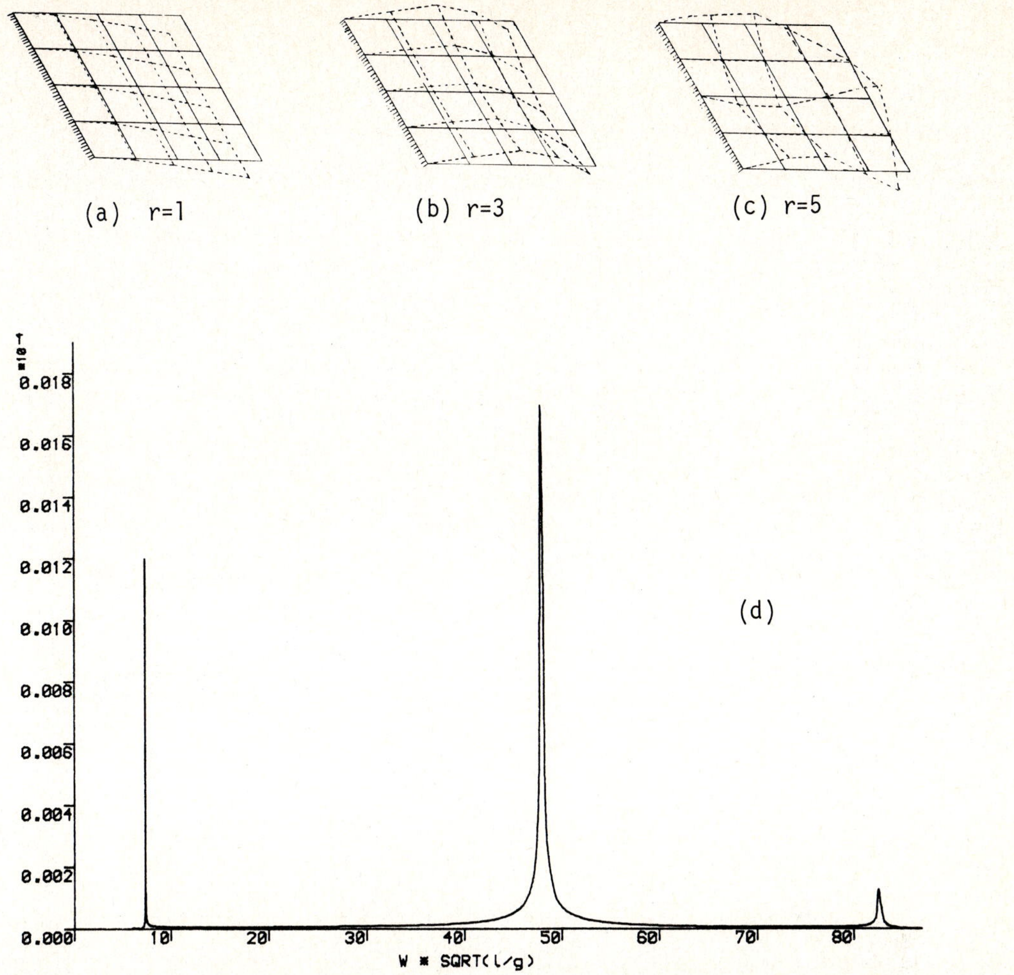

Figure 7. Horizontally cantilevered square plate (10x10x0.1m) submerged at 5m below the free surface. (a-c) denote the first three symmetric dry modes and (d) illustrates the frequency variation of the principal coordinate $|p_3|/a$.

Mode number	Distortion	Natural dry frequency (rad/s)	Resonance wet frequency (rad/s)
1	symmetric	12.94	7.27
2	antisymmetric	31.78	17.93
3	symmetric	80.43	47.4
4	antisymmetric	100.89	80.6
5	symmetric	115.74	81.8

Theoretical and Applied Mechanics
F.I. Niordson and N. Olhoff
Elsevier Science Publishers B.V. (North-Holland)

NON-CLASSICAL MATERIAL CONTINUA

Dominik Rogula

Polish Academy of Sciences
Institute of Fundamental
Technological Research
ul. Swietokrzyska 21
00-049 Warsaw, Poland

A model of material continuum is considered non-classical if it violates or at least does not fully respect one or more assumptions on which the classical continuum mechanics is based. The paper is devoted to discussion of such models from the point of view of fundamental laws of mechanics, with justification or criticism of the models as regards their ability to describe sub-macrolevel phenomena in real material bodies. Conforming to the architecture of the classical continuum theory, the non-classical models are considered according to the modifications they introduce into the relevant kinematical, dynamical, or constitutive assumptions. Particular attention is paid to nonlocal continuum models, i.e. the models that admit some direct interrelation between phenomena occuring at finitely distant (although microscopically close) points of the material continuum. Nonlocal theories of continua involving (a) finite range interaction forces, (b) pseudocontinuum kinematical microstructure, and (c) nonlocally correlated multi-component media are presented.

1. INTRODUCTION

Classical continuum mechanics was very successful in solving numerous problems of large-scale macroscopic behaviour of material bodies. However, if one considers physical phenomena of smaller and smaller characteristic scale of length, classical continuum mechanics appears less and less adequate. It is expected that, under appropriate definitions of the parameters involved, the criterion for validity of classical theory can be expressed in the form of simple inequality

$$\lambda > \ell \tag{1.1}$$

where λ and ℓ are length parameters which characterize the phenomenon considered and the intrinsic scale of length of the material, respectively. The appropriate scale of length in real bodies depends on the underlying intrinsic structure due to micro-defects, finite range of interaction forces, atomic structure, correlation of randomly distributed elements, etc. The relevance of intrinsic length parameters depends on the phenomenon under consideration. As most typical examples one can quote stress concentrations, defect distortions, rapid vibrations, strong polarization, effective meanfield properties of random media.

In the present lecture we shall consider non-classical models of material media, i.e. models which violate or at least do not fully respect one or more assumptions on which the classical continuum mechanics is based. However, we shall confine ourselves to models which do not give up the idea of material continuum, although we include pseudocontinuum theory (section 5) which, in an essential sense, is intermediate between genuine continuum and discrete atomic lattice.

The aim of our present lecture will be two-fold. First, we shall try to sketch a framework of continuum theory which would enable a compact presentation of its non-classical concepts, and to see where the existing modifications of classical continuum mechanics can be placed, as well as to draw some hints concerning new ones. Second, we shall briefly present some most recent trends in the domain of non-classical continua. In our presentation we shall follow mainly (to quote only modern authors) Datta-Gairola [34,35,48], Edelen [10-19,26], Eringen [20-33], Green [19,36], Kröner [48-51], Kunin [53-57], Kaliski [43-45], Iesan [41-42], Mindlin [59-60], Nowacki [61-69], Rivlin [36], Rzewuski [91-92], Rymarz [89-90,105], Stojanović [94], Tiersten [100], Toupin [101-102], Zorski [105-106], although for some views only the present author should be considered responsible.

2. CLASSICAL CONTINUUM

Fundamental concepts of continuum mechanics appropriate for theoretical description of the macroscopic behaviour of deformable material bodies were almost completely formed in the first half of the nineteenth century in the historical works by Navier, Cauchy and Green [1]. These ideas resulted in a scientific theory - the classical continuum mechanics - which unifies theoretical simplicity with incomparable success in applications to practical problems. To understand the status of non-classical refinements it is crucial to answer the question as to whether the experiment confirms all the assumptions of classical continuum mechanics or only a part of them. In the latter case one can reasonably anticipate existence of non-classical models which would be compatible with experiment in classical domains and effective in description of some phenomena which are outside the range of classical theory.

Before discussing refined continuum models of material media we shall very briefly state the basic assumptions of classical continuum mechanics. The phrasing of these assumptions is intended to reflect the main ideas constituting the basic architecture of classical continuum mechanics without much detail or formalization [84]. Basic *kinematical* assumption: a material body is a continuum collection of structureless entities behaving like material points. The kinematics of a body is entirely determined by a map

$$x: \mathcal{B} \times \mathcal{T} \to \mathcal{E} \tag{2.1}$$

where $\mathcal{B}$ $\mathcal{T}$ and $\mathcal{E}$ represent the material continuum, physical time and physical space, respectively. This map represents the configuration of the body as a function of time.

Basic *dynamical* assumption: intrinsic interactions in a material body at each instant are uniquely determined by a field of the symmetric stress tensor. More precisely, this is a four-level hierarchy of the following statements:

(i) There are only *contact* interactions.
(ii) These interactions are given by *force* vectors.
(iii) The force vectors are determined by the *stress* tensor.
(iv) The stress tensor is *symmetric*.

In more detail, the idea of the assumption (i) consists in admitting only mutual interactions between infinitesimally close material particles. Under this assumption the concept of "interaction through a surface element" is meaningful. The interactions through disjoint surface pieces are assumed to be additive.

According to the assumption (iii), the interaction forces acting through surface elements can be derived from the stress vector p_i which is given by the formula

$$p_k = \sigma_{ik} n_i , \qquad (2.2)$$

where the stress tensor σ_{ik} is independent of the surface element. In particular, this means that the stress vector p_i depends linearly on the normal vector n_k. The assumption (iv) says that no net moment of force on a particle can arise from the interaction between particles.

Basic *constitutive* assumption: the stress tensor at a material point X can be entirely determined from the deformation tensor at the same point X.

3. NON-CLASSICAL MODIFICATIONS

The idea of generalizing the fundamental assumptions of mechanics of material media can be traced back already to the works by Cauchy, Voigt and Duhem and other creators of classical continuum mechanics. In recent years the interest in this subject revived. Many authors, guided by various ideas, developed a number of modified theories with the view to improve the results of classical continuum theory, particularly where the phenomena of submacroscopic characteristic scale of length or coupled effects are concerned.

Depending upon the kind and degree of violation of the classical assumptions, various non-classical theories of material continua arise.

The kinematical assumption can be violated by endowing the medium with some kind of local structure defined for each particle of the body. Such a structure, a priori independent of the configuration (2.1), requires some extra fields defined over the body to represent its kinematical state. With extra degrees of freedom resulting from such a structure, the field of the displacement vector is no longer sufficient for that purpose. By specification of the local structure, specific non-classical theories can be obtained.

In the famous work by Cosserat and Cosserat [6] each point of the material continuum is endowed with a rigid rectangular triade whose axes depend on the material point and can vary in time, thus defining a local structure with extra degrees of freedom of the rotational type. In modern history of the problem, some authors guided by physical ideas arrived at non-classical continuous media, like the spinning fluid of Weyssenhoff and Raabe [103] or the fluid of diatomic molecules of Zelazny [107].

From the phenomenological point of view the simplest physically consistent local structure consists in admitting *microrotations* which are independent of local rotations arising from the displacement field. This idea led to the so-called *micropolar* theory [63].

The next step in generalizing the kinematics of a material continuum consists in admitting deformable microstructure, as opposed to rigid microstructure in the micropolar theory. Here Green and Rivlin's [36] *multipolar* theory or Toupin's [102] theory of continuum endowed with an arbitrary system of *directors* are best known.

Further generalization consists in treating particles of material continuum as separate *microcontinua*, kinematically independent of displacements of the usual macro-level continuum. Eringen and Suhubi's [22,23] *micromorphic* theory is based on such an idea. In this approach any point of a material continuum is endowed with infinitely many local degrees of freedom. Frequently, however, it is suitable to constrain the microcontinuum to a finite number of local degrees of freedom, thus reducing it effectively to the former case.

Generally, the state of an enriched kinematical structure of material continuum can be represented by a map of the following form:

$$\chi: \beta \to \xi \times Y \tag{3.1}$$

where Y stands for an appropriate local structure space. Although in the simplest case Y is a vector space, generally such an assumption is not necessary. Even more, there are many physically justified instances of local structure spaces which cannot bear any reasonable vector structure, as examples of finite rotations, saturated spin, or the space of Bravias lattices considered in [75] show. Any material system can serve as a model of local structure. A general procedure of forming the corresponding local structure spaces has been given in [74].

The simplest possible modification of the dynamical assumption consists in rejecting the statement (iv) while retaining the statements (i) - (iii). In this way one admits the nonsymmetric stress tensor. On purely mechanical grounds such an approach was proposed by Bodaszewski [3]. It seems, however, more consistent to introduce an asymmetric stress tensor by combining this modification with other ones, kinematical or constitutive.

The next modification could consist in rejecting the statement (iii) while retaining the statements (i) - (ii). No thorough investigation, however, has been carried out in this direction. It can result in curvature-dependent transmission of forces through surface elements.

By modifying the assumption (ii), a number of non-classical continuum models can be created. The simplest modification of this type consists in retaining the assumption (i) and modifying the assumption (ii) by admitting, apart from the interaction via force vectors, some additional interactions via couple, i.e. moment of force vectors. This leads to the so-called *couple stress* theory. According to this theory, the contact interactions arise from a more general state of stress which can be described by two tensor fields: the usual force-stress tensor σ_{ik} and an additional couple-stress tensor μ_{ik}.

The couple-stress vector is given by the formula

$$m_i = \mu_{ki} n_k \tag{3.2}$$

and the tensor μ_{ki} enters the balance of moment of momentum. Neither σ_{ik} nor μ_{ik} need be symmetric.

Additional interactions introduced above can be assumed more general. Namely, one can introduce hyperstress of higher and higher orders, which describe non-classical interactions via higher order multipoles of forces. In the proper sense, hyperstress define contact interactions that do not enter directly the force and moment balances. However, they always enter the energy balance.

Moreover, the statement (i) can be modified. This results in some kind of dynamically nonlocal models of the integral type [53,49].

Also the classical constitutive assumption can be violated in several ways. It can be modified by taking the stress tensor dependent constitutively not only on the deformation tensor, but also on its gradient or even a number of higher order gradients. In theories with a local structure, the corresponding extra degrees of freedom can also appear in the corresponding constitutive relations like, for example, micropolar elasticity. The stress tensor can also be assumed to depend on temperature and/or electromagnetic fields, what results in a theory of the coupled-field type. And, last but not least, the locality principle contained in the classical constitutive assumption can be rejected by assuming the stress tensor at X dependent on the situation at points Y finitely distant from X. In this way continuum models with a nonlocal stress-strain relation are obtained [48,19,26].

Appropriate combinations of the above modifications, taking into account their mutual interdependence, are also possible and even sometimes necessary. The following sections will be devoted to some examples.

4. NONLOCAL ELASTICITY

The idea of locally determined contact forces was for long time considered a necessary ingredient of continuum theory. It was not until 1960's when it was clearly recognized that the concept of material continuum and the range of interaction forces are independent of each other, and when first papers on nonlocal continuum theory appeared, starting from the paper by Kröner and Datta [48]. The idea of nonlocal continuum has also been formulated and developed by Kunin [53], Edelen [10], Eringen [20] and other authors.

If one considers a material whose intermolecular spacing is a and typical range of intermolecular forces is ℓ, then for characteristic distances λ such that

$$a < \lambda < \ell , \tag{4.1}$$

the local theory is no longer valid, while continuum model can still be useful.

Simple examples of nonlocal continuum models can be constructed starting from the balance of momentum. Under classical assumptions concerning continuity of the medium and the distribution of physical quantities, the corresponding equation can be written in the form

$$\dot{p}_i + (p_i v_k)_{,k} + r_i = f_i \tag{4.2}$$

where r_i stands for the density of forces which arise as a result of interaction between particles of the body. The classical theory of elasticity gives

$$r_i(x) = - \partial_j c_{ijk\ell}(x) \partial_\ell u_k(x) \tag{4.3}$$

If the place of this classical expression an integral relation

$$r_i(x) = \int_\Omega dv' \, \phi_{ij}(x , x')[u_j(x') - u_j(x)] \tag{4.4}$$

is assumed one arrives at the simplest form of integral model, which instead of contact interactions makes use of direct long-range forces between material particles. If one compares (4.4) with the corresponding expression of lattice theory,

$$r_i^A = \sum_{B\varepsilon\Omega} \Phi^{AB} [u_j^B - u_j^A] \tag{4.5}$$

the analogy is immediate.

The Kernel $\Phi_{ij}(x , x')$ appears as a continuum counterpart of the force constant matrix Φ_{ij}^{AB}.

Another simple possibility is the model of nonlocally determined stresses. In constructing such a model one retains the classical concept of contact forces with rejecting local relation between stress and strain. Instead of classical relation

$$\sigma_{ij}(x) = c_{ijkl}(x)\varepsilon_{ij}(x) \tag{4.6}$$

one takes

$$\sigma_{ij}(x) = \int_\Omega dv' \, c_{ijkl}(x , x')\varepsilon_{kl}(x') + D_{ijkl}(x)\varepsilon_{kl}(x) \ . \tag{4.7}$$

Under appropriate circumstances a combination of these two models is adequate.

The above examples represent particular cases of models with a non-local governing equation. An equation

$$Ay = z \tag{4.8}$$

where A stands for a linear operator is called (weakly) local if

$$\text{supp } Ay < \text{supp } y \tag{4.9}$$

for each y in the domain A (containing sufficiently many functions of compact support). Similar definition holds for a nonlinear operator [81]. A theory governed by an equation which is not local is called strictly nonlocal.

However some equations, although local in the above defined mathematical sense, are able to describe relations typical for physical nonlocality. A material medium is nonlocal in this weaker sense if its physical properties determine an intrinsic measure of length, i.e. if the coefficients that fully characterize its physical properties can be combined into a parameter of the dimension of length. A mathematical criterion for nonlocality of this type is provided by scale transformation. If there exists a transformation

$$(x , y , z) \to (x' , y' , z') \quad \text{with} \quad x' = a\, x \tag{4.10}$$

which leaves the equation (4.8) invariant, the corresponding theory is called strictly local. Otherwise it is called (weakly) nonlocal.

It is the property of strict locality that makes classical elasticity unable to tell the difference between small and large. On the other hand, theories like micropolar or strain gradient elasticity, in spite of the fact that they are governed by systems of purely differential, and thus weakly local equations, are endowed with intrinsic length parameters.

In order not to contradict physical facts which support classical theory, any nonlocal model has to be asymptotically compatible with classical theory. By that we understand that if in eq. (1.1) $\lambda \to \infty$ (or, what is equivalent, $\ell \to 0$), the solutions of nonlocal models tend to their classical counterparts. This asymptotic correspondence principle is one of the cornerstones of phenomenological theory of reinfined continuum models.

Based on this principle a general theory of linearly nonlocal elastic media has been constructed [77]. The crucial concept of the theory is that of the singular order of the governing operator L as well as of its inverse G . Intuitively speaking, the material is "singular hard" if the singularity of displacements created by a concentrated force is weak. And if the material is "singular soft" a concentrated force creates a strong singularity in the displacement field. It is proved that the relation

$$s(G) > 6 - s(L) \tag{4.11}$$

always holds. For given L , this relation determines the best limit for the singularity of the fundamental solution. When some properties of solutions are anticipated relation (4.11) imposes necessary restrictions on the choice of the governing operator in the model.

5. PSEUDOCONTINUUM

The models discussed in the previous section exemplify so called dynamical nonlocality due to interaction forces between finitely distant particles, with accepting all the kinematical properties of classical continuum. However, this reminder of the classical approach to material media can also be subject to serious criticism.

If one considers phenomena of spatial scale close to interatomic spacing, continuum theory provides too much - the displacement field defined in the empty space between atoms. The physical meaning of such solutions is rather doubtful. Nevertheless, the behaviour of such solutions on interatomic spacing is not, in continuum theory, a pure ornament, as it contributes to the total energy and therefore influences the solution at physically meaningful points.

The problem of converting lattice theory into a continuum theory has been rigorously solved by Krumhansl [52], who arrived at generalized continuum theory with the energy dependent on displacement gradients up to infinite order. Kunin [53-56] and the present author [76] developed so called pseudocontinuum (or quasicontinuum) theory which completely resolves the above mentioned difficulty.

This theory establishes a one-to-one correspondence between functions f_n defined on some regular lattice and a certain class of functions $f(\underline{x})$ of a continuous argument $\underline{x}$ (provided some natural conditions at infinity are satisfied). This relation has the interpolation property

$$f_n = f(x_n) \tag{5.1}$$

and is most effectively expressed in terms of discrete-continuous Fourier transform and its inverse

$$f_n = \frac{1}{(2\pi)^3} \int_{BZ} e^{i\underline{k}\underline{x}_n} f(\underline{k})\, d^3k \, , \tag{5.2}$$

$$f(\underline{x}) = \frac{1}{(2\pi)^3} \int_{BZ} e^{i\underline{k}\underline{x}} f(k)\, d^3k \, , \tag{5.3}$$

$$\delta(k) = \frac{1}{\Omega} \sum_{\underline{n}} e^{-ik\underline{x}_n} \delta_{\underline{n}} , \qquad (5.4)$$

where the integration extends over the first Brillouin zone of the lattice and Ω denotes the volume of the unit cell.

The functions representable in the form (5.3) are far from being arbitrary: the restrictions following from this representation are sufficiently severe to assure the desired one-to-one correspondence. In consequence the discrete and continuum expressions for the kinetic and potential energies are rigorously equivalent,

$$T = \frac{m}{2} \sum_{\underline{n}} v_{\underline{n}}^2 = \frac{1}{2} \int \tilde{v}^2(\underline{x}) d^3x \qquad (5.5)$$

$$U = \frac{1}{2} \sum_{\underline{n},\underline{n}'} u_{\underline{n}} \Phi_{\underline{nn}'} u_{\underline{n}'} = \frac{1}{2} \int u(\underline{x}) \Phi(\underline{xx}') u(\underline{x}') d^3x d^3x' .$$

As a result, strict equivalence between lattice and pseudocontinuum holds. However, the pseudocontinuum representation of discrete lattice is continuum-like and can easily be transformed in more or less accurate continuum models.

Consider some operators L and L' acting in sufficiently broad class of functions. These operators are equivalent in pseudocontinuum theory provided that

$$Lu = L'u \qquad (5.6)$$

for any field u admissible in this theory. This equivalence gives the pseudocontinuum theory a great flexibility: operators of entirely, different structure - discrete, differential, integral - can be strictly equivalent.

6. A FIELD MODEL OF COHESIVE FORCES AND NONLOCALLY COUPLED MULTI-CONTINUA

In this section we shall very briefly discuss two recently developed ideas in the domain of non-classical continuum models.

The first of them consists in the following. Considered on physical grounds, the stress in material bodies always result in consequence of electron-electron interactions (with appropriate ionic component). It is therefore expected that a unique carrier field for interactions within (as well as between) all materials exist. Such a field can be conceived as a certain generalization of chemical potential. To describe cohesive interactions in solids, the simplest reasonable proposition [88] seems to be a field of a symmetric tensor quantity, which generalizes the scalar chemical potential. From the quantitative point of view the field $\psi_{ij}(\underline{x})$ is connected with the deformation $\varepsilon_{ij}(\underline{x})$ in such a way as the electrostatic potential is connected with the electric charge. It means that the deformation of the body is at the same time a source and a detector of the field $\psi_{ij}(\underline{x})$.

In the linear approximation (with irrelevant body forces omitted) one obtains the following set of field equations

$$(C_{ijk\ell} \beta_{k\ell} + N_{k\ell ij} \psi_{k\ell} + D_{ij})_{,j} = 0 \qquad (6.1)$$

$$M_{ijk\ell mn} \psi_{\ell m,nk} - L_{ijk\ell} \psi_{k\ell} - N_{ijk\ell} \beta_{k\ell} - P_{ij} = 0$$

with the corresponding boundary conditions

$$(C_{ijk\ell}\,\beta_{k\ell} + N_{k\ell ij}\,\psi_{k\ell} + D_{ij})n_j = p_i\ , \tag{6.2}$$

$$M_{ijk\ell mn}\,\psi_{\ell m,n}\cdot n_k = 0\ ,$$

where the tensors $\underline{C}$, $\underline{D}$, $\underline{M}$, $\underline{L}$, $\underline{N}$, $\underline{P}$ depend on the material and, in heterogeneous bodies, may vary with $\underline{x}$. The tensor β denotes the mechanical distortion of the medium; in the case of compatible deformation it can be expressed as the displacement gradient.

The equations (6.1) together with the boundary conditions (6.2) can be solved for various homo- and heterogeneous bodies. It turns out that a ψ - field homogeneous body submitted to homogeneous deformations is experimentally indistinguishable from a classical homogeneous body of the same shape and appropriately renormalized elastic modulae. Inhomogeneous deformation leads to nonlocal $\sigma - \varepsilon$ relation. In the case of a heterogeneous body, extra forces localized at interfaces appear.

The main idea of the multi-continuum model is to apply the concept of interpenetrating continua (as developed by Triersten [100]) to microhomogeneous bodies, not only to micro-heterogeneous ones. An attractive feature of such models is their ability to reproduce micro-oscillations of the displacement field observed in atomic lattices.

A bi-continuum model has been developed and compared with appropriate discrete models [8]. Let u and v denote displacements of the components. By introducing the new field variables

$$\bar{p} = \frac{1}{2}(\bar{u} - \bar{v}) \tag{6.3}$$

$$\bar{q} = \frac{1}{2}(\bar{u} + \bar{v})$$

and making use of the appropriate variational principle one arrives at the following set of equations

$$-\ (\hat{k} + \hat{c})\bar{q} - \hat{h}\bar{p}'' = 0 \tag{6.4}$$

$$-\ (\hat{k} + \hat{d})\bar{p}'' + \hat{h}\bar{q}' + \hat{e}\bar{p} = 0$$

where $\hat{k}$, $\hat{c}$, $\hat{h}$, $\hat{e}$, $\hat{d}$ denote the matrix coefficients, and the dash represents differentation with respect to the independent variable (assumed single for simplicity). Excepting d all these matrices were fitted to force constants of the lattice; where the matrix d is concerned it was chosen so that the best fit for attenuation exponents was obtained.

Within this model the problem of [112] reflection twin boundary in a bcc lattice was solved with making use of the following boundary conditions

$$[\bar{v}] = \text{given}\ , \tag{6.5}$$

$$[\bar{u}] = 0\ ,$$

$$[(\hat{k} + \hat{d})\bar{p}'] = 0\ ,$$

$$[(\hat{k} + \hat{c})\bar{q}' + \hat{h}\bar{p}'] = 0\ .$$

The square brackets denote the discontinuity jump across the interface.

Good agreement with the results of computer simulation [4] has been obtained.

REFERENCES

[1] For references of historical interest see:
Truesdell, C.A. and Toupin, R.A., *The classical field theories*, Handbuch der Physik III/1; Timoshenko, S.P., History of strength of materials (McGraw-Hill 1953).

[2] Banach, Z., *On a certain wave motion in the spatially nonlocal and nonlinear medium*, IJES 19 (1981) 1047.

[3] Bodaszewski, S., *On nonsymmetric state of stress and its application in mechanics of continuous media*, Arch. Mech. 5 (1953) 351.

[4] Bristowe, P.D. and Crocker, A.G., *A computer simulation study of the structures of twin boundaries in body-centred cubic crystals*, Phil. Mag. 31 (1975) 503.

[5] Brulin, O. and Hsieh, R.K.T., *Polar mechanical interactions of lattice defects*, in: New Problems in Mechanics of Continua, Study No. 17 (Univ. of Waterloo Press, 1983), 91.

[6] Cosserat, E. and Cosserat, F., *Théorie des corps déformables* (Hermann 1909).

[7] Crocker, A.G., *Defects in crystalline media*, in: Nonlocal Theory of Material Media, Ed.: D. Rogula, CISM Courses and Lectures 268 (Springer 1982) 1.

[8] Crocker, A.G., Rogula, D. and Sztyren, M., *Bi-continuum model of twin boundary in bcc lattices*; to be published.

[9] Datta Gairola, B.K. and Kröner, E., *The nonlocal theory of elasticity and its application to interaction of point defects*, in: Nonlocal Theory of Material Systems, Ed.: D. Rogula (Ossolineum, Wroclaw, 1976).

[10] Edelen, D.G.B., *Nonlocal variations and local invariance of fields*, (American Elsevier, New York, 1969).

[11] Edelen, D.G.B., *Nonlocal variational mechanics*, IJES 7 (1969) 269,287,373,391,401,677,843 (1970); 13 (1975) 861.

[12] Edelen, G.E.B., *Invariance theory for nonlocal variational principles*, IJES 9 (1971) 741,801,819,921.

[13] Edelen, D.G.B., *A nonlocal variational formulation of the equations of radiative transport*, IJES 11 (1973) 1109.

[14] Edelen, D.G.B., *Irreversible thermodynamics of nonlocal systems*, IJES 12 (1974) 607.

[15] Edelen, D.G.B., *On compatibility conditions and stress boundary value problems in linear nonlocal elasticity*, IJES 13 (1975) 971.

[16] Edelen, D.G.B., *Theories with carrier fields: multiple interaction nonlocal formulation*, Arch. Mech. 28 (1976) 353.

[17] Edelen, D.G.B., *Nonlocal field theories*, in: Continuum Physics, (Academic Press 1976).

[18] Edelen, D.G.B., *A global formulation of continuum physics and the resulting equivalence classes of nonlocal field equations*, in: Nonlocal Theory of Material Systems, Ed.: D. Rogula (Ossolineum, Wroclaw, 1976).

[19] Edelen, D.G.B., Green, A.E. and Laws, N., *Nonlocal continuum mechanics*, ARMA 43 (1971) 36.

[20] Eringen, A.C., *Theory of micropolar continua*, Proc. 9th Midwestern Mech. Congress 1965, (Wiley 1967).

[21] Eringen, A.C., *Linear theory of micropolar elasticity*, J. Math. and Mech. 15 (1966) 909.

[22] Eringen, A.C., *Mechanics of micromorphic materials*, Proc. 11th International Congress of Applied Mechanics 1964 (Springer 1966).

[23] Eringen, A.C. and Suhubi, E.C., *Nonlinear theory of simple microelastic solids I, II*, IJES 2 (1964) 189,389.

[24] Eringen, A.C., *Theory of micropolar fluids*, J. Math. and Mech. 16 (1966) 1.

[25] Eringen, A.C., *Linear theory of micropolar elasticity*, IJES 5 (1967) 191.

[26] Eringen, A.C. and Edelen, D.G.B., *On nonlocal elasticity*, IJES 10 (1972) 233.

[27] Eringen, A.C., *Nonlocal polar elastic continua*, IJES 10 (1972) 1.

[28] Eringen, A.C., *Linear theory of nonlocal elasticity and dispersion of plane waves*, IJES 10 (1972) 425.

[29] Eringen, A.C., *Theory of nonlocal electromagnetic elastic solids*, J. Math. Phys. 14 (1973) 733.

[30] Eringen, A.C., *Foundations of micropolar thermoelasticity*, CISM Courses 27 (Springer 1970).

[31] Eringen, A.C., *Theory of nonlocal thermoelasticity*, IJES 12 (1974) 1063.

[32] Eringen, A.C. and Dixon, R.C., *A dynamical theory of polar elastic dielectrics*, IJES 3 (1965) 359.

[33] Eringen, A.C. and Kim, B.S., *On the problem of crack tip in nonlocal elasticity*, in: Continuum Mechanics Aspects of Geodynamics and Rock Fracture (1974).

[34] Gairola, B.K.D., *The nonlocal theory of elasticity and its application to interaction of point defects*, Arch. Mech. 28 (1976) 393.

[35] Gairola, B.K.D., *The nonlocal continuum theory of lattice defects*, in: Nonlocal Theory of Material Media, Ed.: D. Rogula, CISM Courses and Lectures 268 (Springer 1982) 52.

[36] Green, A.E. and Rivlin, R.S., *Multipolar continuum mechanics*, ARMA 17 (1964) 113.

[37] Holnicki-Szulc, J. and Rogula, D., *Nonlocal continuum models of large engineering structures*, Arch. Mech. 31 (1979) 793.

[38] Holnicki-Szulc, J. and Rogula, D., *Boundary problems in nonlocal continuum models of large engineering structures*, Arch. Mech. 31 (1979) 803.

[39] Hsieh, R.K.T., *Micropolarized and magneticed media*, CISM Courses and Lectures (World Scientific, 1982).

[40] Hsieh, R.K.T., Voros, G. and Kovacs, I., *Stationary lattice defects as sources of elastic singularities in micropolar media*, Physica 101B (1980) 201.

[41] Iesan, D., *Saint-Venants problem for inhomogeneous and anisotropic elastic solids with microstructure,* Arch. Mech. 29 (1977) 419.

[42] Iesan, D., *On the existence and uniqueness of the solutions of the dynamic theory of the linear elasticity with microstructure,* Bull. Acad. Pol. Sci., Ser. Techn. 22 (1974) 329.

[43] Kaliski, S., *Thermo-magneto-microelasticity,* Bull. Acad. Pol. Sci., Ser. Techn. 16 (1968) 7.

[44] Kaliski, S. and Nowacki, W., *Wave-type equations of thermo-magneto-microelasticity,* Bull. Acad. Pol. Sci., Ser. Techn. 18 (1970) 277.

[45] Kaliski, S., Plochocki, Z. and Rogula, D., *Asymmetric stress tensor and the angular momentum conservation law in the equations of combined mechanical and electromagnetic field in a continuous medium,* Proc. Vibr. Probl. 3 (1962) 253.

[46] Kapelewski, J. and Rogula, D., *Pseudocontinuum approach to the theory of interactions between impurity defects and crystal lattice,* Arch. Mech. 31 (1979) 27.

[47] Kotowski, R. and Rogula, D., *Differential pseudocontinua,* Arch. Mech. 31 (1979) 43.

[48] Kröner, E. and Datta, B.K., *Nichtlokale Elastostatik: Abteilung aus der Gittertheorie,* Z. f. Physik 196 (1966) 203.

[49] Kröner, E., *Elasticity theory of materials with long-range cohesive forces,* IJES 3 (1967) 731.

[50] Kröner, E., *Interrelations between various branches of continuum mechanics,* in: Mechanics of Generalized Continua, IUTAM Symp. 1968.

[51] Kröner, E., *The problem of non-locality in the mechanics of solids: review of present status,* in: Fundamental Aspects of Dislocation Theory, Ed.: J.A. Simmon (NBS Special Publ. II1970).

[52] Krumhansl, J.A., *Generalized continuum field representation for lattice vibrations,* in: Lattice Dynamics, Ed.: E. Wallis (Pergamon Press, 1965).

[53] Kunin, I.A., *Model of elastic medium simple structure with spatial dispersion,* Prikl. Math. Mech. 30 (1966) 942, in Russian.

[54] Kunin, I.A., *Theory of elasticity with spatial dispersion. One-dimensional complex structure,* Prikl. Math. Mech. 30 (1966) 866, in Russian.

[55] Kunin, I.A., *Inhomogeneous elastic medium with nonlocal interaction,* Prikl. Mat. Techn. Phys. 3 (1967) 60.

[56] Kunin, I.A., *Theory of elastic media with microstructure. Non-local theory of elasticity,* (Moskow 1975), in Russian.

[57] Kunin, I.A. and Vaisman, A.M., *On problems of the nonlocal theory of elasticity,* in: Fundamental Aspects of Dislocation Theory (NBS Spec. Publ. 1970).

[58] Minagawa, S., *Elastic fields of dislocations and disclinations in an isotropic micropolar continuum,* Lett. Appl. Engng. Sci. 5 (1977) 85.

[59] Mindlin, R.D., *A continuum theory of diatomic dielectrics,* IJES 8 (1972) 7.

[60] Mindlin, R.D. and Tiersten, H.F., *Effects of couple stresses in linear elasticity,* ARMA 11 (1962) 415.

[61] Nowacki, W., *Thermoelasticity,* (Pergamon Press 1962; Addison-Wesley 1962).

[62] Nowacki, W., *Theory of asymmetric elasticity,* (PWN, Warszawa 1970), in Polish.

[63] Nowacki, W., *Theory of micropolar elasticity,* CISM Courses and Lectures 25 (Springer 1970).

[64] Nowacki, W., *Zweidimensionale Probleme der Mikropolaren Elastostatik,* Z. Angw. Math. Mecy. 52 (1972) 268.

[65] Nowacki, W., *Three-dimensional problem of micropolar elasticity,* Bull. Acad. Pol. Sci., Ser. Techn. 22 (1974) 363.

[66] Nowacki, W., *Coupled fields in elasticity,* in: Trends in Applications of Pure Math. to Mechanics, Ed.: G. Fichera, (Pitmon Publ., London 1976).

[67] Nowacki, W., *Some problems of micropolar magneto-elasticity,* Proc. Vibr. Probl. 12 (1971) 105.

[68] Nowacki, W., *Green functions for micropolar thermoelasticity,* Bull. Acad. Pol. Sci., Ser. Techn. 16 (1968) 11.

[69] Nowacki, W., *Couple-stresses in the theory of thermoelasticity,* Proc. IUTAM Symp., Vienna 1966 (Springer 1968).

[70] Nowacki, J.P., *The linear theory of dislocations in Cosserat elastic continuum,* Bull. Acad. Pol. Sci., Ser. Techn. 22 (1974) 611.

[71] Nowacki, J.P., *Theory of disclinations in Cosserat media,* Arch. Mech. 29 (1977) 531.

[72] Olesiak, Z. and Wagrowska, M., *On continua with two kinds of yielding,* in: New Problems in Mechanics of Continua, Study No. 17 (Univ. of Waterloo Press, 1983) 281.

[73] Rogula, D., *Noether theorem for a continuous medium interacting with external fields,* Proc. Vibr. Probl. 7 (1966) 977, 377.

[74] Rogula, D., *Continuum models of structural media,* in: Study No. 12, Continuous Models of Discrete Systems, Proc. Symp. Mont Gabriel 1977, Ed.: J.W. Provan (Univ. of Waterloo Press, 1978).

[75] Rogula, D., *Large deformations of crystals, homotopy and defects,* in: Trends in Application of Pure Mathematical to Mechanics, Proc. Conf. Lecce 1975, Ed.: G. Fichera (Pitman Publ. London 1976).

[76] Rogula, D., *Influence of spatial acoustic dispersion on dynamical properties of dislocation, I, II,* Bull. Acad. Pol. Sci., Ser. Techn. 13 (1965) 337.

[77] Rogula, D., *On nonlocal continuum theories of elasticity,* Arch. Mech. 25 (1973) 233.

[78] Rogula, D., *Some basic solutions in strain-gradient elasticity of an arbitrary order,* Arch. Mech. 25 (1973) 43.

[79] Rogula, D., *Dislocation lines in nonlocal elastic continua,* Arch. Mech. 25 (1973) 967.

[80] Rogula, D., *Quasicontinuum theory of crystals,* Arch. Mech. 28 (1976) 563.

[81] Rogula, D., *Geometrical and dynamical nonlocality*, Arch. Mech. 31 (1979) 65.

[82] Rogula, D., *Introduction to nonlocal theory of material media*, in: Nonlocal Theory of Material Media, Ed.: D. Rogula, CISM Courses and Lectures 268 (Springer 1982) 125.

[83] Rogula, D., *Generalized interactions in nonlocal continua*, in: Continuum Models of Discrete Systems (Univ. of Waterloo Press 1980).

[84] Rogula, D., *Coupled fields and non-classical continua*, J. Techn. Phys. 23 (1982) 69.

[85] Rogula, D. and Kotowski, R., *On the correspondance between representations in the pseudocontinuum theory*, Bull. Acad. Pol. Sci., Ser. Techn. 26 (1978) 555.

[86] Rogula, D. and Sztyren, M., *On the one-dimensional models in nonlocal elasticity*, Bull. Acad. Pol. Sci., Ser. Techn. 26 (1978) 341.

[87] Rogula, D. and Sztyren, M., *Fundamental one-dimensional solutions in nonlocal elasticity*, Bull. Acad. Pol. Sci., Ser. Techn. 26 (1978) 417.

[88] Rogula, D. and Sztyren, M., *A field model of nonlocal heterogeneous material bodies*, to be published.

[89] Rymarz, Cz., *Continuous nonlocal models of a bounded elastic medium*, Proc. Vibr. Probl. 15 (1974) 283.

[90] Rymarz, Cz., *Boundary problems in the nonlocal theory*, Proc. Vibr. Probl. 15 (1974) 355.

[91] Rzewuski, J., *Field theory II*, (Iliffe Books Ltd., London, PWN Warsaw 1969).

[92] Rzewuski, J., *Nonlocal classical and quantum field theories*, in: Nonlocal Theory of Material Systems, Ed.: D. Rogula (Ossolineum, Wroclaw 1976).

[93] Sokolowski, M., *Theory of couple stresses in bodies with constrained relations*, CISM Courses 26 (Udine 1970).

[94] Stojanovic, R., *Mechanics of polar continua*, CISM Courses (Udine 1970).

[95] Sztyren, M., *On the boundary forces in a solvable integral model of nonlocal elastic half-space*, Bull. Acad. Pol. Sci., Ser. Techn. 26 (1978) 537.

[96] Sztyren, M., *Boundary value problems and surface forces for integral models of nonlocal elastic bodies*, Bull. Acad. Pol. Sci., Ser. Techn. 27 (1979) 327.

[97] Sztyren, M., *Boundary value problems and surface forces for integro-differential models of nonlocal elastic bodies*, Bull. Acad. Pol. Sci., Ser. Techn. 27 (1979) 335.

[98] Sztyren, M., *On nonlocal boundary value problems*, in: CMDS3 Proc. (Univ. of Waterloo Press 1980).

[99] Sztyren, M., *On solvable nonlocal boundary value problems*, in: Nonlocal Theory of Material Media, Ed.: D. Rogula, CISM Courses and Lectures 268 (Springer 1982) 224.

[100] Tiersten, H.F., *On the mechanics of interpenetrating solid continua*, in: Continuum Models of Discrete Systems, Study No. 12 (Univ. of Waterloo Press 1978).

[101] Toupin, R.A., *Elastic materials with couple stresses*, ARMA 11 (1962) 385.

[102] Toupin, R.A., *Theories of elasticity with couple-stress*, ARMA 17 (1964) 85.

[103] Weysenhoff, J.W. and Raabe, A., *Relativistic dynamics of spin particles*, Acta Phys. Polon. 9 (1947).

[104] Wozniak, Cz., *On the nonlocal effects in continuum mechanics due to internal constraints*, in: Nonlocal Theory of Material Systems, Ed.: D. Rogula (Ossolineum, Wroclaw 1976).

[105] Zorski, H., Rogula, D. and Rymarz, Cz., *Nonlocal continuum models of discrete systems*, Advances in Mechanics 1 (1979) in Russian.

[106] Zorski, H., *Non-existence of a continuum that models a Newtonian system of interacting particles*, ARMA 56 (1974) 320.

[107] Zelazny, R., *Derivation of hydrodynamic equations of quantum system of diatomic molecules*, Phys. Rev. 117 (1960) 1.

Theoretical and Applied Mechanics
F.I. Niordson and N. Olhoff
Elsevier Science Publishers B.V. (North-Holland)

THREE DIMENSIONAL STABILITY AND BIFURCATION OF STEADY WATER WAVES

P.G. Saffman

Applied Mathematics 217-50
California Institute of Technology
Pasadena, California 91125 U.S.A.

The stability of finite amplitude two-dimensional waves of permanent form on deep water to arbitrary three-dimensional disturbances is considered. The possibility of instability is related to the existence of resonances between linear waves. The regions of instability are calculated. It is shown that spontaneous three-dimensional waves of permanent form can arise, from bifurcation of two-dimensional waves at wave heights and wave lengths for which stationary disturbances exist, and their properties are determined.

INTRODUCTION

Finite amplitude water waves have been an object of scientific study by mathematicians and physicists for over 130 years since the pioneering work of C.G. Stokes in 1847. A vast body of mathematical theory has been constructed, and in fact problems of water waves appear to have been responsible for many of the fundamental techniques developed in applied mathematics and theoretical mechanics, from the method of stationary phase developed by Lord Kelvin to the ingenious techniques (such as inverse scattering, etc) for solving exactly certain nonlinear partial differential equations. In recent years, water waves have proved a challenge to computing techniques, and were the field of application of some of the first uses in continuum mechanics of ideas like Pade approximants. The mathematical richness of the field may seem surprising at first sight, since the governing differential equation of the classical problem is just Laplace's equation, but of course the boundary conditions are nonlinear, and the position of the boundary is one of the unknowns, and this is the source of the wealth of problems. It is also remarkable that in a classical field as well studied as water waves, new physical phenomena are still being discovered, both theoretically and experimentally, and many open questions remain.

For the particular topic of two-dimensional periodic, irrotational surface waves of permanent form propagating under the effect of gravity or surface tension on water of infinite or finite depth, major advances were made in the last two decades by a powerful combination of analytical and numerical techniques. Quite apart from the mathematical questions of the rigorous existence of finite amplitude waves of permanent form (which has attracted and still attracts pure mathematicians) new phenomena of physical interest have been discovered and investigated, and old phenomena have been reinterpreted in new ways.

For example, instability of steady finite amplitude waves to long wave two-dimensional disturbances was predicted by Lighthill (1965) by the use of Whitham's variational method and an approximate Lagrangian. Using Hamiltonian methods, Zakharov (1966,1968) showed that weakly nonlinear gravity waves are unstable for modulations longer than a critical wavelength depending upon the waveheight. Both two- and three-dimensional disturbances were considered. Benjamin & Feir (1967) examined the case of two-dimensional disturbances to weakly non-linear waves employing standard perturbation methods and found instability to two-dimensional disturbances of sufficiently long wavelength. Using the numerical results of Schwartz (1974) for finite amplitude Stokes waves, Longuet-Higgins (1978a,b) investigated the stability of finite amplitude water waves to superharmonic and subharmonic two-dimensional disturbances. This work extended the results of Zakharov and Benjamin & Feir to finite amplitude waves and disturbances of shorter wavelength. It confirmed Lighthill's prediction that the long wave instability disappears when the wave is steep, and gave growth rate values that agree well with the observations of Benjamin & Feir (see Benjamin 1967) and Lake & Yuen (1977). Longuet-Higgins also discovered that when the wave is steep two-dimensional subharmonic disturbances of twice the wavelength of the undisturbed wave are unstable and have growth rates larger than the Lighthill, Zakharov and Benjamin & Feir (LZBF) type.

A second area in which new and at first sight surprising results were obtained in the last five years is the question of the uniqueness of gravity waves of permanent form. Mathematicians agonized for many years over the question of existence, and indeed Lord Rayleigh (1917) in one of his last papers tried to help resolve the issue by calculating terms in the expansion up to and including $O(h/\lambda)^7$. The work of Nekrasov and Levi-Civita in the early 1920's, who proved rigorously existence and uniqueness for sufficiently small two-dimensional waves, put this question to rest, and it appears to have been felt that proving existence and uniqueness for all heights up to the maximum, and proving existence of the highest (120° cusped) two-dimensional wave was only a technical, although probably very hard, problem. Existence has, in fact, now been proved for waves of all heights up to and including the maximum, although it has not yet been shown that the solutions form a continuous family, and the proof of the existence of the extreme wave was achieved only in the last few years (Amick, Fraenkel & Toland 1982). On the other hand, despite the belief that waves are unique, it was not possible to prove this for finite amplitude, except that Garabedian (1965) employed Steiner symmetrization to show that steady periodic gravity waves are unique if their crests and troughs are all the same height. This is just as well, since Chen & Saffman (1980) presented surprising but convincing numerical evidence (subsequently confirmed by other workers and now established beyond reasonable doubt although still awaiting mathematical proof) that sufficiently steep gravity waves are not unique, in the sense that given the spatial period and height (defined as the vertical distance between the highest crest and lowest trough) many different waves may exist with unequal crests and troughs. This phenomenon had already been known for gravity-capillary waves (Wilton 1915), where both surface tension and gravity provide the restoring force, but there a simple explanation exists in terms of resonance where a wave and its first harmonic travel at the same speed. There is no similar explanation known for the pure gravity wave case, and a simple physical explanation is still lacking here. The non-uniqueness of

Wilton's ripples was explained in terms of a bifurcation phenomenon by Chen & Saffman (1979), who also uncovered an extremely rich structure of steady solutions for gravity-capillary waves.

Following these and other advances in two-dimensional water waves, interest has been directed to three-dimensional effects, where it might be anticipated that there is an even greater richness of phenomena. A review of progress here is the purpose of this report. It is now appropriate to introduce an important distinction in three-dimensional phenomena between forced and spontaneous three-dimensionality. The former is where the dependence upon the second horizontal dimension is forced by boundary or initial conditions. This is the case, for example, when one studies the effect of nonlinearity on the interaction of two equal but non-parallel wave trains or the reflection of an obliquely incident wave train on a wall. A special case when the waves are at 180° to each other is the two-dimensional standing wave. A mathematical description of forced three-dimensional waves is to say that they are essentially superharmonic modifications. The basic linear state is

$$\eta(x, z, t) = a \cos(kx \cos\theta + kz \sin\theta - \omega t) + a \cos(kx \cos\theta - kz \sin\theta - \omega t) \tag{1.1}$$

where $\omega = \Omega(k)$ is the linear dispersion relation for waves of wavenumber k and the two wave trains make an angle 2θ with each other. The case $\theta = 0$ is the limit of two-dimensional propagating waves, and the case $\theta = \frac{1}{2}\pi$ is a two-dimensional standing wave. The finite amplitude forced three-dimensional steadily propagating wave of permanent form is a solution of the form

$$\eta(x, z, t) = \sum_0^\infty \sum_0^\infty a_{m,n} \cos[m k \cos\theta\,(x-ct)] \cos[n k \sin\theta\, z] \tag{1.2}$$

where c is the unknown propagation speed of the wave pattern. These wave patterns go under the name of short crested waves. The first calculations of such waves appear to have been made by Fuchs (1952) and Chappelear (1961). The latest work is by Roberts (1983). A significant property of these waves is that they exist in the infinitesimal limit, and can therefore be calculated formally by expansions in powers of wave height h. There are, however, serious doubts whether such expansions converge because of resonances or the small divisor problem and the existence of steady short crested waves is an open question. We shall say more about this difficulty later. The standing wave case ($\cos\theta \to 0$ with $c \cos\theta$ finite) was treated by Penney & Price (1952) and more recently by Schwartz & Whitney (1981). These expansions also suffer from the same existence uncertainties.

The spontaneous three-dimensional waves are of a different origin. They arise from instabilities or bifurcations of a uniform two-dimensional wave train, and in general they cannot be of arbitrarily small amplitude. Mathematically, they can be described or interpreted as subharmonic bifurcations, where a two-dimensional wave of wavelength $2\pi/k$

$$\bar{\eta} = \sum_0^\infty a_\ell \cos k\ell\,(x-ct) \tag{1.3}$$

bifurcates at a critical height into a steadily propagating three dimensional wave of the form

$$\eta = \bar{\eta} + \sum_{0}^{\infty} \sum_{-\infty}^{\infty} \sum_{-\infty}^{\infty} A_{\ell,m,n} \cos\left[(\ell+mp)k(x-ct) + knqz\right] \tag{1.4}$$

where p and q are arbitrary real numbers with $0 < p < 1$. If p is an integer, these would be short crested waves. The critical wave height at which bifurcation occurs depends upon the values of p and q. The surface given by (1.4) is periodic in the transverse direction with wavelength $2\pi/kq$. The longitudinal variation can be thought of as a modulation of the Stokes wave (1.3) by a subharmonic component of wavelength $2\pi/kp$, and is not exactly periodic unless p is rational. The particular value $p = \frac{1}{2}$ corresponds to a doubling of the wavelength in the direction of propagation.

The existence of such waves was first demonstrated by Saffman & Yuen (1980a) for general nonlinear dispersive systems where the nonlinearity is described by four wave interactions and calculated in detail for water waves by Saffman & Yuen (1980b) using the Zakharov equation model. They pointed out that if the medium is isotropic, the bifurcation is degenerate and the solutions on the new branches can be either skew or symmetric about the direction of propagation. In the skew case,

$$A_{\ell,m,n} \neq A_{\ell,m,-n} \tag{1.5}$$

and the wave surface is not symmetric about the direction of propagation. In the symmetric case equality holds, the surface is symmetric about the direction of propagation as in the short crested wave case, and it is described by

$$\eta = \bar{\eta} + \sum_{0}^{\infty} \sum_{-\infty}^{\infty} \sum_{0}^{\infty} A_{\ell,m,n} \cos\left[(\ell+mp)k(x-ct)\right] \cos knqz \tag{1.6}$$

Solutions of this type have been calculated by Meiron, Saffman & Yuen (1982) for gravity waves and by Chen & Saffman (1984) for both capillary and gravity waves using the exact water wave equations.

In the present report, we shall describe the ideas leading to the prediction of spontaneously generated three dimensional gravity waves, discuss some of their properties and list some of the open questions and puzzles. Only a small part of their properties has so far been uncovered, and there is no doubt that the study of three-dimensional waves will be one of the leading areas of research in the next decade. It is worth emphasizing that while the initial theoretical investigation was under way, these waves were discovered and measured experimentally by Ming-Yang Su and colleagues (Su 1982, Su et al 1982) simultaneously and completely independently. Further experimental investigations are described by Melville (1982).

INSTABILITY OF FINITE AMPLITUDE WATER WAVES

We suppose we have a steady water wave on deep water with elevation given by (1.3) and we examine the stability of this flow to general infinitesimal disturbances. Note that since the governing equations are supposed to be the non-dissipative Euler equations, the wave is said to be stable if the disturbance does not grow. The behavior of finite amplitude disturbances is unknown. For simplicity, scales are now normalised so that the wavelength of the undisturbed wave is 2π and $k = 1$. Note also that although we shall only discuss the infinite depth case, there are no difficulties in treating the case of finite depth. The coefficients a_n and wave speed c of the undisturbed wave are supposed to be known functions of the wave height h. The fundamental work and discovery of the new phenomena were carried out by McLean, Ma, Martin, Saffman & Yuen (1981), hereafter referred to as MMMSY, who identified the instabilities as being a parametric instability, with further details supplied by McLean (1982a,b), although it became clear in retrospect that the possibility of the MMMSY type of instability had been suggested by Zakharov (1968) by a resonance argument. There is no known way, however, of demonstrating the actual existence of the instability by general analytic arguments and to date heavy numerical calculations are needed to establish that the resonance is actually an instability when the wave is of finite height. Indeed, although the resonances always give finite amplitude instabilities for gravity waves, the opposite is the case for capillary waves, where the resonances are in general stable, even though the general argument has equal validity.

We move into a coordinate system moving with the undisturbed wave, by writing $x' = x-ct$, and for sake of clarity drop the prime. It follows from Floquet or Bloch wave theory that the general disturbance to the surface of the wave is a sum of terms of the form

$$\eta' = e^{ipx}\, e^{iqz}\, e^{-i\sigma t} \sum_{-\infty}^{\infty} a'_m\, e^{imx} \qquad (2.1)$$

The expression (2.1) is the eigenfunction, and has the form of a periodic function of x, with the same wavelength as the undisturbed wave, multiplied by an oscillatory function of x with wavelength $2\pi/p$, an oscillatory function of the transverse coordinate z with transverse wavelength $2\pi/q$, and an exponential function of time. The numbers p and q are arbitrary real numbers, and σ is an unknown, possibly complex, eigenvalue to be determined by the requirement that the linearized equation for η' and the corresponding disturbance to the velocity potential have a non-trivial solution. For details of the actual equations that η' satisfies, see McLean (1982a) or Chen & Saffman (1984). If σ is real, then the disturbance is said to be stable; if σ is complex, then the disturbance or its complex conjugate is unstable. If σ is real for all values of p and q, the wave is stable. As will be shown, this is never the case, at least for gravity waves and capillary waves on deep water. It should be noted that the expression (2.1) for the eigenfunction is degenerate with regard to the dependence upon p. Clearly, p can be changed by an integer M without changing the eigenfunction if the coefficients a'_m are relabelled $p \to p+M$, $a'_m \to a'_{m-M}$. Thus there would be no loss of generality in restricting p to be in the range $0 \le p \le 1$. However, it proves convenient not to remove the degeneracy and allow p to be quite general. If p is an integer, then the wavelength of the disturbance is the same as that of the undisturbed wave and the disturbance can be called superharmonic. When p is not an integer, then in general the disturbance contains components with a wavelength greater than 2π, and is called subharmonic. It will be seen that the value $p = \frac{1}{2}$ is especially significant; this corresponds to a doubling of the wavelength in which every other crest, say, will be the same height. Because the equations are analytic, it follows that the eigenfunctions will be analytic functions of the disturbance wave numbers p and q, and the wave height h, i.e. $\sigma = \sigma(p,q,h)$.

In the limit $h = 0$, we can write down explicitly the eigenvalues and eigenfunctions, since they are nothing more than the infinitesimal waves on a flat surface with the doppler shift due to the undisturbed flow taken into account. Thus, we have pairs of eigenvalues and eigenvectors

$$\sigma_M^+ (p, q, 0) = \Omega\left[\{(p+M)^2 + q^2\}^{\frac{1}{2}}\right] - (p+M)c_o \quad , \quad a'_m = \delta_{mM}$$

$$\sigma_M^- (p, q, 0) = -\Omega\left[\{(p+M)^2 + q^2\}^{\frac{1}{2}}\right] - (p+M)c_o \quad , \quad a'_m = \delta_{mM} \tag{2.2}$$

where M is a mode number, the ± designates copropagating and counterpropagating relative to the undisturbed flow, and $c_o = \Omega(1)$. For gravity waves, with units chosen so that $g = 1$, $\Omega(k) = k^{\frac{1}{2}}$. For capillary waves, with units chosen so that the coefficient of surface tension is one, $\Omega(k) = k^{3/2}$ These eigenvalues are all real.

To have instability for finite h, we need a complex eigenvalue and its complex conjugate. Since they are analytic functions of h,p and q, this can only happen if two real ones coalesce, i.e. there is a degeneracy. For small h, instability may then happen for modes with p, q values near the points (p,q) where two of the eigenvalues for zero amplitude are equal:

$$\sigma_{M_1}^{s_1} (p, q, 0) = \sigma_{M_2}^{s_2} (p, q, 0) \tag{2.3}$$

where the mode numbers and directions of propagation are not all the same. If the equations (2.3) have real solutions, they provide curves in p,q space which may separate for finite h into the boundaries of regions of instability separating those values of p and q for each h for which disturbances are unstable from those for which disturbances are stable.

It is convenient to separate the possible solutions into two classes, depending upon whether $M_1 - M_2$ is even or odd. The former were called class I by MMMSY, and the latter class II. Because of the arbitrariness in p, we can suppose that $M_1 = -M_2 = M > 0$ for degeneracies of class I type and $M_1 = M$, $M_2 = -M-1$ for class II. The zero amplitude degeneracies are as follows.

For gravity waves, it follows from simple algebra that degeneracy can occur only if S_1 and S_2 are of opposite sign, i.e. the waves are propagating in opposite directions and $M \geq 1$. The class I curves are closed ovals symmetrical about the q-axis, except for $M = 1$ when the resonance curve is a figure of eight. The class II curves are closed ovals symmetrical about $p = \frac{1}{2}$. All curves are symmetrical about $q = 0$.

For capillary waves, the situation is more complicated. For $M \geq 1$ there is degeneracy only if S_1 and S_2 are the same sign, i.e. the waves are copropagating. The loci are also unbounded. For $M = 0$ and $M = 1$, the origin is, however, an exceptional point. Also, there is one closed degeneracy curve for class II, $M = 0$ and counterpropagating waves.

An alternative representation of the degeneracies can be given in terms of resonance conditions. Equations (2.2) can be expressed as

$$\underset{\sim}{k}_1 = \underset{\sim}{k}_2 + N\underset{\sim}{k} \;, \qquad \omega_1 = \omega_2 + N\omega \;, \qquad N = M_1 - M_2 \tag{2.4}$$

which shows that the degeneracy is equivalent to a resonance between two waves and the undisturbed motion, where k_1, k_2 and k are the wave numbers of the perturbations and the undisturbed wave, and the ω are their frequencies relative to fixed coordinates. If finite amplitude of the undisturbed wave makes the resonance unstable, we therefore expect that the growth rate should be proportional to the amplitude raised to the power $M_1 - M_2$, at least for small amplitude. It is found that this is generally true, but appears to be an overestimate for the special case of two-dimensional disturbances.

The question now is the effect of finite amplitude upon the degenerate frequencies. There are two possibilities. Either the frequencies remain real but are perturbed, or they acquire small real parts and the disturbance is unstable. In principle, perturbation methods can be applied, but the algebra is ferocious and no satisfactory way of overcoming the complexity has yet been discovered, except for the long wave LZBF instability which can be handled easily by using the cubic Schrdinger equation. However, numerical methods prove to be feasible and have been employed. These, of course, have the advantage that they are not limited to small h.

Some results for gravity waves are shown in figure 1. In this case, all degeneracies appear to become unstable at finite h and bands of instability in the p,q plane appear. We show in the figure instability bands for the class I and II cases for $M = 1$ for $h = 0.60$ For smaller values of h, the class I instability is dominant and is predominantly two dimensional. For larger values of h, the class II dominates and is predominantly three dimensional. The class I actually disappears as h increases and the values of q for which class II exists decreases. For $h = 0.823$, the class II is unstable for $q = 0$. This new relatively large two-dimensional instability was first discovered by Longuet-Higgins (1978b). But the largest growing instability is always three dimensional and occurs for $p = \frac{1}{2}$, i.e. is wavelength doubling. The dashed lines in the figure are the loci of stationary disturbances with $\sigma = 0$.

These are significant for bifurcation into new classes of steady solutions.

For capillary waves, the situation is rather different. Now, all the degeneracies remain stable at finite amplitude, except for the special case of degeneracy at the origin in p,q space and the oval for class II, M = 0. The instability bands are shown in figure 2. Now the most unstable disturbance is always three-dimensional and again occurs for $p = \frac{1}{2}$. This wavelength doubling instability dominates the LZBF long wave instability which is now confined to a region around the origin and does not extend to shorter wavelengths as for gravity waves.

For further details, see McLean (1980a,b) and Chen & Saffman (1984).

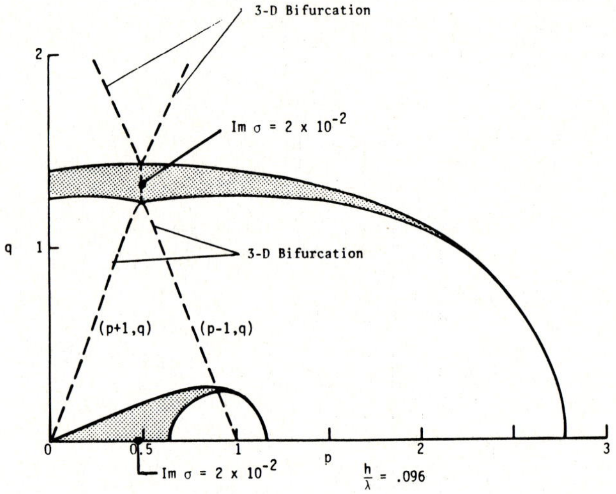

Figure 1.
M = 1 instability bands for class I and class II disturbances for h = 0.60. Dots show values of p and q for maximum growth rate. Dashed lines are loci of stationary disturbances.

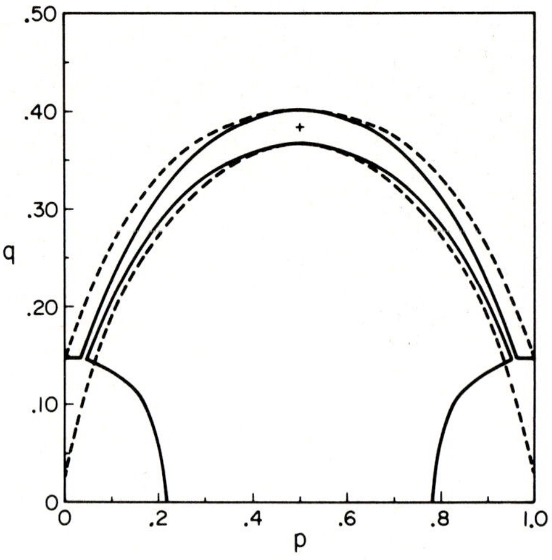

Figure 2.
Instability bands for capillary waves for h = 0.75. Cross marks most unstable disturbance. Dashed lines are loci of stationary disturbances.

SPONTANEOUS THREE DIMENSIONAL WAVES OF PERMANENT FORM

The results summarised in the previous section show that uniform wave trains of gravity and capillary waves are unstable. Gravity waves are most unstable to wavelength doubling three-dimensional instabilities when they are fairly steep, and capillary waves are most unstable to these disturbances for all heights. An open question of great interest is the long time evolution of the instability. A first theoretical step towards understanding the behavior of unstable wave trains and a question of interest in its own right is the investigation of three-dimensional steady states associated with the MMMSY instability. These can arise spontaneously and continuously as bifurcations from the uniform wave train when a stationary infinitesimal disturbance can be superposed on a steady system. For a given value of h, this is possible on an infinite number of loci in the p,q plane. Examples are shown by dashed lines in figures 1 and 2.

Consider a slightly three-dimensional steady wave whose surface in the frame of reference moving with the wave is of the form

$$\eta(x, z) = \bar{\eta}(x) + \varepsilon\, \eta'(x, z) + O(\varepsilon^2) \tag{3.1}$$

where η' is an eigenfunction corresponding to a stationary disturbance and has the form

$$\eta' = \sum_{-\infty}^{\infty} \alpha_j \cos\{(j+p)x\} \cos qy + \sum_{-\infty}^{\infty} \beta_j \sin\{(j+p)x\} \cos qy \tag{3.2}$$

In general, combinations can be chosen which describe waves that are symmetric or skew symmetric about the direction of propagation. Skew solutions were calculated by Saffman & Yuen (1980b) using the Zakharov equation approximation which is valid for small p.

The case $p = \frac{1}{2}$ is special and of particular interest. First, it gives waves of the same form as the most unstable three-dimensional disturbance and second the symmetry associated with this case allows only symmetrical stationary eigenfunctions. The allowable forms of three-dimensional wavelength doubling steady waves are of the form in bottom fixed coordinates

$$\eta = \bar{\eta}(x-ct) + \sum_{0}^{\infty} \sum_{0}^{\infty} B_{m,n} \cos \tfrac{1}{2}m(x-ct) \cos nqz \tag{3.3}$$

or

$$\eta = \bar{\eta}(x-ct) + \sum_{0}^{\infty} \sum_{0}^{\infty} B_{m,n} \sin \tfrac{1}{2}m(x-ct) \cos nqz \tag{3.4}$$

where q can be supposed given and the coefficients and wave speed c are to be determined. The form (3.3) gives waves which are symmetrical in the direction of propagation about the crests. The waves described by (3.4) have symmetry about troughs. It appears from figures 1 and 2 that for $p = \frac{1}{2}$ there are two values of q for infinitesimal bifurcation where the dashed lines of stationary disturbances meet the stability boundaries. The crest symmetric waves (3.3) are associated with bifurcation from the smaller value of q; the trough symmetric waves (3.4) come from the larger value of q. For detailed reasons, see Chen & Saffman (1984)

The coefficients have been calculated numerically using a collocation method for crest symmetric gravity waves by Meiron, Saffman & Yuen (1982) and by Chen & Saffman (1984) for trough symmetric gravity waves and both categories of capillary waves. A typical wave profile is shown in figure 3. This is for a crest symmetric gravity wave.

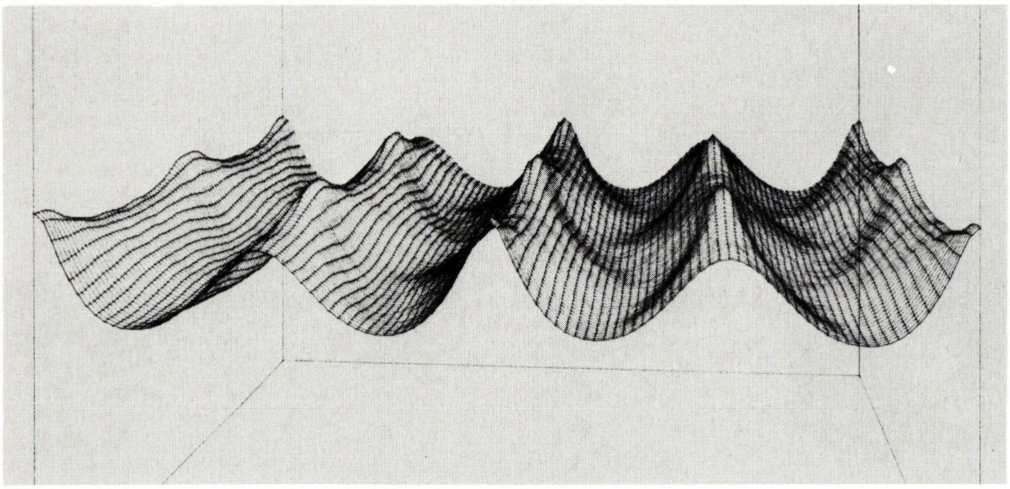

Figure 3.
Perspective plot of three-dimensional crest symmetric gravity wave.

A remarkable feature of the results is the behavior of the total energy of the three-dimensional waves (kinetic + potential) as a function of wave height. For capillary waves, the energy is a monotonically increasing function of the waveheight. In contrast, the energy of three-dimensional gravity waves is a decreasing function of wave height and is less than the energy of a two-dimensional wave of the same height. This suggests that three-dimensional gravity waves are favored by the processes of dissipation and that in a real fluid, where some loss of energy must occur, the three-dimensional components will grow together with the waveheight. This could continue indefinitely, according to the results, but presumably the wave will eventually reach a limiting height at which it breaks and energy dissipation increases. It is not possible to predict at present what will occur, but a plausible speculation is that the wave will 'snap back' into a two-dimensional form. According to the experiments of Su et al (1982), this is exactly what happens. These experiments show that a uniform wave train with h about 0.62 develops three-dimensional instabilities and forms a three-dimensional wave after propagating a distance about 25 wavelengths. Waves of this height have an e-folding time of about 7 periods, so the time for the appearance of the three-dimensional wave is consistent with the three-dimensional growth rates of the MMMSY instability. In 90% of the cases, a quasi-steady three-dimensional wave formed corresponding quite well with the predictions of the steady state three-dimensional calculation, except that the wave crest showed continuous spilling. This, however, is consistent with the idea that energy dissipation causes the waves to steepen and break. With the reasonable speculation that there is a minimum energy below which the wave cannot exist in a three-dimensional state, it is understandable that they will disappear and two-dimensional waves will reappear as seen in the experiment. In 9% of the experiments, bifurcation corresponding to wavelength tripling was seen and in the remaining 1% of the time wavelength quadrupling occurred. The reasons for this are unknown.

An interesting open question is the limiting form of the three-dimensional wave solutions for inviscid fluid. Here, we are not concerned with the problem of proving existence which may be of mathematical interest, but with the real physical problem of the limiting shape. There are two main possibilities for gravity waves. First, it is possible that the wave of greatest height is cusped along a fraction of the crest, tending smoothly to the limiting Stokes wave as $q \to 0$, or second it may be coned with a singularity at only one point. For capillary waves, the limiting wave is probably associated with the surface crossing itself, as for two-dimensional waves.

It is now appropriate to mention a serious mathematical difficulty about the steady three-dimensional waves. The calculations assume that the solutions form a continuous family of isolated solutions, i.e. given q and h, only one solution exists or if there is more than one they are distinct and not arbitrarily close. The problem is that the null space of the wave system at the bifurcation point may have infinite dimension; or expressed physically there may be infinitely many stationary infinitesimal modes on the undisturbed wave of the form

$$\eta' = e^{\frac{1}{2}iM(x-ct)+iNqz} \sum_{-\infty}^{\infty} a'_m \, e^{im(x-ct)} \tag{3.5}$$

We are unable to prove that this is the case, but the analysis for small h suggests that this is likely. If $h \sim 0$, then $c \sim 1$ and for gravity waves $q \sim \surd(45/16)$. Then (3.5) is a solution with $a'_m = 0$ unless $m = 0$ if

$$\frac{45}{4} N^2 + M^2 = \frac{1}{4} M^4 \tag{3.6}$$

This equation has the solution $N = 1$ and $M = 3$, which gives the solution branch calculated by Meiron, Saffman & Yuen. Another solution is $N = 7$ and $M = 7$, and it can be shown (C. Hindman, private communication) that there are infinitely many solutions. The assumption of isolated families is therefore suspicious. However these other values lie outside the truncation range of the calculations and there was no sign in these numerical investigations of any trouble. It may be that for finite h and general irrational q the degeneracy is limited to very large values of N and M so that numerical calculations are little affected and comparisons with experiment are likewise unaltered because in reality viscosity will damp large wave number components. But the infinite degeneracy raises some curious mathematical questions about the uniqueness and existence of solutions which seem to be outside the range of present mathematical understanding and techniques.

This difficulty is, of course, present for the forced short crested waves, where it appears to be more serious since there is no doubt about the infinite degeneracy of the flat surface. In this case, modes with longitudinal wave numbers p and Np, and transverse wave numbers q and Mq, respectively, will be stationary relative to each other in the gravity wave case if

$$M^2p^2 + N^2q^2 = M^4(p^2+q^2) \qquad (3.7)$$

For given p and q, this equation will either be satisfied exactly for infinitely many M and N, or satisfied infinitely closely infinitely often. This means that a power series expansion of short crested waves in powers of wave height will have zero radius of convergence. We repeat, however, that our approach to spontaneous three-dimensional waves is not an expansion in wave height, but the solution by truncation of an infinite system, and the solution may therefore make sense even if not isolated. Clearly this is a challenging problem.

REFERENCES

Amick, C.J., Fraenkel, L.E. & Toland, J.F. On the Stokes conjecture for the wave of extreme form, Acta Math. 148 (1982) 193-214.
Benjamin, T.B. Instability of periodic wavetrains in nonlinear dispersive systems, Proc. Roy. Soc. A299 (1967) 59-75.
Benjamin, T.B. and Feir, J.E. The disintegration of wavetrains on deep water. Part 1. Theory, J. Fluid Mech. 27 (1967) 417-430.
Chappelear, J.E. On the description of short crested waves, Core of Engineers U.S. Army. Tech. Memo. no. 125 (1961)
Chen, B. and Saffman, P.G. Steady gravity-capillary waves on deep water I. Weakly nonlinear waves, Stud. Appl. Math. 60 (1979) 183-210.
Chen, B. and Saffman, P.G. Numerical evidence for the existence of new types of gravity waves of permanent form on deep water, Stud. Appl. Math. 62 (1980) 1-21.
Chen, B. and Saffman, P.G. Three dimensional stability and bifurcation of capillary and gravity waves on deep water, Stud. Appl. Math. (1984) to appear.
Fuchs, R.A. On the theory of short-crested oscillatory waves, Gravity waves, Natl. Bur. Standards Circ. 521 (1951) 187-200.
Garabedian, P.R. Surface waves of finite depth, J. D'Anal. Math. 14 (1965) 161-169.
Lake, B.M. and Yuen, H.C. A note on some nonlinear water wave experiments and the comparison of data with theory, J. Fluid Mech. 83 (1977) 75-81.
Lighthill, M.J. Contributions to the theory of waves in nonlinear dispersive systems, J. Inst. Math. Appl. 1 (1965) 269-306.
Longuet-Higgins, M.S. The instabilities of gravity waves of finite amplitude in deep water I. Superharmonics, Proc. Roy. Soc. A360 (1978a) 471-488.
Longuet-Higgins, M.S. The instabilities of gravity waves of finite amplitude in deep water II. Subharmonics, Proc. Roy. Soc. A360 (1978b) 489-505.
McLean, J.W. Instabilities of finite-amplitude water waves, J. Fluid Mech. 114 (1982a) 315-330.
McLean, J.W. Instabilities of finite-amplitude gravity waves on water of finite depth, J. Fluid Mech. 114 (1982b) 331-341.
McLean, J.W., Ma, Y.C., Martin, D.U., Saffman, P.G. and Yuen, H.C. Three-dimensional instability of finite amplitude water waves, Phys. Rev. Lett. 46 (1981) 817-820.

Meiron, D.I., Saffman P.G. and Yuen, H.C. Calculation of steady three-dimensional deep water waves, J. Fluid Mech. 124:109-121 (1982) 109-121.
Melville, W.K.The instability and breaking of deep-water waves, J. Fluid Mech. 115 (1982) 165-185.
Penney, W.G. & Price, A.J. Finite periodic stationary gravity waves in a perfect liquid, Phil. Trans. A 224 (1952) 254-284.
Rayleigh, Lord On periodic irrotational waves at the surface of deep water, Phil Mag. 23 (1917) 381-389.
Roberts, A.J. 1983 Highly nonlinear short-crested water waves, J. Fluid Mech. 135 (1983) 301-321.
Saffman, P.G. & Yuen, H.C. Bifurcation and symmetry breaking in nonlinear dispersive waves, Phys. Rev. Let. 44 (1980a) 1097-1100.
Saffman, P.G. and Yuen, H.C. A new type of three-dimensional deep water wave of permanent form, J. Fluid Mech. 101 (1980b) 797-808.
Schwartz, L.W. Computer extension and analytic continuation of Stokes Stokes expansion for gravity waves, J. Fluid Mech. 62 (1974) 553-478.
Schwartz, L. & Whitney, A.K. A semi-analytic solution for nonlinear standing waves in deep water, J. Fluid Mech. 107 (1981) 147-171.
Su, M-Y. Three-dimensional deep-water waves. Part 1. Experimental measurement of skew and symmetric wave patterns, J. Fluid Mech. 124 (1982) 73-108.
Su, M-Y., Bergin, M., Marler, P. & Myrick, R. Experiments on nonlinear instabilities and evolution of steep gravity-wave trains, J. Fluid Mech. 124 (1982) 45-72.
Wilton, J.R. On ripples, Phil. Mag. 29 (1915) 688-700.
Zakharov, V.E. The instability of waves in nonlinear dispersive media, Sov. Phys. J.E.T.P. 51 (1966) 1107-1114.
Zakharov,V.E. Stability of periodic waves of finite amplitude on the surface of a deep fluid, J. Appl. Mech. Tech. Phys. 2 (1968) 190-194.

ACKNOWLEDGEMENT

The author's work was supported by the Office of Naval Research and the National Science Foundation.

Theoretical and Applied Mechanics
F.I. Niordson and N. Olhoff
Elsevier Science Publishers B.V. (North-Holland)

YIELD-STRENGTH OF ANISOTROPIC SOILS

Jean Salençon

Ecole Polytechnique, Ecole Nationale des Ponts et Chaussées
Laboratoire de Mécanique des Solides
91128 Palaiseau Cedex - France

Natural soils often exhibit a significant mechanical anisotropy. Taking it into account in structure stability analyses requires defining the procedure and the interpretation of experimental tests properly, describing the strength-characteristics of the soil through a mechanically correct criterion, and devising adequate stability analysis methods. Soil mechanical reinforcement has been developped and is now commonly used : the composite material so obtained at the structure level can be modelized as a homogeneous continuum with an anisotropic yield-strength, allowing anisotropic stability analysis methods to be used.

1 - THE ORIGIN OF A CONCEPT

Since he started to improve his environment, man has come up against the problem of the stability of the structures he did build on the ground, sometimes with the very soil of that ground or with borrow material (buildings, bridges, dikes, dams, ...). He has also been concerned with the safety of the natural relief (e.g. slopes, embankments, ...) and with that of mines, pits and quarries he did develop in it. Soil mechanics can therefore be considered as a rather old discipline, but its historical appearance is usually connected with Coulomb's celebrated memoir [1], notwithstanding , as pointed out by Chandler [2], Morton's early contribution to the study of slope stability which does not refer to mechanics.

Coulomb's memoir is devoted both to structures and to soil mechanics : it deals with the stability of pillars and vaults, and with the problem of earth pressure on a retaining wall. As a matter of fact, the range of problems which were to be the main concern of soil mechanics for more than 150 years was defined there : active and passive earth-pressure calculations, stability analyses of slopes, bearing capacity calculations, through yield-design methods. Though we do not intend to produce an exhaustive list, let us recall some famous names in that domain of soil mechanics after Coulomb : Massau, Rankine, Kötter, Caquot, Sokolovski, Mandel.

That approach is based upon some fundamental concepts :

• soil is treated as a *continuous medium*, in spite of its being both particulate as clays, or granular as sands, and polyphasic (particles, water, air) ;
• this medium is defined through a *strength-criterion* which corresponds to failure or yielding when reached ;
• the analysis is performed by investigating if the equilibrium of the studied structure is possible under the applied loads (dead and active) with the limitation of the strength-criterion.

Yield-strength appears then as the first concept introduced by engineers to modelize "soil" as a material through a constitutive law, though incomplete since it is only a limitation imposed on the internal force intensity derived from the risk of failure. The introduction of that concept is certainly to be related to the very patterns of the failure mechanisms. As shown in the highly documented papers by Desrues [4] and Habib [5], one frequently observes that, when failure occurs, the deformation of the material seems to be confined in very thin soil layers (figure 1) which can be treated as surfaces. This localization may be noticed in the failure mechanisms of structures (e.g. landslides or breakings of earth fills), or when performing classical laboratory tests. It has induced engineers to apply intuitive reasoning based on the concept of mobilized strength on a failure plane, which unfortunately reveal some deficiencies from the mechanical point of view.

Anyhow it must be understood that having retained yield-strength as the working concept for stability analyses for more than 150 years shows its good adequation to practical applications.

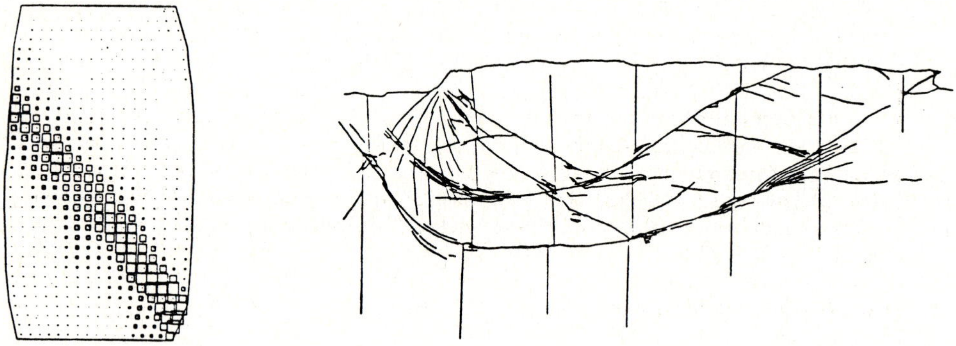

Figure 1 : Localization of deformation at failure : distorsion map for a plane strain compression test (from [4]), plane strain punch indentation test (from [5]).

2 - STABILITY ANALYSIS METHODS

As apparent in the title of the memoir, Coulomb's original study is carried out within the frame of Statics.

Taking as an example the compression of a cylindrical block between the plates of a press assumed to be perfectly smooth (a more precise definition of these boundary conditions is not useful for what follows), the following necessary condition is stated :

Let the specimen be separated in two parts 1 and 2 by an arbitrary plane (P) ; so that failure of the specimen do not occur it is necessary that, whatever (P), the global equilibrium of block 1 or 2 (from the Statics of rigid bodies point of view) be possible under the action of the vertical force Q imposed by the plate, and of the resisting forces developped by the material along the section by plane (P).

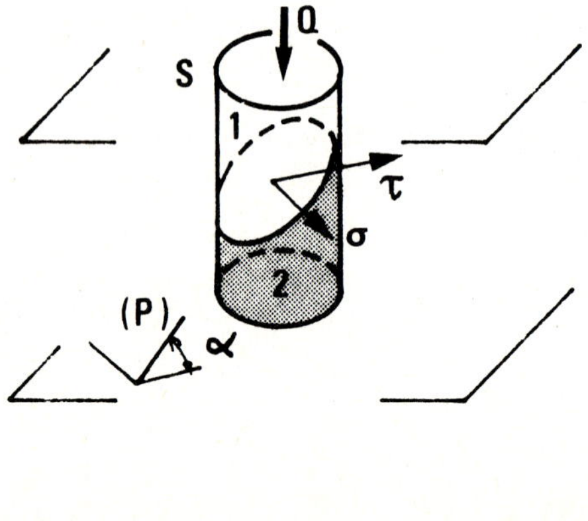

Figure 2 : Compression test of a cylindrical specimen.

The yield-strength of the material being characterized by a cohesion C and a friction angle φ, the Coulomb strength condition indicates that, whatever (P) and whatever the point in (P), the normal stress σ and the shear stress τ must verify the inequality :

$$|\tau| \leqslant C - \sigma \tan\varphi \tag{2.1}$$

(tensile stresses are counted positive). It then follows that, whatever $\alpha>\varphi$, the equilibrium of block 1 is compatible with (2.1) only if :

$$Q \leqslant SC\,(1 + \tan^2\alpha) \,/\, (\tan\alpha - \tan\varphi)\ , \tag{2.2}$$

a condition on Q whose minimum is obtained when $\alpha = (\pi/4 + \varphi/2)$ and writes :

$$Q \leqslant 2SC \tan(\pi/4 + \varphi/2). \tag{2.3}$$

So it turns out that an upper bound for the loads Q which can be applied to the specimen without causing its failure, is obtained through this reasoning and is given by (2.3).

This example deserves some comments.

• The performed reasoning does not completely make use of the main idea included in the necessary stability condition, namely the compatibility between the studied system equilibrium and the material yield-strength. In fact this condition could be checked within the frame of continuous mechanics instead of restricting the study to a partition of the specimen into two blocks and checking global equilibrium. Therefore it will be necessary to adopt continuous mechanics formulations and to express the Coulomb strength-condition as a criterion on the stress tensor $\underline{\underline{\sigma}}(\underline{x})$ at current point $\underline{x}$. σ_i being the principal values of $\underline{\underline{\sigma}}(\underline{x})$ Coulomb's criterion is derived from (2.1)[1]:

$$f(\underline{\underline{\sigma}}(\underline{x})) = \sup_{i,j=1,2,3} \left\{\sigma_i(1 + \sin\varphi) - \sigma_j(1 - \sin\varphi) - 2C\cos\varphi\right\} \leqslant 0 \tag{2.4}$$

which is equivalent to (2.1) expressed at point $\underline{x}$ for any orientation of plane (P).

• Bringing together the schema of figure 2 used for the static reasoning, and the often observed failure mechanisms where deformation seems to be localized and rigid blocks to slip with respect to one another (figure 1) suggests the introduction of an intuitive kinematic reasoning : taking the partition of figure 2 again, it is assumed that failure of the specimen will occur anytime the work of the external forces Q exceeds the work of the resisting forces mobilized on (P). Stated that way, unless the material be purely cohesive ($\varphi = 0$), the method cannot be accepted for it is not mechanically consistent with the original static approach. Nevertheless the idea of a kinematic method must be retained, but it needs being built up by dualizing the static approach through the principle of virtual work.

This points out the fact that the significance of the *yield-strength* of the material must be clearly defined, especially by indicating the stability analysis method it is associated with. Moreover it must be recalled that the yield-strength of a soil is a mechanical characteristic which is to be determined from experimental tests ; therefore the interpretation of those tests must be made in accordance with the method to be used later for the analysis of structures. Asking for a sound mechanical reasoning may seem nothing but banal, yet practical examples, dealing with isotropic or anisotropic soils as well, have shown that this demand has often not been satisfied. Let us add for a fairness' sake that experimental results do not, as a rule, enjoy the same purity as that given by theoreticians to their analyses !

From the mechanical point of view, when setting up and using the concept of yield strength, soil mechanicians can rely on two general theories :

• the theory of representations of isotropic tensorial functions, for what concerns the strength criterion of the soil (cf. many documented studies by Boehler such as [6], and also [7]) ;

• the theory of yield design, when devising consistent static and kinematic approaches for stability analyses.

3 - OUTLINES OF THE YIELD-DESIGN THEORY

3.1 - Setting of the problem

We shall restrict ourselves to a presentation of the main results of the theory of yield design ; comments and discussions may be found in [8].

Given V and S respectively the volume and the boundary of the studied mechanical system ;
$\underline{\underline{\sigma}}$ denotes a stress-field and $\underline{\underline{\sigma}}(\underline{x})$ its value at point $\underline{x}$;
$\underline{v}$ is a velocity-field, $\underline{v}(\underline{x})$ its value at point $\underline{x}$;
$\underline{\underline{d}}$ is the associated strain-rate field ; $[\![\underline{v}(\underline{x})]\!]$ denotes the jump of field $\underline{v}$ at point $\underline{x}$ when crossing the discontinuity surface Σ following its normal $\underline{n}(\underline{x})$.

The system is loaded according to a loading process depending linearly on n scalar parameters Q_j, the components of a loading-vector $\underline{Q}$; $\dot{\underline{q}}$ is the kinematic vector associated with $\underline{Q}$ when expressing the work of the external forces in a velocity field $\underline{v}$.

The material of the system (which needn't be homogeneous) is only known through its strength characteristics, given at any point $\underline{x}$ of V by a domain $G(\underline{x})$ in $\mathbb{R}^6$ where $\underline{\underline{\sigma}}(\underline{x})$ must stay. As a rule $G(\underline{x})$ contains $\underline{\underline{\sigma}}(\underline{x}) = 0$ and is star-shaped with respect to 0 ; it is often convex. $G(\underline{x})$ is usually given by means of a strength-criterion (such as (2.4)) :

(3.1) $$f[\,\underline{x}\,;\,\underline{\underline{\sigma}}(\underline{x})\,] \leq 0 \quad (\text{resp.} > 0) \iff \underline{\underline{\sigma}}(\underline{x}) \text{ in (resp. out of) } G(\underline{x})$$

The problem to be solved is to determine whether the system so defined will be "stable" under a given load $\underline{Q}$ of $\mathbb{R}^n$.

3.2 - Static approach "from inside"

As already seen on the example of chapter 2, a necessary stability condition for the system is the

(3.2) $$\text{compatibility} \begin{cases} \text{equilibrium under } \underline{Q} \\ \text{material strength capacities} \end{cases}$$

The following definitions are then adopted :

(3.3) $$\left.\begin{matrix} \underline{Q} \text{ is potentially safe} \\ \text{for the system} \end{matrix}\right\} \iff \left\{\begin{matrix} \text{system is potentially} \\ \text{stable under } \underline{Q} \end{matrix}\right.$$
$$\Updownarrow$$
$$\exists\, \underline{\underline{\sigma}} \text{ S.A. (statically admissible) with } \underline{Q}$$
$$\text{such that } \underline{\underline{\sigma}}(\underline{x}) \in G(\underline{x}) \quad \forall\, \underline{x} \in V$$

The set of all potentially admissible loads is denoted by K. As a consequence of the properties of $G(\underline{x})$, it can be shown that : K contains $\underline{Q} = 0$ and is star-shaped with respect to 0 ; K is convex if $G(\underline{x})$ is convex $\forall\, \underline{x} \in V$. The loads at the boundary of K in $\mathbb{R}^n$ are called the *extreme loads* for the system.

The static approach is just the application of definition (3.3) to construct points in K. It is clear that this construction can be simplified by taking advantage of the star-shape of K, or of its convexity when valid (figure 3). One shall notice that in the case of a unique positive loading parameter Q, as is fairly often encountered, the static approach will lead to a lower bound for the extreme value $Q^\star$ of Q.

3.3 - Kinematic approach "from outside"

The kinematic approach of K from outside is derived by dualizing (3.3) through the principle of virtual work.

Introducing function $\pi(.)$ defining $G(\underline{x})$ by duality (more precisely : G's convex envelope)

$$\pi\,(\underline{x}\,;\underline{\underline{d}}(\underline{x})) = \text{Sup}\,\{\,\underline{\underline{\sigma}}(\underline{x}) : \underline{\underline{d}}(\underline{x}) \mid \underline{\underline{\sigma}}(\underline{x}) \in G(\underline{x})\,\}\,, \tag{3.4}$$

and function

$$\pi\,(\underline{x},\,\underline{n}(\underline{x})\,;\,[\![\underline{v}(\underline{x})]\!]) = \text{Sup}\,\{[\![\underline{v}(\underline{x})]\!] . \underline{\underline{\sigma}}(\underline{x}) . \underline{n}(\underline{x}) \mid \underline{\underline{\sigma}}(\underline{x}) \in G(\underline{x})\}\,, \tag{3.5}$$

makes it possible to define functional $P(\underline{v})$ for any velocity field $\underline{v}$

$$P(\underline{v}) = \int_V \pi\,(\underline{x}\,;\,\underline{\underline{d}}(\underline{x}))dV + \int_\Sigma \pi\,(\underline{x},\,\underline{n}(\underline{x})\,;\,[\![\underline{v}(\underline{x})]\!])\,d\Sigma\,. \tag{3.6}$$

We then get the following result :

$$\left\{\begin{array}{l} \forall\,\underline{v}\ \text{kinematically admissible (K.A)} \\ K \subset \{\,\underline{Q} \mid \underline{Q} . \underline{\dot{q}}(\underline{v}) - P(\underline{v}) \leqslant 0\,\} \end{array}\right. \tag{3.7}$$

and the approach of K from outside as sketched on figure 3. In the case of a unique positive loading parameter Q, this kinematic approach leads to an upper bound for $Q^\star$.

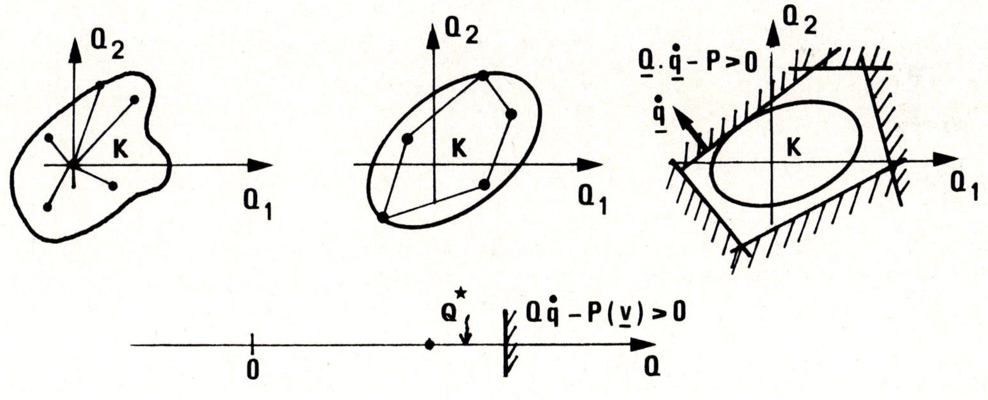

Figure 3 : Approaches from inside and from outside in the theory of yield design.

4 - YIELD-STRENGTH ANISOTROPY OF NATURAL SOILS

4.1 - Introduction

Although most geotechnical materials exhibit anisotropy, be it slight or strong, classical stability analyses have been performed usually under the assumption of isotropic soils. This simplifying hypothesis is justified as long as precise anisotropic yield-strength characteristics cannot be drawn from the identification tests, and as long as the experimental validation of the corresponding stability analyses remains good enough.

From a survey of the available experimental results about the anisotropy of natural soils, Hueckel and Nova [9] state that anisotropy has been observed for rocks (slates, shales, ...), for sands (with elongated grains) and for clays ; yet, as noticed by Ung Sen [10], the anisotropic effects that can be evidenced for the friction angle of sands are most often of the same order of magnitude as the accuracy of the measures. As a matter of fact, except for the works by Boehler et al. dealing with rocks, all the studies of anisotropic yield-strength which have been carried out as far as stability analyses of structures in soil mechanics, are concerned with purely cohesive soils (i.e. with a zero friction angle) ; we shall focus our attention on this kind of materials.

4.2 - Inherent and induced anisotropy

The distinction introduced by Casagrande and Carrillo [11] between inherent and induced anisotropy is now usually referred to. Inherent anisotropy pre-existing in a soil would be evidenced if it were possible to examine that soil without alteration ; induced anisotropy is generated by the testing procedures (sampling and laboratory test conditions ; remoulding for *in-situ* tests). This distinction is worth keeping, for it points out that much care must be taken when realizing the tests and interpreting them ; but it is clear that, at the end, soil mechanicians have to rely on measured characteristics which retain a part of induced anisotropy.

Yield-strength anisotropy of soils originates essentially in the existence of preferential orientations at the fabric level. Those may proceed from the grain or particle shapes (anisotropy is stronger for soils with elongated particles that tend to orientate themselves perpendicularly to the direction of the highest tectonic pressures), from the structure (floculated or loose structure of clays), or from the preferred orientations of heterogeneities (organic matters, change of materials : cf. [12] for the case of diatomite). These considerations, although the scales are different, can be compared with those developped in chap. 6 dealing with reinforced soils where applying the homogenization theory leads to an anisotropic model material (cf. also [13]).

4.3 - Test techniques and experimental results

As far as the principle is concerned, the determination of yield-strength characteristics for an anisotropic soil only requires generating a known and controlled uniform stress fiels in a sample, with the possibility of any orientations of its principal directions with respect to that of the sample.

Classical triaxial compression tests can then be performed on samples drilled out of the ground at various inclinations, but the interpretation of this simple procedure is not so easy, as recalled in [14], due to the non-coincidence of the principal directions of stresses and strains during the experiment as a consequence of anisotropy. Broms et al. [15, 16], Saada et al. [17, 18], have developped hollow cylinder test techniques for anisotropic soils ; for other anisotropic materials Boehler has reàlized precise experiments evidencing side sway under hydrostatic pressure, that could be worth adapting in soil mechanics.

As has been said before this paper will be concerned with purely cohesive anisotropic soils : for those, in the common range of pressures, the undrained compression triaxial test shows that the applied stresses only appear through their difference $(\sigma_3 - \sigma_1)$ as regards failure. The value of this difference at failure will be plotted as a function of the sample orientation. Due to the already mentioned origin of the anisotropy, such soils are usually transversally isotropic around the axis of the major tectonic pressures which is often vertical. The test results will therefore yield the value of $(\sigma_3 - \sigma_1)$ at failure as a function of α, the inclination of the axis of the sample with respect to the orthotropy axis :

(4.1) $$\sigma_3 - \sigma_1 = 2\,C\,(\alpha).$$

Most often the deformation of the sample at failure is localized in a plane ; the inclination of that plane may also be measured as a function of α. Duncan and Seed [20] have found this inclination with respect to the sample axis to be practically constant : $\psi = 30°$ (Lo [21] : $\psi = 35°$).

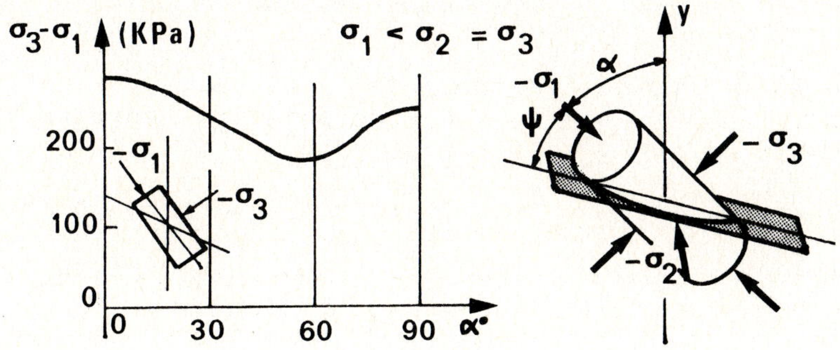

Figure 4 : Triaxial compression test [20].

Figure 5 gives examples of the results obtained through such undrained triaxial compression tests, which are plotted in the form of polar diagrams : the vector radius with a modulus $C(\alpha)$ is inclined at angle α on the vertical direction and indicates the orientation of the sample since the transverse isotropy axis is vertical too.

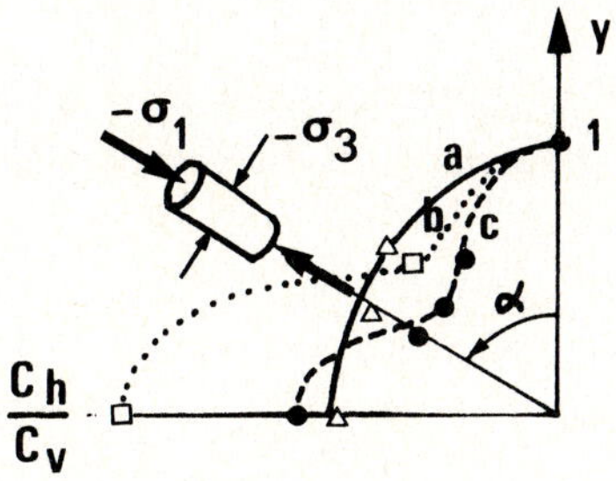

Figure 5 : Polar diagrams : polar representation of $C(\alpha)/C_v$ for the results by Duncan and Seed [22] a, Bishop [23] b, Lo and Milligan [24] c.

Two types of anisotropy can be distinguished from this picture :
Type I (curve a) : normally consolidated soil whose fabric appears homogeneous ; the polar diagram has an "elliptic pattern"
Type II (curves b and c) : those soils demonstrate a minimum cohesion for $\alpha \# 45°$, which is significantly lower than C_h and C_v corresponding to the vertical and horizontal orientations ; the fabric is oriented, due to high overconsolidation of the soil or to thin layers of stratified materials.

Practically, laboratory tests have produced anisotropy ratios within the ranges

$$0.6 \leqslant C_h/C_v \leqslant 1.6 \quad \text{and} \quad 0.4 \leqslant C_{45}/C_v \leqslant 1.2$$

(actually, in some cases of soils with complex fabrics, one may come across values of $C_{45}/C_v > 1$).

The first formula for describing $C(\alpha)$ was proposed by Casagrande and Carrillo [11] (as it seems, without any experimental support) :

$$C(\alpha) = C_v \cos^2\alpha + C_h \sin^2\alpha \quad ; \tag{4.2}$$

it may be seen that it only accounts for type I anisotropy. From his own experimental results on London clay, Bishop [23] suggested a formula with one parameter more :

$$C(\alpha) = C_v(\cos^2\alpha + K_1\sin^2\alpha)\,[\cos^2 2\alpha + 2K_2\sin^2 2\alpha/(1 + K_1)]$$

(4.3) with

$$K_1 = C_h/C_v \quad \text{and} \quad K_2 = C_{45}/C_v \, .$$

This formula has been found to fit many experimental data fairly well.

5 - STABILITY ANALYSIS OF ANISOTROPIC NATURAL SOIL STRUCTURES

5.1 - Strength-criteria

Yield-design problems of purely cohesive anisotropic soil structures have always been treated as two-dimensional (Lo, Chen, Menzies), using methods which were adapted from those for isotropic soils by introducing the anisotropic cohesion derived from triaxial test experimental data. Therefore the concept of strength criterion of the soil considered as a three-dimensional continuous medium, had never to be referred to in those analyses.

Besides, the studies about strength criteria for anisotropic soils are so numerous that it would be vain to quote them all. The concept of anisotropy-tensor has been used by Caquot [25], Goldenblatt [26], Sobotka [27], and by Boehler and Sawczuk [28] who formulated Mises' and Drucker-Prager's type criteria in the case of anisotropic soils. Nova and Sacchi [29] also made use of that concept after having tried another approach [30]. Relying on the studies by Pipkin, Rivlin, Smith, Spencer, and Wang, about the representations of isotropic tensorial functions, Boehler [31] have produced a most relevant contribution in that field. His three-dimensional analysis makes it possible to study plane problems on sound mechanical bases : the obtained equations have been used by Pastor and Turgeman [32] for finite element calculations.

The data derived from undrained compression triaxial tests, as represented by formula (4.1), are not sufficient to determine a strength criterion valid for any true triaxial stress state ; Salençon and Tristán-López [33] have suggested to construct that criterion under the complementary hypothesis that it should not depend on the intermediate principal stress [1]: the so-obtaned criterion fulfils the invariance-conditions established by Boehler (cf. [34]), and writes :

$$f(\underline{\underline{\sigma}}(\underline{x})) = \sigma_3 - \sigma_1 - 2C(\alpha) \leqslant 0$$

(5.1) $\sigma_1 \leqslant \sigma_2 \leqslant \sigma_3$ principal stresses,

$$\alpha = (\underline{O}y, \sigma_1)$$

(tensile stresses are counted positive)

where Oy is the direction of the transverse isotropy axis. With that strength-criterion, yield-design of structures in the case of plane strain parallel to the transverse isotropy axis is proved to reduce to a plane problem the corresponding two-dimensional strength criterion is nothing but (5.1) applied to the two-dimensional stress tensor in the plane ; the later result matches the procedure used in the available methods of stability analysis for anisotropic soils, it confirms that the assumption explicitly made here on the strength-criterion, was implicitly formulated by their authors.

Bishop's formula (4.1) has been retained by Salençon and Tristán-López to express $C(\alpha)$ in (5.1).

5.2 - π- functions

Two problems of yield design in plane strain conditions for anisotropic soils with criterion (5.1, 4.3) will be presented. Solving them through the kinematic approach requires knowing the two-dimensional π - functions for plane strain parallel to the transverse isotropy axis Oy. We get :

- for $\pi(\underline{\underline{d}}(\underline{x}))$:

  let d_1 be the major principal value of two-dimensional $\underline{\underline{d}}(\underline{x})$, and $\beta = (\underline{Oy},\underline{d_1})$; then :

  $$tr(\underline{\underline{d}}(\underline{x})) \neq 0 \Longrightarrow \pi(\underline{\underline{d}}(\underline{x})) = +\infty$$

  $$tr(\underline{\underline{d}}(\underline{x})) = 0 \Longrightarrow \pi(\underline{\underline{d}}(\underline{x})) = \underset{\alpha}{\text{Max}}\ (-2C(\alpha)\ d_1 \cos 2(\beta-\alpha))$$

  i.e. $$\pi(\underline{\underline{d}}(\underline{x})) = C_v\ d_1\ \pi(\beta) \quad .$$

- for $\pi(n(x), [\![\underline{v}(\underline{x})]\!])$:

  define unit vector $\underline{t}(\underline{x})$ in the plane by $(\underline{n}(\underline{x}), \underline{t}(\underline{x})) = \pi/2$; denote by v_n and v_t the corresponding components of $[\![\underline{v}(\underline{x})]\!]$ and let $\varepsilon = (\underline{Oy}, \underline{n}(\underline{x}))$; then:

  $$v_n \neq 0 \Longrightarrow \pi(\underline{n}(\underline{x}), [\![\underline{v}(\underline{x})]\!]) = +\infty$$

  $$v_n = 0 \Longrightarrow \pi(\underline{n}(\underline{x}), [\![\underline{v}(\underline{x})]\!]) = \underset{\alpha}{\text{Max}}\ (C(\alpha)\ v_t\ (\underline{x})\ \sin 2(\varepsilon-\alpha))$$

  i.e. $$\pi(\underline{n}(\underline{x}), [\![\underline{v}(\underline{x})]\!]) = \begin{cases} C_v |v_t| \pi_+(\varepsilon) & \text{if} \quad v_t > 0 \ , \\ C_v |v_t| \pi_-(\varepsilon) & \text{if} \quad v_t < 0 \ . \end{cases}$$

Functions $\pi(\beta)$ and $\pi_{\pm}(\varepsilon)$ have been computed in the case of criterion (5.1,4.3).

5.3 - Bearing capacity of a footing on an anisotropic cohesive soil

Davis and Christian [35] appear to have been the first to investigate the effects of cohesive soil anisotropy on the bearing capacity of foundations. As a strength-criterion for the material they adopted a simple polynomial form, depending on three parameters, that generalizes the one previously written by Hill [36]. Thus, they could determine the bearing capacity of strip footings by applying the theory of plane limit equilibriums (method of characteristics) since they remarked that a simple change of variable reduced the equations to those solved by Hill. In their paper Davis and Christian also retained the validity of Bishop's formula for a correct description of the available experimental data, but they noticed that no structure stability analysis had been performed using it.

The determination of the bearing capacity of strip-footings with Bishop's formula has been carried out by Salençon and Tristán-López [33] using both approaches of para. 3.2 & 3.3 ; they constructed static and kinematic solutions depending on several parameters which were optimized (figure 6).

With the notations of figure 6, the loading parameter for this problem is the axial force Q applied to the footing (per unit transversal length). The extreme value of this positive parameter is $Q^\star$ which writes (after a dimensional analysis) :

(5.2) $$Q^\star = B\ C_v\ \varphi(K_1,K_2)$$

where φ is a scalar function of K_1 and K_2. For isotropic soils, $K_1 = K_2 = 1$ in (4.3), the solution is known : $\varphi(K_1,K_2) = \pi + 2$.

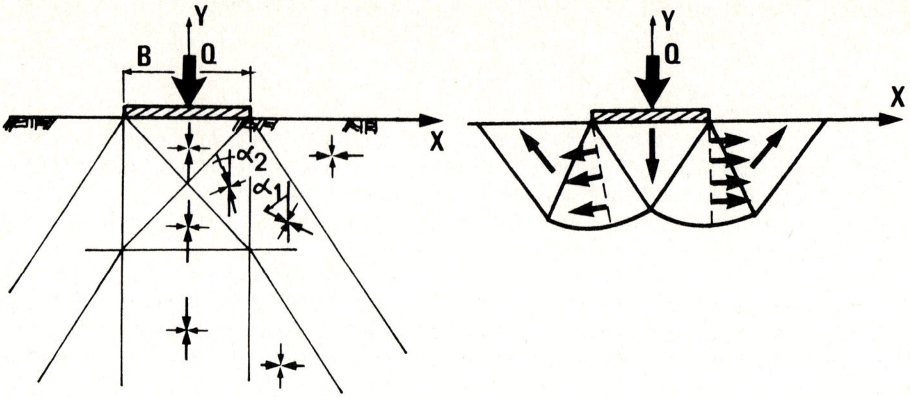

Figure 6 : Bearing capacity of strip-footings on anisotropic cohesive soils : static and kinematic solutions [33].

Results of this analysis are presented on figure 7, for three values of K_2, showing the thin zones delimited by the upper and lower bounds of the correction factor :

(5.3) $$n(K_1,K_2) = \varphi(K_1,K_2)/(\pi + 2) \ .$$

Complete charts are available in [34], and a detailed critical analysis of these results from the practical point of view may be found in [37]. Looking at correction factor (5.3) immediatly gives an estimation of the error committed on the bearing capacity of a footing on an anisotropic soil, if it is calculated assuming the soil to be isotropic from triaxial tests on vertically drilled samples. Although allowing for the usual tolerances in soil mechanics, it appears that the correction may be important either in the conservative or in the non-conservative sense ; moreover, simple empirical rules, as that of a mean cohesion suggested by Meyerhof [38], cannot be considered as sufficient (this one is proved to hold only in the case of Casagrande and Carrillo's soils).

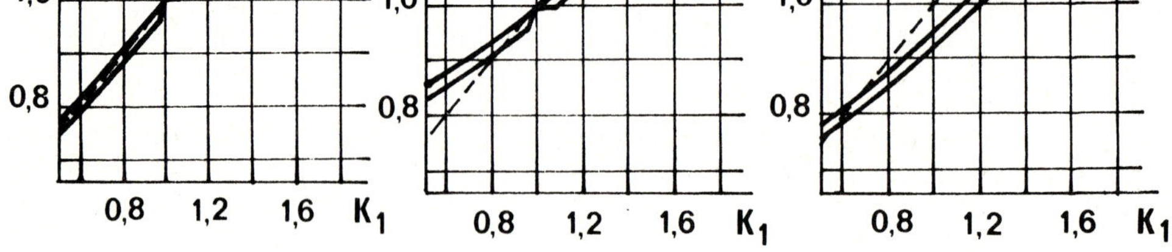

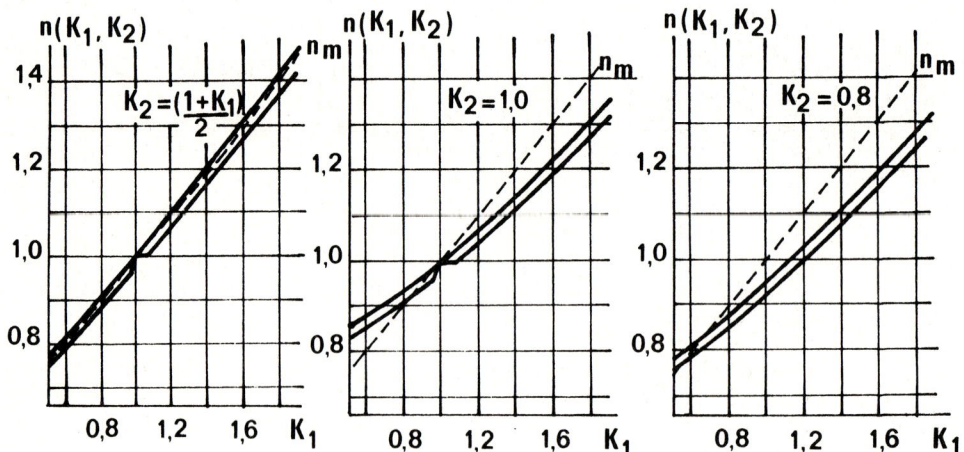

Figure 7 : Bearing capacity of a foundation on a cohesive anisotropic soil [37] ; $K_2 = (1 + K_1)/2$ corresponds to Casagrande-Carrillo's soils; n_m is the empirical result obtained assuming the soil to be isotropic with a cohesion $(C_v + C_h)/2$.

Comparing those results with Davis and Christian's ones shows that both strength-criterion (5.1, 4.3) and Davis and Christian's are convenient for practical applications.

5.4 - Stability of slopes for anisotropic cohesive soils

Let h and β be the geometrical characteristics of the studied slope (figure 8). The constitutive soil with voluminal weight γ, is anisotropic with strength-criterion (5.1, 4.3). Dimensional analysis shows that the stability of this structure is governed by the non-dimensional loading-parameter $N = \gamma h/C_V$; its extreme value is a function of β, K_1, K_2 : $N^{\star}(\beta, K_1, K_2)$.

The construction of static solutions in order to apply the yield-design approach "from inside" has proved difficult and poorly efficient for this kind of problems; therefore they are usually dealt with by means of the yield-design approach "from outside".

Such a stability analysis has been carried out by Tristán-López [34], applying Eq. (3.7) with the velocity-fields defined by the rigid body rotation of a block sliding along a circular line. Thus he established the first charts for slope stability evaluation in the case of an anisotropic cohesive soil whose yield-strength is defined by three parameters (i.e. two anisotropy ratios). An example is given at figure 8.

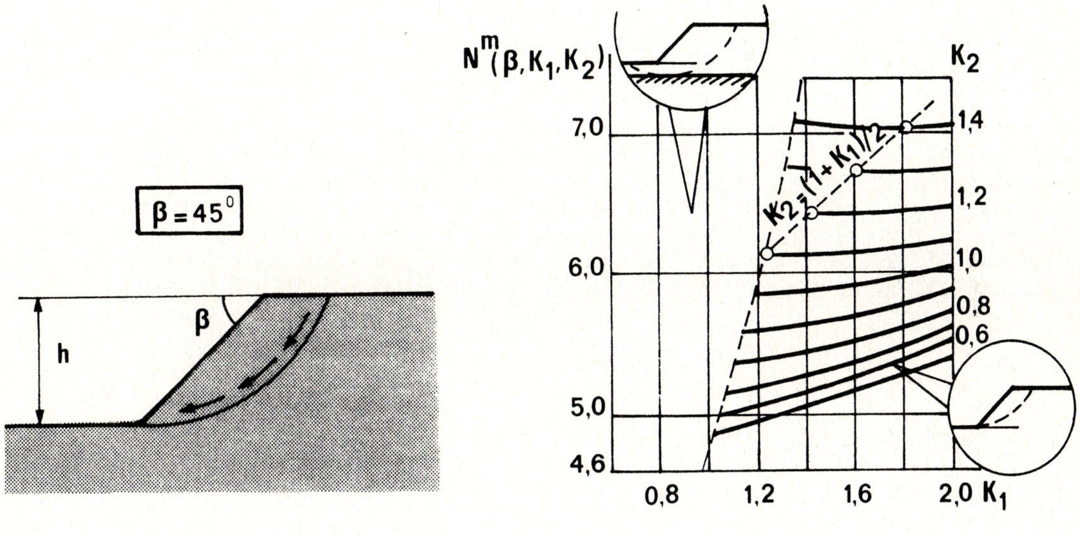

Figure 8 : Slope stability for anisotropic cohesive soil [34] : $\gamma h/C_V \leq N^m(\beta, K_1, K_2)$.

As mentioned before, Bishop's formula reduces to Casagrande and Carrillo's when $K_2 = (1 + K_1)/2$; stability analyses were already available in that case (Lo [21], Chen [39]) whose results can be compared with those obtained by Tristán-López (figure 9).

This comparison deserves a few comments from the mechanical point of view. As a matter of fact, Lo's and Chen's analyses are performed using the same velocity fields as in Tristán-López'; nevertheless, though the strength-criterion is the same for the three authors, the obtained results are different. This discrepancy which originates in the way the approach from outside is realized, emphasizes the importance of a sound mechanical reasoning. Tristán-López' analysis relies on

Eq. (3.7) and definition (3.5) of function $\pi(\ )$ to calculate $P(\underline{v})$: therefore it is consistent with the definition of the soil strength-criterion from the mechanical view-point. Lo's analysis is made following the idea mentioned in chap.2: the work of the external forces is compared with the work of the resisting ones ; in order to evaluate the later, the slip-circle in the velocity field is considered as an envelope of failure planes in triaxial tests (figure 9) and the mobilized shear stress on any of these planes (inclined at $\psi = 35°$ on the sample axis) is said to be equal to the corresponding $C(\alpha)$. Since $\psi \neq \pi/4$, this reasoning is indeed not in accordance with the procedure and the interpretation of the triaxial test (uniform cylindrical stress-field around σ_1). The same ideas have been followed by Reddy and Srinivasan, Chen, and Menzies, assuming other values for ψ ($\psi = 0$ in some cases). Due to the values of ψ adopted by the authors on the example of figure 9, the discrepancy between the results obtained that way, and through the mechanical dualization, is slight and can be considered negligible for practical applications ; but the question should not be discarded.

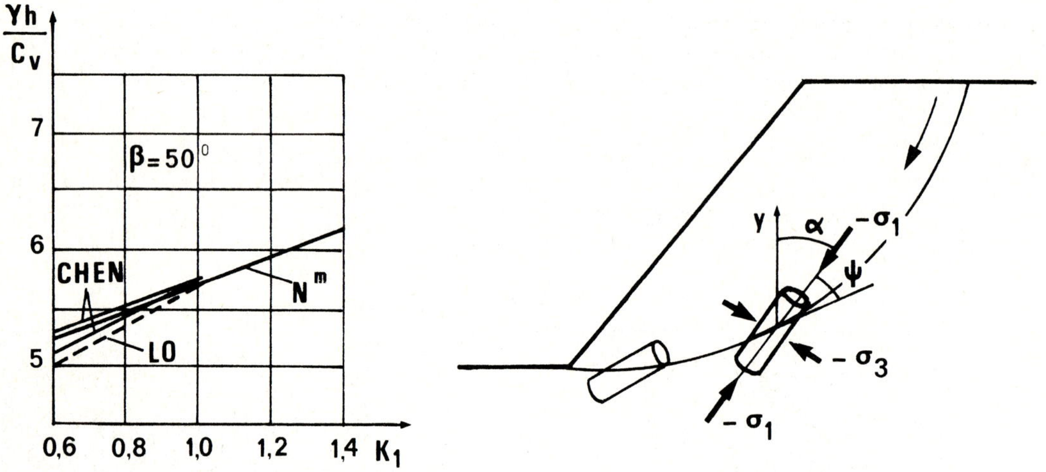

Figure 9 : Casagrande and Carrillo's soil : comparison between Tristán-López', Lo's, and Chen's results.

6 - SOIL REINFORCEMENT AND ANISOTROPY

6.1 - Soil reinforcement

Mechanical soil-reinforcement for building earth-structures has been used since antiquity, but it has enjoyed an extensive development during the late twenty years generated by important technical research. A few examples can illustrate typical techniques :

- spreading of regularly disposed reinforcement in the soil (the principle of reinforced earth),
- substitution of a better soil for a part of the native soil, following a regular mesh (the principle of ballast columns and trenches),
- improvement of the strength-characteristics of the native soil, following a regular mesh (the principle of lime columns),

and it may be noticed that the regularity of the set up reinforcing material, and the existence of preferential orientations, are features common to these processes.

Practical stability analyses for reinforced soil structures (e.g. slope stability, bearing capacity of surface footings) apply two-dimensional yield-design methods adapted from those classical in the case of homogeneous isotropic soils ; therefore any new development on this topic will be welcome. Here, we intend to give a first survey of the methods and results that can be derived from the application of the yield-design theory to the stability analysis of reinforced soil structures.

The characteristic scale of any of the abovementioned reinforcements, the usual civil engineering one, appears macroscopic when compared with that of the soil particles. But it looks like a fabric scale when compared with the scale of the structures themselves. This suggests that it should be possible to analyse the stability of a reinforced-soil structure by referring to a homogeneous "reinforced-soil" material whose definition and use should be clearly specified. From the preferential orientations of the reinforcement one may expect the "reinforced-soil" material to exhibit anisotropic strength properties.

The theoretical developments about homogenization in yield-design and limit analysis, by Suquet [40,41] and de Buhan [42], will not be presented ; the general reasoning and an example of the obtained results will be exposed on a typical problem.

6.2 - Stability of a reinforced soil slope

The considered slope is made of a purely cohesive isotropic native soil with cohesion C_1, reinforced by vertical layers of another purely cohesive anisotropic soil with cohesion $C_2 > C_1$. The stability of this structure will be studied in plane Oxy (figure 10).

Let e_i denote the layer thickness for the material with cohesion C_i $(i = 1,2)$; let $e = e_1 + e_2$ and $\lambda_i = e_i/e$ $(i = 1,2)$. The contact between the native soil and the reinforcement one is assumed to be perfectly adhesive. As a simplification, both soils will be supposed to have the same voluminal weight γ.

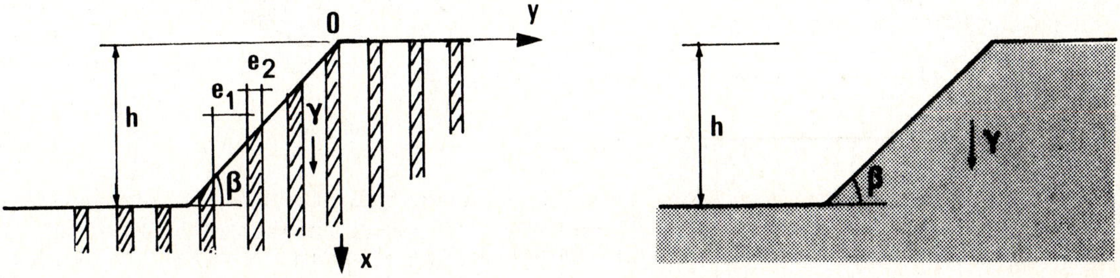

Figure 10 : Stability analysis of a reinforced-soil slope and associated homogeneous structure.

Dimensional analysis applied to the yield-design of that structure shows that the non-dimensional factor $N = \gamma h/C_1$ acts as the loading parameter for the problem ; its extreme value is a function of β, C_2/C_1, λ_2/λ_1, e/h :

$$(6.1) \qquad N = \gamma h/C_1 \quad , \quad N^{\star}(\beta, C_2/C_1, \lambda_2/\lambda_1, e/h) .$$

The value of $N^{\star}$ is looked for when the thickness of the layers (the scale of the reinforcement) can be considered small with respect to the height of the slope (the scale of the structure), i.e. when $e/h \to 0$. For simplicity we introduce :

$$(6.2) \qquad N_o^{\star}(\beta, C_2/C_1, \lambda_2/\lambda_1) = \lim N^{\star}(\beta, C_2/C_1, \lambda_2/\lambda_1, e/h) \quad \text{as} \quad e/h \to 0.$$

The determination of $N_o^{\star}$ through the approaches of yield-design proves rather difficult. Due to the two layered materials present in the system, the stress-or velocity-fields to work with soon become sophisticated, except for the velocity-fields defined by the rigid body rotation of a block sliding along a circular line (as in para. 5.4) which remain convenient. These ones are used in the classical methods [44] : thus, applying the kinematic approach with this family of velocity-fields leads to an upper bound N_o^m for $N_o^{\star}$:

(6.3) $$N_o^m(\beta,\ C_2/C_1,\ \lambda_2/\lambda_1) \geqslant N_o^\star(\beta,\ C_2/C_1,\ \lambda_2/\lambda_1)\ .$$

It may be proved that N_o^m depends only on β and on the factor $r = [1 + (\lambda_2/\lambda_1)(C_2/C_1)] / (1 + \lambda_2/\lambda_1)$, and writes :

(6.4) $$N_o^m(\beta,\ C_2/C_1,\ \lambda_2/\lambda_1) = r\ N^m(\beta)\ ,$$

where $N^m(\beta)$ is the result obtained for the stability of a cohesive soil slope by the slip circle method. In other words, the slope with the "composite" soil when $e/h \to 0$, is perceived through the slip-circle analysis in just the same way as if it were made of a homogeneous isotropic soil with the cohesion rC_1 i.e. the ponderated mean value of both constitutive soil cohesions.

6.3 - Homogenization and yield-design

Homogenization in yield-design introduces a homogeneous structure associated to the composite soil structure when $e/h \to 0$, which is geometrically identical and is submitted to the same loading process as that one (figure 10). The stability of this associated homogeneous structure is then analysed.

The homogenized material is only defined for this application. A first yield-design problem is solved on the representative elementary volume of the "composite" material : it corresponds to the change of scales and produces the strength criterion to be adopted for the "reinforced-soil" homogenized material from the characteristics C_1, C_2/C_1, λ_2/λ_1, through the static and kinematic approaches.

In the present state of the theory it is proved that, with this definition of the homogenized material, the stability analysis of the associated homogeneous structure gives an over-estimation of the stability of the composite one (Suquet [41]). A stronger result would indeed better fit the intuitive conjecture made in para. 6.1, but the value of this already available one will be demonstrated through our following example.

6.4 - Strength-criterion of reinforced soil

For the reinforced soil structure describe in para. 6.2, the homogenized material is thus obtained in the form of a purely cohesive soil, transversally isotropic around Oy. For the two-dimensional problems in plane strain parallel to Oy, its strength-criterion writes : let σ_1 and σ_3 be the principal stresses ($\sigma_1 \leqslant \sigma_3$) and $\alpha = (Oy,\sigma_1)$

(6.5) $$\sigma_3 - \sigma_1 - 2C(\alpha) \leqslant 0\ ,$$

(6.6) $$C(\alpha) = C_1\ \rho(\alpha,\ C_2/C_1,\ \lambda_2/\lambda_1)\ ,$$

where the non-dimensional factor ρ can be explicitly calculated [42]. Figure 11 shows ρ as a function of α.

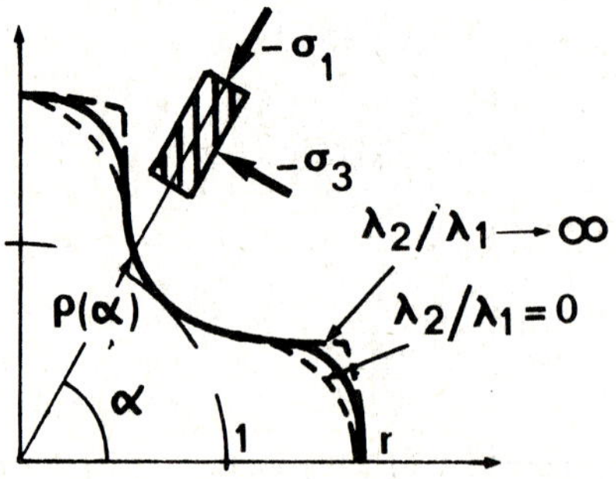

Figure 11 : Polar diagrams for the homogenized soil : $\rho(\alpha) = C(\alpha)/C_1$.

It exhibits the following properties :

$\alpha = \pi/4 : \rho = 1, \quad C_{45} = C_1$ ("weak" soil cohesion)

(6.7) $\left.\begin{array}{l}\alpha = 0\\ \alpha = \pi/2\end{array}\right\} : \rho = r, \quad C(\alpha) = rC_1$ (ponderated mean value of the soil cohesions)

$0 \leqslant \alpha \leqslant \pi/2 : \rho(\alpha) = \rho(\pi/2 - \alpha) \qquad 1 \leqslant \rho(\alpha) \leqslant r$

6.5 - Stability analysis of the associated homogeneous structure

With the results (6.5, 6.6) the stability analysis of the associated homogeneous slope (figure 10) proves similar to the problem already studied in para. 5.4 for anisotropic natural soils ; here, formula (6.6) is substituted for Bishop's formula in expressing $C(\alpha)$. The corresponding π- functions for plane strain parallel to Oy have been determined [45].

Once again the velocity-fields defined by the rigid-body rotation of a block sliding along a circular line are used in the kinematic approach. An upper bound $N^m_{hom}(\beta, C_2/C_1, \lambda_2/\lambda_1)$ is thus obtained for the extreme value $N^\star_{hom}$ of the loading parameter $\gamma h/C_1$ for the associated homogeneous structure :

(6.8) $$N^m_{hom}(\beta, C_2/C_1, \lambda_2/\lambda_1) \geqslant N^\star_{hom}(\beta, C_2/C_1, \lambda_2/\lambda_1) .$$

Applying Suquet's theorem, the loading parameters being identical for both structures, we can also state that :

(6.9) $$N^m_{hom}(\beta, C_2/C_1, \lambda_2/\lambda_1) \geqslant N^\star_o(\beta, C_2/C_1, \lambda_2/\lambda_1) .$$

Figure 12 presents the obtained results in the case $\beta = 90°$, as functions of parameters λ_2/λ_1 and r.

N^m_{hom} enjoys the following properties whatever β :

- for $r = 1$ (homogeneous soil with cohesion C_1), $N^m(\beta)$ is found again :

(6.10) $$r = 1, \forall \lambda_2/\lambda_1 : N^m_{hom}(\beta, C_2/C_1, \lambda_2/\lambda_1) = N^m(\beta) ;$$

- for $r > 1$, N^m_{hom}/r decreases as r increases when λ_2/λ_1 is fixed

  N^m_{hom}/r increases with λ_2/λ_1 when r is fixed,

  thence, taking Eq. (6.10) into account, we get :

(6.11) $$N^m_{hom}(\beta, C_2/C_1, \lambda_2/\lambda_1) \leqslant r\, N^m(\beta).$$

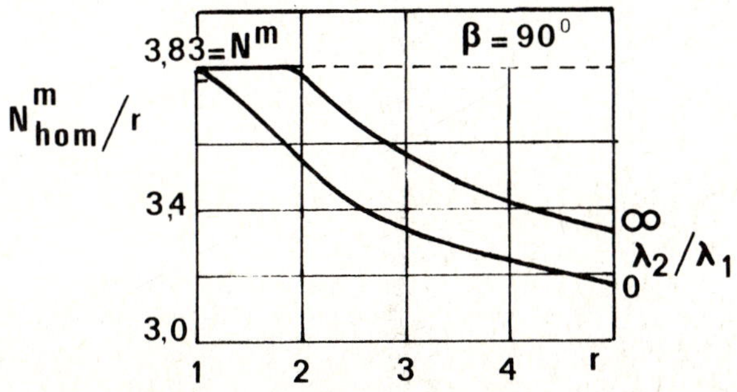

Figure 12 : Stability analysis of a reinforced soil slope : $\gamma h/C_1 \leqslant N^m_{hom}$.

6.6 - Comments : homogenization and anisotropy

Putting together formulae (6.3, 6.4, 6.9 and 6.11), or looking at figure 12, it is easy to compare the results obtained through the application of the slip-circle method : first to the "composite"-soil slope ($e/h \to 0$), then to the associate slope made of a homogenized soil with anisotropic cohesion. It appears then that, working on the same class of simple velocity-fields, and applying rigorously the kinematic approach of yield-design, we get significantly different estimations of the composite-soil slope stability : since it is lower, the value obtained through homogenization shall be retained. A similar conclusion has been drawn by de Buhan [46] when studying the bearing capacity of surface footings, within the same framework.

The property is general and its explication is close to mechanical intuition. It was shown in para. 6.2 that the classical method leads to the same evaluation of the slope stability as if the constitutive soil were homogeneous and isotropic with the cohesion ponderated mean value ; stability analysis after homogenization, thanks to the two successive applications of yield-design approaches it implies, can account for the macroscopic anisotropy of reinforced-soil (and for this one, cohesion $C(\alpha)$ has been proved to be always smaller than the cohesion ponderated mean value).

7 - CONCLUSION

Due to techonological incitements and to progresses in computational means, solid mechanicians have taken increasing interest into the study of anisotropic materials, during the late decades. General fundamental researches have been carried out as well as more specific investigations in many disciplines of applied mechanics.

In soil mechanics, constitutive laws are now formulated, based on modelizations which take anisotropy into account ; numerous papers have been devoted to that subject (a recent general lecture by Hueckel and Nova [9] includes some 250 references !). We have chosen to present here an aspect of the reflexions on soil anisotropy which is directly connected with practical applications : the yield-strength of anisotropic soils from the point of view of the stability analysis of structures.

The necessity of sound mechanical bases to rely on, as well when setting up or interpreting experimental tests, as when devising stability analysis methods, has been evidenced : ambiguities, which we can remain unaware of in the case of isotropy, may lead to inconsistency when the material is anisotropic. The theory of representation of isotropic tensorial functions and that of yield-design proved highly valuable for that purpose. From the practical view point the obtained results are relevant.

Apart from the anisotropy of natural soils, the macroscopic anisotropy of mechanically reinforced soils can also be taken into account. Applying the homogenization process to yield-design makes it possible to formulate efficient methods for the stability analysis and the design of reinforced-soil structures. New developments should appear in this field during the next few years, which are more likely to be of an applied character, in order to simulate the real reinforcement techniques better.

REFERENCES :

[1] Coulomb, C.A., Essai sur une application des règles de Maximis et Minimis à quelques problèmes de statique relatifs à l'architecture, mémoire présenté à l'Académie Royale des Sciences (Paris, 1773).

[2] Chandler, R.J., Two early contribution to the study of slopes, *Géotechnique*, 31, 4, (1981), 553-554.

[3] Morton, J., *The natural history of Northamptonshire* (London, 1712).

[4] Desrues, J., *La localisation de la déformation dans les matériaux granulaires*, th. Dr Sc., Univ. Grenoble, (1984).

[5] Habib, P., Les surfaces de glissement en mécanique des sols, *Revue Française de Géotechnique*, 27, (1984), 7-21.

[6] Boehler, J.P., Lois de comportement anisotrope des milieux continus, *J. de Mécanique*, 17, 2, (1978), 153-190.

[7] Wang, C.C., A new representation theorem for isotropic functions, Part. I & II, *Arch. Rat. Mech. An.*, 36, (1970), 166-223, Corrigendum, 43, (1971), 392-395.

[8] Salençon, J., *Calcul à la rupture et analyse limite*, Presses de l'E.N.P.C., Paris, (1983).

[9] Hueckel, T., Nova, R., Anisotropic failure criteria for geotechnical materials, recent trends, *Int. Coll. CNRS, "Failure criteria of structured media"*, Villard-de-Lans, (1983).

[10] Ung Sen, Y., *Etude de l'influence des paramètres d'essais sur la mesure de la résistance au cisaillement des argiles raides*, th. Dr Ing., E.N.P.C., Paris, (1975).

[11] Casagrande, A., et Carrillo, N., Shear failure of anisotropic materials, *J. Boston Soc. Civil Eng.*, 31, 4, (1944), 74-87.

[12] Allirot, D., Boehler, J.P., Evolution de l'anisotropie d'une roche sous compression cyclique isotrope, *Bul. Ac. Pol. Sc.*, série Sciences et Tech., 24, 9, (1976), 405-409.

[13] Goguel, J., La prise en compte de l'anisotropie dans la mécanique des déformations tectoniques, *Bul. Soc. Géol., Fr.*, 18, 6, (1976), 1489-1495.

[14] La Rochelle, P., & Marsal, R.J., Slope stability, Gen. Rep. 11, *Proc. 10th I.C.S.M.F.E.*, Stockholm, (1981), 141-162.

[15] Broms, B.B., et Ratnam, M.V., Shear strength of an anisotropically consolidated clay, *Journal of the Soil Mech. and Found. div.*, Proc. A.S.C.E., 92, SM6, (nov. 1963), 1-26.

[16] Broms, B.B., et Casbarian, A.O., Effects of rotation of the principal stress axes and of the intermediate principal stress on the shear strength, *Proc. 6th I.C.S.M.F.E.*, Montreal, (1965), 1, 179-183.

[17] Saada, A.S., et Zamani, K.K., The mechanical behavior of cross anisotropic clays, *Proc. 7th I.C.S.M.F.E.*, Mexico, (1969), 1, 351-359.

[18] Saada, A.S., Discussion of "Anisotropy in heavily overconsolidated Kaolin", *Jnl Geotech. Eng. Div.*, A.S.C.E., 102, GT 7, (1976), 823-824.

[19] Boehler, J.P., Raclin, J., Critères de résistance des composites verre-résine sous états de contrainte complexes, *C.R. JNC 3*, Paris, (1982), 309-319.

[20] Duncan, J.M., et Seed, H.B., Anisotropy and stress reorientation in clays, *Jnl Soil Mech. and Found. Div.*, Proc. A.S.C.E., 92, SM6, (sept. 1966), 21-50.

[21] Lo, K.Y., Stability of slopes in anisotropic soils, *Jnl Soil Mech. and Found. Div.*, Proc. A.S.C.E., 91, SM4, (july 1965), 85-106.

[22] Duncan, J.M., et Seed, H.B., Strength variations along failure surfaces in clay, *Jnl Soil Mech. and Found. Div.*, Proc. A.S.C.E., 92, SM6, (nov. 1966), 81-104.

[23] Bishop, A.W., The strength of soils as engineering materials, Géotechnique, 16, 2, (1966), 89-130.

[24] Lo, K.Y., et Milligan, V., Shear strength properties of two stratified clays, *Jnl Soil Mech. and Found. Div.*, Proc. A.S.C.E., 93, SM1, (jan. 1967), 1-15.

[25] Caquot, A., Conférence à l'Académie des Sciences d'Italie, 1953, (quoted in Caquot, A., & Kérisel, J., *Traité de mécanique des sols*, Gauthier-Villars, Paris, (1966).

[26] Goldenblat, I.I., *Some problems of mechanics of deformable media*, Noordhoff, Groningen (1962).

[27] Sobotka, Z., The fundamental relations of theory of plasticity and a new concept of the plastic potential for anisotropic materials, Stavebnicky Casopis, S.A.V. 14, (1966), 377-380.

[28] Boehler, J.P., et Sawczuk, A., Equilibre limite des sols anisotropes, *J. de Mécanique*, 9, 1, (1970), 5-33.

[29] Nova, R., Sacchi, G., A model of the stress-strain relationship of orthotropic geological media, *J. Meca. Th. Appl.*, 1, 6, (1982), 927-949.

[30] Nova, R., Sacchi, G., A generalized failure condition for orthotropic solids, *Proc. Euromech 115, Coll. CNRS "Comportement mécanique des solides anisotropes"*, Villard-de-Lans, (1979).

[31] Boehler, J.P., *Contributions théoriques et expérimentales à l'étude des milieux plastiques anisotropes*, Th. Dr Sc., Grenoble, (1975).

[32] Pastor, J., Turgeman, S., Approches numériques des charges limites pour un matériau orthotrope de révolution en déformation plane, *J. Méca. Th. Appl.*, 2, 3, (1983), 393-416.

[33] Salençon, J., et Tristán-López, A., Force portante des semelles filantes sur sols cohérents anisotropes homogènes, *C.R. Ac. Sc. Paris*, 292, II, (1981), 1097-1102.

[34] Tristán-López, A., *Stabilité d'ouvrages en sols anisotropes*, Th. Dr -Ing. E.N.P.C., Paris, (1981).

[35] Davis, E.H., et Christian, J.T., Bearing capacity of anisotropic cohesive soils, *Jnl of the Soil Mech. and Found. Div.*, Proc. A.S.C.E., 97, SM5, (may 1971), 753-769.

[36] Hill, R., *The mathematical theory of Plasticity*, Clarendon Press, Oxford, (1950).

[37] Salençon, J., Tristán-López, A., Calcul à la rupture en mécanique des sols : cas des sols cohérents anisotropes, *Ann. I.T.B.T.P.*, 413, (1983), 53-83.

[38] Meyerhof, G., Discussion of stability of slopes in anisotropic soils, *J. Soil Mech. and Found. Div.*, Proc. A.S.C.E., 91, SM6, (1965), 132.

[39] Chen, W.F., *Limit analysis and soil plasticity*, Elsevier Sc. Publ. Cy., Amsterdam, (1975).

[40] Suquet, P., *Plasticité et homogénéisation*, Th. Dr Sc. Univ. Pierre et Marie Curie, Paris, (1982).

[41] Suquet, P., Analyse limite et homogénéisation, *C.R. Ac. Sc. Paris*, 296, II, (1983), 1355-1358.

[42] de Buhan, P., Homogénéisation en calcul à la rupture, *C.R. Ac. Sc. Paris*, 296, II, (1983), 933-936.

[43] de Buhan, P., Salençon, J., Analyse de stabilité d'ouvrages en sols renforcés, to appear in *Proc. 13rd I.C.S.M.F.E.*, San Francisco, (1985).

[44] Lime Column Method., Linden-Alimak, Sweden, (1980).

[45] Zghal, A., *Stabilité d'ouvrages en matériaux composites*, Mémoire de D.E.A., Laboratoire de Mécanique des Solides, Palaiseau, (1983), E.N.I.T., (Tunis).

[46] de Buhan, P., Détermination de la capacité portante d'une fondation sur sol renforcé par une méthode d'homogénéisation, to appear in *Proc. Coll. Int. "Renforcement des sols et des roches"*, Paris, 1984.

(1) and that the orientation of the stress-tensor should appear only through the inclination of the major principal pressure towards the orthotropy axis.

Theoretical and Applied Mechanics
F.I. Niordson and N. Olhoff
Elsevier Science Publishers B.V. (North-Holland)

VEHICLE SYSTEM DYNAMICS

Werner O. Schiehlen

Institute B of Mechanics
University of Stuttgart
Stuttgart, F.R.G.

The dynamical analysis of vehicles results in complex systems including the excitation process due to the guideway, the vehicle itself and the evaluation of the passenger response to mechanical motion. The vehicle is modeled as a linear or nonlinear multibody system, the road represents a stochastic process of white or colored noise and the passenger's frequency response follows from biomechanics. The resulting dynamical system is investigated by the covariance analysis.

INTRODUCTION

The main topics in vehicle system dynamics are characterized by the performance of the vehicle's longitudinal motion, the handling and the stabilization of the lateral motion, and the ride quality related to the vertical vibrations. The scientific progress during the last years is well documented in the proceedings of the International Association of Vehicle System Dynamcis (IAVSD) edited by Pacejka [1], Slibar and Springer [2], Willumeit [3], Wickens [4] and Hedrick [5]. A survey on recent research activites is given in the following:

i) Longitudinal motion (Performance);

- control of the driver-vehicle-road system,
- vibrations of the engine-transmission system,
- anti-lock bracking systems,
- accelerating and braking of motorcycles.

ii) Lateral motion (Handling, stability);

- dynamic stability of railway cars,
- curving behavior of wheel/rail vehicles,
- contact problems wheel-rail,
- stability of trailers,
- handling of four wheel drive vehicles,
- motorcycle control.

iii) Vertical motion (Ride characteristics);

- active suspension of railway cars,
- modeling of guideway surfaces,
- random vehicle vibrations,
- damping and friction in passenger car,
- human sensation of vibration.

The mathematical methods used include dynamics of multibody systems, bifurcation theory and covariance analysis of stochastic systems.

This broad variety of problems cannot be covered in one paper. Therefore, the multibody system approach, valid for all kinds of vehicles and motions, will be presented first. Then, random vehicle vibrations will be treated, resulting in a global stochastically excited linear dynamical system, ready for the application of the covariance analysis.

DYNAMICS OF MULTIBODY SYSTEMS

Multibody systems consisting of rigid bodies with inertia interconnected by springs, dampers and bearings without inertia are well qualified for the dynamical analysis for vehicles. Usually the method of multibody systems is restricted to motions with frequencies less than 50 Hz. For high frequency problems such as car body vibrations and acoustics, respectively, the method of finite element systems or the method of continuous systems has to be applied. The method of multibody systems is in a continuous development, see Magnus [6], Wittenburg [7]. The recent state of the art is given by Haug [8].

The kinematics of a multibody system, Fig. 1, are defined by the 3x1-translation vector

$$r_i(y,t) = r_R + r_{Ri}(y,t) \quad , \quad i = 1(1)p \quad , \tag{1}$$

and the 3x3-rotation tensor

$$S_i(y,t) = S_R + S_{Ri}(y,t) \quad , \quad i = 1(1)p \quad , \tag{2}$$

of each body K_i. The absolute quantities r_i, S_i with respect to the inertial frame I may also be expressed by the relative quantities r_{Ri}, S_{Ri} with respect to a moving reference frame R. The translations and rotations of the total system are uniquely given by the fx1-position vector $y(t)$ summarizing the generalized coordinates according to the f degrees of freedom of the system.

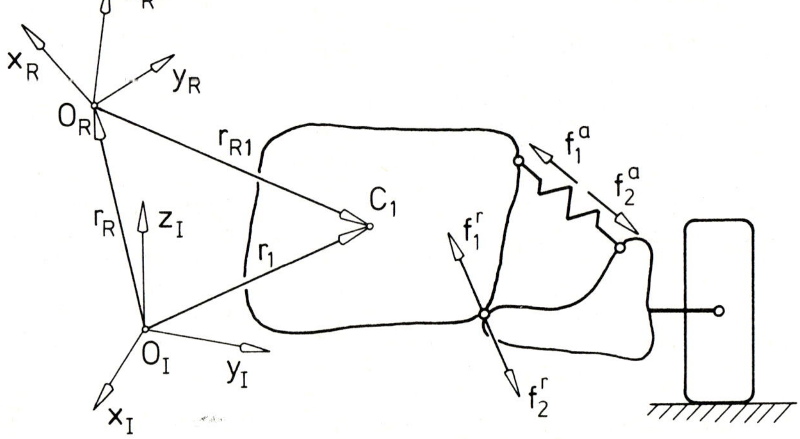

Figure 1
Multibody System

The kinetics of a multibody system are controlled by ideal applied forces and torques, by reaction forces and torques, and by nonideal applied forces and torques, respectively. The corresponding 3x1-forces vectors read for each body

$$f_i^{ia} = f_i(y,\dot{y},w,t) \quad , \tag{3}$$

$$f_i^r = F_i(y,t)\ g(t) \quad , \tag{4}$$

$$f_i^{na} = f_i(g,y,\dot{y},t) \quad , \quad i = 1(1)p \quad . \tag{5}$$

Thus, the ideal forces f_i^{ia} depend only on position variables of the total sytem, position integrals represented by the ρx1-vector $w(t)$ and time. The reaction forces f_i^r are proportional to the qx1-reaction vector $g(t)$ summarizing the general constraint forces according to the q holonomic constraints of the system. And the nonideal forces f_i^{na} due to contact and friction are in general a nonlinear function of reaction and position variables. The relations (3) to (5) are also valid for torques. A typical nonideal torque is the friction torque in a bearing, which is a nonlinear function of the corresponding generalized reaction forces and derivatives of position.

The equations of motion of a multibody system are obtained from Newton's and Euler's equations for each body and D'Alembert's principle as

$$M(y,t)\ \ddot{y}(t) + k(y,\dot{y},t) = q(y,\dot{y},w,g,t) \tag{6}$$

where M is the fxf-inertia matrix and k and q are fx1-vectors of the generalized coriolis and applied forces. Obviously, this equation cannot be solved; it has to be completed by the ρx1-vector differential equation

$$\dot{w} = \dot{w}(y,\dot{y},w,t) \tag{7}$$

of the position integrals appearing in (3), and by the equations of reaction of a multibody system

$$N(y,t)\ g(t) + \hat{q}(y,\dot{y},w,g,t) = \hat{k}(y,\dot{y},t) \quad . \tag{8}$$

Here, N is the qxq-reaction matrix and $\hat{q}$ and $\hat{k}$ are qx1-vectors of applied and coriolis forces.

Eqs. (6) to (8) represent nonlinearly coupled differential and algebraical equations. They have to be solved simultaneously by an integration procedure and a nonlinear equation solver. However, in vehicle system dynamics the contact forces are often proportional to the normal or constraint forces. Then, eqs. (5) , (6) and (8) are also linear in $g(t)$ and - at least formally - the generalized constraint forces can be eliminated in (6) resulting in decoupled equations of motion. For more details see [9].

Eqs. (6) to (8) does not depend on the choice of the moving reference frame introduced in (1) and (2). However, in vehicle dynamics, it is often very useful to apply a moving reference frame: the defintion of generalized coordinates, the formulation of forces and the generation of the equations may be strongly simplified. A typical reference frame is a guideway frame moving with vehicle speed.

For the analysis of vehicle vibrations in lateral or vertical direction, eqs. (6) to (8) may be linearized. Then, the generalized constraint forces can be eliminated again. The remaining decoupled equations of motion read as

$$M\,\ddot{y}(t) + P\,\dot{y}(t) + Q\,y(t) + R\,w(t) = h(t), \tag{9}$$

$$\dot{w}(t) = T\,w(t) + U\,\dot{y}(t) + V\,y(t) + j(t)\,, \tag{10}$$

where P, Q, R, T, U, V, h, j, are matrices and vectors of appropriate dimension. Further, it is assumed that these matrices are time-invariant. Introducing the nx1-state vector

$$x(t) = [y^T(t) \quad \dot{y}^T(t) \quad w^T(t)]^T \tag{11}$$

eqs. (9) and (10) can be summarized in the state equation

$$\dot{x}(t) = A\,x(t) + b(t) \tag{12}$$

where A means the nxn-system matrix and b(t) is the nx1-excitation vector. Eqs (12) represents the standard form of linear time-invariant vibration systems. The methods of solution are available for all kinds of excitation functions in literature, see e.g. [10].

The generation of equations of motion of complex vehicles is a nontrivial problem. Therefore, formalisms have been developed for computer-aided engineering (CAE). The numerical formalisms, e.g. MULTIBODY [11] require the repeated generation of equations of motion for each set of parameters. On the other hand, the symbolical formalisms like NEWEUL [12] are characterized by the first and final generation of equations of motions. Thus, symbolical formalisms are more adequate to computer-aided design (CAD).

GUIDEWAY MODELING

The guideway may have a rigid or a flexible surface. Rigid surfaces result in deterministic or stochastic excitation functions while flexible surfaces have to be modeled as mechanical systems. A typical rigid guideway is a highway or road on a rigid ground. Flexible guideways are primarily found in railway engineering, especially represented by bridges and other elevated constructions, see e.g. Popp [13]. With the random vibration problem in mind, only rigid stochastic guideways will be treated.

The guideway roughness of a rigid road results with respect to the four wheels of a two-axle vehicle in four random processes summarized in a 4x1-vector excitation process

$$\xi(t) = [\xi_{fr}(t) \quad \xi_{fl}(t) \quad \xi_{rr}(t) \quad \xi_{rl}(t)]^T \tag{13}$$

where the space coordinate $s = vt$ of the longitudinal motion is replaced by time t under assumption of a constant vehicle speed v, Fig. 2. Since the 2x1-vector process $\xi_r(t)$ of the rear axle is only delayed due to the axle distance $\Delta s = v\ \Delta t$ the 2x1-vector process $\xi_r(t)$ of the front axle is discussed in more detail.

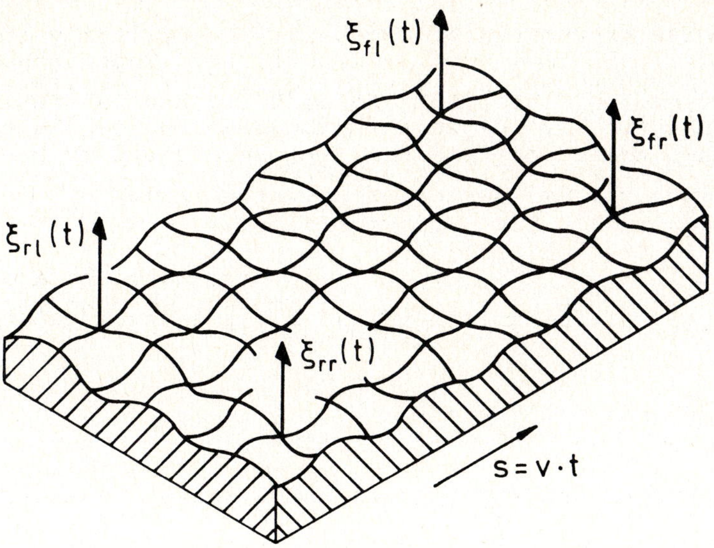

Figure 2
Road Roughness and Stochastic Excitation Processes

Numerous measurements have shown that the road roughness can be characterized by a Gaussian, ergodic and stationary process with zero mean value:

$$m_\xi = E\{\xi(t)\} = 0 \quad , \qquad (14)$$

$$R_\xi(\tau) = E\{\xi(t)\ \xi^T(t-\tau)\} \quad , \qquad (15)$$

$$S_\xi(\omega) = \frac{1}{2\pi} \int_{-\infty}^{+\infty} R_\xi(\tau)\ e^{-i\omega\tau}\ d\tau \quad , \qquad (16)$$

$$P_\xi = R_\xi(0) = \int_{-\infty}^{+\infty} S_\xi(\omega)\ d\omega \ , \qquad (17)$$

where m_ξ is the mean value, $R_\xi(\tau)$ the correlation matrix, $S_\xi(\omega)$ the spectral density matrix and P_ξ the covariance matrix.

Usually, the stochastic properties of guideway processes are found by measurements and they are presented as spectral densities. For the vibration analysis the measurements are approximated in the frequency domain by polynomials or in the time domain by shape filters, respectively. A first order model of a scalar process reads for example as

$$S_\xi(\omega) = \frac{Q}{2\pi} \frac{1}{\omega_0^2 + \omega^2} \qquad (18)$$

or

$$\frac{d\xi}{dt} = -\omega_0\,\xi(t) + w(t) \qquad (19)$$

where ω_0 is a constant filter coefficient and Q is the intensity of a white noise process $w(t)$ with $R_w(\tau) = Q\,\delta(\tau)$.

In vehicle dynamics, the right and left trace of a spatial road represent stochastically dependent processes, while the medium trace ξ_M and the trace difference $2\xi_D$ are found to be independent processes. With the 2x1-vector of the front axle

$$u(t) = [\xi_{fM}(t) \quad \xi_{fD}(t)]^T \qquad (20)$$

the kinematical relations in the road cross-section can be expressed as

$$\xi_f(t) = H\,u(t) \quad , \quad H = \begin{bmatrix} 1 & 1 \\ 1 & -1 \end{bmatrix} \quad . \qquad (21)$$

Then, the corresponding first order shape filter equations read as

$$\frac{du}{dt} = F\,u(t) + G\,w(t) \qquad (22)$$

where F and G are constant diagonal 2x2-matrices and $w(t)$ means a stochastically independent 2x1-white noise process with a diagonal 2x2-intensity matrix. Even with a simple first order shape filter as used in (22), guideway irregularities can be modeled very well as Rill [14] has shown.

For the rear axle, the shape filter approach (21), (22) can also be applied, if the time delay Δt is introduced in the white noise excitation process. Then it remains

$$\xi(t) = [\xi_f^T(t) \quad \xi_r^T(t)]^T = [\xi_f^T(t) \quad \xi_f^T(t-\Delta t)]^T \quad . \qquad (23)$$

This is also true for an arbitrary number of axles.

Some special cases are now easily obtained. For $F = 0$ it is obvious from (22) that the excitation process is white velocity noise, a very handy approximation. With $\xi_D(t) \equiv 0$ it follows from (21) that the traces are parallel, $\xi_{fr}(t) = \xi_{fl}(t)$, $\xi_{rr}(t) = \xi_{rl}(t)$.

If a variable vehicle speed $v(t) \neq$ const is considered, the time domain representation (22) remains consistent even if the coefficients become time-variant and the stochastic processes prove to be nonstationary. For more details see Rill [14].

Now, the nx1-excitation vector appearing in (12) can be specified as

$$b(t) = B_f\,H\,u(t) + B_r\,H\,u(t-\Delta t) \qquad (24)$$

where (21) and (23) have been used. The nx2-excitation matrices B_f and B_r characterize the front and rear axle, respectively; the state equations (12) have to be completed by shape filter equations (22). Due to the stochastic excitation vector process $u(t)$, the response of the linear vibration system (12) will be a random state vector process $x(t)$ with characteristics corresponding to (14) to (17).

HUMAN SENSATION OF VIBRATIONS

The road roughness generates mechanical vibrations of the vehicle acting on the human body. The evaluation of the human sensation depends on the state and excitation vector processes which are assumed Gaussian, ergodic and stationary.

The objectively measurable mechanical vibrations affecting the guideway and the vehicle can be evaluated by scalar vibration variables

$$w_k(t) = c_k^T x(t) + d_k^T \xi(t) \quad , \quad k = 1, 2, \ldots \tag{25}$$

where c_k and d_k are the corresponding weighting vectors. Typical vibration variables in vehicle dynamics are the car body acceleration and dynamical wheel load variation. By definition, the scalar vibration variables $w_k(t)$ are also Gaussian, ergodic and stationary processes characterized by the mean value

$$m_{wk} = c_k^T m_x + d_k^T m_\xi = 0 \tag{26}$$

and the variance

$$\sigma_{wk}^2 = c_k^T P_x c_k + 2c_k^T P_{x\xi} d_k + d_k^T P_\xi d_k \quad . \tag{27}$$

The covariance matrices P are defined by (17) with respect to the processes $x(t)$ and $\xi(t)$. The covariance matrix $P_{x\xi}$ represents the coupling of these processes. If the excitation follows from white noise guideway roughness, $P_\xi \to \infty$, then only vibration variables with $d_k = 0$ are admissible. This may be a restriction for the dynamical wheel load computation in the ride safety analysis.

The human sensation of mechanical vibrations differs from the objectively measureable vibrations. Numerous physiological investigations have shown that the human sensation can be characterized by scalar perception variables $\bar{w}_k(t)$ depending on the passenger's own dynamics and on the objectively measurable vibration variables $w_k(t)$. The relation between the perception variables $\bar{w}_k(t)$ and the vibration variables $w_k(t)$ can be represented by shape filters using frequency response methods or by differential equations in the time domain. However, the perception variables $\bar{w}_k(t)$ are also stochastic processes to be characterized by the corresponding variance $\sigma_{\bar{w}k}^2$ identical with the measure K of the perception, see national standard [15].

With respect to vehicle vibrations only the vertical acceleration a(t) will be used. Then, the perception follows as

$$K = \sigma_{\bar{a}}^2 = \alpha^2 \int_{-\infty}^{+\infty} |f_a(\omega)|^2 S_a(\omega)\, d\omega \tag{28}$$

where $\alpha = 20\ s^2/m$ is a constant, $f_a(\omega)$ is the frequency response of the vertical sensation and $S_a(\omega)$ is the spectral density of the vertical acceleration a(t) . The frequency response $f_a(\omega)$ is given by the international standard [16] and shown in Fig. 5. In the time domain the frequency response is replaced by a second order shape filter

$$\bar{a}(t) = \bar{h}^T \bar{v}(t) \quad , \quad \dot{\bar{v}}(t) = \bar{F}\,\bar{v}(t) + \bar{g}\, a(t) \quad , \tag{29}$$

where $\bar{v}(t)$ is the 2x1-filter state vector, $\bar{F}$ the 2x2-filter matrix, $\bar{g}$ the 2x1-input vector and $\bar{h}$ the 2x1-output vector. A frequency response of the filter (29) is shown in Fig. 3. Now, the perception reads as

$$K = \sigma_{\bar{a}}^2 = \alpha^2\, \bar{h}^T P_{\bar{v}}\, \bar{h} \tag{30}$$

where $P_{\bar{v}}$ is the 2x2-covariance matrix of the shape filter process $\bar{v}(t)$.

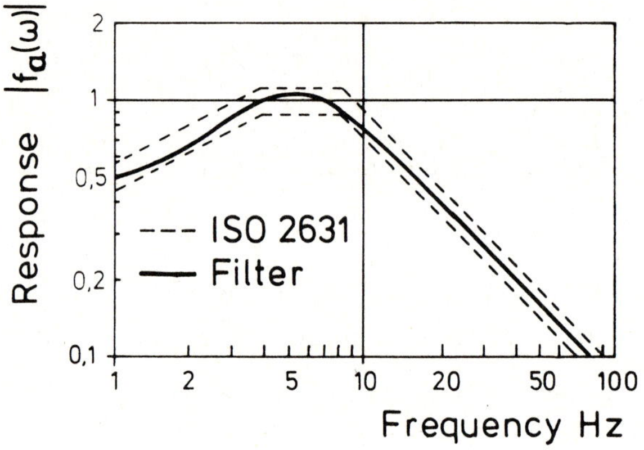

Figure 3
Frequency Response of Human Sensation

The final results (28) and (30) are given in the frequency domain and the time domain, respectively. It turns out that an infinite integral has to be evaluated in the frequency domain while in the time domain only an algebraic matrix operation is required. Therefore, the covariance analysis using the time domain is prefered in this paper. The spectral density analysis using the frequency domain may also be applied in vehicle dynamics, however, it is less accurate and less economic for complex vehicle sstems, see, e.g. [17].

COVARIANCE ANALYSIS OF STOCHASTIC SYSTEMS

The modeling of guideway roughness, vehicle dynamics and human sensation results in a strongly coupled global system under stochastic excitation. Since the excitation and sensation shape filters are linear subsystems, together with the linearized equations of motion a linear global system is obtained.

The extended state equation reads as

$$\dot{\hat{x}}(t) = \hat{A}\,\hat{x}(t) + \hat{B}_f\, w(t) + \hat{B}_r\, w(t-\Delta t) \tag{31}$$

where

$$\hat{x}(t) = [u_f^T(t) \quad u_r^T(t) \quad x^T(t) \quad \bar{v}^T(t)]^T \tag{32}$$

is the extended nx1-state vector and $\hat{A}$, $\hat{B}_f$, $\hat{B}_r$ are corresponding matrices following from (22), (12), (24) and (29). It has to be mentioned, that the vectors $u_f(t)$ and $u_r(t)$ according to (22) differ only in the argument t and $(t-\Delta t)$ of the excitation vector process w. Now, an extended $\hat{n}x\hat{n}$-covariance matrix $\hat{P} = P_{\hat{x}}$ can be found from the algebraic Ljapunov matrix equation

$$\hat{A}\,\hat{P} + \hat{P}\,\hat{A}^T + \hat{B}_f\, Q\, \hat{B}_f^T + \hat{B}_r\, Q\, \hat{B}_r^T$$
$$+ \hat{\Phi}(\Delta t)\, \hat{B}_r\, Q\, \hat{B}_f^T + \hat{B}_f\, Q\, \hat{B}_r^T\, \hat{\Phi}^T(\Delta t) = 0 \tag{33}$$

where

$$\hat{\Phi}(\Delta t) = e^{\hat{A}\Delta t} = \hat{E} + \hat{A}\Delta t + \frac{1}{2}\hat{A}^2(\Delta t)^2 + \ldots \tag{34}$$

may be obtained by series expansion. The Ljapunov equation (33) results immediately in the covariance matrix, without any numerical integration. This is the great advantage of the covariance analysis as discussed in [18].

In a second step the variances of the vibration variables (25) are obtained by the algebraical matrix operations (27). Further, the unknown matrix in (30) is immediately available as submatrix,

$$\hat{P} = \left[\begin{array}{c|c} ---- & ---- \\ \hline ---- & P_{\bar{v}} \end{array}\right] . \tag{35}$$

For the numerical solution of the Ljapunov equation (33) the method of Smith [19] may be applied, since the matrix $\hat{A}$ is asymptotically stable by definition.

RANDOM VIBRATIONS OF A COMPLEX VEHICLE

In Fig. 4 a complex vehicle is shown consisting of 4 mass points and 7 rigid bodies subject to 35 constraints resulting in $f = 19$ degrees of freedom. In addition, there are two serial spring-damper-configurations at the engine characterized by $\varrho = 2$ position integrals. Further, 4 first order excitation shape filters are considered while the sensation shape filter will be neglected. Then, the global system has the order $\hat{n} = 44$. The human sensation of mechanical vibration will be discussed only with respect to the vertical acceleration of the car body. Then, there is $k = 1$ and the only vibration variable reads as

$$w_1(t) = c_1^T \hat{x}(t) = \ddot{ZK}(t) + D\,\ddot{AK}(t) - C\,\ddot{BK}(t) \qquad (36)$$

where (C,D) characterizes a location in the plane of the car body. The 44x1-vector c_1 is defined by (36) under consideration of (9).

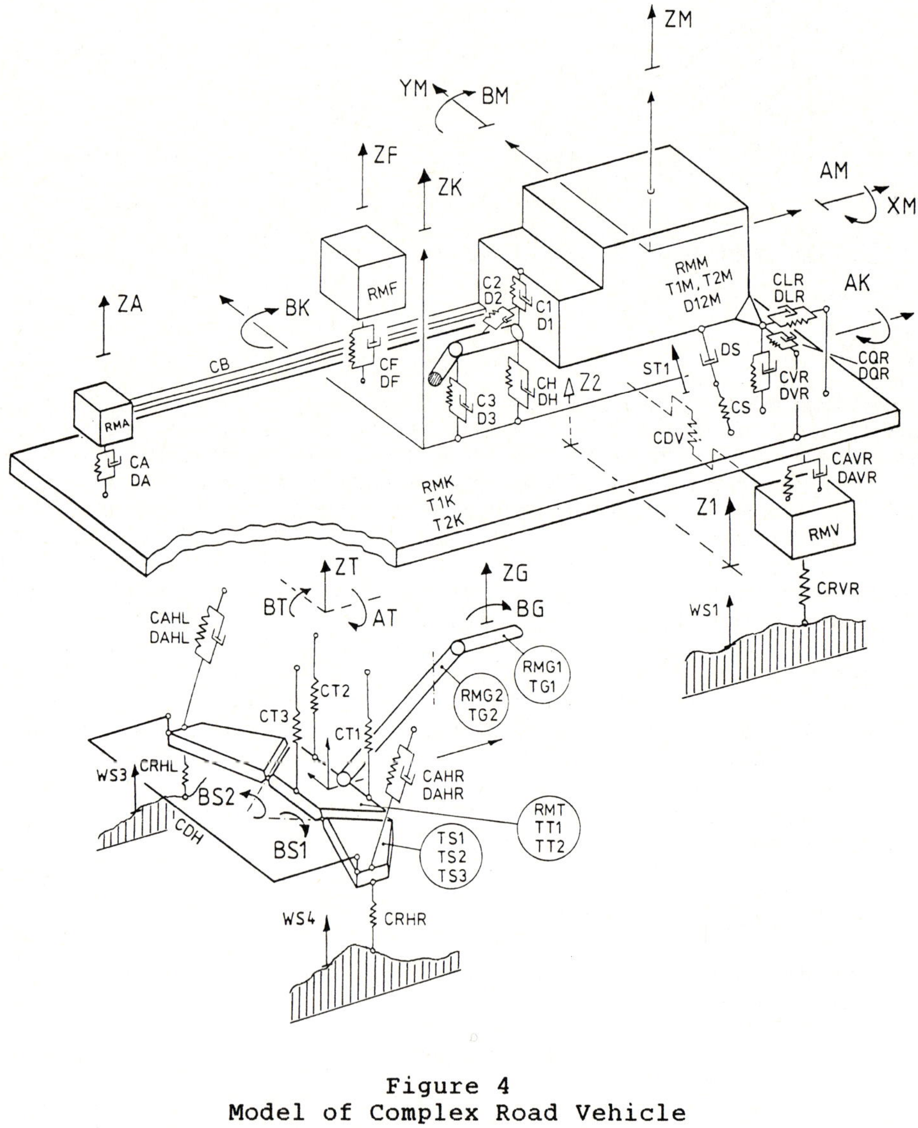

Figure 4
Model of Complex Road Vehicle

Numerical results for this complex vehicle have been published by Kreuzer and Rill [20]. Fig. 5 shows the RMS value or standard deviation, respectively, of the vertical acceleration normalized by the earth acceleration for each location (C,D) on the car body. It turns out that the optimal ride characteristics are found in the middle of the car body. Thus, qualitative experience and quantitative measurements are confirmed by theoretical system analysis very well.

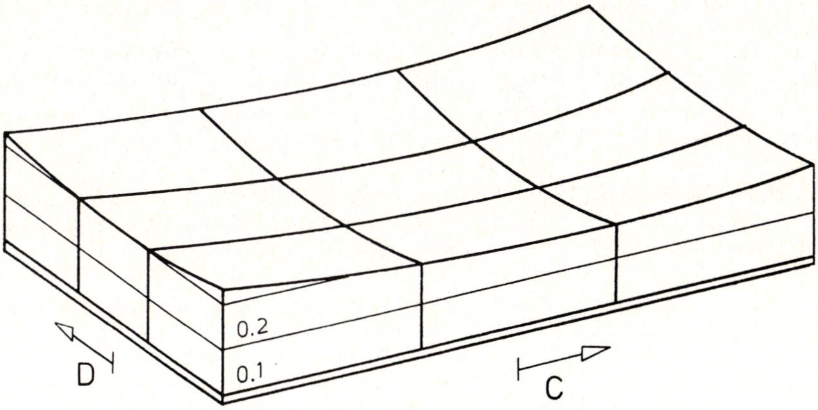

Figure 5
Vertical RMS Acceleration Normalized by Earth Acceleration

CONCLUSION

The established high standards in vehicle engineering were primarily obtained by car racing, by experiments on roads and tracks, and by measurements on test circuits. For a long time theoretical results have not been applicable to real vehicle problems. The extremely increasing costs of experimental research on the one hand and the great scientific progress in applied dynamics on the other hand lead to an increasing importance of the computer-aided vehicle engineering. Thus, vehicle system dynamics will serve the engineering community more and more.

REFERENCES

[1] Pacejka, H.B. (ed.), The Dynamics of Vehicles (Swets and Zeitlinger, Amsterdam, 1976).

[2] Slibar, A., and Springer, H. (eds.), The Dynamics of Vehicles (Swets and Zeitlinger, Amsterdam, 1978).

[3] Willumeit, H.-P. (ed.), The Dynamics of Vehicles (Swets and Zeitlinger, Lisse, 1980).

[4] Wickens, A.H. (ed.), The Dynamics of Vehicles (Swets and Zeitlinger, Lisse, 1982).

[5] Hedrick, J.K. (ed.), The Dynamics of Vehicles (Swets and Zeitlinger, Lisse, 1984).

[6] Magnus, K. (ed.), Dynamics of Multibody Systems (Springer-Verlag, Berlin, 1978).

[7] Wittenburg, J., Dynamics of multibody systems, in: Rimrott, F.P.J. and Tabarrock, B. (eds.), Theoretical and Applied Mechanics, Preprints (North-Holland, Amsterdam, 1980).

[8] Haug, E.J. (ed.), Computer Aided Analysis and Optimization of Mechanical System Dynamics (Springer-Verlag, Berlin, 1984).

[9] Schiehlen, W., Modeling of complex vehicle systems, in: [5].

[10] Müller, P.C. and Schiehlen, W.O., Linear Vibrations (Martinus Nijhoff, The Hague, 1985).

[11] Schwertassek, R., Der Roberson/Wittenburg - Formalismus und das Programmsystem MULTIBODY zur Rechnersimulation von Mehrkörpersystemen. (DFVLR, Oberpfaffenhofen, 1978).

[12] Kreuzer, E.J. and Schiehlen, W.O., NEWEUL, A Software Package for Symbolical Equations of Motion, CCG-Course V1.08, Carl-Cranz-Ges., Oberpfaffenhofen (April 1984).

[13] Popp, K., Stochastic and elastic guideway models, in: Schiehlen, W.O. (ed.), Dynamics of High-Speed Vehicles (Springer-Verlag, Wien, 1982).

[14] Rill, G., Instationäre Fahrzeugschwingungen bei stochastischer Erregung, Ph. D. Thesis, University of Stuttgart (February 1983).

[15] VDI-Richtlinie 2057, Beurteilung der Einwirkung mechanischer Schwingungen auf den Menschen, Verein Dt. Ing. Düsseldorf (1975-1979).

[16] International Standard ISO 2631, Guide for the Evaluation of Human Exposure to Whole-Body Vibrations, Int. Org. Standardization (1974).

[17] Schiehlen, W.O., Probabilistic analysis of vehicle vibration, in: Huang, T.C. and Spanos, P. (eds.), Random Vibrations (ASME, New York, to appear).

[18] Müller, P.C., Popp, K. and Schiehlen, W.O., Berechnungsverfahren für stochastische Fahrzeugschwingungen, Ing. Arch. 49 (1980) 235-254.

[19] Smith, R.A., Matrix equation XA + BX = C, SIAM J. Appl. Math. 16 (1968) 198-201.

[20] Kreuzer, E. and Rill, G., Vergleichende Untersuchung von Fahrzeugschwingungen an räumlichen Ersatzmodellen, Ing. Arch. 52 (1982) 205-219.

Theoretical and Applied Mechanics
F.I. Niordson and N. Olhoff
Elsevier Science Publishers B.V. (North-Holland)

FLOW FIELD VISUALIZATION

Sadatoshi Taneda

Research Institute for Applied Mechanics
Kyushu University
Kasuga, Fukuoka 816
Japan

Flow visualization is the most useful research tool for determining flow structures. Streamlines, streaklines, particle paths, timelines, flow directions, flow separation, limiting streamlines, density distributions, temperature distributions, stress distributions, etc. are revealed using flow visualization techniques. However, vorticity distributions can not be made visible. Since the flow pattern depends on the visualization technique and on the reference frame from which the flow is observed, great care is required in determining the flow structure from the flow pattern.

INTRODUCTION

Flow visualization has played a very important role in improving our physical understanding of complicated flow phenomena. The structure of flow can be quantitatively determined by means of flow visualization techniques. In this article a brief survey of the flow visualization techniques is described together with some remarks about the difficulties in interpreting the flow pattern. Further details of the individual visualization techniques and the examples of the flow photographs can be found in the references [1],[2],[3],[4],[5],[6],[7],[8] at the end of this article.

Figure 1
Surface waves around an oscillating sphere

VISUALIZATION OF STREAMLINES

Streamlines are the lines at any point of which the direction of the tangential line coincides with the instantaneous direction of the flow velocity at the corresponding point. When a lot of fine tracer particles are uniformly suspended in a flow and their motion is photographed with a proper exposure time, the streamline pattern is obtained. Aluminum flakes, glass beads, polystyrene beads, polyethyrene beads, oil droplets, air bubbles, etc. are used as the tracer particles in water, and zinc stearate powder, metaldehyde flakes, etc. in air.

Figures 2 and 3 show the streamline patterns in a steady flow and in an unsteady flow respectively.

It is desirable that the density of tracer particles is the same as that of the fluid medium. It should be noted, however, that even when the tracer particles and the fluid have the same density, the particles, if they are large, suffer lift forces in shear flows. Therefore, for observing the true streamlines, the tracer particles should be as small as possible. If the size of the tracer particles is sufficiently small, the particles follow the flow correctly even when their density is different from the fluid density, because the inertial and gravitational forces (bodily forces) decrease rapidly with decreasing particle sizes, and the drag forces (surface forces) become dominant.

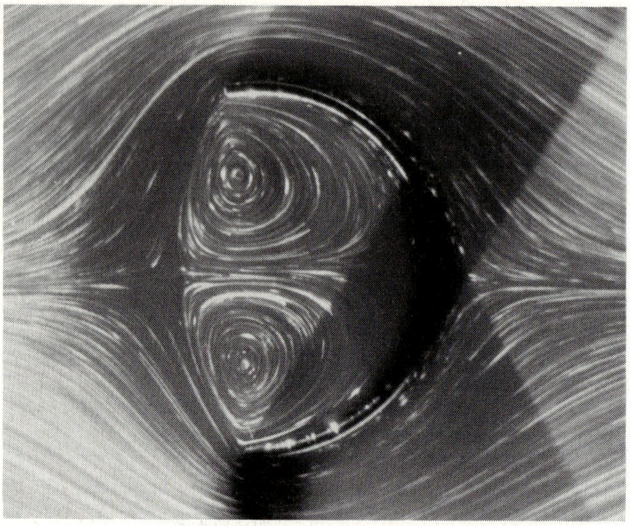

Figure 2
Streamlines in steady flow around a semi-circular arc at Reynolds number 3.1×10^{-2} (aluminum flake method)

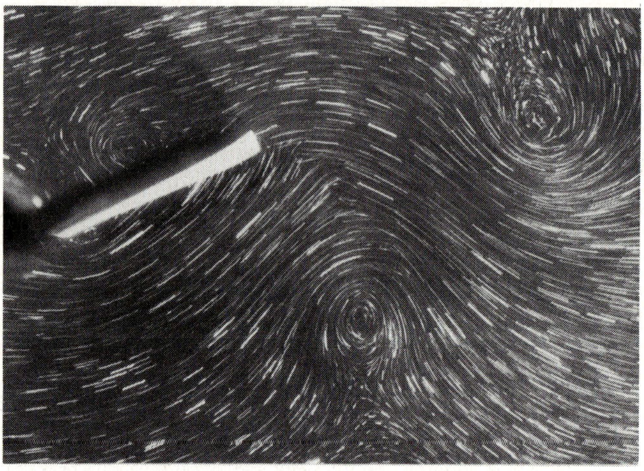

Figure 3
Streamlines in unsteady flow around a plate oscillating about its leading edge (aluminum flake method)

VISUALIZATION OF PARTICLE PATHS

A particle path is the path of a particular fluid particle in the flow field. A way of obtaining particle paths is to suspend a small number of fine tracer particles in the fluid and photograph their motion with a long exposure time. The particle paths starting from particular points of the flow field are obtained by introducing tracer particles into the flow at the points and taking a long-time exposure photograph of the motion of the tracer particles. Oil droplets, air bubbles, illuminant particles, etc. are used as the tracer particles in water, and soap bubbles, metaldehyde flakes, etc. in air.

VISUALIZATION OF STREAKLINES

A streakline is the instantaneous loci of all fluid particles which have passed through a given point in the flow field. In practice, streaklines are observed by continuously injecting tracer material into the flow at selected positions. If the flow is steady, the streamline, particle path and streakline coincide. In unsteady flows, however, these three types of lines are different one another. For observing streaklines, the electrolytic precipitation method, hydrogen bubble method, pH indicator method, tellurium method, dye injection method, etc. are used in water, and the smoke injection method, smoke-wire method, etc. in air. Figure 4 shows an example of the streakline pattern obtained by means of the electrolytic precipitation method. Figure 5 shows the streamline and streakline pattern of the unsteady flow behind a circular cylinder. The photograph was taken using the aluminum flake method and the electrolytic precipitation method simultaneously. The aluminum flakes show the streamline pattern and white dye produced by electrolytic precipitation method shows the streaklines. It will be seen that the streamlines do not coincide with the streaklines, and that the vortex centers and saddle points in the streamline pattern are not located on the streaklines.

Figure 4
Streaklines in unsteady flow around a plate oscillating about its leading edge (electrolytic precipitation method)

It should be noted here that the hydrogen bubble method and the smoke-wire method have a drawback [9]. These methods need fine metal wires from which tracer particles are released. When a tracer-generating wire is located on the upstream

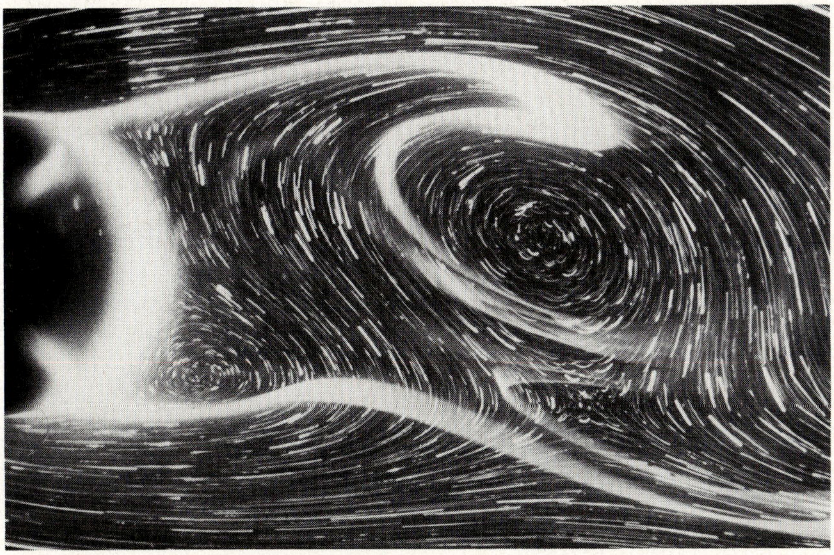

Figure 5
Streamlines and streaklines in unsteady flow behind a circular cylinder at Reynolds number 100

side of an obstacle, the wake of the wire induces a pair of horse-shoe vortices around the obstacle. Consequently, the tracer particles do not indicate the true flow pattern. Figure 6 shows the streakline pattern of the flow around a circular cylinder obtained by means of the electrolytic precipitation method. The gap between the obstacle and the sheet of the tracer particles increases as the diameter of the wire is increased and the distance between the wire and the obstacle is decreased.

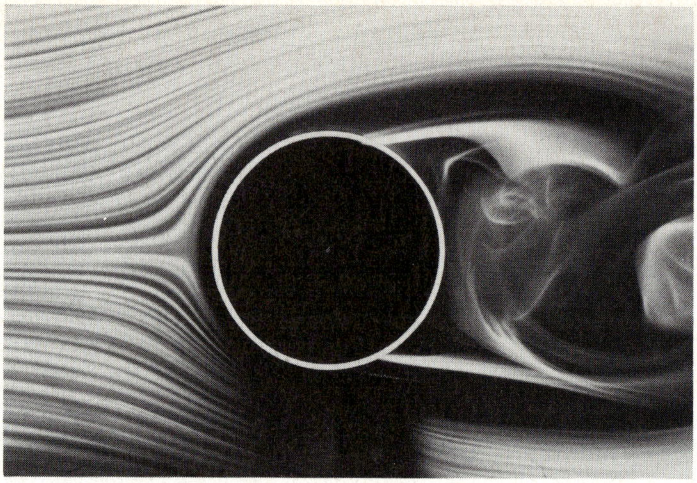

Figure 6
Streakline pattern around a circular cylinder at Reynolds number 650 (electrolytic precipitation method)

VISUALIZATION OF VORTEX MOTIONS

Figure 7 shows the streamlines and the lines of constant vorticity around a circular cylinder calculated using the numerical method. The Reynolds number is 40 and the flow is steady. What is important is that the position of the vortex centers in the streamline pattern are quite different from the positions where the vorticities are concentrated. However, it should be noted here that as yet no technique has been developed for visualizing the vorticity distribution.
Figure 8 shows an example of the variation of streamline and streakline patterns with time for the case when the velocity of a circular cylinder is increased rapidly from one steady

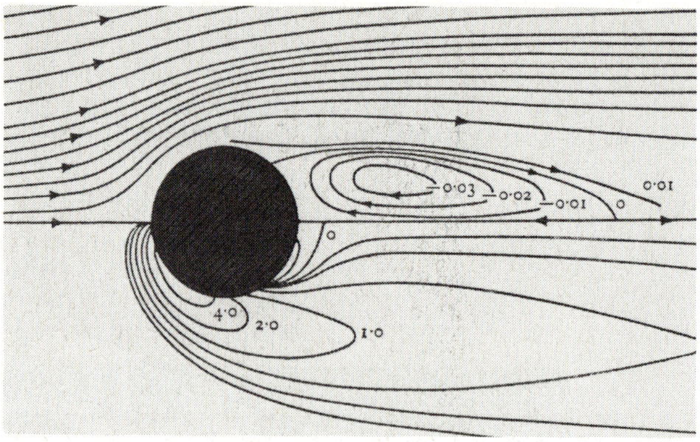

Figure 7
Calculated distributions of stream function and vorticity in flow past a circular cylinder at Reynolds number 40

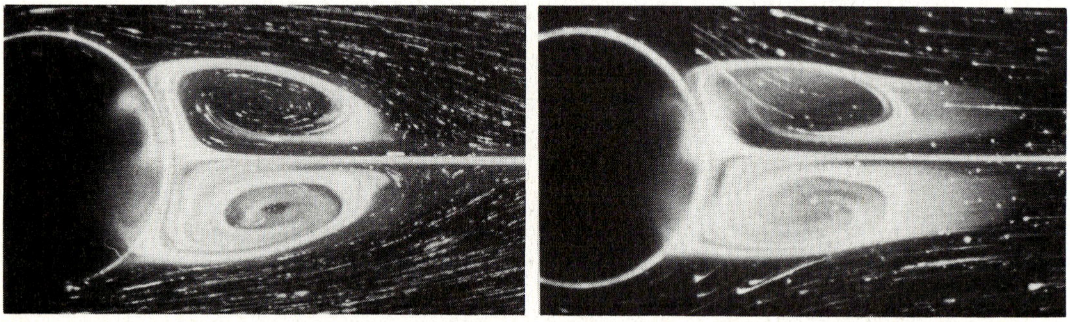

(a) steady flow at Reynolds number 22 (b) immediately after velocity change
Figure 8
Streamlines and streaklines around a circular cylinder accelerated abruptly (electrolytic precipitation method and aluminum flake method)

velocity to another higher velocity. The photographs were taken using the aluminum flake method and the electrolytic precipitation method simultaneously. It will be seen that at small times after the velocity change the primary separation bubble disappears in the streamline pattern, while it remains unaltered in the streakline pattern. Since the streamline, streakline and particle path do not coincide one another, great care should be taken in determining the flow structure. Hama (1962) [10] has discussed the shape of streamlines and streaklines in a sinusoidally perturbed shear flow, and found that the streamlines with respect to the reference frame moving with the wave velocity take on the so called cat's-eye streamlines and the streakline near the critical layer exhibits an apparent amplification and rolling up in spite of the fact that there is no amplification of the wave. Figure 9 shows schematic sketches of the flow patterns in a boundary layer flow with travelling instability waves.

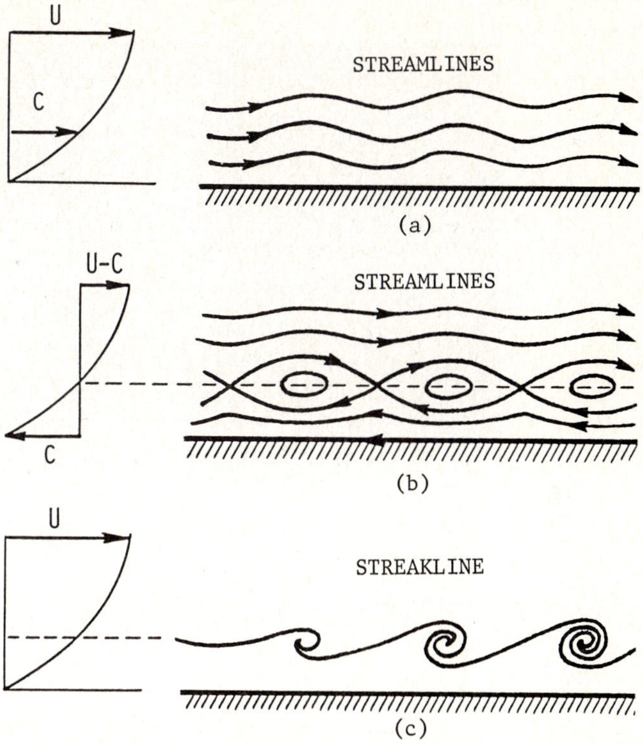

Figure 9
Streamlines and streaklines in a boundary layer with travelling instability waves; C is the wave velocity

If an instability wave grows in a shear layer, the vorticity distribution in the shear layer becomes periodical. Consequently, a row of vortices are formed. For example, the Tollmien-Schlichting wave forms a row of vortices, and the Karman vortex street, the Taylor vortices, the Goertler vortices, etc. are a manifestation of the presence of instability waves.
Figure 10 shows the wake behind a flat plate at zero incidence. The photograph was taken using the aluminum flake method and the condensed milk method simultaneously. It will be seen that the streamline pattern clearly indicates the existence of the double row of vortices, while the streakline pattern shows the appearance of a progressive wave motion whose amplitude increases in the downstream direction.

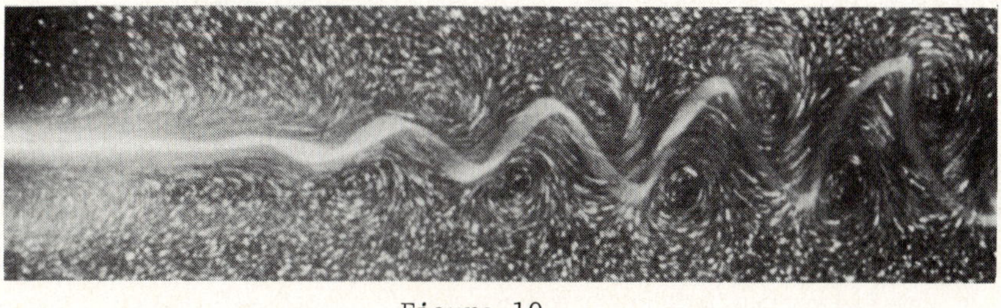

Figure 10
Streamline and streakline patterns behind a flat plate at zero incidence (aluminum flake method and condensed milk method)

VISUALIZATION OF TIMELINES

A timeline is the curve formed by fluid particles which have started simultaneously from a straight line across the flow. The timeline indicates the local velocity

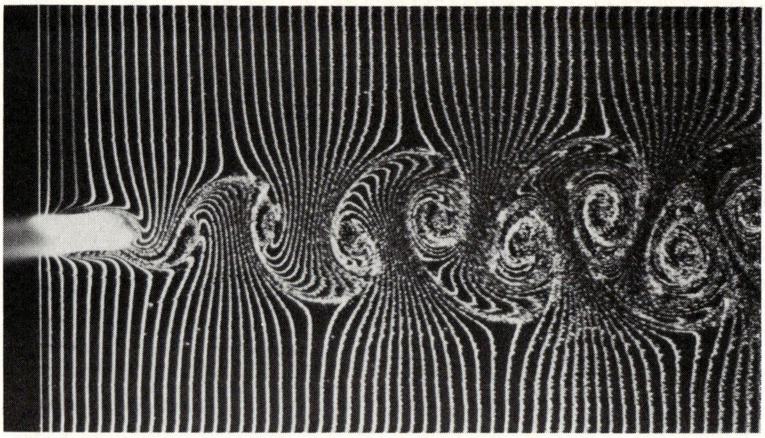

Figure 11
Timelines behind a circular cylinder at Reynolds number 152 (hydrogen bubble method)

profile. Timelines are obtained using the hydrogen bubble method, electrolytic pH indicator method, pulse luminescence method, etc. in water, and the smoke-wire method, spark tracer method, etc. in air. Figure 11 shows the timelines in the wake behind a circular cylinder.

VISUALIZATION OF FLOW DIRECTIONS

For visualizing the flow direction the tuft method and the aluminum flake method are convenient.
The tuft method and the tuft-grid method have the advantage that they can be used in a wide range of flow velocity both in air and in water, and that the photograph of the tufts are easily taken with a usual illumination.
The tracer particles, such as aluminum flakes, mica platelets, graphite powder, etc., orient themselves parallel to the direction of local flow. Consequently,

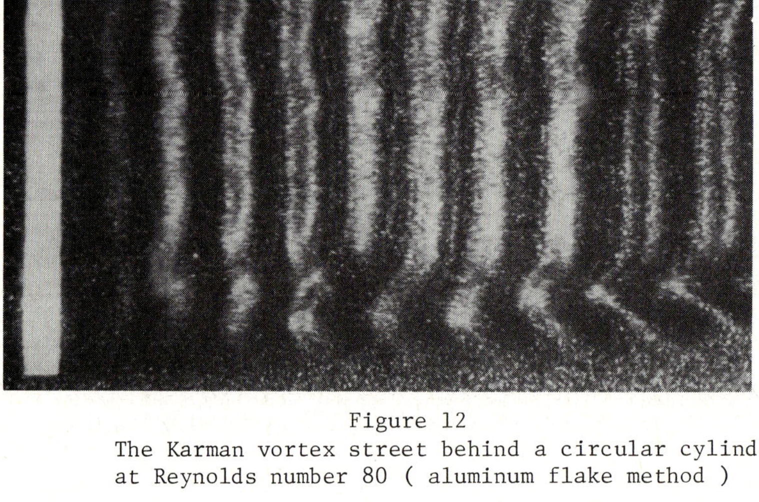

← direction of illumination

Figure 12
The Karman vortex street behind a circular cylinder at Reynolds number 80 (aluminum flake method)

the brightness of reflected light varies according to the flow direction. Since these tracer particles respond very rapidly to the change in flow direction, information contained in a single photograph is information about the current state of the motion. Figure 12 shows an example of the photographs obtained by this technique.

VISUALIZATION OF FLOW SEPARATION

As is well known, in the case of steady two-dimensional flow over a fixed wall the separation point can be determined using the Prandtl condition, or the vanishing of the wall shear. It should be noted that in the case of two-dimensional flow over a fixed wall the Prandtl separation point coincides with the streamline separation point and the streakline separation point. However, the Prandtl condition is meaningless in the cases of two-dimensional unsteady flow, two-dimensional steady flow over a moving wall and three-dimensional flow. Figure 13 shows the streamline pattern of the steady flow around a circular cylinder rotating about its center in a uniform flow. It will be seen that although a large isolated dead water region exists some distance downstream from the cylinder the flow does not separate from the surface of the cylinder. However, the flow around a rotating circular cylinder can separate three-dimensionally. Figure 14 shows the flow around a circular cylinder rotating in still water. The tracer particles generated on the whole surface of the cylinder by means of the electrolytic precipitation method separate from many lines spaced regularly on the cylinder surface, and form a series of ring-shaped vortices.

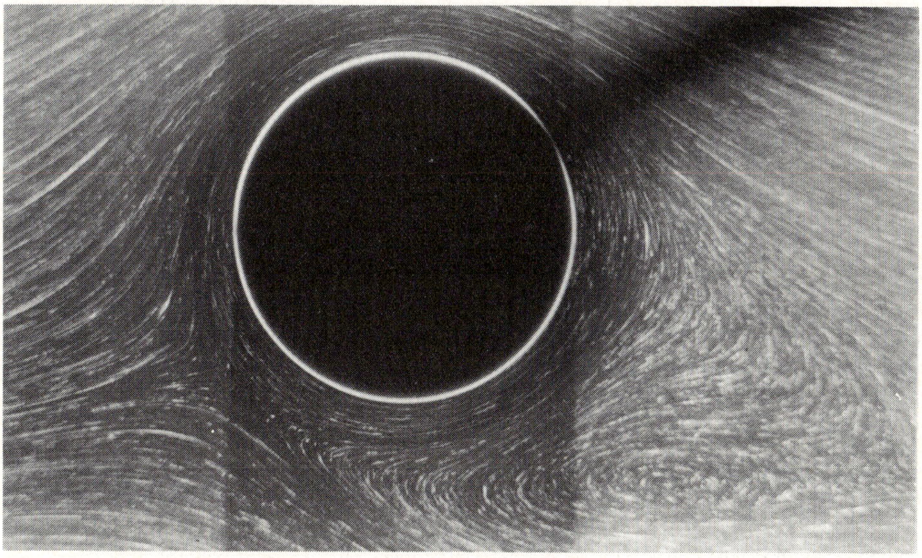

Figure 13
Streamlines around a circular cylinder rotating in a uniform flow; Reynolds number 68 and V/U = 2.1, where V is the circumferential velocity and U the velocity of undisturbed flow (aluminum flake method)

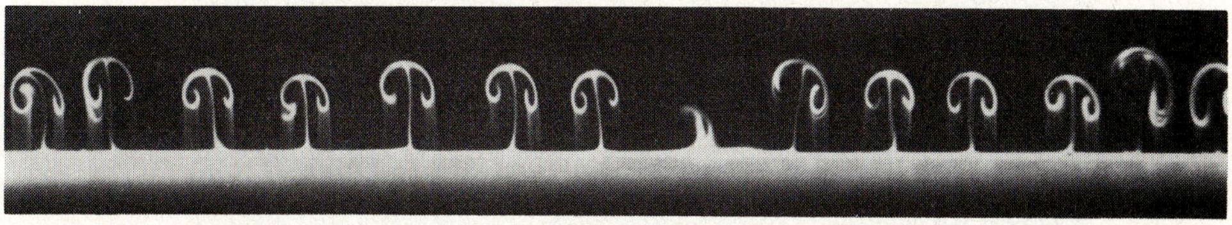

Figure 14
Sheets of tracer particles separated from the surface of a rotating circular cylinder (electrolytic precipitation method)

It seems that the separation of fluid particles from the wall is the most meaningful definition of flow separation for all kinds of flows. This definition coincides with the Prandtl condition in the case of two-dimensional flow over a fixed wall. The only way to observe the behavior of fluid particles close to the

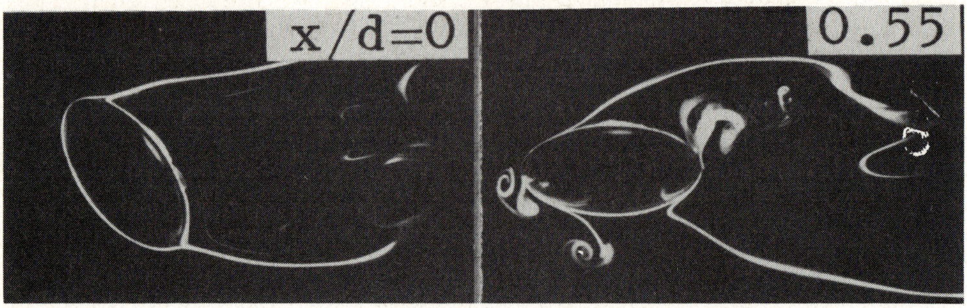

Figure 15
Integrated streaksheets around an elliptic cylinder whose angle of incidence has been abruptly decreased; x is the distance from the start of unsteady motion (electrolytic precipitation method)

body surface is to place tracer material on the whole surface of the body. The streaklines started from all the points on the whole surface separate from the surface in the form of thin sheets. If we call the sheets composed of all fluid particles which come out of the whole surface the "integrated streaksheets", flow separation can be detected by observing the integrated streaksheets. Figure 15 shows the integrated streaksheets around an elliptic cylinder after the angle of incidence has been abruptly decreased from 60° to 0°. It will be seen that the integrated streaksheets are separated from many lines on the surface of the elliptic cylinder immediately after the angle of incidence has been changed. When the flow is strongly unsteady, it is generally difficult to determine the integrated streaksheet separation lines, because the thickness of the newly developed boundary layer is extremely small.

Figure 16 shows the integrated streaksheets around an elliptic cylinder performing a rotatory oscillation about its center in a uniform flow. What is interesting is that the integrated streaksheet separates from the elliptic cylinder in the form of a single sheet, and that an isolated large dead-water region is formed some distance downstream of the cylinder.

The integrated streaksheets behind an oil drop falling through still water can be easily made visible by means of a dye in the drop. The dye is scrubbed from the drop as it passes through the water. Figure 17 shows the wake of a carbon tetrachloride drop.

Integrated streaksheets can be rendered visible using the electrolytic precipitation method, condensed milk method, kalium iodite starch method, electrolytic pH indicator method, etc. in water, and the titanium tetrachloride method etc. in air.

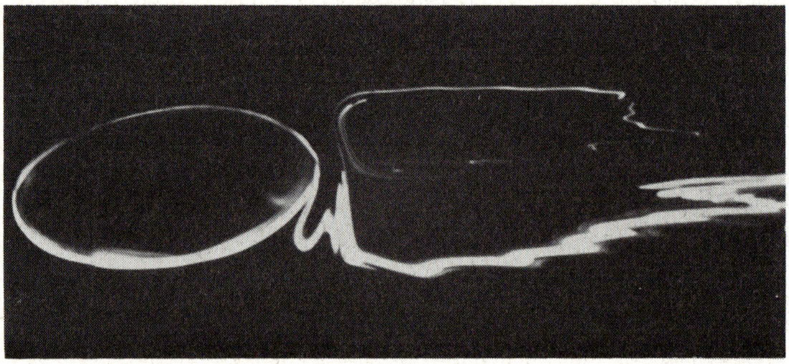

Figure 16
Integrated streaksheets around an elliptic cylinder performing a rotatory oscillation about its center (electrolytic precipitation method)

VISUALIZATION OF SURFACE PATTERNS

Various methods are available for revealing the state of the flow close to a solid surface. The oil-flow method, sublimation method, evaporation method, china-clay method, luminescent lacquer method, etc. are used in air, and the oil paint method, hydroquinon diacetate method, etc. in water. The surface indicator techniques can indicate whether the boundary layer is laminar or turbulent, attached or separated. At some time after the start of the test, the coating will have been removed in the region of the turbulent boundary layer whilst remaining in the laminar layer.

Oil-flow patterns, however, need cautious interpretation, because the oil particles are moved by the actions of their own gravitational, and viscous forces and the external pressure forces and surface shears. Figure 18 shows an example of the oil-flow pattern on the rear surface of a sphere at Reynolds number 4.7×10^5.

To visualize an instantaneous temperature field close to the wall temperature-sensitive liquid crystals can be used. If the wall is coated with a very thin film of such liquid crystals, the color of the film changes with the temperature on the wall.

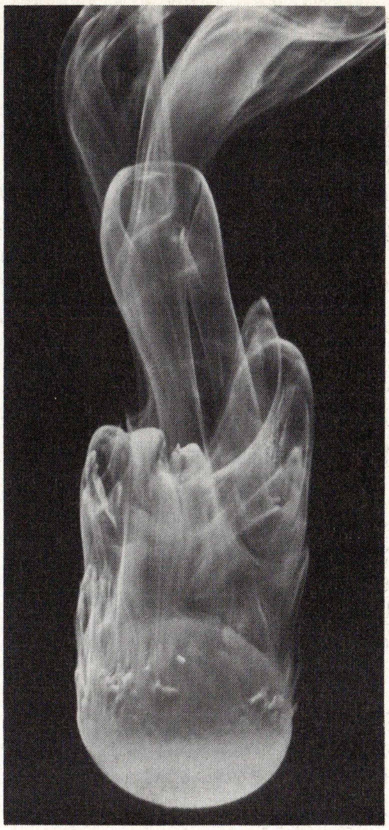

Figure 17
Integrated streaksheets behind an oil drop falling through still water

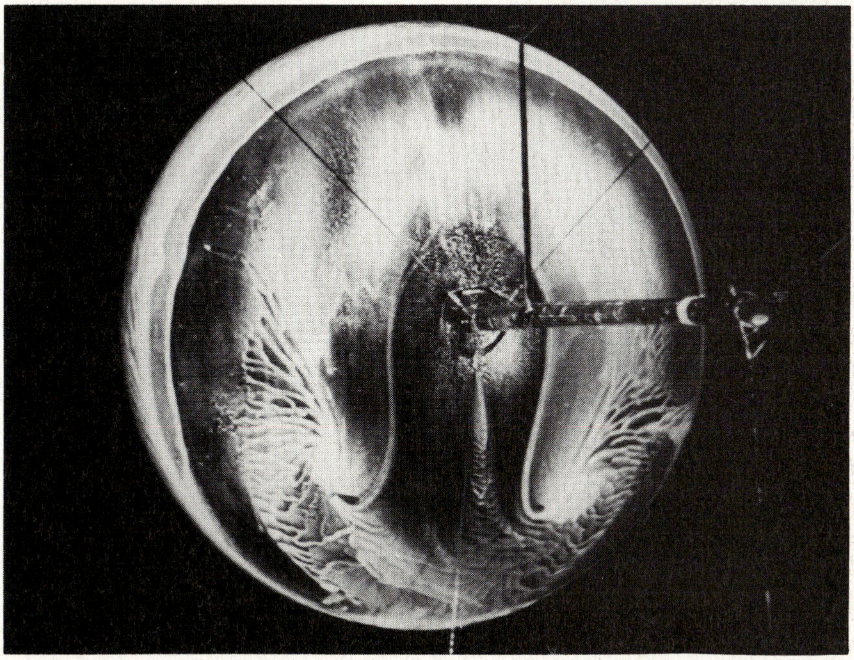

Figure 18
Oil-flow pattern on the rear surface of a sphere at Reynolds number 4.7×10^5

VISUALIZATION OF DENSITY DISTRIBUTIONS

The visualization of density distribution in flows is an important means of understanding high-speed flows, density stratified flows, mixing of two flows with density differences, thermal convection, temperature stratified flows, etc.
The optical methods (shadowgraph method, schlieren method, Mach-Zehnder interferometer method, holographic method, etc.), electron beam method, glow discharge method, etc. are used to visualize the density distribution in fluids.
Figure 19 shows the shadowgraph of free convection flow over a point heat source. The distortions produced in a set of diagonal lines are viewed through the water tank.

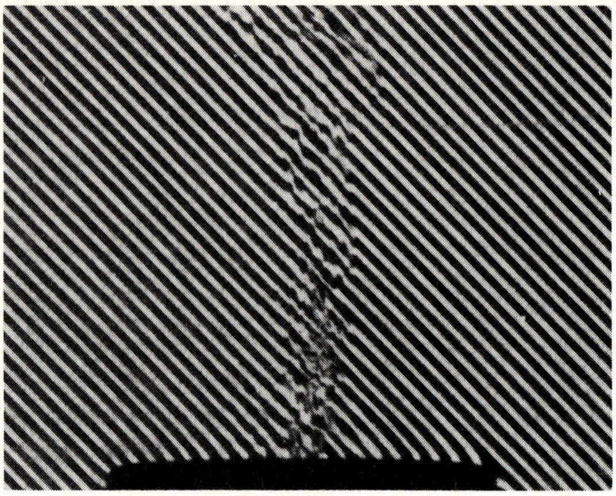

Figure 19
Shadowgraph of free convection flow

VISUALIZATION OF STRESS DISTRIBUTIONS

Certain pure liquids, homogeneous solutions and colloidal suspensions are isotropic when at rest, but become birefringent when a preferred orientation is imposed on the molecules or suspended particles by an external force. This makes it possible for an analysis to be made using polarized light as in the analogous subject of photo-elasticity. However, the method has not been extensively used as a research tool because the phenomenon is still not fully understood.
It is a property of certain concentrations of glycerine-water solutions that the planes of equal shear in the liquid become visible in ordinary light, if viewed along a path tangent to the shear plane. However, it is not known at present just why the lines are visible [11]. Figure 20 shows an example of the photograph obtained by means of this method.

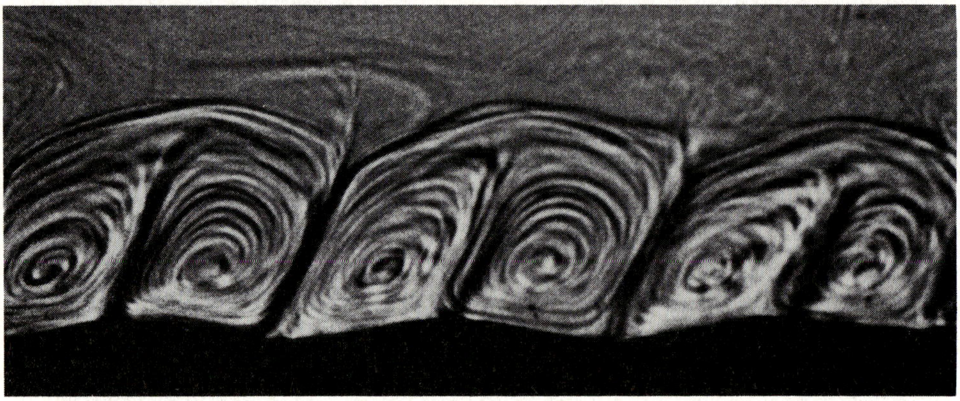

Figure 20
Steady streaming induced by oscillatory flow over a wavy wall [12]

VISUALIZATION OF TURBULENT FLOWS

Many flow visualization experiments have been carried out to examine the structure of turbulent flows. Nowadays it is well known that turbulent flows are characterized by a remarkable degree of order. It seems that a turbulent flow consists of many kinds of coherent structures. For determining the structures of turbulent flows, the smoke injection method, smoke-wire method, shadowgraph method, schlieren method, etc. are used in air, and the dye injection method, aluminum flake method, hydrogen-bubble method, etc. in water. Figures 21 and 22 show the flow patterns of the turbulent wake behind a circular cylinder. It will be seen that the wake contains at least two kinds of instability waves, two-dimensional instability waves (Karman vortex street type waves) and three-dimensional instability waves (Goertler vortex type waves), and that the wavelength of the two-dimensional instability waves increases in the downstream direction.

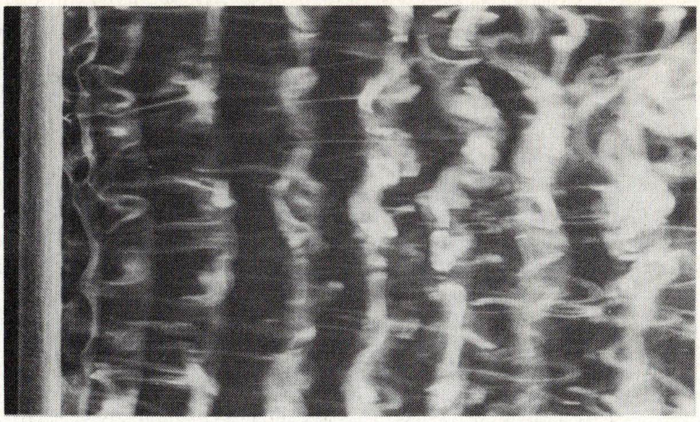

Figure 21
Two-dimensional and three-dimensional instability waves in the wake behind a circular cylinder at Reynolds number 348 (electrolytic precipitation method)

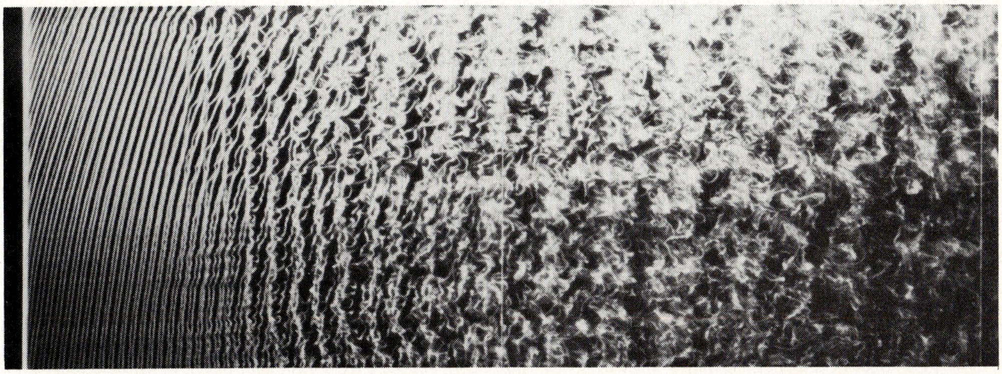

Figure 22
Variation of wavelength of two-dimensional instability waves with distance from the cylinder at Reynolds number 360 (smoke-wire method)

Figures 23, 24 and 25 are the flow patterns of the turbulent boundary layer on a flat plate. From the photographs it may be suggested that the turbulent boundary layer contains at least three kinds of instability waves. They are large-scale two-dimensional waves (Tollmien-Schlichting type waves), large-scale three-dimensional waves (Goertler vortex type waves in the outer region of the boundary layer) and small-scale three-dimensional waves (Goertler vortex type waves in the near wall region).

Figure 23
Two-dimensional instability waves in the turbulent boundary layer on a flat plate at $R_{\delta *}$= 5000 [side view] (smoke-wire method)

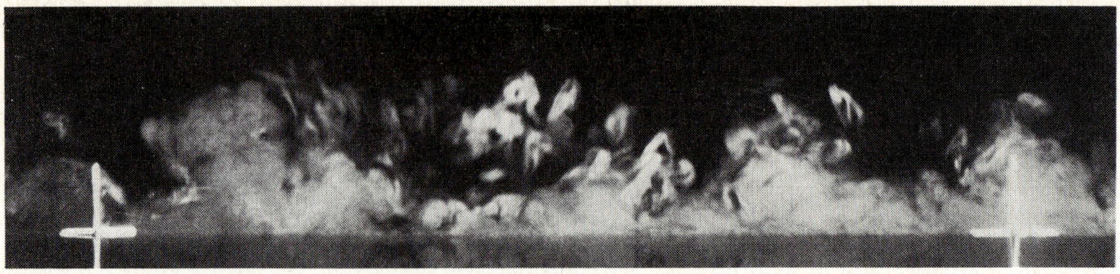

Figure 24
Large-scale three-dimensional instability waves in the turbulent boundary layer at $R_{\delta}*= 4600$ [transverse section] (titanium tetrachloride method)

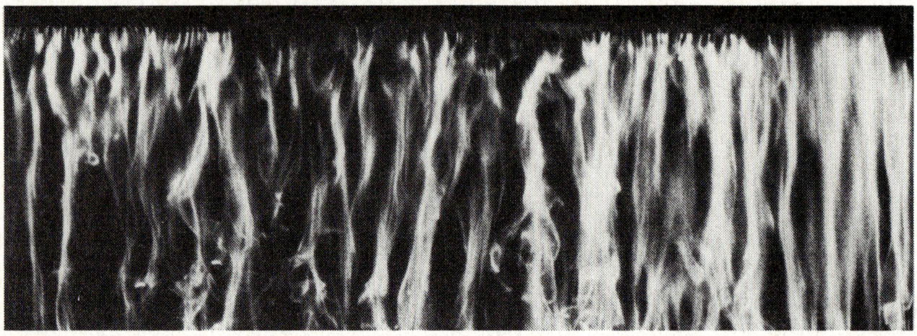

Figure 25
Small-scale three-dimensional instability waves in the near-wall region of the turbulent boundary layer at $R_{\delta}*= 2100$ (smoke-wire method)

CONCLUDING REMARKS

Flow visualization is the most useful research tool for obtaining information on the flow structure. Recently many new techniques, such as the laser technique, holographic interferometry, digital image processing, computer-generated color graphics, computed tomography, nuclear magnetic resonance, etc. have been developed in the field of flow visualization. Flow visualization may continue to play a crucial role in the development of our understanding of the flow phenomena.

REFERENCES

[1] Pankhurst, R.C. and Holder, D.W., Wind-tunnel technique (Pitman, London, 1965).
[2] Clayton, B.R. and Massey, B.S., Flow visualization in water: a review of technique, J. Sci. Instrum. 44 (1967) 2-11.
[3] Werle, H., Hydrodynamic flow visualization, Annu. Rev. Fluid Mech. 5 (1973) 361-382.
[4] Merzkirch, W., Flow visualization (Academic, New York, 1974).
[5] Asanuma, T. (ed.), Flow visualization (Hemisphere, Washington, 1979).
[6] Merzkirch, W. (ed.), Flow visualization II (Hemisphere, Washington, 1982).
[7] Preprints of the third international symposium on flow visualization, September 6-9, 1983, Ann Arbor, USA.
[8] Van Dyke, M., An album of fluid motion (Parabolic, Stanford, 1982).
[9] Taneda, S., Honji, H. and Tatsuno, M., The behaviour of tracer particles in flow visualization by electrolysis of water, J. Phys. Soc. Japan 37 (1974) 784.
[10] Hama, F.R., Streaklines in a perturbed shear flow, Phys. Fluids 5 (1962) 644-650.
[11] Hagerty, W.W. and Mich, A.A., Use of an optical property of glycerine-water solutions to study viscous fluid-flow problems, J. Appl. Mech. 17 (1950) 54-58.
[12] Kaneko, A. and Honji, H., Double structures of steady streaming in the oscillatory viscous flow over a wavy wall, J. Fluid Mech. 93 (1979) 727-736.

LIST OF CONTRIBUTED PAPERS PRESENTED AT THE CONGRESS

ABEYARATNE, Rohan and TRIANTAFYLLIDIS, N.: On the Localization of Deformation in a Foam Type Rubber using the Homogenization Theory Approach.

ABOUDI, J.: See BENVENISTE, Y.

ADLER, P.M. and BRENNER, Howard: Spatially Periodic Suspensions of Convex Particles.

ADVANI, Sunder, TOROK, J., LEE, J.K., and KHATTAB, H.: Some Exact and Numerical Solutions Associated with the Mechanics of Hydraulic Fracturing.

AKYLAS, T.R. and DEMURGER, J.-P.: Finite-Amplitude Stability of Slowly-Rotating Pipe Flow.

ALTENBACH, J.: A Generalized Semi-Moment Vlassow Shell-Theory for Prismatic and Nonprismatic Thinwalled Structures.

ALWIS, W.A.M.: See WANG, C.-M.

AMBARTSUMIAN, S.A., BELUBEKIAN, M.V., and KAZARIAN, K.B.: Magnetoelastic Surface Waves on an Interface between two Electro-Conducting Solids.

AMBARTSUMIAN, S.A. and MINASSIAN, M.M.: On the Model of Bodies with the Mechanical Properties Depending on the Strain Rate.

AMBERG, G., DAHLKILD, A., and BARK, F.: Sediment Transport in some Unsteady Settling Processes.

ANDERSON, C.: See BUCKMASTER, J.

ANDERSSON, Lars-Erik: See LUNDBERG, Bengt.

ANTAR, Basil N.: Finite Amplitude Baroclinic Waves of Eady's Model.

APPLEBY, J.C. and BRINDLEY, J.: The Effects of Modal Truncation on Unsteady Solutions in a Geophysical Model.

ARAI, M.: See YAMAMOTO, Yoshiyuki.

ARBOCZ, Johann, VERMEULEN, P.G., and GEER, J. van: The Buckling of Axially Compressed Imperfect Shells with Elastic Edge Supports.

AREF, Hassan and TRYGGVASON, Gretar: Interface Dynamics by the Vortex-in-Cell Method.

ARYA, V.K.: On the Creep Deformation, Bulge Behaviour and Failure of Zircaloy-Tubes.

ATADAN, A.S. and HUSEYIN, Koncay: Bifurcation Analysis of Aircraft Pitching Motions and Brusselator via an Intrinsic Perturbation Procedure.

ATANACKOVIC, T.M.: See DJUKIC, Djordje S.

ATANASIU, N.E. and IRIMESCU, B.R.: Fatigue Crack Growth in the High-Rate Regime for a Medium Strength Steel.

ATASSI, H. and GRZEDZINSKI, Janusz: Three-Dimensional Gust Acting on Joukowski Airfoil.

ATTEN, P., CAPUTO, J.G., MALRAISON, B., and GAGNE, Y.: Experimental Determination of Attractors Dimension for Various Flows.

AZOUNI, Maherzia Aza: See BOLTON, Edward W.

BABCOCK, Charles D.: Time Scales in Dynamic Buckling.

BAER, M.R.: See NUNZIATO, Jace W.

BAHAR, Leon Y. and KWATNY, Harry G.: Extension of Noether's Theorem to Non-Conservative Dynamical Systems.

BAI, Y.-L.: See CHENG, C.-M.

BAJAJ, Anil K.: Bifurcations in a Parametrically Excited Nonlinear Oscillator.

BAKKER, P.G.: The Flow around an External Corner as a Bifurcation Problem.

BALSER, Jeffrey R.: See COWIN, Stephen C.

BAMMANN, Douglas J. and JOHNSON, George C.: Descriptions of Inelastic Behavior in Finite Deformation Plasticity.

BANIOTOPOULOS, C.C.: See PANAGIOTOPOULOS, P.D.

BANKS-SILLS, Leslie and BORTMAN, Yaacov: Reappraisal of the Quarter-Point Quadrilateral Element in Fracture Mechanics.

BARAL, G.C. and SHARMA, R.N.: Transition Process in Pulsating Tube Flows. An Experimental Investigation.

BARBER, J.R.: A Two-dimensional Transient Thermoelastic Contact Problem.

BARK, F.: See AMBERG, G.

BARTHES-BIESEL, Dominique and SGAIER, Hedi: Behaviour of a Capsule Suspended in Simple Shear Flow Role of Membrane Viscoelasticity.

BASDEVANT, C.: See CHOMAZ, J.M.

BASSANI, John L.: Crack Tip Stresses in a Creeping Solid.

BAUER, Jacek and SOKOLOWSKI, Jan: On the Unilateral Problem of a Rigid Punch Pressed on an Elastic Arch.

BEARMAN, P.W., DOWNIE, M.J., and GRAHAM, J.M.R.: Vortex Induced Roll Damping of a Barge in Waves.

BECHTEL, S.E. and BOGY, David B.: On the Stability of Initially Unrelaxed Non-Newtonian Capillary Jets.

BEEK, P. van: A Unifying Approach to Renormalization and Effective Medium for Media with Random Inclusions.

BEGUIER, C.: See CALAMOTE, J.

BEHRENDT, Lars, JONSSON, Ivar G., and SKOVGAARD, Ove: A Finite Element Model for the Calculation of Wind Wave Diffraction.

BELTZER, Abraham I. and BRAUNER, Neima: Waves of an Arbitrary Frequency in Random Composites via the K-K Relations.

BELUBEKIAN, M.V.: See AMBARTSUMIAN, S.A.

BELYAEV, Yu.N.: See YAVORSKAYA, I.M.

BELYAEV, Yu.N., MONAKHOV, A.A., SCHERBAKOV, S.A., and YAVORSKAYA, I.M.: Transition Sequences to Stochasticity in a Spherical Couette Flow and the System's Attractors Properties.

BENDIKSEN, Oddvar O.: Aeroelastic Stabilization by Disorder and Imperfections.

BENDSØE, Martin, OLHOFF, Niels, and SOKOLOWSKI, Jan: Sensitivity Analysis of Problems of Elasticity with Unilateral Constraints.

BENDSØE, Martin: Optimal Design of Rib-Reinforced Plates.

BENVENISTE, Y. and ABOUDI, J.: The Effect of Debonding on the Mechanical Behaviour of Fiber-Reinforced Composites.

BERGAMASCHI, S. and LUCCHETTI, F.: A Preliminary Investigation on the Natural Frequencies of Monopiles.

BERGAN, Pål G., MOLLESTAD, Egil, and SANDSMARK, Nils: Nonlinear Static and Dynamic Response Analysis of Floating Off-Shore Structures.

BERGER, S.A.: Instability of a Liquid Jet: The Initial-Value Problem.

BEVILACQUA, Luis and NORONHA, Danro B.: Bending Free Shells under an Axial Force.

BICANIC, N.: See WILLAM, Kaspar J.

BIDINI, Gianni and MARTELLI, F.: Numerical Calculation of Condensing Flows.

BIESHEUVEL, A. and GORISSEN, W.C.M.: On Void Fraction Waves in Bubbly Liquids.

BLYTHE, P.A., DANIELS, P.G., and SIMPKINS, P.G.: High Rayleigh Number Cavity Flows Driven by Horizontal Differential Heating.

BODNER, Sol R.: See SHACHAM, Itzhak.

BODUROGLU, Hasan: Dynamic Stability of Polar Orthotropic Annular Plates.

BÖHM, Juhani von: See RATY, Raimo.

BOER, R. de and EHLERS, W.: On the Problem of Fluid- and Gas-Filled Elasto-Plastic Porous Continua.

BOGACZ, R. and MAHRENHOLTZ, O.: On Design of Structures under Circulatory Load.

BOGY, David B.: See BECHTEL, S.E.

BOGY, David B. and YANG, H.J.: Elastic Wave Scattering from an Interphase Crack in a Layered Half Space.

BOLTON, Edward W., AZOUNI, Maherzia Aza, and BUSSE, Friedrich H.: Drifting Convection Columns in a Rotating Annulus.

BORTMAN, Yaacov: See BANKS-SILLS, Leslie.

BOS, H.J.: Solutions of the Two-Dimensional Wave Equation of Non-Integer Degree of Homogeneity.

BOSZNAY, Adam: Improvable Frequency Constraint Inequalities.

BOUKARY, M.S.: See LEBON, G.

BOULALA, M. and DAMOU, M.: Optimization of Mechanical Vibration Isolation Systems with Multi-Degrees of Freedom.

BRADY, John F.: See KOCH, Donald L.

BRÆSTRUP, Mikael W.: Pull-Out and Punching of Concrete - Lower-Bound Plastic Analysis.

BRAUNER, Neima: See BELTZER, Abraham I.

BRENNER, Howard and NADIM, Ali: Antisymmetric Stresses Induced by the Rigid-Body Rotation of Suspensions of Dipolar Particles in an External Field: Vortex Flows.

BRENNER, Howard: See ADLER, P.M.

BRINDLEY, J.: See APPLEBY, J.C.

BRØNS, Morten: A Mathematical Analysis of Surge in Compressor Systems.

BRORSEN, Michael and LARSEN, Jesper: Open Boundaries in Gravity Wave Simulations with the Boundary Integral Equation Method.

BROWN, E.H. and BURGOYNE, C.J.: Non-Uniform Elastic Torsion.

BROWN, Susan N., CHENG, Hsien K., HAFEZI, Hamid: The Evolution of Wave and Anti-Cyclonic Patterns above a Thin Obstacle in a Rotating Stratified Fluid.

BUCHIN, V.A., LYUBIMOV, G.A., and TRIFONOV, V.D.: Experimental Relization of Unstable Regimes in Flow Reactors.

BUCKINGHAM, Alfred C.: See HALL, Mary.

BUCKMASTER, J., ANDERSON, C., and NACHMAN, A.: A Model for Intumescent Paints.

BUCKMASTER, J. and PETERS, N.: The Infinite Candle and its Stability - A Study of Buoyancy-Induced Flows.

BUDIANSKY, Bernard, HUTCHINSON, John W., and EVANS, Anthony G.: Matrix Fracture in Fiber-Reinforced Ceramics.

BURDESS, J.S. and METCALFE, A.V.: Active Control of Vibration Caused by Harmonic Exitation in Structures with Negligible Natural Damping.

BURGOYNE, C.J.: See BROWN, E.H.

BURTON, Tom D. and RAHMAN, Z.: On the Development of Chaos in the Duffing Oscillator.

BUSSE, Friedrich H.: See BOLTON, Edward W.

BUX, S.L.: See ROBERTS, John W.

CAFLISCH, Russel: Non-Linear Oscillation of a Bubbly Liquid.

CAHOUET, J., EUVRARD, D., GUTTMANN, C., LAHALLE, D., and LENOIR, M.: Some Recent Progress towards a Numerical Solution to the Problem of Ship-Wave Resistance.

CALAMOTE, J. and BEGUIER, C.: Destruction of the Karman Vortex Shedding by the Rotation of the Cylinder.

CALLADINE, C.R.: See PELLEGRINO, S.

CALLIAS, Constantine and MARKENSCOFF, Xanthippi: The Near Field of a Moving Dislocation Loop.

CAMOTIM, Dinar and ROORDA, John: On the Effect of Residual Stresses in the Plastic Postbuckling of Structures.

CAMPOS, L.M.B.C.: On the Propagation of Acoustic-Gravity Waves in Atmospheric Temperature Gradients.

CAPUTO, J.G.: See ATTEN, P.

CARLSSON, Janne, KAISER, Sten, and SONNERLIND, Henrik: The J-Integral Applied to Cases of Non-Proportional Loading.

CARROLL, M.M.: Expansion or Compaction of Elastic and Elastic-Plastic Hollow Spheres and Cylinders.

CARTMELL, M.P.: See ROBERTS, John W.

CARVALHO, Sonia Pinto de: See KOILLER, Jair.

CASAS-VAZQUEZ, J.: See LEBON, G.

CASEY, J.: Strength-Differential Effect and Plastic Volume Expansion.

CERCIGNANI, Carlo: On the Speed of Propagation of Disturbances in a Gas.

CHAN, Eng-Soon and MELVILLE, W.K.: Deep Water Breaking Wave Forces on Surface Piercing Structures.

CHANDRA, Abhijit and MUKHERJEE, Subrata: A Beam Analysis of Large Strain Viscoplasticity Problems Including the Effects of Induced Material Anisotropy.

CHANG, Jeng-Shian: See LIBOVE, Charles.

CHAO, Ching Cheng and WANG, Chih Cheng: Stresses in Composite Foam Sandwich Cylinders with Multi-Stiffeners.

CHAVES, Humberto, SPECKMANN, Hans-Dieter, MEIER, Gerd E.A., and THOMSON, Philip A.: Splitting of Condensation Shock Waves and Evaporation Waves in Fluids of High Specific Heat.

CHEN, C.F. and THANGAM, S.: Convective Stability of a Variable-Viscosity Fluid in a Vertical Slot.

CHEN, Jingyn: See ZHAO, Ling-Cheng.

CHEN, Sixiong and OGILVIE, T. Francis: Solution of the Water Waves Generated by a Slowly Moving Two-Dimensional Body.

CHEN, Xiang-Jun: See ZHANG, Jin.

CHENG, C.-M., BAI, Y.-L., and YU, S.-B.: On Evolution of Thermo-Plastic Shear Band.

CHENG, Gengdong: See ZHONG, Wanxie.

CHENG, H.S.: See MILLER, G.R.

CHENG, Hsien K.: See BROWN, Susan N.

CHENG, Shun and HE, F.B.: An Accurate Theory and Simple Fourth Order Governing Equations for Orthotropic and Composite Cylindrical Shells.

CHIANG, C.R. and WENG, George J.: The Dependence of Polycrystal Plasticity on Single-Crystal Properties.

CHIEN, Wei-Zang: Further Study on Generalized Variational Principles in Elasticity.

CHIRIACESCU, Sergiu T., MANGERON, Dumitru, and CRACIUNAS, Aurora: Stability of Multivariable Dynamic Systems with Time-Lags. Application to Stability Analysis of Machine-Tool Chatter.

CHOMAZ, J.M., COUDER, Y., BASDEVANT, C.: Observation and Numerical Simulation of Modulated Waves in a Circular Shear Zone.

CHOSSAT, P. and IOOSS, G.: Primary and Secondary Bifurcations in the Couette-Taylor Problem.

CHRISTENSEN, Richard M. and FENG, William W.: Nonlinear Compressive Deformation of Viscoelastic Porous Materials.

CHRISTIANSEN, Peter Leth, IF, F., SØRENSEN, M.P., and SKOVGAARD, O.: Solitons and Chaos in Dynamical Systems.

CHUDNOVSKY, Alexander, DOLGOPOLSKY, A., and KACHANOV, M.: Stresses Near a Crack Tip Surrounded by Microcracks.

CINQUINI, Carlo, CONTRO, Roberto, and CORRADI, Leone: Finite Element Approach to Elastic-Plastic Structural Optimization.

CLEMENS, H.: See WAUER, J.

CLIFTON, Rodney J., KLOPP, R.W., and LI, C.H.: Pressure-Shear Impact and the Dynamic Viscoplastic Response of Metals.

COENE, R.: On the Froude Efficiency of Propulsion in Waves.

COLLET, B.: Nonlinear Wave Propagation in Deformable Dielectrics.

COMNINOU, Maria: The Thermoelastic Hertz Problem with Pressure Dependent Contact Resistance.

COMNINOU, Maria: See HILLS, David A.

COMO, M. and GRIMALDI, A.: Collapse Analysis of Masonry Structures under Vertical and Horizontal Loads.

CONTRO, Roberto: See CINQUINI, Carlo.

COOPER, John M.: See WALTON, Otis R.

CORRADI, Leone: See CINQUINI, Carlo.

COUDER, Y. and RABAUD, M.: Two Dimensional Turbulence in Thin Liquid Films.

COUDER, Y.: See CHOMAZ, J.M.

COWELL, R.G.: See HUNT, Giles W.

COWIN, Stephen C.: Mechanical Modeling of the Shrinking and Swelling of Porous Elastic Solids.

COWIN, Stephen C., HART, Richard T., KOHN, David H., and BALSER, Jeffrey R.: Surface Remodeling in Long Bone; a Comparison of Model Predictions with Animal Experiments.

CRACIUNAS, Aurora: See CHIRIACESCU, Sergiu T.

CRAMER, M.S., KLUWICK, A., WATSON, L.T., and PELZ, W.: The Propagation of Viscous, Nonlinear Waves in Fluids having both Positive and Negative Nonlinearity.

CRANDALL, Stephen H. and PARISEANU, George: Limit-Cycle Vibrations of a Rolling Cylinder.

CROCCO, Luigi and ORLANDI, Paolo: Advanced Calculation of the Non Local Interactions in Isotropic Turbulence.

CURRIE, I.G.: See VIOLA, J.

DAHAN, M. and PREDELEANU, M.: Closed Form Solutions for an Anisotropic Solid Weakened by a Ductile Crack.

DAHLKILD, A.: See AMBERG, G.

DAMOU, M.: See BOULALA, M.

DANIELS, P.G.: See BLYTHE, P.A.

DATTA, Subhendu K., WONG, K.C., and SHAH, A.H.: Dynamic Amplification of Surface Displacements Due to Subsurface Cavities.

DATTA, Subhendu K. and LEDBETTER, Hassel M.: Physical-Mechanical Properties of a Particle-Reinforced Composite: Measurements and Modeling.

DAVIES, John T.: Drop Sizes in Emulsions Related to Turbulent Energy Dissipation Rates.

DAVIS, S.H.: See MÜLLER, U.

DAVYS, J.W., HOSKING, R.J., and SNEYD, A.D.: Waves on Floating Ice Plates.

DEBIEVE, Jean-Francois and GAVIGLIO, Jean: Some Aspects of a Shock Wave Turbulence Interaction.

DEGER, Y.: Coupling Effects on Free Longitudinal, Torsional and Flexural Vibrations of Prismatic Bars.

DEHGHANYAR, Tejav J., MASRI, Sami F., and MILLER, Richard K.: Semi-Active Control of Nonlinear Flexible Structures.

DEMIRAY, H.: Incremental Modulus for Soft Biological Tissues.

DEMS, K.: See MROZ, Z.

DEMURGER, J.-P.: See AKYLAS, T.R.

DETEMPLE, E. and ECKELMANN, H.: The Influence of Sound Waves on Karman Vortex Streets.

DEVIENNE, R.: See SCIROCCO, V.

DIAZ, F.: See GIRALT, Francesc.

DIETERMAN, Harm A.: An Analytically Derived Lumped-Impedance Model for the Dynamic Behaviour of a Watertower.

DIETSCHE, C.: See MÜLLER, U.

DINGEMANNS, M.W.: See RADDER, A.C.

DJUKIC, Djordje S. and ATANACKOVIC, T.M.: An Extremum Variational Principle and Error Bounds.

DOLGOPOLSKY, A.: See CHUDNOVSKY, Alexander.

DONG, Ming: See ZHANG, Ruqing.

DOWELL, Earl H.: Observation and Evolution of Chaos for an Autonomous System.

DOWNIE, M.J.: See BEARMAN, P.W.

DOYLE, John F.: Dynamic Analysis of Curved Line Elements.

DUNAYEVSKY, Victor: The Effect of Strain Hardening on the Field Near a Rapidly Propagating Crack Tip.

DURBAN, David: Drawing and Extrusion of Composite Sheets, Wires and Tubes.

DVORAK, George J. and LAWS, Norman: Analysis of Damage Processes in Composite Laminates.

DYBBS, Alexander and EDWARDS, Robert V.: Darcy and Beyond - A New Look at Flow in Media.

EASSON, William James and GREATED, Clive: Impact of Breaking Waves on Horizontal Cylinders.

ECKELMANN, H.: See DETEMPLE, E.

EDWARDS, Robert V.: See DYBBS, Alexander.

EHLERS, W.: See BOER, R. de.

EL CHAZLY, Nihad M.: Formability of 3004, H-19 Alloy for Rigid Containers.

EL FATMI, R. and LADEVEZE, P.: Mechanics of Spotwelded Joints.

EL-ASHKAR, I.: See NOVAK, M.

ELISHAKOFF, Isaac: Probabilistic Methods in Buckling and Vibration.

ELLERMEIER, Wolfgang: Resonant Surface Waves over a Wavy Bottom.

ELLYIN, Fernand and KUJAWSKI, D.: Crack Growth Rate Model for Cyclic Loading.

ELZANOWSKI, Marek and EPSTEIN, Marcelo: Uniformity Characterization in Hyper-elasticity.

ENFLO, Bengt: Saturation of a Nonlinear Cylindrical Sound Wave Generated by a Sinusoidal Source.

EPSTEIN, J.S.: See SMITH, C.W.

EPSTEIN, Marcelo: See ELZANOWSKI, Marek.

ESSLINGER, Maria and SCHMIDT, Herbert: Theoretical and Experimental Investigations of Thin-Walled Shells of Revolution in the Elastoplastic Range.

EUVRARD, D.: See CAHOUET, J.

EVANS, Anthony G.: See BUDIANSKY, Bernard.

FAN, Jinghong and VALANIS, K.C.: Endochronic Elastoplastic Constitutive Equation and its Experimental Verification in Spatially Varying Strain Fields.

FAN, Y.: See KRASINSKI, J. de.

FAUVE, S., LAROCHE, C., and PERRIN, B.: Competing Instabilities in a Rotating Layer of Mercury Heated from Below.

FENG, Willian W.: See CHRISTENSEN, Richard M.

FENG, William W.: Inflation of a Plane Circular Viscoelastic Membrane.

FENTON, R.G.: See TABARROK, B.

FISHER, Martin J. and HERRMANN, George: Acoustoelastic Measurements of Elastic-Plastic and Residual Stresses.

FOMINA, N.I.: See YAVORSKAYA, I.M.

FOSDICK, R.L. and MAC SITHIGH, Gearoid P.: Minimum-Energy Conditions for Constrained Finite Elasticity with a Non-Convex Stored Energy.

FREDSØE, Jørgen and JUSTESEN, Peter: Drag Forces on Large Circular Cylinders in Oscillatory Flow.

FREUND, L.B.: The Mechanics of Cleavage Fracture Initiation in Carbon Steel.

FRISCH, Uriel: Turbulent Transport of Temperature, Magnetic Field and Momentum.

FUJINO, Masataka: See FUKASAWA, Toichi.

FUKASAWA, Toichi and FUJINO, Masataka: On the Slamming Response of a Large Bulk Carrier.

FUKUMOTO, Y.: See HASIMOTO, H.

GAGNE, Y.: See ATTEN, P.

GAO, Y.C.: The Asymptotic Solution to the Dynamic Crack-Tip Field in a Strain-Damage Material.

GAUL, Lothar: Modal Properties of Structures Coupled by Joints and Springs: Analysis and Measurement.

GAUTESEN, A.K.: Scattering of a Rayleigh Wave by an Elastic Quarter Space.

GAVALDA, J.: See GIRALT, Francesc.

GAVIGLIO, Jean: See DEBIEVE, Jean-Francois.

GEER, J. van: See ARBOCZ, Johann.

GEIER, B. and ROHWER, K.: Behavior of Curved Anisotropic Panels beyond the Bifurcation Buckling Load.

GHOSH, K.: See HANAGUD, S.

GIRALT, Francesc, GAVALDA, J., DIAZ, F., KAWALL, J.G., and KEFFER, J.F.: Development and Structural Characteristics of a Distorted Turbulent Wake Generated by the Interacting of a Rotating Cylinder and a Stationary Cylinder.

GLADWELL, Graham M.L.: The Inverse Problem for the Vibrating Beam.

GLASS, Irvine I.: Explosive-Driven Hemispherical Implosions and their Application.

GLENDINNING, P. and SPARROW, C.: Bifurcation near Homoclinic Orbits.

GORISSEN, W.C.M.: See BIESHEUVEL, A.

GORLOV, Alexander M.: Disaster of the Mianus River Bridge. Where could Lateral Vibrations come from?

GRABITZ, G. and JUNGOWSKI, W.M.: Mechanism of some Self-Sustained Flow Oscillation.

GRAHAM, J.M.R.: See BEARMAN, P.W.

GRAICHEN, K.: See HOFFMEISTER, M.

GREATED, Clive: See EASSON, William James.

GREELEY, Ronald: See IVERSEN, James D.

GREGORY, R. Douglas and WAN, Frederic Y.M.: Correct Boundary Conditions for Thin and Thick Plate Theories.

GRIMALDI, A.: See COMO, M.

GRIMSHAW, R. and PULLIN, D.I.: Finite-Amplitude Interfacial Gravity Waves; Wave Profiles and Stability.

GRISTCHAK, V.Z.: See MOSSAKOVSKY, V.I.

GROSS, R.J.: See NUNZIATO, Jace W.

GRYSA, Krzysztof: Associated Integral Equations for Helmholtz Equation and their Application to the Inverse Heat Transfer and Wave Motion Problems.

GRZEDZINSKI, Janusz: See ATASSI, H.

GUJ, G.: See MATTEIS, G. de.

GUNNESKOV, Ole: See KRENK, Steen.

GUO, Shang Ping, HUANG, Yan Zhang, ZHOU, Juan, and KUANG, Pei Qiong: Microscopic Research of the Flow of Physical Chemistry Fluid through Porous Media.

GUSTAFSON, C.W.: See JOHNSON, R.E.

GUTTMANN, C.: See CAHOUET, J.

HADDAD, Y.M.: A Micromechanical Approach to the Rheology of Randomly Structured Fibrous Systems.

HADDOW, Alan G., NINH, HaQuangt, and MOOK, Dean T.: Experimental and Analytical Study of Subharmonic Resonance in a Structure.

HAFEZI, Hamid: See BROWN, Susan N.

HAGEN, Deborah A.: See WALTON, Otis R.

HALL, Mary and BUCKINGHAM, Alfred C.: Calculation of the Interaction between a Compliant Material and an Unsteady Flow.

HALL, Philip: The Görtler Vortex Instability Mechanics in Three-Dimensional Boundary Layers.

HANAGUD, S. and GHOSH, K.: Interaction of a Moving Shell Structure and a Plane Wave in a Fluid Medium.

HANDKE, E.: See OBERMEIER, F.

HANSEN, J.S.: See TENNYSON, R.C.

HARRIS, David: The Solution of Differential Equations Governing the Double-Shearing Deformation of Granular Materials.

HARRIS, John G. and POTT, John: Scattering of an Acoustic Gaussian Beam from a Fluid-Solid Interface.

HART, Richard T.: See COWIN, Stephen C.

HASIMOTO, H., ISHII, Katsuga, MIYAZAKI, T., and FUKUMOTO, Y.: Stokeslet in the Presence of Boundaries and its Application.

HAVNER, Kerry S. and SUE, Ping-Liang: Theoretical Analysis of Freely Deforming or Partially Constrained Crystals in Multiple Slip.

HE, F.B.: See CHENG, Shun.

HE, Li-Nan: See YU, Mao-Hong.

HE, You-Sheng: See LU, Chuan-Jing.

HENDRIKS, Ferdinand: Chaotic Motion of a Fast Impact Printing Hammer.

HENRIKSEN, Mogens and HOGAN, Harry A.: A Parametric Study of Micropolar Theory for Applications in Biomechanics - Fluid Flow.

HEO, H.: See IBRAHIM, R.A.

HERRMANN, George: See FISHER, Martin J.

HILLS, David A. and COMNINOU, Maria: An Analysis of Fretting Fatigue Cracks.

HINCHEY, M.J. and SULLIVAN, P.A.: Heave Stability of Hovercraft Hovering over Water.

HÖGFORS, Christian: Autobalancing.

HOFFMEISTER, M. and GRAICHEN, K.: Properties of the Flow over a Swept, Wall-Bounded Bar with a Square Cross Section.

HOGAN, Harry A.: See HENRIKSEN, Mogens.

HOLMES, Philip: Knotted Orbits and Bifurcation Sequences in Periodically Forced Oscillations.

HOSKING, R.J.: See DAVYS, J.W.

HSIEH, R.K.T.: Magnetic Forces in Ferrofluids.

HSIEH, R.K.T. and ZHOU, S.A.: A Statistical Theory of Elastic Materials with Micro-Defects.

HSU, C.S. and KIM, Myun C.: A Study of Statistical Properties of Strange Attractors by Generalized Cell Mapping.

HSU, Chen-Chi and LIAKOPOULOS, A.: On a Class of Compressible Laminar Boundary-Layer Flows and their Solution Behaviour Near Separation.

HUANG, T.C. and SHEN, K.S.: Dynamic Interaction of a Floating, Moving, and Flexible Structure and Fluid with Deterministic Surface Waves.

HUANG, Yan Zhang: See GUO, Shang Ping.

HUNT, Giles W., WILLIAMS, K.A.J., and COWELL, R.G.: Hidden Symmetry Concepts in the Elastic Buckling of Axially-Loaded Cylinders.

HUSEYIN, Koncay: See ATADAN, A.S.

HUSSAINI, N. Yousuff: See LAKIN, William D.

HUTCHINSON, James R. and ZILLMER, Scott D.: On the Transverse Vibration of Beams of Rectangular Cross-Section.

HUTCHINSON, John W.: See BUDIANSKY, Bernard.

HYCA, Milan: An Approximate Theory of Shear-Lag in Open Cross-Section Beams under Torsion.

IBRAHIM, R.A., SOUNDARAJAN, A., and HEO, H.: An Improved Non-Gaussian Closure for Stochastic Nonlinear Systems.

IF, F.: See CHRISTIANSEN, Peter Leth.

IGEL, E.A.: See NUNZIATO, Jace W.

IIDA, Kunihiro: See YAMAMOTO, Yoshiyuki.

ILANKAMBAN, R.: See KRAJCINOVIC, Dusan.

INMAN, Daniel J.: Modes and Critical Damping in Asymmetric Linear Dynamic and Systems.

IOOSS, G.: See CHOSSAT, P.

IRIMESCU, B.R.: See ATANASIU, N.E.

ISARIE, Illie: See SARBU, Nicolae.

ISHII, Katsuga: See HASIMOTO, H.

ISHII, Katsuya and KUWAHARA, Kunio: Computation of Compressible Flow around a Circular Cylinder.

ISHIKAWA, H. and LIPPMANN, H.: Plastic Flow Rule for Cyclic Loading.

ISOMAKI, Heikki M.: See RATY, Raimo.

IVANOV, Tsolo P.: Proof of Koiter's Problem for the Second Variation in the Theory of Elastic Stability.

IVERSEN, James D. and GREELEY, Ronald: A Space Station Wind Tunnel for Studying Sediment Transport.

JACOB, K.I.: See LEISSA, Arthur W.

JANSEN, A.J.M.: General Solutions of the Inviscid- and Viscous Fluid Motions Induced by an Electric Current and its Associated Magnetic Field.

JENKINS, James T. and RICHMAN, Mark W.: A Kinetic Theory for Rapidly Deforming Granular Materials Consisting of Identical, Rough, Inelastic, Circular Discs.

JIMENEZ, Javier and ZUFIRIA, Juan A.: Boundary Layer Models for Nonlinear Convection.

JIN, Wenlu: Spectral Resolving Methods for Analyzing Nonlinear Random Vibrations.

JOHNSON, George: The Effective Second-Order and Third-Order Elastic Constants of Rolled Plates.

JOHNSON, George C.: See BAMMANN, Douglas J.

JOHNSON, Lee W.: See PLAUT, Raymond H.

JOHNSON, Millard W. and SUHLING, Jeffrey C.: Thin Plate Theories for Orthotropic Nonlinear Elastic Materials.

JOHNSON, R.E. and GUSTAFSON, C.W.: Fluid Mechanics of Hydraulic Fracture.

JONAS, Pavel and PICHAL, Miroslav: A Contribution to the Investigation of the Turbulent Boundary Layer Structure.

JONES, Frederick L.: See MIKSAD, Richard W.

JONES, Norman and SONG, Boquan: Shear and Bending Response of a Rigid-Plastic Beam to Partly Distributed Blast-Type Loading.

JONKER, J.B.: See WERFF, K. van der.

JONSSON, Ivar G.: See BEHRENDT, Lars.

JONSSON, Mikael, KARLSSON, Lennart, and LINDGREN, Lars-Erik: Plate Motion and Thermal Stresses in Root-Bead Butt-Welding and One-Pass Butt-Welding of Two Different Steels.

JUNGOWSKI, W.M.: See GRABITZ, G.

JUSTESEN, Peter: See FREDSØE, Jørgen.

KAAS-PETERSEN, Chr.: Chaos in a Railway Bogie.

KACHANOV, M.: See CHUDNOVSKY, Alexander.

KAFKA, V.: A Mathematical Model of the Mechanical Response of Heterogeneous Structures.

KAISER, Sten: See CARLSSON, Janne.

KAMBE, T.: Acoustic Wave Emission by a Vortex Ring Passing by a Sharp Edge.

KANAGASUNDARAM, S.: See KARIHALOO, Bhushan Lal.

KANT, Rishi and O'SULLIVAN, Timothy C.: The Symmetric and Anti-Symmetric Vibrations of Thin-Walled Rectangular Tubes.

KAPLAN, Paul: Flat Bottom Slamming Impact of Advanced Marine Vehicles in Waves.

KARIHALOO, Bhushan Lal and KANAGASUNDARAM, S.: Minimum-Weight/Maximum-Strength Design of Plane Structural Frames.

KARIM-PANAHI, K.: Fluid-Structure Wave Motion of Eccentric Cylindrical Shells Filled with a Viscous Compressible Fluid.

KARLSSON, Lennart: See JONSSON, Mikael.

KATSUBE, Noriko: Mechanical Response of Fluid-Filled Poroelastic Solids.

KAWALL, J.G.: See GIRALT, Francesc.

KAWAMURA, Tetuya and KUWAHARA, Kunio: Direct Simulation of Turbulent Flow in a Duct.

KAWASHIMA, Koichiro and YOSHIDA, Takafumi: A New Stress-Strain Law for the Transient and Stable Cyclic Plasticity.

KAZA, K.R.V.: See SUBRAHMANYAM, K.B.

KAZARIAN, K.B.: See AMBARTSUMIAN, S.A.

KEER, L.M.: See MILLER, G.R.

KEFFER, J.F.: See GIRALT, Francesc.

KELLY, Robert E.: See PEARLSTEIN, Arne J.

KEMA, Kasuyuki: See SEGUSCHI, Yasuyuki.

KERR, Arnold D.: Recent Advances in Railroad Track Mechanics.

KHATAAN, H.A.: See TABARROK, B.

KHATTAB, H.: See ADVANI, Sunder.

KIDA, Shigeo: Interaction between Turbulence of Different Scales over Short Times.

KIKUCHI, Noboru: See TAYLOR, John E.

KIM, Myun C.: See HSU, C.S.

KIOUSIS, P.D.: See VOYIADJIS, G.Z.

KISKE, S. and VASANTA-RAM, V.I.: A Study of the Propagation of Disturbances in Turbulent Shear Flow.

KITAGAWA, Hiroshi and MATSUSHITA, Hisashi: Flow Localization in Elastic-Plastic Material Developing from Stress Free Surface.

KIYA, Masaru and SASAKI, Kyuro: Turbulence Structure in a Separation-Reattachment Flow.

KLARBRING, Anders: Contact Problems with Friction - Using a Finite Dimensional Description and the Theory of Linear Complementarity.

KLEPACZKO, Janusz R.: Fracture Initiation under High Loading Rates - New Experimental Methods and Results.

KLOPP, R.W.: See CLIFTON, Rodney J.

KLUWICK, A.: See CRAMER, M.S.

KNIGHT, J.H.: See PHILIP, J.R.

KNUDSEN, Thomas S.: Dynamic Buckling of Cylindrical Shells Subjected to Axial Step-Loadings.

KOBAYASHI, Yasunori and MATSUMOTO, Tsuyoshi: Two-Dimensional Condensing Vapour Flow in the Enclosure.

KOCH, Donald L. and BRADY, John F.: Dispersion in a Dilute Fixed Bed.

KOGA, Tasuzo: Asymptotic Solutions for the Free Vibrations and the Buckling under External Pressure of Circular Cylindrical Shells.

KOHN, David H.: See COWIN, Stephen C.

KOHN, Robert V.: Optimal Bounds for Effective Moduli as a Tool for Structural Optimization.

KOILLER, Jair and CARVALHO, Sonia Pinto de: Chaotic Motions and Nonintegrability on a Restricted Problem of Four Point Vortices.

KOSAROV, M.: Thermal Buckling of Thin-Walled Orthotropic Cylindrical Shells Heated along an Axial Strip.

KOSCHMIEDER, E.L.: Surface Tension Driven Benard Convection.

KOSKY, Michael S.: See SU, Tsung-Chow.

KOWALEWSKI, Tomasz A.: Experimental Studies of the Flow of Suspension through the Tube.

KRAJCINOVIC, Dusan and ILANKAMBAN, R.: Continuous Damage Mechanics of Rock Materials.

KRASINSKI, J. de and FAN, Y.: Some Visco-Elastic Aspects of Liquid Foams of a High Dryness Fraction and Possibilities of Application.

KRAWIETZ, Arnold: Upper and Lower Bounds to the Overall Incremental Stiffness of Prestressed Composites.

KRENK, Steen and GUNNESKOV, Ole: Theory and Computation of Turbine Blade Vibrations.

KREUZER, Edwin J.: Analysis of Chaotic Systems using the Cell Mapping Approach.

KRISTENSEN, Hans Saustrup: See XU, Hong Quing.

KUANG, Pei Qiong: See GUO, Shang Ping.

KUJAWSKI, D.: See ELLYIN, Fernand.

KURANISH, Masatsugu: See NISHIMURA, Tetsu.

KUWAHARA, Kunio: See ISHII, Katsuya.

KUWAHARA, Kunic: See KAWAMURA, Tetuya.

KWATNY, Harry G.: See BAHAR, Leon Y.

KYRIAKIDES, Stelios and SHIAU, Guan-Jon: Propagating Buckles in Long Confined Cylindrical Shells.

LABISCH, F.K.: On the Dual Determination of the Morphology of Non-Unique Solutions in Non-Linear Elastostatics.

LADEVEZE, P. and LEMAITRE, J.: Damage Effective Stress in Quasi-Unilateral Conditions.

LADEVEZE, P.: See EL FATMI, R.

LADRIERE, P.: See LEDUC, B.

LAHALLE, D.: See CAHOUET, J.

LAKIN, William D. and HUSSAINI, N. Yousuff: Stability of the Laminar Boundary Layer in a Streamwise Corner.

LAROCHE, C.: See FAUVE, S.

LARSEN, Jesper: See BRORSEN, Michael.

LARSEN, Poul Scheel and SØRENSEN, Søren K.: Effect of Secondary Flows and Turbulence on Electrostatic Precipitator Efficiency.

LARSEN, Poul Scheel: See XU, Hong Quing.

LAWS, Norman: See DVORAK, George J.

LEBON, G., BOUKARY, M.S., and CASAS-VAZQUEZ, J.: Extended Irreversible Thermodynamics and New Perspectives in Fluid Dynamics.

LEBOUCHE, M.: See SCIROCCO, V.

LEDBETTER, Hassel M.: See DATTA, Subhendu K.

LEDUC, B., NAGEL, Ph., and LADRIERE, P.: Experimental Study of Condenser Tubes Vibrations in Relation to the Supporting Conditions.

LEE, E.H.: Finite Deformation Effects in Plasticity Analysis.

LEE, J.K.: See ADVANI, Sunder.

LEE, Jon: Free Vibration of Lowest-Order Models of a Large Amplitude Deflected Plate.

LEGROS, J.Cl., LIMBOURG, M.Cl., and PETRE, G.: On Thermocapillarity in a Microgravity and in a 1-g Environment.

LEGUILLON, D. and LENE, F.: Damage as a Consequence of a Microglide Mechanism in Composite Materials.

LEHMANN, Theodor: On a Generalized Constitutive Law in Thermo-Plasticity Taking into Account Different Yield Mechanisms.

LEISSA, Arthur W. and JACOB, K.I.: Three-Dimensional Vibrations of Twisted, Cantilevered Parallelepipeds.

LEISSA, Arthur W.: See SUZUKI, K.

LEKSZYSKI, Tomasz: On Optimal Design Problems of Overdamped Structures.

LEMAITRE, J.: See LADEVEZE, P.

LENE, F.: See LEGUILLON, D.

LENOIR, M.: See CAHOUET, J.

LETELIER SOTOMAYOR, M.F.: See VIOLA, J.

LETELIER SOTOMAYOR, M.F. and LEUTHEUSSER, H.J.: Laminar Flow in Conduits of Unconventional Shape.

LEUTHEUSSER, H.J.: See VIOLA, J.

LEUTHEUSSER, H.J.: See LETELIER SOTOMAYOR, M.F.

LEVINSON, Mark: Free Vibrations of a Simply Supported, Rectangular Plate: An Exact Elasticity Solution.

LI, C.H.: See CLIFTON, Rodney J.

LI, Guang-Xuan: See MOON, Francis C.

LI, Hao: See QING, Jiang.

LI, Yongchi: See TING, T.C.T.

LIAKOPOULOS, A.: See HSU, Chen-Chi.

LIBOVE, Charles and CHANG, Jeng-Shian: Buckling and Post-Buckling Behavior of Sparsely Connected Build-up Columns.

LIBRESCU, Liviu: Transient Response of Elastic Thin Panels Undergoing Arbitrary Small Motions in a Supersonic Flow Environment.

LIMBOURG, M.Cl.: See LEGROS, J.Cl.

LIN, S.R.: See LIN, T.H.

LIN, T.H. and LIN, S.R.: Dislocations, Slip Band and Fatigue Crack Nucleation.

LINDGREN, Lars-Erik: See JONSSON, Mikael.

LING, Fuhua: See ZHANG, Wen.

LING, Fuhua: Periodic, Almost Periodic and Chaotic Motions of Forced Self-Sustained Oscillators.

LIPPMANN, H.: Velocity Field Equations and Strain Localization.

LIPPMANN, H.: See ISHIKAWA, H.

LITEWKA, Andrzej: Overall Mechanical Properties of Materials with Oriented Damage.

LIU, Cheng-Qun: Integral Invariant of Nonholonomic Conservative Systems and its Applications.

LIU, Peng: Combination Method of Generalized Transfer Matrix and Finite Element.

LIZZIO, R.: See STAGNI, L.

LU, Chuan-Jing and HE, You-Sheng: Unsteady Theory of the Hydrofoil of Finite Span.

LUCCHETTI, F.: See BERGAMASCHI, S.

LUN, C.K.K.: See SAVAGE, Stuart B.

LUNDBERG, Bengt and ANDERSSON, Lars-Erik: Illustration of Some Fundamental Properties of Impedance Transitions by Means of Microcomputer Simulation.

LUONGO, Angelo, REGA, Giuseppe, and VESTRONI, Fabrizio: Nonplanar Nonresonant Finite Motion of Inextensional Beams.

LYUBIMOV, G.A.: See BUCHIN, V.A.

MAC SITHIGH, Gearoid P.: See FOSDICK, R.L.

MAGNUSON, Allen H.: Ship Wave Perturbation Analysis using Numerical Surface Singularity (Panel) Methods.

MAHANTI, P.K.: Construction of New Element in the Finite Element Method.

MAHONY, J.J.: See PHILIP, J.R.

MAHRENHOLTZ, O.: See BOGACZ, R.

MAJERUS, John N.: A Hybrid Model for Analyzing Bodies Subjected to Intense Impact.

MAKER, B.N.: See TRIANTAFYLLIDIS, N.

MALRAISON, B.: See ATTEN, P.

MANGERON, Dumitru: See CHIRIACESCU, Sergio T.

MANNEVILLE, Paul, PUMIR, Alain, and TUCKERMAN, L.: Intrinsic Stochasticity with Many Degrees of Freedom.

MARCUS, Philip: See TUCKERMAN, Laurette.

MARGOLIS, Stephen B. and MATKOWSKY, Bernard J.: Flame Propagation in Channels: Secondary Bifurcation to Quasi-Periodic Pulsations.

MARKENSCOFF, Xanthippi: See CALLIAS, Constantine.

MARTELLI, F.: See BIDINI, Gianni.

MARUO, Hajime: A Contribution to the Slender Body Theory in Ship Hydrodynamics.

MASRI, Sami F.: See DEHGHANYAR, Tejav J.

MASRI, Sami F. and MILLER, Richard K.: A Time-Domain Method for the Identification and Modeling of Nonlinear Vibrating Structures.

MATKOWSKY, Bernard J.: See MARGOLIS, Stephen B.

MATSUMOTO, Tsuyoshi: See KOBAYASHI, Yasunori.

MATSUSHITA, Hisashi: See KITAGAWA, Hiroshi.

MATTEIS, G. de, GUJ, G., and PIVA, R.: Anisotropic Porous Medium for Modelling Thermal Flows through a Cylinders Array.

MAUGIN, Gerard A.: See TOURATIER, Maurice.

MAUGIN, Gerard A.: See POUGET, Joel.

MCCOMB, W.D. and SHANMUGASUNDARAM, V.: Renormalisation Methods Applied to the Calculation of the Subgrid-Scale Eddy Viscosity for Isotropic Turbulence.

MCINTYRE, Michael E. and MOBBS, Stephen D.: On the "Pseudomomentum Rule" for Wave-Induced Mean Forces.

MEECHAM, W.C.: Low and Moderate Frequency Rough Surface Scattering using a Wiener-Hermite Random Process Representation.

MEIER, Gerd E.A.: See CHAVES, Humberto.

MELVILLE, W.K.: See CHAN, Eng-Soon.

MENENDEZ, A.N. and RAMAPRIAN, B.R.: Quasi-Steady Modeling of Unsteady Turbulent Boundary Layers.

MERTENS, Robert: See OTTOY, Jean-Piere.

METCALFE, A.V.: See BURDESS, J.S.

MIKSAD, Richard W., JONES, Frederick L., and POWERS, Edward, J.: The Generation of Low-Frequency Fluctuations by Nonlinear Interactions during Transition to Turbulence.

MIKSIS, Michael J. and TING, Lu: Wave Propagation in a Bubbly Liquid with Finite Amplitude Asymmetric Oscillations.

MILLER, G.R., KEER, L.M., and CHENG, H.S.: On the Mechanics of Fatigue Crack Growth due to Contact Loading.

MILLER, Richard K.: See DEHGHANYAR, Tejav J.

MILLER, Richard K.: See MASRI, Sami F.

MINASSIAN, M.M.: See AMBARTSUMIAN, S.A.

MITCHELL, Thomas P.: Axisymmetric Creeping Flows in Shear-Thinning Fluids.

MITSOTAKIS, Konstantin: See ZAUNER, Erwin.

MIYATA, Masafumi: See NAKAMURA, Ikuo.

MIYAZAKI, T.: See HASIMOTO, H.

MIZUSHIMA, Jiro and SAITO, Yoshio: Statistical Properties of Burgers' Turbulence for Various Initial Conditions.

MIZUSHIMA, Jiro: See SAITO, Yoshio.

MOBBS, Stephen D.: See MCINTYRE, Michael E.

MOLLESTAD, Egil: See BERGAN, Pål G.

MONAKHOV, A.A.: See BELYAEV, Yu.N.

MOOK, Dean T.: See HADDOW, Alan G.

MOON, Francis C. and LI, Guang-Xuan: The Fractal Dimension of the Two-Well Potential Strange Attractor.

MØRCH, Knud Aage: The Life-Cycle of Acoustically Induced Cavity Clusters.

MOSER, Alfred: Computation of Nonlinear Gas Oscillations by an Iterated Mapping and Interpretation of Chaos.

MOSSAKOVSKY, V.I. and GRISTCHAK, V.Z.: Buckling and Vibration of the Nonhomogeneous Constructions.

MROZ, Z. and DEMS, K.: A New Class of Conservation Rules in Elasticity and their Application.

MÜLLER, U., DAVIS, S.H., and DIETSCHE, C.: Pattern Selection in Single-Component Systems Coupling Benard Convection and Solidification.

MUKHERJEE, Subrata: See CHANDRA, Abhijit.

MULLER, Michael R. and SHANG, Paul C.: Properties of Explosive Resonant Interactions between Non-Singular Modes of Internal Gravity Waves in Moving Layered Fluids.

MURAKAMI, Sumio: See TANAKA, Eiichi.

NACHMAN, A.: See BUCKMASTER, J.

NADIM, Ali: See BRENNER, Howard.

NAGEL, Ph.: See LEDUC, B.

NAGHDI, Paul M. and VONGSARNPIGOON, L.: The Downstream Flow Beyond an Obstacle.

NAKAMURA, Ikuo, SAKAI, Yasuhiko, and MIYATA, Masafumi: Diffusion of Matter by a Non-Buoyant Plume in Grid-Generated Turbulence.

NARASIMHA, Roddam: Statistical Models for Transitional Intermittency.

NAYROLES, B. and NGUYEN, Q.S.: Infrared Thermography and Solid Mechanics.

NEALE, Kenneth W. and TUGCU, P.: Analysis of Necking and Neck Propagation in Polymeric Materials.

NEEDLEMAN, Alan and TVERGAARD, Viggo: An Analysis of Ductile Rupture in Notched Bars.

NEWMAN, Barry G.: Multiple Actuator-Disc Theory for Wind Turbines.

NEWMAN, J. Nicholas: Algorithms for the Free-Surface Green Function.

NGUYEN, Q.S.: See NAYROLES, B.

NINH, HaQuangt: See HADDOW, Alan G.

NISHIMURA, Tetsu and KURANISH, Masatsugu: The Yield Condition on Linear Combination of the Invariants J3/2 and J2/3.

NORONHA, Danro B.: See BEVILACQUA, Luis.

NOVAK, M. and EL-ASHKAR, I.: Response of Cable Roofs to Wind.

NUNZIATO, Jace W., BAER, M.R., IGEL, E.A., and GROSS, R.J.: One-Dimensional Flame Propagation in a Reactive Granular Material.

O'SULLIVAN, Timothy C.: See KANT, Rishi.

OBERMEIER, F. and HANDKE, E.: Some Theoretical Results on Mach-Reflection of Moderately Strong Shock Waves.

OCHOA, O.O.: See REDDY, J.N.

OGILVIE, T. Francis: See CHEN, Sixiong.

CHAYON, Roger: Vibration Analysis of Periodic Structures and Brillouin Zones Computation. Finite Element Applications to Stiffened Cylinders.

OLAOSEBIKAN, O.: See SMITH, C.W.

OLHOFF, Niels: See BENDSØE, Martin.

ONISHI, Yoshimoto: See SONE, Yoshio.

OOKA, Masahiro: See TANAKA, Eiichi.

ORGILL, Gary: See WILSON, James F.

ORLANDI, Paolo: See CROCCO, Luigi.

OSHIMA, Koichi and OSHIMA, Yuko: Experimental and Numerical Study of Two- and Three-Dimensional Vortex Interactions.

OSHIMA, Nobunori: Theoretical Estimate of Effective Elastic Moduli of Dispersely Reinforced Composite Materials.

OSHIMA, Yuko: See OSHIMA, Koichi.

OTTOY, Jean-Piere and MERTENS, Robert: On the Adiabatic and Exact Invariants of some Special Hamiltonian Systems.

PALM, Enok, SKOGVANG, Arnljot, and TVEITEREID, Morten: On the Transition to Chaotic Motion in a Low Prandtl Number Field.

PANAGIOTOPOULOS, P.D. and BANIOTOPOULOS, C.C.: Nonconvex Superpotentials, Hemivariational Inequalities and Applications in Thermoelasticity.

PAO, Y.H.: See ZIEGLER, Franz.

PAPATZACOS, Paul: Fluid Flow in a Porous Medium due to a Fractured Well with Infinite Conductivity.

PARISEANU, George: See CRANDALL, Stephen H.

PARKINSON, G.V.: On Modelling Separated Potential Flows using Wake Singularities.

PARLAND, H.: On the Rocking Oscillations of a Non-Monolithic Portal Frame.

PARNES, Raymond: Steady State Ring Load Pressure on a Bore Hole Surface.

PAULUN, J. and PECHERSKI, Ryszard: A New Evolution Equation for Kinematic Hardening in Finite Deformation Plasticity.

PEARLSTEIN, Arne J. and KELLY, Robert E.: Effect of a Heterogeneous Chemical Reaction on the Onset of Convective Instability in a Doubly Diffusive Fluid Layer.

PECHERSKI, Ryszard: See PAULUN, J.

PELLEGRINO, S. and CALLADINE, C.R.: Mechanics of Pin-Jointed Assemblies, Applied to Cable Net Structures.

PELZ, W.: See CRAMER, M.S.

PERALTA-FABI, Ricardo: Micromechanics of Creep in Clay Soils.

PEREGRINE, Dennis H.: Refraction and Focussing of Finite-Amplitude Water Waves.

PERRIN, B.: See FAUVE, S.

PERZYNA, Piotr: Dependence of Fracture Phenomena upon the Evolution of Constitutive Structure of Solids.

PETERS, N.: See BUCKMASTER, J.

PETRE, G.: See LEGROS, J.Cl.

PHILIP, J.R., KNIGHT, J.H., and MAHONY, J.J.: Mechanics of Collodial Suspensions with Applications to Stress Transmission, Volume Change, and Cracking in Clay Soils.

PIAN, Theodore H.H.: Finite Elements Based on Consistently Assumed Stresses and Displacements.

PICHAL, Miroslav: See JONAS, Pavel.

PICUGA, Alija: Minimum-Cost Synthesis of Multi-Purpose Beams.

PIGNATARO, Marcello, RIZZI, Nicola, and TATONE, Amabile: On the Postbuckling Behaviour of Beams with Curved Axis.

PINDERA, Jerzy T.: Plane and Generalized Elastic and Photoelastic Isodynes - Theories and Applications.

PIVA, R.: See MATTEIS, G. de.

PLATTEN, J.K.: See VILLERS, D.

PLAUT, Raymond H. and JOHNSON, Lee W.: Optimal Forms of Shallow Elastic Shells.

POMAZI, L.: Investigations on the Stability of Constructionally Anisotropic Multilayered (Multi-Sandwich) Plates.

POTT, John: See HARRIS, John G.

POUGET, Joel and MAUGIN, Gerard A.: Nonlinear Elastic Waves Generated by Solitons in Ferroelectric Crystals.

POWERS, Edward J.: See MIKSAD, Richard W.

PREDELEANU, M.: See DAHAN, M.

PULLIN, D.I.: See GRIMSHAW, R.

PUMIR, Alain: See MANNEVILLE, Paul.

PYLKKAENEN, Jaakko V.: Collective Collapse of Spherical Cloud Cavity.

PYRZ, Ryszard: On the Liapunov Stability of Viscoelastic Perfect Column.

QING, Jiang and LI, Hao: New Variational Principles in Linear and Nonlinear Theories of Elasticity, Thermoelasticity and Viscoelasticity.

RABAUD, M.: See COUDER, Y.

RADDER, A.C. and DINGEMANNS, M.W.: Canonical Equations for Almost-Periodic, Weakly-Nonlinear Surface Waves.

RAHMAN, Z.: See BURTON, Tom D.

RAKOWSKI, Jerzy and SYGULSKI, Ryszard: On the Analysis of Nonlinear Cable Networks using the Boundary Element Method.

RAMAPRIAN, B.R.: See MENENDEZ, A.N.

RASHED, Ahmed A.: Three-Dimensional Dynamic Behaviour of Gravity Dam-Reservoir Systems.

RATY, Raimo, BOEHM, Juhani von, and ISOMAKI, Heikki M.: Development of Chaos of Duffing's Oscillator.

RAVN-JENSEN, Kim: Influence of Residual Stresses on the Postbuckling Behaviour of Plate and Shell Structures.

REDDY, J.N. and OCHOA, O.O.: A Refined Laminate Theory with Transverse Shear Deformation.

REGA, Giuseppe: See LUONGO, Angelo.

REIFSNIDER, Kenneth L.: See TALREJA, Ramesh.

REINHALL, Per Gustaf and STORTI, Duane William: Periodic and Chaotic Motion of a Drill in the Drilling of Thin Plates.

REMENYIK, Carl J.: See TRAUGOTT, Stephen C.

RICHMAN, Mark W.: See JENKINS, James T.

RIMROTT, F.P.J.: Dissipative Rigids.

RIZZI, Nicola: See PIGNATARO, Marcello.

ROBERTS, John W., BUX, S.L., and CARTMELL, M.P.: Some Effects of Internal Resonance on the Vibration of Beam System.

ROHWER, K.: See GEIER, B.

ROORDA, John: See CAMOTIM, Dinar.

ROZVANY, George I.N.: New Applications of Prager's Theory of Optimal Structural Layouts.

RU, Chong-Qing: See WANG, Ren.

SABINA, Federico J.: Low Frequency Acoustic Scattering by a Soft Body.

SACHSE, Wolfgang: Studies with Simulated and Controlled Acoustic Emission Signals in Materials.

SAITO, Yoshio: See MIZUSHIMA, Jiro.

SAITO, Yoshio and MIZUSHIMA, Jiro: Statistical Properties of Strong MHD Turbulence for Various Magnetic Prandtl Number.

SAKAI, Yasuhiko: See NAKAMURA, Ikuo.

SANBONGI, Shigeo: See TODA, Susumu.

SANDSMARK, Nils: See BERGAN, Pål G.

SANO, Masaki, SATO, S., and SAWADA, Y.: Development of Strange Attractors with Increasing Spatial Degrees of Freedom in Rayleigh-Benard System.

SARBU, Nicolae and ISARIE, Illie: On Vibrations in Shafts with Variable Mass.

SARPKAYA, Turgut: Hydrodynamic Resistance of Smooth and Rough Cylinders Subjected to Wave-Current Combination.

SARPKAYA, Turgut: Effect of Orbital Motion on Wave Forces on Horizontal and Vertical Cylinders.

SASAKI, Kyuro: See KIYA, Masaru.

SATO, S.: See SANO, Masaki.

SAVAGE, Stuart B. and LUN, C.K.K.: Kinetic Theory of Rapid Flows of Dense Concentrations of Dispersed Solid-Fluid Mixtures.

SAWADA, Y.: See SANO, Masaki.

SAYED, Sayed M.: Cylindrical Cavity Expansion in Nonlinear Dilatant Media.

SCHERBAKOV, S.A.: See BELYAEV, Yu.N.

SCHMIDT, Henrik: Reflection of Narrow Ultrasonic Beams from a Stratified Solid Half Space.

SCHMIDT, Herbert: See ESSLINGER, Maria.

SCHMIDT, Rüdiger: Contributions to the Nonlinear Theory of Thin Elastic Shells.

SCHNEIDER, Wilhelm: See ZAUNER, Erwin.

SCHOUTEN, G.: The Vector Potential in Branched Space.

SCHREYER, Howard L.: A Viscoplastic Model of Frictional Materials.

SCHWAB, A.L.: See WERFF, K. van der.

SCHWIEGER, H.: A New Application of Differential Interferometry for Dynamic Stress Analysis.

SCIROCCO, V., DEVIENNE, R., and LEBOUCHE, M.: Heat Transfer for a Pseudoplastic Fluid Flowing in the Thermal Entry Region of a Tube.

SCLAVOUNOS, Paul D.: Wave Forces on Stationary Slender Bodies.

SEGUCHI, Yasuyuki and TANAKA, Masao: Optimal/Adaptive Incremental Sequence in Nonlinear Finite Element Analysis.

SEGUCHI, Yasuyuki, TADA, Yukio, and KEMA, Kasuyuki: Shape Determination of Non-Conservative Structural Systems by the Inverse Variational Principle.

SELVADURAI, A.P.S.: Rigid Elliptical Inhomogeneity at a Transversely Isotropic Bi-Material Elastic Interface.

SGAIER, Hedi: See BARTHES-BIESEL, Dominique.

SHACHAM, Itzhak, WELLER, Tanchum, and BODNER, Sol R.: Whipping Response of a Ship Hull due to Wave Impact.

SHAH, A.H.: See DATTA, Subhendu K.

SHANG, Paul C.: See MULLER, Michael R.

SHANMUGASUNDARAM, V.: See MCCOMB, W.D.

SHARMA, R.N.: See BARAL, G.C.

SHAW, Steven W.: Forced Vibrations of Mechanical Systems having Amplitude Constraints: Subharmonic and Chaotic Motions.

SHEN, Hui-Li et al.: The Prediction and Analysis for the Response of a Turbojet Engine to the Blast-Wave.

SHEN, K.S.: See HUANG, T.C.

SHIAU, Guan-Jon: See KYRIAKIDES, Stelios.

SHIH, Tien-Mo: See TRAUGOTT, Stephen C.

SIGINER, Aydeniz: Some New Results Concerning Free Surface Phenomena in Rheological Fluid Mechanics.

SIMMONDS, James G.: The Derivation, Implication, and Application of Stress-Strain Relations for General, Rubber-Like Shells.

SIMPKINS, P.G.: See BLYTHE, P.A.

SKOGVANG, Arnljot: See PALM, Enok.

SKOVGAARD, Ove: See BEHRENDT, Lars.

SKOVGAARD, Ove: See CHRISTIANSEN, Peter L.

SMITH, C.W., EPSTEIN, J.S., and OLAOSEBIKAN, O.: Boundary Layer Effects in Cracked Bodies; An Engineering Assessment.

SNEYD, A.D.: See DAVYS, J.W.

SÖDERHOLM, Lars: On Gases, Kinetic Theory, and Material Frame-Indifference.

SOKOLOWSKI, Jan: See BAUER, Jacek.

SOKOLOWSKI, Jan: See BENDSØE, Martin.

SOMMERIA, Joel: An Experimental Investigation of the Two-Dimensional Inverse Energy Cascade.

SONE, Yoshio and ONISHI, Yoshimoto: Kinetic Theory Analysis of a Two Phase System of a Gas with Condensable and Noncondensable Components and its Condensed Phase.

SONG, Boquan: See JONES, Norman.

SONG, Ling-Yu: See YU, Mao-Hong.

SONNERLIND, Henrik: See CARLSSON, Janne.

SØRENSEN, M.P.: See CHRISTIANSEN, Peter Leth.

SØRENSEN, Søren K.: See LARSEN, Poul Scheel.

SOUNDARAJAN, A.: See IBRAHIM, R.A.

SPARROW, C.: See GLENDINNING, P.

SPECKMANN, Hans-Dieter: See CHAVES, Humberto.

SPENCE, D.A.: Self-Similar Solutions for Elastohydrodynamic Cavity Flow.

STADLER, W.: Nonexistence of Solutions in Optimal Structural Design.

STAGNI, L. and LIZZIO, R.: The Elastic Field within an Internally Stressed Infinite Strip.

STEKETEE, J.A.: On a Transformation of Staniukovich for the Rectilinear Unsteady Flow of a Non-Homentropic Gas.

STORTI, Duane William: See REINHALL, Per Gustaf.

STRANG, Gilbert: Optimal Design and Specific Cost Functions for Layered Composites.

STUMPF, Helmut: General Concept of Energy-Consistent Elastic Shell Theories with Buckling and Post-Buckling Analysis.

STURE, S.: See WILLAM, Kaspar J.

SU, Tsung-Chow and KOSKY, Michael S.: A Numerical Study on Shallow Water Wave Breaking and Wave-Structure Interactions.

SUBRAHMANYAM, K.B. and KAZA, K.R.V.: Improved Methods of Vibration Analysis of Pretwisted Airfoil Blades.

SUE, Ping-Liang: See HAVNER, Kerry S.

SUHLING, Jeffrey C.: See JOHNSON, Millard W.

SULLIVAN, P.A.: See HINCHEY, M.J.

SUMI, Yoichi: Computational Crack Path Prediction.

SUMNER, Eric E.: Thermoelastic Attenuation in Composites.

SUQUET, P.M.: Plasticity of Highly Heterogeneous Media.

SUZUKI, K. and LEISSA, Arthur W.: Exact Solutions for the Free Vibrations of Non-circular Cylindrical Shells having Circumferentially Varying Thickness.

SYGULSKI, Ryszard: See RAKOWSKI, Jerzy.

TABARROK, B., KHATAAN, H.A., and FENTON, R.G.: Plane Strain Plastic Flow Analysis using the Method of Principal Stress Lines.

TADA, Yukio: See SEGUCHI, Yasuyuki.

TAKAHASHI, Kazuo: Dynamic Stability of a Rectangular Plate Subjected to Inplane Forcing.

TALBOT, D.R.S.: The Effective Sink Strength of a Random Array of Parallel Edge Dislocations.

TALREJA, Ramesh and REIFSNIDER, Kenneth L.: Stiffness-Damage Relationships for Composite Materials.

TANAKA, Eiichi, MURAKAMI, Sumio, and OOKA, Masahiro: Strain-Hardening under Multiaxial Non-Proportional Cyclic Plastic Deformation.

TANAKA, Masao: See SEGUCHI, Yasuyuki.

TARNAI, Tibor: Bifurcation of Equilibrium and Bifurcation of Compatibility.

TATONE, Amabile: See PIGNATARO, Marcello.

TAYLOR, John E. and KIKUCHI, Noboru: Optimal Fail-Safe Structural Design: A Multicriteria Formulation.

TENNYSON, R.C. and HANSEN, J.S.: Buckling Analysis of Composite Cylinders.

THANGAM, S.: See CHEN, C.F.

THOMSON, Philip A.: See CHAVES, Humberto.

THORNTON, C.: A Numerical Simulation Approach to the Micromechanics of Particulate Material.

TING, Lu: See MIKSIS, Michael J.

TING, T.C.T. and LI, Yongchi: Simple Waves and Shock Waves Generated by an Incident Shock Wave in Two-Dimensional Hyperelastic Materials.

TODA, Susumu and SANBONGI, Shigeo: Buckling of Sinusoidally Corrugated Plates in Shear.

TOROK, J.: See ADVANI, Sunder.

TOURATIER, Maurice and MAUGIN, Gerard A.: Elastic Propagation in Two-Phase Heterogeneous Wave Guides of Rectangular Cross Section.

TRAUGOTT, Stephen C., SHIH, Tien-Mo, and REMENYIK, Carl J.: Influence of an Interfacial Film on the Hydrodynamics of Large Confined Bubbles.

TREVINO, G.: The Bispectrum of Turbulence.

TRIANTAFYLLIDIS, N.: See ABEYARATNE, Rohan.

TRIANTAFYLLIDIS, N. and MAKER, B.N.: On the Comparison between Microscopic and Macroscopic Instability Mechanisms in a Class of Fiber Reinforced Composites.

TRIFONOV, V.D.: See BUCHIN, V.A.

TRYGGVASON, Gretar: See AREF, Hassan.

TUCKERMAN, L.: See MANNEVILLE, Paul.

TUCKERMAN, L. and MARCUS, Philip: Formation of Taylor Vortices in Spherical Couette Flow.

TUGCU, P.: See NEALE, Kenneth W.

TUMAY, M.T.: See VOYIADJIS, G.Z.

TVEITEREID, Morten: See PALM, Enok.

TVERGAARD, Viggo: See NEEDLEMAN, Alan.

URSELL, F.: A Note on Uniqueness in the Linear Theory of Two-Dimensional Water Waves.

VALANIS, K.C.: See FAN, Jinghong.

VALID, R.: Static and Dynamic Analysis of Cyclically Symmetric Structures - Extensions.

VASANTA-RAM, V.I.: KISKE, S.

VAUGHAN, Henry: Three-Dimensional Motions of Ships and Platforms in Waves.

VERMEULEN, P.G.: See ARBOCZ, Johann.

VERRON, Jacques: Mesoscale Topography Effects in a Quasigeostrophic Ocean.

VESTRONI, Fabrizio: LUONGO, Angelo.

VILLERS, D. and PLATTEN, J.K.: Rayleigh-Benard Instability in Systems Presenting a Minimum in Surface Tension.

VINOGRADOV, A.M.: Concepts of Buckling and Bifurcation in the Creep Stability Analysis.

VIOLA, J., LETELIER SOTOMAYOR, M.F., LEUTHEUSSER, H.J., and CURRIE, I.G.: Unsteady Turbulent Flow in Pipes.

VONGSARNPIGOON, L.: See NAGHDI, Paul M.

VOYIADJIS, G.Z., KIOUSIS, P.D., and TUMAY, M.T.: A Lagrangean Formulation Theory for some Large Deformation Problems in Geomechanics.

VULLIERME-LEDARD, M.: The Limiting Amplitude Principle Applied to Linear Naval Hydrodynamics.

WALENTA, Zbigniew Andrzej: Quasi-Stationary, Irregular Reflection of Shocks - Formation and Structure.

WALTON, Otis R., HAGEN, Deborah A., and COOPER, John M.: Interparticulate Force Models for Computational Simulation of Granular Solids Flow.

WAN, Frederic Y.M.: See GREGORY, R. Douglas.

WANG, C.-M. and ALWIS, W.A.M.: On Optimization of Archgrids.

WANG, C.Y.: Load Capacities of Floating Elastic Sheets.

WANG, Chang-Yi: Hydrodynamic Braking.

WANG, Chih Cheng: See CHAO, Ching Cheng.

WANG, Ren and RU, Chong-Qing: Energy Criteria for the Plastic Buckling of Circular Cylindrical Shell under Axial Impulsive Loading.

WANG, Wen-Liang: See ZHANG, Jin.

WATSON, L.T.: See CRAMER, M.S.

WAUER, J. and CLEMENS, H.: On the Dynamics of Radially Loaded Circular Rings.

WEAVER, Richard L.: A Variational Principle for Multiple Scattering of Waves in Random Media.

WEBER, Hans Ingo: Motion Analysis of a Two Degrees of Freedom Rotor about a Fixed Point Driven through a Universal Joint.

WEBER, Herbert: On a Nonlinear Constitutive Equation of Photoviscoelasticity and its Application to Stress Analysis.

WEICHERT, D.: Shakedown at Finite Displacements and Small Strains.

WELLER, Tanchum: See SHACHAM, Itzhak.

WENG, George J.: See CHIANG, C.R.

WERFF, K. van der, JONKER, J.B., and SCHWAB, A.L.: Dynamics of Flexible Mechanisms.

WILLAM, Kaspar J., STURE, S., and BICANIC, N.: Continuous Fracture Computations and Strainlocalization in Cementitious Solids Subjected to Tension and Shear.

WILLIAMS, K.A.J.: See HUNT, Giles W.

WILLIS, John A.B.: Wind Flow over Laboratory Water Waves.

WILSON, James F. and ORGILL, Gary: Optimal Cable Configurations for Offshore Compliant Towers.

WONG, K.C.: See DATTA, Subhendu K.

WU, Han-Chin and YANG, C.C.: Fatigue Life of 304 Stainless Steel with Non-Proportional Axial-Torsional Strain-Paths.

XU, Hong Quing, KRISTENSEN, Hans Saustrup, and LARSEN, Poul Scheel: Pressure Drop Reduction in Cyclone.

YAMAMOTO, Yoshiyuki, IIDA, Kunihiro, and ARAI, M.: Structural Damage Analysis of a Fast Ship due to Bow Flare Slamming.

YANG, C.C.: See WU, Han-Chin.

YANG, H.J.: See BOGY, David B.

YAVORSKAYA, I.M., FOMINA, N.I., and BELYAEV, Yu.N.: Convection in a Rotating Spherical Layer as a Model for Global Motion in Giant Planet Atmospheres.

YAVORSKAYA, I.M.: See BELYAEV, Yu.N.

YEH, Kai-Yuan: See YU, Huan-Ran.

YOSHIDA, Takafumi: See KAWASHIMA, Koichiro.

YU, Huan-Ran and YEH, Kai-Yuan: Optimal Design of a Thin Elastic Solid Annular Plate under Arbitrarily Distributed Load.

YU, Mao-Hong, HE, Li-Nan, and SONG, Ling-Yu: Twin Shear Stress Criterion and its Generalization.

YU, S.-B.: See CHENG, C.-M.

ZAUNER, Erwin, SCHNEIDER, Wilhelm, and MITSOTAKIS, Konstantin: Turbulent Jets with Buoyancy, Swirl and Mutual Hindering.

ZHANG, Jin, WANG, Wen-Liang, and CHEN, Xiang-Jun: Natural Mode Analysis of N Blades Disc Coupled System --- Modal Synthesis of Symmetric Structure with CNv Group.

ZHANG, Ruqing and DONG, Ming: Two Methods of Solving Dynamic Response by Discreting of Gurtin Variational Principle.

ZHANG, Wen and LING, Fuhua: On the Dynamic Stability of the Rotating Shafts Made of Boltzmann Viscoelastic Solid.

ZHAO, Ling-Cheng and CHEN, Jingyn: The Impedance Matrix Formulation and Error Estimation of Finite Dynamic Element.

ZHONG, Wanxie and CHENG, Gengdong: The Second Order Sensitivity Analysis of Multimodal Eigenvalue and Related Optimization Technique.

ZHOU, Juan: See GUO, Shang ping.

ZHOU, Pei-Yuan: On the Condition of Pseudo-Similarity and the Theory of Turbulence.

ZHOU, S.A.: See HSIEH, R.K.T.

ZIEGLER, Franz and PAO, Y.H.: Transient Elastic Waves in a Wedge-Shaped Layer.

ZILLMER, Scott D.: See HUTCHINSON, James R.

ZIMNY, Josef: See ZYCZKOWSKI, Michal.

ZUFIRIA, Juan A.: See JIMENEZ, Javier.

ZYCZKOWSKI, Michal and ZIMNY, Josef: Plastic Removal of Undesirable Plastic Deformations.